Tissue
Stem Cells

Tissue Stem Cells

edited by

Christopher S. Potten
EpiStem Limited
Manchester, U.K.

Robert B. Clarke
Division of Cancer Studies,
University of Manchester, Paterson Institute
Manchester, U.K.

James Wilson
Centre for Gastroenterology, Barts and The London,
Queen Mary's School of Medicine and Dentistry
London, U.K.

Andrew G. Renehan
Christie Hospital
Withington, Manchester, U.K.

CRC Press
Taylor & Francis Group
Boca Raton   London   New York

CRC Press is an imprint of the
Taylor & Francis Group, an **informa** business

CRC Press
Taylor & Francis Group
6000 Broken Sound Parkway NW, Suite 300
Boca Raton, FL 33487-2742

First issued in paperback 2019

© 2010 by Taylor & Francis Group, LLC
CRC Press is an imprint of Taylor & Francis Group, an Informa business

No claim to original U.S. Government works

ISBN-13: 978-0-8247-2899-1 (hbk)
ISBN-13: 978-0-367-39092-1 (pbk)

Visit the Taylor & Francis Web site at
http://www.taylorandfrancis.com

and the CRC Press Web site at
http://www.crcpress.com

A CIP record for this book is available from the British Library.
Library of Congress Cataloging-in-Publication Data available on application

We dedicate this book to the memory of our friend and colleague
Lez Fairbairn
who died unexpectedly on 26th July 2005 at the age of 46.

Preface

Tissue stem cells and their medical applications have become a major focus of research over the last 10 years. The ethical issues surrounding the therapeutic use of stem-cell transplantation, and in particular embryonic stem-cell transplantation, have also been the subject of much attention, particularly in the popular media. In this book, we have sought to provide a thorough and up-to-date summary of the current position of scientific knowledge with regard to stem cells from a range of adult tissues, including the skin, the intestine, and the liver. All the reviews are written by internationally regarded experts in the field. The different chapters focus on a variety of aspects of stem-cell research including the molecular biology of stem-cell regulation, experimental models of stem-cell function, and the clinical applications of stem-cell biology. They also seek to address controversial issues, such as the debate between stem-cell transdifferentiation and stem-cell fusion, and the mathematical modeling of stem cells in tissues. We hope *Tissue Stem Cells* will provide an essential reference for all those working in this field. We also hope that this book, with its broad outlook and high-quality authorship, will be of interest to all researchers and students and not just stem-cell biologists.

Christopher S. Potten
Robert B. Clarke
James Wilson
Andrew G. Renehan

Contents

Contributors

Ilaria Bellantuono   Academic Unit of Bone Biology, The Sheffield Medical School, Sheffield, U.K.

Robert B. Clarke   Division of Cancer Studies, University of Manchester, Paterson Institute, Manchester, U.K.

Mark d'Inverno   Centre for Agent Technology, Cavendish School of Computer Science, University of Westminster, London, U.K.

Joanne Ewing   Birmingham Heartlands Hospital, Bordesley Green East, Birmingham, U.K.

Leslie J. Fairbairn[†]   Cancer Research U.K. Gene Therapy Group, Paterson Institute for Cancer Research, Manchester, U.K.

Mark E. Furth   Institute for Regeneratine Medicine, Wake Forest Medical Center, Winston Salem, North Carolina, U.S.A.

Joerg Galle   Interdisciplinary Centre for Bioinformatics, University of Leipzig, Leipzig, Germany

David Gerber   Department of Surgery, UNC School of Medicine, Chapel Hill, North Carolina, U.S.A.

Trevor Graham   Molecular and Population Genetics Laboratory, Cancer Research U.K., London Research Institute, London, U.K.

Sanjeev Gupta   Departments of Medicine and Pathology, Albert Einstein College of Medicine, Bronx, New York, U.S.A.

Pritinder Kaur   Epithelial Stem Cell Biology Laboratory, Peter MacCallum Cancer Centre, Melbourne, Victoria, Australia

Nigel L. Kennea   Institute of Reproductive and Developmental Biology, Imperial College, London, London, U.K.

[†]Deceased.

Jun Kohyama   Department of Physiology, Keio University School of Medicine, Tokyo and Core Research for Evolutional Science and Technology (CREST), Japan Science and Technology Agency, Saitama, Japan

Markus Loeffler   Institute for Medical Informatics, Statistics, and Epidemiology, University of Leipzig, Leipzig, Germany

E. Georg Luebeck   Public Health Sciences Division, Fred Hutchinson Cancer Research Center, Seattle, Washington, U.S.A.

Randall E. McClelland   Department of Cell and Molecular Physiology, UNC School of Medicine, Chapel Hill, North Carolina, U.S.A.

Stuart A. C. McDonald   Digestive Diseases Centre, University Hospitals Leicester, Leicester, U.K.; Histopathology Unit, Cancer Research U.K., Lincoln's Inn Fields and Department of Histopathology, Bart's and the London School of Medicine and Dentistry, London, U.K.

Huseyin Mehmet   Institute of Reproductive and Developmental Biology, Imperial College, London, London, U.K.

Aloa Melhem   Department of Cell and Molecular Physiology, UNC School of Medicine, Chapel Hill, North Carolina, U.S.A.

Rebecca J. Morris   Department of Dermatology, Columbia University Medical Center, New York, New York, U.S.A.

Hiroyuki Ohba   Department of Physiology, Keio University School of Medicine, Tokyo and Core Research for Evolutional Science and Technology (CREST), Japan Science and Technology Agency, Saitama, Japan

Hideyuki Okano   Department of Physiology, Keio University School of Medicine, Tokyo and Core Research for Evolutional Science and Technology (CREST), Japan Science and Technology Agency, Saitama, Japan

Hirotaka James Okano   Department of Physiology, Keio University School of Medicine, Tokyo and Core Research for Evolutional Science and Technology (CREST), Japan Science and Technology Agency, Saitama, Japan

Christopher S. Potten   Epistem Ltd. and School of Biological Sciences, University of Manchester, Manchester, U.K.

Jane Prophet   CARTE, University of Westminster, London, U.K.

Richard P. Redvers   Epithelial Stem Cell Biology Laboratory, Peter MacCallum Cancer Centre, Melbourne, Victoria, Australia

Lola M. Reid   Department of Cell and Molecular Physiology, Department of Biomedical Engineering and Program in Molecular Biology and Biotechnology, UNC School of Medicine, Chapel Hill, North Carolina, U.S.A.

Andrew G. Renehan   Department of Surgery, Christie Hospital NHS Trust, Manchester, U.K.

Adnan Z. Rizvi   Department of Surgery, Oregon Health and Science University, Portland, Oregon, U.S.A.

Ingo Roeder   Institute for Medical Informatics, Statistics, and Epidemiology, University of Leipzig, Leipzig, Germany

Masanori Sakaguchi   Department of Physiology, Keio University School of Medicine, Tokyo and Core Research for Evolutional Science and Technology (CREST), Japan Science and Technology Agency, Saitama, Japan

Eva Schmelzer   Department of Cell and Molecular Physiology, UNC School of Medicine, Chapel Hill, North Carolina, U.S.A.

James L. Sherley   Division of Biological Engineering, Center for Environmental Health Sciences, Biotechnology Process Engineering Center, Center for Cancer Research, Massachusetts Institute of Technology, Cambridge, Massachusetts, U.S.A.

Takuya Shimazaki   Department of Physiology, Keio University School of Medicine, Tokyo and Core Research for Evolutional Science and Technology (CREST), Japan Science and Technology Agency, Saitama, Japan

Gilbert H. Smith   Mammary Biology and Tumorigenesis Laboratory, Center for Cancer Research, National Cancer Institute, National Institutes of Health, Bethesda, Maryland, U.S.A.

Yvonne Summers   Northern Ireland Cancer Centre, Belfast City Hospital, Belfast, U.K.

Neil Theise   Division of Digestive Diseases, Departments of Medicine and Pathology, The Milton and Carroll Petrie Division of Beth Israel Medical Center, New York, New York, U.S.A.

Akinori Tokunaga   Department of Physiology, Keio University School of Medicine, Tokyo and Core Research for Evolutional Science and Technology (CREST), Japan Science and Technology Agency, Saitama, Japan

Ian P. M. Tomlinson   Molecular and Population Genetics Laboratory, Cancer Research U.K., London Research Institute, London, U.K.

William S. Turner   Department of Biomedical Engineering, UNC School of Medicine, Chapel Hill, North Carolina, U.S.A.

Reaz Vawda   Institute of Reproductive and Developmental Biology, Imperial College, London, London, U.K.

Eliane Wauthier   Department of Cell and Molecular Physiology, UNC School of Medicine, Chapel Hill, North Carolina, U.S.A.

Melissa H. Wong   Department of Dermatology, Cell and Developmental Biology, Oregon Health and Science University, Portland, Oregon, U.S.A.

Nicholas A. Wright   Histopathology Unit, Cancer Research U.K., Lincoln's Inn Fields and Department of Histopathology, Bart's and the London School of Medicine and Dentistry, London, U.K.

Hsin-lei Yao   Department of Biomedical Engineering, UNC School of Medicine, Chapel Hill, North Carolina, U.S.A.

Lili Zhang   Department of Infectious Diseases, Nanjing Medical University, Nanjing, China

1

Mathematical Modeling of Stem Cells: A Complexity Primer for the Stem-Cell Biologist

Mark d'Inverno
Centre for Agent Technology, Cavendish School of Computer Science, University of Westminster, London, U.K.

Neil Theise
Division of Digestive Diseases, Departments of Medicine and Pathology, The Milton and Carroll Petrie Division of Beth Israel Medical Center, New York, New York, U.S.A.

Jane Prophet
CARTE, University of Westminster, London, U.K.

INTRODUCTION

Many of us have been fascinated by the straight line of ants stretching from food sources to the anthills we found in our gardens. Taking a closer look, we found that this straight line was made from hundreds of individual, industrious ants, each behaving energetically. If we next focus on an individual ant's behavior (from a modeling perspective, looking at the individual elements of a system is often called the *micro view*), it is very easy to interpret the behavior of the individual as unfocused and chaotic, and it is certainly very difficult to interpret its behavior as being *purposeful* when taken in isolation. It is only when we take a step back and look at the behavior of the entire group (called the *macro view*) that we can observe a purposeful global system of behavior. This purpose, bringing food back to the anthill, *emerges* from the total of the individual behaviors and interactions of the apparently undirected individual. Somehow, the sum of the local interactions of each individual—responding only to their local environment—produces a stable, surviving system, even though individual ants get lost or die.

This system is a commonly cited example of a large class of systems known as *complex adaptive systems*. In this paper, we provide details of several examples of such systems that illustrate some fundamental qualities and issues, before describing the properties of such systems in general. We will argue that this is precisely the paradigm to understand the global behavior of stem cells. We also discuss some of the details of our current project to model and simulate stem-cell behaviors using this paradigm.

Simulating Ants

Using techniques from computer science, we can build an artificial model of ants to simulate the essential behavior of an ant colony by defining the behavior of each ant using the same simple computational algorithm. Some readers might be unclear on what an algorithm is, but conceptually it is no more than an automatic process, a clearly defined sequence of explicit instructions. This process defines how the ant should behave, precisely and completely. Let us consider one possible way to codify an individual ant's behavior as detailed in the rules that follow. (This example is adapted from many examples of the ant algorithm. The interested reader is encouraged to look at education.mit.edu/starlogo/and cognitrn.psych.indiana.edu/rgoldsto/complex/for some excellent examples of complex systems such as ant colonies.) There are three ordered rules that determine behavior:

Rule 1. Wander around randomly.

Rule 2. If you do happen to find some food, take it back to the anthill and leave a trail of pheromones that will evaporate over time. Once you have done this go back to Rule 1.

Rule 3. If you find a pheromone trail, then follow it in the direction that takes you away from the anthill until either:

(i)   You find food and can perform Rule 2, or

(ii)   The trail disappears, in which case, go back to Rule 1.

It is a simple matter to set up this simulation. First, we build an artificial model of an environment that is typically a two-dimensional space. This environment will contain the ant colony and several sources of food scattered around within it. We then program each of our ants with exactly the same algorithm described earlier. Lastly, we provide some initial conditions for this simulation as follows:

1.   Place the anthill at some location.
2.   Place the food sources at other locations.
3.   Create a reasonably large (typically greater than 100) number of ants.
4.   Position each individual ant at some point close to the anthill.

The program can then be *run*. This means that each ant starts processing the rules; it searches randomly and changes its behavior if it discovers entities in the environment, such as food or pheromones. What is extraordinary about the first time most observers see such an experiment is just how *intelligent* and *sophisticated* the simulated ant colony behaves as a whole. They soon find the food sources and bring them back, bit by bit, to the anthill. Some sources are found and depleted more quickly than others but, in most cases, all the food is discovered and brought back to the anthill.

This example is often used as an exemplar of a *complex system* (1). A complex system is often described as one in which the overall behavior of a system is somehow qualitatively different from that of the individual parts; the parts themselves behave as though they are not goal-directed, whereas the global behavior of the system is *meaningful*. In our current example, it is a simple matter to perceive the goal of the colony as searching for, and returning home with, food. This meaning only arises or *emerges* when looking at the whole system and comes from the collective behavior and interactions of a set of simple elements called *agents* (2). Because each individual agent's behavior or interaction may have significant ramifications for the entire future of the system, the global behavior of any such system cannot be predicted. Currently, there is no mathematical system that can model complex systems in a sufficiently detailed way for such predictions to be made. Only by *running* the simulation can we see what will happen in the future, given a specific initial state. It is practically impossible to predict what global behavior will arise in general. Before we discuss agents, complex systems, and emergence in more detail, we will first give descriptions of two other examples of systems where sophisticated global behavior arises through the very simple behavior of a large number of interacting computational entities.

Simulating Populations: The Game of Life

One of the first and, at the time, most extraordinary examples of an algorithm-based system that is often described as an example of a complex system is the so-called *Game of Life* by John Conway (3), which he first proposed in the late 1960s. The aim of constructing this artificial system was to show how the simplest set of rules for individual agents could be used to generate sophisticated global behavior that modeled some very basic principles of birth, death, and survival in a homogenous population. By homogenous, we mean that all agents (as with our simulated ants) had the same set of behavioral rules. In this simulation, agents would survive or die depending on the conditions of their *local environment*. If they were isolated or overcrowded, the agent would die. If neither was the case, then the agent would survive and, furthermore, in some appropriate conditions where there was neither overcrowding nor isolation, agents would be born into empty locations.

The rules of this system were chosen such that the behavior of each population was unpredictable and, after some revisions, the following set of criteria were developed:

1. There should be no initial configuration for which there is a simple proof (law) that the population can grow without limit.
2. Even without any such laws some initial configurations can grow without limit.
3. Depending on the initial conditions, the population would either:
   (i) change and grow but ultimately die,
   (ii) settle into a stable equilibrium that would never change, or
   (iii) enter an oscillating phase in which they repeat a cycle of two or more periods endlessly.

Once again, there are simple agents (called counters) that can inhabit locations in a two-dimensional grid environment. On each successive *clock tick* of the system (sometimes called a *move* and sometimes a *generation* in the system's life history) there are very simple rules about whether counters survive or die or whether new ones are born. These rules are applied simultaneously as follows:

1. Survival. Every counter with either two or three neighboring counters survives for the next generation.
2. Death.
   (i) Each counter with four or more neighbors dies (i.e., is removed) because of overcrowding.
   (ii) Every counter with one or zero neighboring counter dies from isolation.
3. Birth. Every empty cell adjacent to *exactly* three neighbors is a birth cell and a counter is created in this space on the next move.

What was so extraordinary about this system was that it was one of the first to show how complex global behavior could emerge through the behavior and interaction of a number of very simple agents. Depending on the initial conditions (which grid squares had counters at the beginning of time, referred to as $t = 0$), whole societies would seemingly rise then die or oscillate between different stable states, cycling endlessly. The interested reader is encouraged to visit (4) to download and experiment with the game itself.

Engineering Complex Systems: Robotic Rock Collection on Mars

Rather than wondering in amazement at the emergent global behavior of various systems, it is often the case that we wish to *design* a system that has key global properties that arise from a set of simple interacting agents. One of the best examples of this was in the work of Steels (5), that was subsequently described excellently by Wooldridge (6). The problem Steels set himself can be paraphrased as follows:

> Suppose we want to collect samples of a particular type of rock on another planet such as Mars. We don't know where it is, but it's typically clustered together. A number of vehicle agents are available that can drive around the planet and later re-enter a mother spaceship and go back to earth. There is no detailed map of the planet but it is known that there are rocks, hills, and valleys that prevent the agents from communicating.

In a solution to this problem, Steels makes use of an agent architecture, first proposed by Brooks, called the subsumption (7,8) architecture. In the subsumption architecture, the different behaviors of an agent are layered with those lower-level behaviors that are most critical to the agent and that take precedence over any higher-level behaviors. At the time, Brooks' work was revolutionary because most researchers believed that the only way to build robots capable of sophisticated behavior was to use techniques from artificial intelligence such as symbolic representation and reasoning. He suggested that intelligent behavior (from the perspective of the individual or group) does not necessarily require agents to have a sophisticated model of the world. He argued that intelligence could emerge from simple systems that responded to the environment by stimulus-response rules, as in the two examples we have discussed previously. In such systems, the environment provides a stimulus that causes a rule to fire and the agent responds with some behavior that in turn affects the environment and so (possibly) the future behavior of itself and other agents sharing the same environment.

He pioneered the idea that intelligent systems could be engineered in this way and that intelligent behavior is an emergent phenomenon arising from the interaction of *societies of nonintelligent systems* (7) as stated earlier. Subsumption architecture comprises eight task-achieving behaviors, each of which is implemented separately. The hierarchy of layers reflects how specific the behavior is—the more specific the task, the higher the level. In the case of the mobile robot, there are eight levels from zero to seven that relate to contact avoidance, wandering, exploring, building maps, noticing change, distinguishing objects, changing the world according to goals, and reasoning about the behavior of others.

The first step in the agent's construction is to build the zeroth control level and, once this has been tested, to build the first control level on top of the zeroth level. The first level has access to the data in level zero and can also supply its own inputs to this layer to suppress the normal activity of the zeroth layer. The zeroth level continues to execute, unaware that there is a higher level intermittently influencing its behavior. This process is then repeated for each successive layer. Subsequently, each layer competes to control the behavior of the robot.

Before Steels designed his subsumption architecture for each of the agents, he introduced a gradient field, so that the agents could always locate the mother ship. For the agent to find the mother ship, all it needed to do was move up the gradient. Then he programmed the agent as follows. The first rule at level zero in the subsumption architecture and, therefore, the rule with the highest priority was concerned with obstacle avoidance. The other rules can then be described in decreasing order of priority as follows. Note that there are five levels.

Rule 1. If you detect an obstacle, then avoid it.
Rule 2. If you are carrying samples and at the mother ship, drop the samples.
Rule 3. If you are carrying samples and not at the base, then travel up the gradient.
Rule 4. If you detect a sample, pick it up.
Rule 5. Move randomly.

Using this set of rules, the agents are *noncooperative*. There are no interactions between them and, there is certainly nothing that looks like global emergent system intelligence. However, inspired by the ant colony example described earlier, he introduced a new mechanism. The idea was that agents would carry radioactive crumbs that could be dropped, picked up, and detected by passing agents. Using this simple technique, sophisticated cooperation between the agents could now take place. The rules are almost identical except that Rule 3 is altered and there is a new level (Rule 5) introduced just before the highest level behavior to obtain a six-level architecture. These new rules introduce *cooperation* between the agents.

Rule 3. If you are carrying samples and you are not at the base, then *drop 2 crumbs* and travel up the gradient.
Rule 5. If you sense crumbs, then pick up one crumb and travel down the gradient.

What was extraordinary about this work was that it showed that near optimal performance could occur with a collection of very simple agents. Along with Brooks, Steels was one of the first to show how intelligent systems could be designed from the emergent behavior of simple interacting agents to achieve real tasks. From an engineering perspective, the solution was also significant because it was cheap on computational resources (these agents are really *very* simple!) and it was robust. And as with our ants, one or two agents breaking down would not impact significantly on the overall system performance.

In many senses, this was an attempt to harness the power of complex systems. However, there is a very important but subtle point to realize here: the complexity was predetermined; agents were specifically engineered to achieve an overall system behavior. However, we can recapture the emergence by realizing that if the simple specification (rule set) of each of the agents was shown to an observer, all but the most experienced programmer would not anticipate the optimal, collective, cooperative system behavior. To the observer then, the system would be displaying emergence. In short, there is some sense here of emergence being a personal phenomenon, that "emergence is in the eye of the beholder" if you like. This is quite important, and we shall discuss emergence and other issues relating to complex systems that we have touched on in our three examples in the next section.

EMERGENCE

The systems we discussed earlier are some of the key original examples of computational *complex adaptive systems* (9,10). There are many other noncomputational systems that exhibit the same kind of emergent self-organization including (11) economies, social organisations [human (12) or animal], embryologic development, the weather, traffic, ecologies, growth of cities, the rise and extinction of species, and the diversity of immune system responses. One key factor that is common to all these systems is that there is an *emergent* self-organization arising on the macro-scale from micro-scale interactions of the individuals constituting the system.

There has been a great deal of debate about what constitutes such systems, but it is essentially the notion that some kind of *order* or *structure* or *intelligence* occurs that is not predetermined. Perhaps, the best definition of emergence (and certainly one of the most cited) is given by Cariani (13). He first describes emergence as involving "the creation of qualitatively new structures and behaviors, which cannot be reduced to those already in existence" and then goes on to describe three kinds of emergence: computational, thermodynamic, and "relative to a model."

Computational Emergence

The three examples we first described in this chapter can be seen as examples of *computational emergence* in which complex global behaviors or structures arise from local computational interactions. Cariani makes the point that such emergence occurs only because of the observational frame through which the system is considered (i.e., the kind of person the observer is, and the degree of technical understanding of the system's underlying algorithms they have). He argues that because there is simply a set of initial conditions and behavioral rules, there is a sense that everything is predetermined and the consequence of this is that nothing is emergent. In Cariani's view, therefore, the game of life is not displaying emergence, because as soon as you encode it in a program, you have by default defined the set of possible states for that program. Introducing stochastic (random) elements into the program does not help either, he argues, because even random elements are at some lower level of the computational process deterministic. That is, the random parameters are themselves actually generated by deterministic algorithms.

A good way to understand how something at one level can be random but at another lower level completely determined is in the tossing of a coin before a football game. The home captain tosses the coin into the air and the opposition captain calls. From the perspective of both captains, as the coin is spinning in the air, whether the coin ends up heads or tails is totally up to chance. Half the time it will be heads and half the time tails, and that is all both captains know. If, however, the opposition captain was possessed with extraordinary powers of perception and mathematical ability, they could work out the rate of spinning, gravitational pull, air resistance, wind velocity, and so on and calculate with all certainty whether the coin will land on its head or not. At this lower level then, the tossing of a coin is a deterministic process completely decided by the laws of physics. As soon as we return to the higher level of everyday human modeling and perception, we lose this determinism, and nondeterminism is reintroduced.

In order to reintroduce emergence into work from multi-agent systems and related disciplines such as Artificial Life, Cariani then moves toward introducing a more pragmatic definition of emergence as being *relative to a model*.

Emergence Relative to a Model

While there is no "system emergence" with any computational system, there is clearly emergence occurring from the point of view of the observer, and so Cariani provides this very pragmatic view. (Recall our previous comment "emergence is in the eye of the beholder.") Emergence arises "relative to a model," because an observer of a computational system does not typically have a detailed view of the processes that occur inside the system. That is, the observational frame is incomplete and the observed emergent behaviors arise because they are based on issues that are *outside of this frame*. A more succinct definition of this type of emergence is the "deviation of the behavior of a physical system from an observer's model of it." Cariani summarizes this category as follows:

> The emergence-relative-to-a-model view sees emergence as the deviation of the behavior of a physical system from an observer's model of it. Emergence then involves a change in the relationship between the observer and the physical system under observation. If we are observing a device which changes its internal structure and consequently its behavior, we as observers will need to change our model to track the device's behavior in order to successfully continue to predict its actions.

Thermodynamic Emergence

This category is a much stronger, physical view of emergence and is essentially characterized as the emergence of order from noise in the physical environment. Again though, it is the phenomenon where nondeterministic processes at the micro-level lead to structures or behaviors at the macro-level. The typical example of this type of emergence is when considering a particular gas such as oxygen. The nondeterministic behavior of electrons, atoms, and molecules somehow leads to a stable gas with well-defined properties relating to pressure, temperature, and volume at a higher level.

However, there is also some notion here of emergence relative to a model, even though Cariani sees fit to distinguish it from computational emergence. Gas is only an emergent property of molecules, because we do not fully understand how that process works. If we could understand all the laws of the universe, then getting gas from molecules might seem pretty obvious to us. Taking this view, it would seem that the only way to understand the phenomenon of emergence is by having a model of the individuals (observers) who are perceiving it.

COMPLEX ADAPTIVE SYSTEMS WITH MULTIPLE INTERACTING AGENTS

Although there is no agreed upon definition of exactly what constitutes either a complex system (14), emergence (15), or even what an agent is (16), there are a number of general properties that we list and outline here that have been instrumental in guiding our work in modeling the society of stem cells:

- Order is emergent rather than predetermined.
- The system's future is, in general, unpredictable.
- The basic entities of a complex system are agents. There is a huge debate that has been raging for years about what constitutes an agent (16) but, in this context, they are autonomous or semi-autonomous entities that seek to maximize some measure of usefulness (this could relate to a goal, motivation, or utility) by responding to the local environment according to a set of rules that define their behavior. (This is sometimes referred to in more economically-biased accounts as the agent's *strategy*.)
- The individuals are not aware either of the larger organization or its goals and needs. Clearly any single agent within the system cannot know the state and current behavior of every other agent and, as a result, cannot determine its behavior based on such complete global system information. Instead, the behavior of agents is governed by rules based on the local environment.

- Agents typically follow reactive rules that are typically of the form: if *condition* then fire *action*. For example, a possible rule for an agent might be *if there's a space next to me* and *I am currently too hot*, then *move into an empty space*.
- Agents can perceive aspects of the environment they are in and can act so as to change the state of the environment. Critically, in a complex system, agents must affect the environment in such a way that the environmental change can (*i*) be perceived by others and (*ii*) affect the behavior of others. (Recall the difference between the noncooperative and cooperative versions of the robot vehicle rock collectors.) That is, the agents must have a reasonable degree of *interaction* beyond that of, say, simple obstacle avoidance (17).
- The rules of an agent will often contradict and there must be some mechanism (possibly nondeterministic) for selecting from competing behaviors/rules. In general, the behavior of an agent will not be deterministic.
- Agents may be equipped with the ability to adapt and learn rules so as to have a more effective way of maximizing their usefulness in given situations. New rules would try, for example, to make the agent more able to act effectively in a wider variety of environmental situations.
- Rules may compete for survival. The more a rule is used in determining behavior, the greater the chance that it has of surviving in the future. Rules that are seldom or never used will have less chance of survival. Rules may change randomly or intentionally and may be integrated for more sophisticated action.
- Agents are resource-bounded and can only perceive (or experience) their *local* environment. Typically, they will also have the ability to determine what to do next when there is incomplete or contradictory sensory information about their local environment. It is also possible that agents might have access to some global information.
- A few individuals or agents will not make a sustainable complex system. It is only when the number of agents reaches a certain critical threshold that the system will exhibit global, meaningful behavior.

Using these basic principles, we have built and are currently implementing a model of stem cells and cell lineages. There are also several others who have done similar work that we discuss briefly here.

MODELING STEM CELLS AND CELL LINEAGES AS COMPLEX ADAPTIVE SYSTEMS

Although mathematical modeling of stem-cell lineage systems is critical for the development of an integrated attempt to develop ideas in a systematic manner, it has not been a research area that has received a large amount of attention. Over the last year or so, there has been a noticeable climate change in this respect, and there is now a growing awareness of the need to use mathematical modeling and computer simulation to understand the processes and behaviors of stem cells in the body. Some reasons have been pointed out by Viswanathan and Zandstra (18) in an excellent survey of mathematical techniques for predicting behaviors of stem cells. We summarize the key points here:

- In the adult body, stem cells cannot be distinguished morphologically from other primitive nondifferentiated cell types.
- Extracting stem cells from an embryo means sacrificing it, posing serious ethical difficulties.

- There is no way to determine whether any individual isolated cell is a stem cell and to be able to model what its potential behavior might be. It is not possible to make any definite statements about this cell. At best, it can be tracked and its behavior observed, though clearly this behavior is simply one of many possible paths. The notion of a stem cell refers to the wide-ranging set of potential behaviors that it might have that are influenced by internal, environmental, and stochastic processes.
- The number of possible interactions and behaviors of a large number of stem cells makes the system extremely complex in all the senses described earlier. Theoretical simplifications are key to understanding fundamental properties.

There is, thus, a need for new theoretical frameworks and models that can be directly mapped to a computer simulation and that look at the dynamic self-organization of stem cells.

Before introducing a summary of our own work, we will consider some related theoretical investigations. We first introduce the recent work of Agur et al. (19) as it is very similar, algorithmically, to the game of life introduced earlier in this paper.

A Simple Discrete Model of Stem Cells

In their recent work, Agur et al. used a model very similar to that of the game of life to understand what mechanism might be employed for maintaining the number of stem cells in the bone marrow and producing a continuous output of differentiated cells. This work is important, because it is one of the few examples where a mathematical model has been used to show what properties of stem cells might be required to enable the maintenance of the system's homeostasis.

Essentially, they model a niche as having the ability to maintain a reasonably fixed number of stem cells, to produce a supply of mature (differentiated) cells, and to be capable of returning to this state even after very large perturbations that might occur through injury or disease. The behavior of a cell is determined by both internal (intrinsic) factors (a local clock) and external (extrinsic) factors (the prevalence of stem cells nearby), as stated by the authors as follows:

1. Stem-cell behavior is determined by the number of its stem-cell neighbors. This assumption is aimed at simply describing the fact that cytokines secreted by cells into the micro-environment are capable of activating quiescent stem cells into proliferation and differentiation.
2. Each cell has internal counters that determine stem-cell proliferation, stem-cell transition into differentiation, and the transit time of a differentiated cell before it migrates to the peripheral blood.

The niche is modeled as a connected, locally finite, undirected graph, but for most intents and purposes, we can visualize this as a two-dimensional space made up of grid squares as in the game of life. Their model certainly applies for this topology.

Any grid square is either empty, or it is occupied by either a stem cell or a differentiated cell. A stem cell is able to interpret messages from neighboring locations (horizontal or vertical, not diagonal) such that it knows what is at those locations. Stem cells can divide into two stem cells (called proliferation) or become determined cells (no division takes place). Determined cells stay in the niche for a period and then eventually leave to enter the bloodstream.

There are three constant values (let us call them N_1, N_2, and N_3) that are used to reflect experimental observation. The first constant (N_1) represents the time taken for a differentiated cell to leave the niche. The second (N_2) represents the cycling phase of a stem cell; a certain number of ticks of the clock are needed before the cell is ready to consider dividing. Finally, the third (N_3) represents the amount of time it takes for an empty space that is continuously neighbored by a stem cell to be populated by a descendent from the neighboring stem cell. The rules of the model, expressed in simple English, are as follows:

Rule for determined cells

1. If the internal clock has reached N_1, then leave the niche. Reset local clock to 0.
2. If the internal clock has not yet reached N_1 then increment is 1.

Rule for stem cells

1. If the counter at a stem-cell location has reached N_2 and all stems are neighbors, then become a differentiated cell. Reset the clock to 0.
2. If the counter of a stem cell is equal to N_2, but not all the neighbors are stem cells, then do nothing. Leave clock unchanged.
3. If the counter has not reached N_2, then do nothing except increment the clock.

Rule for empty spaces

1. If the counter at an empty grid has reached N_3 and there is a stem-cell neighbor, then introduce (give birth to) a stem cell in that location. Reset clock.
2. If the counter at an empty grid has not reached N_3 and there is a stem-cell neighbor, then increase the clock.
3. If there are no stem-cell neighbors at all, then reset the clock to 0.

A move is just as it is in the game of life; the *next state* of the system is a function of the clock, the state of the cell, and the state of the neighboring cells. All locations are then updated simultaneously as before. As with the game of life, there are no stochastic elements. The only real difference is that the agents have a local state (clock) that is not present in the game of life. What is remarkable here is that this simple model allows for sophisticated global behaviors to arise. All the basic *common sense ground rules* of stem cells to proliferate, to remain quiescent, and to produce continuous supplies of differentiated cells can be found in all the possible behaviors of this systems (That is, if you discount extreme situations, such as where all the grids are occupied by stem cells.). Moreover, there is always a sufficient density of stem cells in the niche and the system never dies out.

Although there are a number of difficulties with this work, in particular the fact that it requires spaces in the niche to have counters (rather than the cells as originally set out in the text), it is one of the few attempts to capture observable qualities of stem-cell systems in a simple mathematical model that can be simulated computationally. It is a very simple model and, as a result, the authors were able to mathematically *prove* many properties of their system, and their results are extremely important for paving the way for a more sophisticated analysis of various stem-cell-like properties in the future.

What is perhaps most extraordinary about this work is how similar the basic algorithms are to that of the game of life and the fact that it took 30 years to get from a cute mathematical game to cutting-edge work on the theoretical modeling of stem cells!

Plasticity and Reversibility in Stem-Cell Properties

From a biological viewpoint, the model of Agur et al. does not allow any reversibility or plasticity in the basic properties of cells. For example, once a cell has differentiated, it cannot become a stem cell again (or, in a more continuous view, more plastic). Moreover, once a cell has left the niche, it cannot return. A recent example, an approach that uses a more sophisticated model and addresses these issues, is that of Loeffler and Roeder at the University of Leipzig, who model hematopoietic stem cells using various (but limited) parameters including representing both the growth environment within the marrow (one particular stem-cell niche) and the cycling status of the cell (20–22). The ability of cells to both escape and re-enter the niche and to move between high and low niche affi-nities (referred to as within-tissue plasticity) is stochastically determined. The validity of their model is demonstrated by the fact that it produces results in the global behavior of the system that exactly matches experimental laboratory observations. The point is that the larger patterns of system organization emerge from these few simple rules governing vari-ations in niche-affinity and coordinated changes in cell cycle.

Another example, also from Loeffler, working with colleagues Potten and Meineke, models movement and differentiation of small intestinal stem cells from the stem-cell niche to the villous tip in a two-dimensional lattice-free cylindrical surface (23). In this model, cells interact by viscoelastic forces. Simulations were compared directly with experimental data obtained from observations of cells in tissue sections. These showed that the model is consistent with the experimental results for the spatial distribution of labeling indices, mitotic indices, and other observed phenomena using a fixed number of stem cells and a fixed number of transit cell divisions. Moreover, the model suggested a gradient, perhaps a diffusable protein, which could explain differen-tiation of cells as they moved up the villus. Thus, not only did the model fit experimental data already in hand, but it made predictions that could form the basis of new investigations.

An Agent-Based Approach to Modeling Stem Cells

In our current work, we are building a more comprehensive formal model of cells as reac-tive agents responding to local environmental factors that can maintain some balance of cells under various conditions, using the criteria outlined in Section 3. The intention is to provide a toolkit for researchers and students to investigate behaviors of stem-cell systems, given a set of rules, environmental influences, and so on. As with Roeder and Loeffer (22), we will also allow reversibility, plasticity, and nondeterminism, but we model a greater number of internal and environmental parameters. As with the related work we have discussed in this chapter, there is something in common here with the "reverse engineering" approach of Steels and his design of robots to achieve an overall system behavior. What we wish to do is build our model of a stem cell in such a way that the overall system behavior has many of the observable qualities viewed in current medical experiments.

Currently, all cells are modeled as agents with *identical* abilities, perceptual capa-bilities, and rules. In line with Roeder and Loeffer (22), we see "stemness" not as a "yes" or "no" quality of any given cell, but as a continuum of potential behaviors. The more a cell has stemness, the more likely a cell is to behave in a stem-like way. The agent model details how the internal state, the local environment (proteins, populations, fluid pressure, and so on) affect the probabilities of behaving in certain ways, such as moving to or from a niche and cell division.

Their state will include information on how many divisions can occur, how likely the cell is to stay in the niche, whether it's more or less likely to divide, whether division is symmetric or asymmetric, how sensitive the cell is to protein signaling in the micro-environment, how likely it is to react to them once they are sensed, and so on. They can also perceive local environmental conditions, such as the relative concentrations of other stem cells and cells at various stages in the set of available lineages we model, as well as various signaling proteins such as SDF1. Their behavior will be nondeterministic and based on their current state and the state of the local environment. Emerging from this nondeterministic micro view, we expect a stable dynamic system that can be re-instantiated even after traumatic events.

Suppose, for example, that a certain stem cell is in a given environment. At any stage, it will have some probability of dividing and, if it does, there will be some probability of producing a daughter cell along one lineage and another probability of producing a cell along an entirely different lineage, and so on. Because we cannot say for certain what will happen—sometimes one action will happen and, in exactly the same situation, sometimes something else might happen—we introduce randomness or, more formally speaking, nondeterminism into the system. As we stated earlier, most commentators argue that some degree of nondeterminism is needed for a system to be a complex adaptive one. Self-organization fails to emerge in completely determined systems.

Once we simulate these agents and run the system so that it is in a kind of stable equilibrium (of the kind exemplified by the game of life example we described earlier), we can then consider the effects of disease and life-threatening environments. It will be possible to model and investigate what kinds of behavioral changes, even of a single stem cell because of a chance mutation in its rule base, might lead to system imbalance or collapse. In other words, do precisely defined rule changes mimic known clinical conditions?

This conceptual approach thus focuses on the fact that complex, adaptive systems typically have multiple equilibrium states where, for example, the number of agents of a particular kind may be kept constant. An equilibrium can be more or less stable: a very stable equilibrium needs massive events (either internal or external) to affect it while a nonstable one can be upset by relatively small events. These less-stable equilibria are more dangerous for the safety of a system, as a tiny event may lead to massive system change or total system collapse. Commonly cited examples are mass extinctions of species, collapse of stock markets, and the demise of cultures and civilizations. It is often changes in the interactions or behavior at the micro-level that affect phenomena such as mass extinctions.

Analogously, the failure of stem-cell systems is sometimes not merely due to the size of the internal or external change: it may be simply a necessary result of the generally high durability and sustainability, but complexity, of the cell system. Aplastic anemia, for example, a complete failure of the hematopoietic system, may not have a specific precipitating event. Likewise, acute hepatitis A is usually benign and self-limited, but a very few infected people suffer massive hepatic necrosis leading to death or the need for transplant. The unpredictability of these events may relate to our limited understanding of pathogenesis, but it might instead be inherent because the stem-cell system is a complex one.

In addition, we can use the formal model of our stem-cell complex system to build a computer simulation of a large system of stem cells and progeny. As we described earlier in this chapter, computer simulations have been very successful in showing how emergent, global, self-organizing properties can arise through very simple descriptions of individual behavior. However, the computational demands required to model large systems are enormous, and we will have to use a grid cluster to perform this simulation of hundreds of interaction agents.

The formal specification of our model is now being used as the blueprint to build the simulation. Using logic we can prove that the simulation implements the model exactly and completely. As far as we are aware, this has not been achieved before and it means that observed events produced by running the simulation can be carefully interpreted within the semantic context of the formal structured model. Careful statistical analysis of our simulated clinical events may shed a new and very different light on these dire occurrences.

RAMIFICATIONS FOR CURRENT THINKING

A new perspective from which to debate some of the key contested issues of adult stem-cell research arises when one considers cell lineages from a complex systems approach. For example, the longstanding debate as to whether stem-cell lineages are determined or stochastic processes becomes clear (22). As in the work of Loeffler and colleagues, the stochastic elements are required to obtain the observable results. Indeed, the increasing number of articles on the reversibility of gene restriction makes the stochasticity of lineage fate unavoidable in conceptualizing issues of cell plasticity (24,25). These theoretical notions are backed up by the results of both clinical studies (26) and single-cell culture and gene-expression experiments (27,28) where a greater variability of gene-expression pathways is revealed than would be expected from a complete determinism.

In this paper, we do not go into a detailed discussion of these findings, but we wish to make it clear that the evidence now strongly indicates a nondeterministic view. This is crucial for our complex-system interpretation to be appropriate: if we conceive of cell lineages as complex and adaptive, then stochasticity is implicit because fluctuations are necessary for self-organizing systems to explore new possibilities.

Another current controversy concerning adult stem-cell lineages relates the often low engraftment from bone marrow into other system organs: often less than five percent, sometimes less than one percent, in the absence of overt, severe injury (29,30). Some have argued that even if bone marrow plasticity can be demonstrated, such low levels of engraftment from the blood are physiologically trivial and insufficiently robust to be of relevance to tissue maintenance (31).

However, if we consider these alternate lineage phenomena as parts of a complex adaptive system, it reveals to us that the converse is more likely to be true. The documented low level of apparently random fluctuation, this "quenched disorder" that we mentioned earlier, is precisely what allows the system to be adaptive. Going back to our ants example, it is only the small percentage of ants straying from the main path that enables the formation of new paths to food in the event that the current line becomes interrupted or the food source runs out. From the complex system point of view, the low-level engraftment fluctuations are critical: without them, robust responses to injury might not be so efficient or even possible.

It is precisely this intermediate level of stochastic variation, somewhere between a fully determined system (where all events can be predicted; the behavior of each element is simply a function of the current state of the whole system) and a totally nondetermined system where any event can happen at any time (referred to as "chaos") that makes cell lineage systems, and therefore our own bodies, complex, adaptive, and alive.

CONCLUSION

We believe that recent experimental evidence makes it clear that it is increasingly necessary to use formal, computational models to investigate the nature of stem-cell systems rather than stem cells in isolation. There are several key reasons. First, adult stem cells cannot be easily isolated; indeed, it may be that it is only by looking at their behavior in a system, not in isolation, that we can tell what kind of cell we were originally looking at. Second, even if we were able to track the behavior of a cell in the body, it would only tell us about one of the possible behaviors of the original cell; it tells us nothing about the potentially infinite array of behaviors that may have been possible if the environment and the chance elements had been different. Third, there is evidence to suggest that mechanical forces on cells are critical in determining stem-cell behavior. If this is the case, then any act of withdrawing cells from the original system would potentially affect that cell irrevocably (32). Fourth, by removing a cell from its original and natural habitat, the new environmental conditions will influence future behavior and lead to misleading results. Fifth, it is the totality of the stem cells as a *system* in the human body that is important. A key quality of the system is its ability to maintain exactly the right production of cells in all manner of different situations.

In response, therefore, we have developed a formal model that reflects many of the key experimental and recent theoretical developments in stem-cell research. Using techniques from multi-agent systems, we are currently building a complex, adaptive system to simulate stem-cell systems in order to provide a testbed from which to be able to investigate their key properties in general and to formulate new experiments to identify the underlying physiological mechanisms of tissue maintenance and repair.

ACKNOWLEDGMENTS

The team of collaborators in this project (entitled CELL) also included the curator Peter Ride and the A-life programmer Rob Saunders (who was instrumental in helping us form some of the views here regarding emergence), both from from the University of Westminster.

REFERENCES

1. Luck M, d'Inverno M. A formal framework for agency and autonomy. In: Proceedings of the First International Conference on Multi-Agent Systems. San Francisco, CA: AAAI Press/MIT Press, 1995:254–260.
2. Odell J. Agents and complex systems. J Object Technol 2002; 1(2):35–45.
3. Gardner M. The fantastic combinations of John Conway's new solitaire game, 'Life.' Sci Am 1970; 223(4):120–123.
4. http://www.bitstorm.org/gameoflife/
5. Steels L. Cooperation between distributed agents through self-organization. In: Proceedings of the First European Workshop on Modelling Autonomous Agents in a Multi-Agent World. Holland: Elsevier Science Publishers, 1990:175–196.
6. Wooldridge M. An Introduction to MultiAgent Systems. New York, NY: Wiley, 2002.
7. Brooks RA. Intelligence without reason, Proceedings of 12th International Joint Conference on Artificial Intelligence. Sydney, Australia, 1991:569–595.
8. Brooks RA. Robust layered control system for a mobile robot. IEEE J Rob Autom 1986; 2(1): 14–33.
9. Lewin R. Complexity: Life at the Edge of Chaos. 2nd ed. Chicago IL: University of Chicago Press, 2002.
10. Johnson S. Emergence. New York, NY: Scribner, 2001.
11. Walker A, Wooldridge M. Understanding the emergence of conventions in multi-agent systems. In: ICMAS95 1995:384–389.
12. Gilbert N, Conte R. Artificial Societies: The Computer Simulation of Social Life. London: UCL Press, 1995.

13. Cariani P. Emergence and artificial life. In: Langton CG, Taylor C, Farmer JD, Rasmussen S, eds. Artificial Life II. Boston, MA: Addison-Wesley Longman Publishing Co., 1991:775–797.
14. Holland J. Emergence: From Chaos to Order. Oxford: Oxford University Press, 2000.
15. Shoham Y, Tennenholtz, M. On the emergence of social conventions: modeling, analysis, and simulations. Artif Intell 1997; 94(1):139–166.
16. d'Inverno M, Luck M. Understanding Agent Systems. 2nd ed. Berlin Heidelberg, New York: Springer, 2003.
17. d'Inverno M, Luck M. Understanding autonomous interaction. In: Wahlster W, ed. ECAI '96: Proceedings of the 12th European Conference on Artificial Intelligence. Budapest, Hungary: John Wiley and Sons, 1996: 529–533.
18. Viswanathan S, Zandstra PW. Toward predictive models of stem cell fate. Cytotech Rev 2004; 41(2/3):1–31.
19. Agur Z, Daniel Y, Ginosar Y. The universal properties of stem cells as pinpointed by a simple discrete model. J Math Biol 2002; 44(1):79–86.
20. Loeffler M, Roeder I. Tissue stem cells: definition, plasticity, heterogeneity, self-organization and models—a conceptual approach. Cells Tissues Organs 2002; 171:8–26.
21. Roeder I. Dynamical Modelling of Hematopoietic Stem Cell Organisation. Ph.D. dissertation, Leipzig University, 2003. http://people.imise.uni-leipzig.de/ingo.roeder/diss_ingo.pdf (accessed July 2004).
22. Roeder I, Loeffler M. A novel dynamic model of hematopoietic stem cell organization based on the concept of within-tissue plasticity. Exp Hematol 2002; 30:853–861.
23. Meineke FA, Potten CS, Loeffler M. Cell migration and organization in the intestinal crypt using a lattice-free model. Cell Prolif 2001; 34(4):253–266.
24. Theise ND, Krause DS. Toward a new paradigm of cell plasticity. Leukemia 2002; 16:542–548.
25. Theise ND. New principles of cell plasticity. C R Biol 2003; 325:1039–1043.
26. Thornley I, Sutherland R, Wynn R, Nayar R, Sung L, Corpus G, Kiss T, Lipton J, Doyle J, Saunders F, et al. Early hematopoietic reconstitution after clinical stem cell transplantation: evidence for stochastic stem cell behavior and limited acceleration in telomere loss. Blood 2002; 99:2387–2396.
27. Madras N, Gibbs AL, Zhou Y, Zandstra PW, Aubin JE. Modeling stem cell development by retrospective analysis of gene expression profiles in single progenitor-derived colonies. Stem Cells 2002; 20:230–240.
28. Krause DS, Theise ND, Collector MI, Henegariu O, Hwang S, Gardner R, Neutzel S, Sharkis SJ. Multi-organ, multi-lineage engraftment by a single bone marrow-derived stem cell. Cell 2001; 105:369–377.
29. Wagers AJ, Sherwood RI, Christensen JL, Weissman IL. Little evidence for developmental plasticity of adult hematopoietic stem cells. Science 2003; 297:2256–2259.
30. Weissman IL, Anderson DJ, Gage F. Stem and progenitor cells: origins, phenotypes, lineage commitments, and transdifferentiations. Annu Rev Cell Dev Biol 2001; 17:387–403.
31. Eastwood M, Mudera V, McGrouther D, Brown R. Effect of mechanical loading on fibroblast populated collagen lattices: Morphological changes. Cell Motil Cytoskel 1998; 40:13–21.
32. Ogawa M. Stochastic model revisited. Int J Hematol 1999; 69:2–5.

2

Theoretical Concepts of Tissue Stem-Cell Organization

Ingo Roeder
Institute for Medical Informatics, Statistics, and Epidemiology, University of Leipzig, Leipzig, Germany

Joerg Galle
Interdisciplinary Centre for Bioinformatics, University of Leipzig, Leipzig, Germany

Markus Loeffler
Institute for Medical Informatics, Statistics, and Epidemiology, University of Leipzig, Leipzig, Germany

INTRODUCTION

Many recent experimental findings on heterogeneity, flexibility, and plasticity of tissue stem cells are challenging the classical stem-cell concept of a pre-defined, cell-intrinsic developmental program. Moreover, a number of these results are not consistent with the paradigm of a hierarchically structured stem-cell population with a unidirectional development. Nonhierarchical, self-organizing systems provide a more elegant and comprehensive alternative to explain the experimental data.

Within the last decade, our modeling attempts in stem-cell biology have evolved considerably and now encompass a broad spectrum of phenomena, ranging from the cellular to the tissue level. On the basis of our results, we advocate abandoning the classical assumption of a strict developmental hierarchy and, instead, understanding stem-cell organization as a dynamic, self-organizing process. Such a concept makes the capabilities for flexible and regulated tissue function based on cell–cell and cell–environment interactions the new paradigm. This would permit the incorporation of context-dependent lineage plasticity and generation of stem-cell heterogeneity, as a result of a dynamically regulated process. This perspective has implications for a prospective characterization of tissue stem cells, for example, regarding gene expression profiles and genetic regulation patterns.

To be validated, such concepts need a rigorous examination by quantitative and predictive modeling of specific, biologically relevant tissues. In this chapter, we provide some general ideas on how to proceed with such theories and illustrate this with a working model of hematopoietic stem cells applied to clonal competition processes. Furthermore, we give an example of how to include the possible effects of a spatial arrangement of cells into the proposed new stem-cell paradigm.

DEFINING TISSUE STEM CELLS

"Is this cell a stem cell?" This frequently posed question implies the idea that one can decide about the capabilities of a selected cell without relating it to other cells and without testing its capabilities functionally. We argue that this is a very naive and unrealistic point of view. To explain this perspective, let us start by taking a look at the definition of tissue stem cells, which has been extensively discussed elsewhere (1,2). Stem cells of a particular tissue are a (potentially heterogeneous) population of functionally undifferentiated cells, capable of (*i*) homing to an appropriate growth environment (GE), (*ii*) proliferation, (*iii*) producing a large number of differentiated progeny, (*iv*) self-renewing their population, (*v*) regenerating functional tissues after injury, and (*vi*) having flexibility and reversibility in the use of these options. Within this definition, stem cells are defined by virtue of their functional potential and not by an explicit, directly observable characteristic.

This choice of a functional definition is inherently consistent with the biological role of a stem cell particularly linked to the functional tissue-regeneration feature. This kind of definition, however, imposes difficulties as, in order to identify whether or not a cell is a stem cell, its function has to be tested. This inevitably demands that the cell be manipulated experimentally while subjecting it to a functional bioassay. This, however, alters its properties. Here, we find ourselves in a circular situation. In order to answer the question whether a cell is a stem cell, we have to modify it. In doing so, we unavoidably lose the original cell and, in addition, may only see a limited spectrum of responses. In analogy to the *Heisenberg's uncertainty principle* in quantum physics we call this the *uncertainty principle of stem-cell biology*. In simple terms, this principle states that the very act of measuring the functional properties of a certain system always changes the characteristics of that system, hence, giving rise to a certain degree of uncertainty in the evaluation of its properties. We believe that this analogy holds true for the functional tissue stem cells in a very fundamental sense. Therefore, all statements that we can make about stem cells will necessarily be probabilistic statements about the future behavior under particular conditions.

CONCEPTUAL CHALLENGES IN TISSUE STEM-CELL BIOLOGY

One essential aspect of the given definition of tissue stem cells is the flexibility criterion. There is accumulating, experimental evidence for flexibility and reversibility. We would like to highlight a few of these, preferably related to the hematopoietic system.

It is now widely accepted that tissue stem cells are heterogeneous with respect to functional properties such as cycling activity, engraftment potential, or differentiation status, and to the expression of specific markers such as adhesion molecules or cell-surface antigens. However, recent experimental evidence is accumulating that these properties are able to reversibly change (3–12). Many authors have described the variability in the proliferative status of hematopoietic stem cells. One important finding in this respect is the fact that primitive cells may leave the cell cycle for many days and even months, but that almost all re-enter cycling activity from time to time. Consequently, there is no pool of permanently dormant stem cells (13,14). Experimental evidence is also provided for reversible changes of the stem-cell phenotypes involving differentiation profiles, adhesion protein expression, and engraftment/homing behavior associated with the cell-cycle status or the point in the circadian rhythm (6,15). There is increasing evidence that the expression of cell-surface markers (e.g., CD34) on hematopoietic stem cells is not constant but may fluctuate. The property can be gained and lost without affecting the stem-cell quality (5,16). Other groups investigated hemoglobin switching of hematopoietic stem cells in the blastocyst GE. Geiger et al. (17) showed that the switch from embryonic/fetal-type to adult-type globin is reversible. Furthermore, there is a lot of indirect evidence for fluctuations in the stem-cell population based on the clonal composition of functional cells. Chimerism induced by transplantation studies in cats and mice has been shown to fluctuate with time (18–22), indicating variations in the composition of active and inactive tissue stem cells. For the intestinal crypt, there is good evidence for a competition process of tissue stem cells within the individual crypts. This competition leads to a fluctuation of the clonal composition with a dynamic instability leading to crypt fission (23,24). Similar observations were made following retroviral marking of individual stem-cell clones that highlight the relative differences of inheritable cellular properties between stem-cell clones and their impact on the competitive potential (25–29). Another level of flexibility was found for lineage specification within the hematopoietic tissue. It is possible to bias the degree of erythroid, granuloid, or lymphoid lineage commitment by several maneuvers altering the growth conditions in different culture systems (4,30). The present concept of explaining the fluctuations observed in lineage specification is based on a dynamic network of interacting transcription factors (31–37). Cross and Enver (38) put forward the concept of fluctuating levels of transcription factors with threshold-dependent commitment.

Moreover, there is a rapidly growing library of literature that tissue stem cells specified for one type of tissue (e.g., hematopoiesis) can be manipulated in such a way that they can act as tissue stem cells of another tissue (e.g., neuronal, myogenic) (39–43). As suggested by experimental observations on these tissue plasticity phenomena, microenvironmental effects seem to play an essential role in directing cellular development. Very clearly this tissue plasticity represents a particular degree of flexibility consistent with the above definition. On the other hand, this phenomenon explains the necessity to include the homing to a specific GE into the stem-cell definition.

Motivated specifically by these experimental results on stem-cell plasticity, a debate, whether the view of a strict, unidirectional developmental hierarchy within tissue stem-cell populations is still appropriate, has been initiated (8,44–50). Although the general existence of tissue-plasticity properties is widely accepted, the underlying mechanisms (e.g., trans-differentiation, de-differentiation, or cell fusion) and the relevance of this plasticity potential in normal in vivo systems or even in clinical settings is still unclear. Furthermore, high-throughput analysis of genomic data (e.g., gene-expression profiling) and signaling studies offer the chance to extend our knowledge on tissue stem cells to the molecular level (32,51–53). Because classical stem-cell concepts are not able to explain all these experimental findings consistently, new conceptual approaches and theoretical models are required.

PREDICTIVE THEORIES AND QUANTITATIVE MODELS

Within the natural sciences, a *model* is understood as a simplifying abstraction of a more complex construct or process. In contrast to *experimental models* (e.g., animal or in vitro models), we will focus on *theoretical models*. Theoretical models in biology include *qualitative concepts*, that is, descriptive representations, and *quantitative models*, that is, mathematical representations, of a biological process. In contrast to qualitative concepts, quantitative models allow for an analytical, numerical, or simulation analysis.

The more we realize that we cannot prospectively determine stem cells directly, the more we need theoretical approaches to cope with the complexity. We believe that there is a tremendous need for general and specific theoretical concepts of tissue stem-cell organization, as well as for related quantitative models, to validate the concept by comparison of model predictions and experimental results. Such a theoretical framework of tissue stem-cell functioning will have several advantages: The model predictions can assist biologists to select and design experimental strategies, and they help to anticipate the impact of manipulations to a system and its response. Modeling is able to discriminate similar and to link different phenomena. Specifically, models originating from the same principles adapted to different systems (i.e., tissues or cell types) may help to understand common construction and regulation principles. Furthermore, they contribute to the understanding of latent mechanisms or crucial parameters of biological processes and may predict new phenomena; subsequently, we give a list of general requirements, which quantitative models should fulfill, in order to be suitable to serve as the bases for a theoretical framework of tissue stem-cell organization. The model cells must consistently fulfill the criteria listed in the definition of tissue stem cells. This has the following implications:

- The models based on the populations of individual cells to follow clonal development conform with the uncertainty principle, and enable the considerations of population fluctuations.
- They must consider GEs and the interactions between the cells.
- The system has to be dynamic in time and possibly in space.
- The system requires assumptions on mechanism to regulate proliferation, cellular differentiation, and cell–cell/cell–GE interactions.
- The model concept must be comprehensive in the sense of being applicable to the normal unperturbed in vivo homeostasis as well as to any in vivo or in vitro assay procedure. This criterion requests that system-measurement interactions must be consistently considered.

A NEW PERSPECTIVE ON STEM-CELL SYSTEMS

The basic concept of a functional definition of tissue stem cells (see above) has proven useful. This definition implies that one does not require *stemness* as an explicit attribute of cells, but rather considers it as a functional endpoint. Therefore, any concept on tissue stem cells has to specify assumptions about the mechanisms that potentially control the regenerative and proliferative potential of these cells such as proliferation, differentiation, maturation, lineage specification, and homing. Hence, the task is to design a dynamic process that drives and controls the cellular attributes. The leitmotifs here are the aspect of capabilities (i.e., actual and potential expression of cellular properties), of flexibility, and of reversibility. Apparently these aspects are controlled by the genetic and epigenetic status of the cells and by the activity of the signal transduction pathways including the transcription factor networks. Clearly, it is presently impossible to describe these processes in any reasonable detail. It will, therefore, be necessary to propose a simplified basic scheme of the cellular dynamics.

One possibility to consistently explain the variety of experimental phenomena without explicitly assuming a predefined *stemness* property of the cells has been developed by our group recently. This approach radically differs from other concepts presented so far in the literature. It strictly avoids assumptions that conclude with direct or indirect labeling of particular cells as stem cells, a priori. We rather attribute to all model cells only functional properties (e.g., proliferating or not having an affinity for homing to a particular GE, sensitivity to particular growth factors, etc.) and request that the system behavior changes these properties such that the population fulfills the functional criteria of the stem-cell definition.

To explain our conceptual approach, let us consider the activity of genes relevant for the behavior of tissue stem cells. There may be circumstances when sets of genes are insensitive to activation despite the availability of regulatory molecules. This is the case if, for example, epigenetic constellations prevent accessibility or if key regulator molecules such as transcription factor complexes are lacking (54–57). Therefore, we will conceptually distinguish two levels of gene activity control. Level 1 is qualitative and decides whether a gene is accessible for activation or not (sensitive or insensitive). Level 2 is quantitative and describes the degree of gene expression in a sensitive gene. Within this concept of a two-level control, a gene may not be expressed for two very different reasons. It may either not be sensitive (level 1 dynamics), or it may be sensitive but there is no—or only minor—activation due to lack of challenge (level 2 dynamics). State-transition graphs can be used to characterize this two-level dynamics. If they contain only self-maintaining and irreversible acyclic transitions between states, a population can be self-maintaining but not self-renewing (Fig. 1A). In contrast, Figure 1B and 1C illustrates state-transition graphs, which are characterized by reversible transitions.

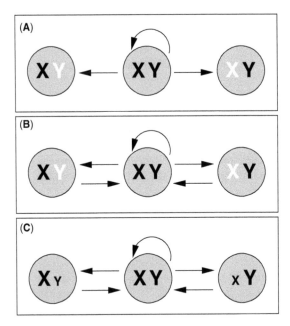

Figure 1   Examples of simple state-transition graphs according to level 1 and 2 dynamics. X and Y illustrate certain genes or functionally related gene clusters. Whereas the color is coding for the level 1 dynamics status (*black*: sensitive, *white*: insensitive), the font size illustrates the quantitative expression level according to level 2 dynamics. (**A**) Irreversible loss of cellular properties due to permanent level 1 inactivation. Only self-maintenance of XY state is possible. (**B**) Due to reversible changes (plasticity) with respect to level 1 dynamics (sensitive, insensitive), true self-renewal of XY state is possible. (**C**) Reversibility (plasticity) of XY state due to changes with respect to quantitative level 2 dynamics.

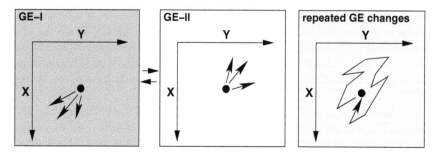

Figure 2 Dependency of cellular development on GE. This figure illustrates the actual position of a cell (•) and the preferred developmental directions (*arrows*) with respect to level 2 dynamics of cellular properties X and Y (e.g., gene expression) depending on the actual GE. Alternating between different GEs can induce fluctuating expression of cellular properties (quantitative plasticity), as illustrated in the right-most panel by one possible example trajectory. *Abbreviation*: GE, growth environment.

This would imply the property of true self-renewal, in the sense that cellular properties can be re-established even if they had been lost or down-regulated before.

Furthermore, we assume that the preferred direction of cellular development is dependent on GE-specific signals. Therefore, alternating homing to various GEs would yield a rather fluctuating development. In such a setting, the influence of the environments would be considerable, in particular, the frequency of transitions between them. For example, Figure 2 illustrates how signals from different GEs can influence the cellular fate, that is, the trajectories of cells within a property (e.g., gene expression) space, with respect to level 2 dynamics. Although only explained for level 2 dynamics, growth environmental signals could also affect transient or permanent inactivation of genes, that is, the level 1 dynamics.

Taken together, such a general concept of GE-dependent dynamics of reversibly changing cellular properties is a possibility to explain processes of self-renewal and differentiation in tissue stem-cell systems.

In the following section, we will demonstrate how this concept, implemented into a quantitative, mathematical model, has been applied to one specific tissue stem-cell system to explain dynamical processes of clonal competition in the hematopoietic system.

MODELING OF THE DYNAMICS OF CLONAL COMPETITION IN HEMATOPOIETIC STEM CELLS

Applying the principles described in the previous section to the hematopoietic stem-cell system leads to the concept of *within-tissue plasticity* (2,58), which will be described subsequently. We assume that cellular properties of hematopoietic stem cells can reversibly change within a range of potential options. The direction of cellular development and the decision whether a certain property is actually expressed depend on the internal state of the cell and on signals from its GE. Individual cells are considered to reside in one of two GEs (GE-A or GE-Ω). The state of each cell is characterized by its actual GE, by its position in the cell cycle (G_1, S, G_2, M, or G_0), and by a property (a), which describes its affinity to reside in GE-A. Cells in GE-Ω gradually lose this affinity, but cells in GE-A are able to gradually regain it (level 2 dynamics). Furthermore, cells in GE-A are assumed to be non-proliferating (i.e., in G_0), whereas cells in GE-Ω are assumed to proliferate with an average generation time τ_c. The transition of cells between the two GEs is modeled as a stochastic process. The corresponding transition intensities (probabilities of GE change per time step α and ω) depend on the current value of the affinity a and on the number of stem cells residing in GE-A and GE-Ω, respectively. If the attachment affinity a of an individual cell has fallen below a certain threshold (a_{min}), the potential to home to GE-A is inactivated (level 1 dynamics). These cells are released from the stem-cell compartment and start the formation of clones of differentiated cells. Figure 3 gives a graphical illustration of the model structure and of the cell number dependency described in the model by the transition characteristics f_α and f_ω.

A mathematical representation of this concept has been implemented in a computer program. Using extensive simulation studies we could demonstrate that this model can describe a large variety of observed phenomena, such as heterogeneity of clonogenic and repopulation potential (demonstrated in different types of colony formation and repopulating assays), fluctuating clonal contribution (observed in chimeric animals or in individual clone tracking experiments), or changing cell-cycle activity of primitive progenitors (described by the use of different S-phase labeling studies) (22,58,59). One of these phenomena—the competition of different stem-cell populations in mouse chimeras—will subsequently be used as an example to illustrate the potential of mathematical modeling in describing and explaining biological observations.

In order to apply the model to a mouse chimera setting, that is, to the coexistence of cells from two different mouse strain backgrounds (DBA/2 and C57BL/6) in one

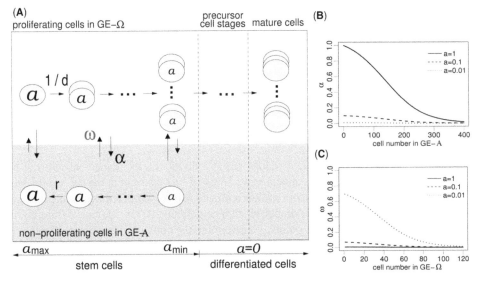

Figure 3 Schematic representation of the model concept. (**A**) The lower part represents GE-A and the upper part GE-Ω. Cell amplification due to proliferation in GE-Ω is illustrated by growing cell numbers (cell groups separated by *vertical dots* represent large cell numbers). The attachment affinity a decreases by a factor $1/d$ per time step in GE-Ω, but it increases by a factor r per time step in GE-A. The actual quantity of the affinity a is sketched by different font sizes. If a fell below a critical threshold a_{min}, the cell lost its potential to switch to GE-A, and a is set to zero (represented by *empty cells*). Transition between GE-A and Ω occurs with intensities $\alpha = (a/a_{max})f_\alpha$ and $\omega = (a_{min}/a)f_\omega$, which depend on the value of a (represented by the differently scaled *vertical arrows*) and on the cell numbers in the target GE. Typical profiles of the cell number-dependent transition intensities f_α and f_ω for different values of attachment affinity a shown in panels (**B**) and (**C**). *Abbreviation*: GE, growth environment.

common host, we consider two populations of cells within one model system. These populations potentially differ in their model parameters d, r, τ_c, f_α, or f_ω. This approach allows the analysis of the influence of these model parameters on the competitive behavior of the two cell types and, therefore, on the dynamics of chimerism development.

Simulation studies lead to two major qualitative predictions for the chimeric situation: first, the model predicts that small differences in model parameters may cause unstable chimerism with a slow but systematic long-term trend in favor of one clone; second, it is predicted that the chimerism development depends on the actual status (i.e., cell numbers) of the entire system. For example, system perturbations by stem-cell transplantation after myeloablative conditioning, cytokine, or cytotoxic treatment, are expected to result in significant changes of chimerism levels at a short timescale. These predictions are also supported by previously reported experimental results on the contribution of DBA/2 (D2) cells to peripheral blood production in C57BL/6 (B6)-D2 allophenic mice (18). In these animals, the D2 contribution declines over time, but can be reactivated by a bone marrow transplantation into lethally irradiated B6-D2-F1 (BDF1) mice.

To subject our qualitative model predictions to an experimental test and to investigate whether these phenomena could be explained consistently by one single parameter configuration of the model, a specific set of experiments was performed. To quantitatively compare experimental data and simulation results, we investigated the chimerism kinetics in primary and secondary B6-D2 radiation chimeras. The detailed experimental procedure has been described elsewhere (22). Briefly, primary irradiation chimeras were constructed by transplantation of fetal liver (FL) cells isolated from B6 and D2 mice into lethally irradiated BDF1 mice. To measure chimerism levels, blood samples were drawn from each chimera at various time points after transplantation. The percentage of leukocytes derived from D2, B6, and BDF1 was assessed by flow cytometry. To determine the effect of serial bone marrow transplantation on the chimerism dynamics, secondary transplantations were performed. Herein, bone marrow cells from individual chimeric donors at different time points after primary transplantation of FL cells were transplanted into cohorts of 5 and 12 lethally irradiated female BDF1 mice, respectively. Identical to primary hosts, the chimerism was determined by repeated peripheral blood samples in these secondary chimeras.

To simulate the chimeric development of individual mice, the actual status of each stem cell, characterized by its attachment affinity (a), its position in the cell cycle, and its current GE (GE-A, GE-Ω, or pool of differentiated cells), is updated at discrete time steps (22). Additionally, the actual number of stem cells in GE-A, GE-Ω, and of differentiated cells is recorded at these time points. To determine the number of peripheral blood leukocytes in the simulations, the pool of mature cells (Fig. 3A) is used. Hereby, it is assumed that the number of mature leukocytes is proportional to the number of cells released from the stem-cell compartment. Details of amplification, differentiation, and maturation within the precursor cell stages are neglected in the current model version. Chimerism levels are obtained by calculating the D2 proportion among model cells within the mature leukocyte compartment.

Due to the assumed stochastic nature of the GE transition of stem cells, individual simulation runs produce different chimerism levels even though identical parameter sets are used; therefore, to determine the mean chimerism levels under a specific parameter set, repeated simulation runs are performed. To illustrate the average behavior, the mean chimerism levels are determined at each time step.

Starting from a parameter configuration previously demonstrated to consistently explain a variety of experimental phenomena in the nonchimeric situation, we fit the simulation outcome to the observed chimerism development in primary irradiation chimeras initiated with a 1:4 ratio of transplanted D2 and B6 fetal liver cells. Due to the documented difference between D2 and B6 cells with respect to their cycling activity, we assumed different average generation times. However, solely assuming this difference is not sufficient to explain the observed biphasic chimerism development. Therefore, we performed a sensitivity analysis of the model parameters controlling the cellular development, that is, the differentiation coefficient (d), the regeneration coefficient (r), and the transition characteristics f_α and f_ω. We found that only differences in the transition characteristics induce the observed biphasic pattern. Although the qualitative chimerism development was primarily determined by the transition characteristics, the maximally reached D2 levels are dependent on the ratio of initially engrafting D2 and B6 cells. Optimal parameter values of the initial D2 proportion of engrafting stem cells and of the shape parameters of the transition characteristics f_α and f_ω have been determined by fitting simulation results to experimental data using an evolutionary strategy. For technical details of the fitting procedure and for a description of the specific form of the transition characteristics, we refer to the work of Roeder et al. (22).

The data points in Figure 4A show the experimentally observed chimerism development in unperturbed radiation chimeras together with an average simulation using the fitted set of model parameters. Without any further change of the model parameters, our simulations demonstrate that the experimentally observed heterogeneity of chimerism development in different experiments can be explained by variations in the initial D2:B6 ratio (Fig. 4B). To test whether these parameter configurations (obtained for the competition situation in chimeric systems) are also able to explain differences in the reconstituting behavior of nonchimeric D2 and B6 systems, we simulated the reconstitution of nonchimeric systems using the D2 and the B6 parameter sets, respectively. It could be shown (22) that the simulations are able to reproduce the differences in the time scales of reconstitution between D2 and B6, which had been observed experimentally.

Furthermore, using the same parameter configuration, simulations predict that a reduction of the total stem-cell pool size, as assumed for the transplantation setting, induces an initial elevation of the D2 contribution in the host (compared to donor chimerism prior to transplantation) followed by a gradual D2 decline (Fig. 4C). This is consistent with the experimental results obtained by the transplantation of bone marrow cells from a primary radiation chimera at day 133 after first transplantation into secondary cohorts of lethally irradiated BDF1 mice, which clearly show a reactivation of the D2 contribution in the peripheral blood (data points in Fig. 4C).

These results provide an experimental test of our novel concept of tissue stem-cell organization based on the within-tissue plasticity idea for the situation of competitive hematopoiesis. Using a parameter configuration obtained by fitting the model to one specific data set, the mathematical model made several predictions for the situation of clonal competition and unstable chimerism. We demonstrated that this single parameter configuration can explain the majority of the presented phenomena in the chimeric situations and is also consistent with the variety of further phenomena analyzed before (22,58,59). It should be noted that parameter adjustments for the simulation of each individual data set would provide even better model fits. However, it was our main goal to validate the model by the application of one parameter configuration to several independent data sets.

Our results suggest that chimerism levels, observed in the peripheral blood, depend on the actual dynamic status of the stem-cell system. The simulation studies reveal that variations in strain-specific cellular properties of stem cells, which sensitively affect the competitive behavior in a chimeric situation, do not necessarily influence their growth

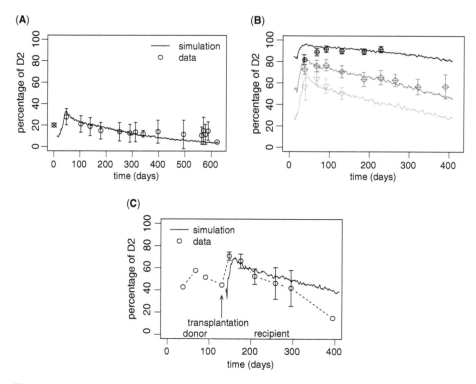

Figure 4 Simulation results on chimerism development. (**A**) Data points (*open circles*) represent the observed chimerism levels (mean ± SD) in primary radiation chimeras with ⊗ illustrating the initial D2:B6 ratio in the transplant. The solid line shows the simulated chimerism of mature model leukocytes (average of 100 simulation runs). (**B**) Effect of the initial D2:B6 ratio: data points represent the results (mean ± SD) from three independent experiments using different D2 proportions of the transplant. Solid lines represent corresponding average simulation results using identical parameter sets but different initial D2 proportions: 85%—black, 50%—dark gray, 30%—light gray. (**C**) The circles show the experimentally observed peripheral blood leukocyte chimerism in a primary radiation chimera (single values) and in a corresponding cohort of secondary host mice (mean ± SD). The solid lines show average simulations for the chimerism development in the secondary chimeras.

and repopulating potential in a nonchimeric system. These findings point to the relative nature of stem cells and their repopulating potential in general. Therefore, stem-cell potential must not be regarded as an isolated cellular property, but has to be understood as a dynamic property taking into account the individual cellular potential, the cell–cell and the cell–microenvironment interactions. This has potentially important implications for the treatment of clonal disorders, gene therapeutic strategies, or tissue engineering processes where is the goal to control the competitive potential of a specific cell type or clone.

SPATIO-TEMPORAL STEM-CELL ORGANIZATION

The assumption of different GEs suggests that a spatial component might also influence tissue stem-cell organization. This hypothesis is supported by several experimental findings (60–63); however, it is ignored in the stem-cell model discussed so far. In the following, we show that the spatial arrangement of cells in a stem-cell compartment and the related effects on the system behavior can consistently be incorporated into the concepts described earlier.

First of all, an extension of the described model to incorporate spatio-temporal dynamics requires an explicit physical representation of the cells. As real cells, the model cells need to have a shape, a volume, and specific biomechanical properties. Furthermore, they need to be able to detect shape and stress changes within their local environment by sensing the degree of their own extension or compression. Thereby, these models need to describe a link between shape changes and functional processes such as proliferation, differentiation, and apoptosis. As a consequence, basic effects of tissue organization can be attributed to cell contact formation between the basic individual cells and their local GE.

Due to recent experimental advances (64–66), the possibilities to collect new information on biophysical parameters of cells and tissues are rapidly improving. Utilizing this information, a specific class of so-called "individual cell-based biomechanical models (ICBMs)," is now available. Recently, we have shown that this model class is capable of explaining the complex spatial growth and pattern formation processes of epithelial stem-cell populations growing in vitro (67). ICBMs permit one to model the growth and pattern formation of large multi-cellular systems as they tie properties averaged on the length scale of a cell to the macroscopic behavior on the cell population and tissue level. Consequently, they allow for an efficient simulation and, therefore, permit the analysis of spatial arrangements of large cell populations on large timescales. Thus, ICBMs enable approaches to cell differentiation, maturation, and lineage specification accounting for tissue formation and regeneration (68,69). A number of different individual-based models of cell populations have been studied so far (70).

In the following list, we describe the basic properties of a lattice-free ICBM, which have been introduced to extend our concepts on stem-cell organization to more general spatio-temporal dynamics.

- In the spatial model, we assume that an isolated cell adopts a spherical shape. As the cell comes into contact with other cells or with the substrate, its shape changes. Cells in contact form adhesive bonds. With decreasing distance, their contact areas increase and so does the number of the adhesive contacts.
- The attractive cell–cell and cell–substrate interaction is assumed to be dominated by receptor–ligand interactions. We assume homogeneously distributed receptors/ligands on the cell surfaces and the substrates. Accordingly, the strength of attraction is proportional to the product of the size of the contact area A_C, the number of receptor–ligand complexes, and the strength of a single bond.
- Contact formation is accompanied by cell deformations. These deformations lead to stress in the cell membranes and cytoskeletons resulting in repulsive interactions. In our model, we approximate a cell by a homogeneous, isotropic, elastic object.
- Furthermore, we consider a subdivision of the cell cycle into two phases: the interphase and the mitotic phase. During the interphase, a proliferating cell doubles its mass and its volume. We model the cell growth process by increasing an intrinsic (target) volume V_T of the cell by stochastic increments. After the V_T reached twice the standard volume V_0, the cell enters the mitotic phase and is split into two daughter cells of equal target volume V_0.

In order to enable the model cells to couple shape changes to processes such as proliferation, differentiation, and apoptosis, we consider a hierarchy of different regulation

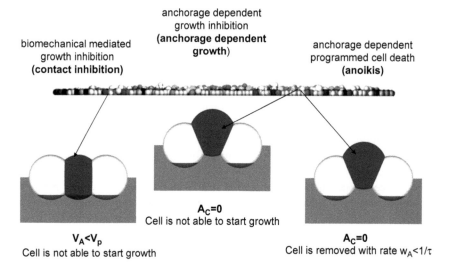

biomechanical mediated
growth inhibition
(contact inhibition)

anchorage dependent
growth inhibition
**(anchorage dependent
growth)**

anchorage dependent
programmed cell death
(anoikis)

$A_C=0$
Cell is not able to start growth

$V_A<V_p$
Cell is not able to start growth

$A_C=0$
Cell is removed with rate $w_A<1/\tau$

Figure 5 Cellular regulation mechanisms controlled via cell–cell and cell–substrate contacts and cell deformation/compression. A_C is the contact area to the substrate; V_A the actual cell volume; V_p a threshold volume.

mechanisms (Fig. 5), namely, (*i*) a biomechanical-mediated form of growth inhibition (contact inhibition), (*ii*) an anchorage-dependent growth inhibition (anchorage-dependent growth), and (*iii*) an anchorage-dependent programed cell death (anoikis).

In simulation studies, we have investigated the consequences of modifying the parameters for cell–substrate adhesion, the cell-cycle time, and have studied how this affects the morphology, the biomechanics, and the kinetics of the growing cell population (67). We found that in particular the cell-substrate anchorage has a significant impact on the population morphology (Fig. 6). For instance, cells within a monolayer undergo contact inhibition of growth only for strong cell–substrate anchorage. Thus, anoikis (anchorage-dependent programed cell death) only substantially contributes to growth control in case of low cell–substrate anchorage, or if contact inhibition is deficient. Whether a variation of the substrate anchorage can initialize the formation of self-organized and spatially

(A)   **(B)**

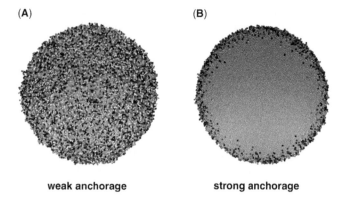

weak anchorage   strong anchorage

Figure 6 Top view of the macroscopic morphology of growing cell populations with $N = 10,000$ cells. Cell anchorage strength: (**A**) 200 mN/m and (**B**) 600 mN/m. The shaded value of the cells is a marker of the cell target volume V_T. Dark shaded cells indicate imminent cell division.

structured clonogenic units (cell niches), which are able to reproduce themselves, remains an open question.

Our model analysis on epithelial cell layers predicts that weak substrate anchorage is accompanied by a continuous cell shedding out of the basal layer and consequently by an ongoing self-renewal of the population (Fig. 6A). In contrast, strong anchorage results in stable growth and an aging population (Fig. 6B). However, the property of self-renewal is also conserved in the latter case and perturbations, for example, emanating from the induced death of cells, would be followed by an immediate re-growth of the population.

The proposed ICBM links properties of individual cells and the substrate on a small spatial scale to the macroscopic spatio-temporal dynamics of a cell population. All cells were assumed to be capable of proliferation and able to produce an unlimited number of progeny. Thus, each cell has the potential to self-maintain the population and to regenerate (self-renew) after injury. In this respect, the cells comply with the stem-cell criteria introduced earlier. However, the capabilities to differentiate and to undergo lineage specification are not yet included in our model representation at the moment. The challenge is to develop a generic theoretical framework of cell–environment interactions, which is controlling these processes. For that purpose, one may allow for cell-specific parameters, which fluctuate due to varying interactions of the cells with their local environment. In other words, one may consider reversibly changing biophysical properties of the cells, combining the general concept of within-tissue plasticity and the concept of spatial effects of tissue stem-cell organization.

How does the cell microenvironment actually influence the cell properties? Experimental studies demonstrate that cells adapt their shape to micro-patterned structures (71,72) and sense their stiffness (73,74) and composition (60,75), thereby changing their growth and differentiation properties. This may include changes of their own specific gene expression. Models of tissues with spatio-temporal organized stem-cell compartments, such as the intestinal mucosa might have to consider all these effects and will be a considerable challenge.

CONCEPTUAL NOVELTY AND ACHIEVEMENTS

The concepts proposed earlier may change the paradigm of the thinking about stem cells. Rather than assuming that these cells are specialized in the first place, we suggest that they are selected and modified due to interactions with the GE. Their properties are considered to fluctuate permanently so that some cells meet a situation of expansion and growth. Therefore, tissue stem cells are conceived as cells capable of behaving in a variety of ways and, hence, it is their potential and the flexibility to use this potential that matters.

We argue that it is conceptually misleading to consider *stemness* as a specific property that can be determined at one point in time without putting the cells to functional tests. The potential of stem cells relates rather to the complexity of the state-transition graphs describing the potential dynamics of gene/protein activation than to the actual activity status in one of these states. This has implications for attempts to define tissue stem cells, for example, by gene- or protein-profiling (76–81). There are several problems that we envisage. First, molecular profiles obtained by high-throughput technologies (e.g., micro-arrays) are mostly measured on cells obtained from negative selection procedures leading to a heterogeneous mixture of cells. Second, the assays typically represent snapshots at one point in time. However, such snapshots give little insight into the potentials and the dynamic responses of a (stem) cell population. It would be essential to track the molecular profiles over time in various experimental settings putting the system under various modes of stress. Such an approach is necessary to sketch the topology of gene/protein activity networks and to identify (potentially reversible) developmental and regulatory pathways. Third, to conform with the functional definition of tissue stem cells, it will be crucial to correlate the molecular activity network to the functional capabilities of the cells in functional assays. Hence, all techniques based on snapshot measurements of some surface markers or gene activity patterns must be considered as surrogate techniques. At present we cannot see the possibility for a molecular definition of tissue stem cells, disregarding functional aspects as a reference point. Thus, we are reluctant to believe that tissue stem cells can be defined by a "tissue-stem-cell chip." Such an approach would ignore the two basic aspects of stem-cell potentiality and of cell growth–environment interaction. Furthermore, the uncertainty principle discussed would still apply and all statements could only be made in a probabilistic sense. However, gene-/protein-profiling approaches are still a possibility to select cells with properties required for (potential) stem cells and one can expect a more detailed insight into the mode of stem-cell operation by investigating the underlying mechanisms. In particular, one can hope for test procedures to screen functional capabilities of tissue stem cells.

There are a number of further predictions arising from the proposed mathematical models. One basic prediction is that two twin cells originating from the same mother cell put into different GEs will take different development paths. This is, however, also predicted if they are placed into identical GEs. The ongoing fluctuations will eventually lead to different fates. Another prediction concerns clonal evolution. All our model simulations presented are based on a simultaneous activity of several co-existing tissue stem cells. They generate several clones and the situation is polyclonal at any given point in time. This should always be evident shortly after introducing some genetic markers (e.g., retro- or lentiviral marking). However, there are fluctuations and some active stem cells become silent (or get lost) and others are activated. Thus, the clones contributing to tissue formation change with time. Actually, in the long run, the pattern is predicted to change. If one could label all cells in a tissue with a unique marker, our simulations would predict that coexistence is impossible in the long run and that descendents from one clone will eventually generate all active stem cells in the tissue. This conversion to long-term monoclonality is a consequence of fluctuations. It would, however, not be possible to know in advance which clone will be the winner. Hence, we predict that depending on the time scale of measurement, it is equally valid to argue that stem-cell systems are polyclonal (actual activity) and monoclonal (descendent status) at the same time. A detailed understanding of the long-term dynamic features will be important in gene therapy based on random insertion of genes into tissue stem cells. A third important model prediction concerns the role of self-renewal. If one has a stem-cell system with a homogenous population of cells, self-renewal and self-maintenance are actually equivalent. In stem-cell systems with heterogeneity the distinction is very important. One can prove that systems that are only capable of self-maintenance can live for a long time but will, with certainty, die out at some point in the future. The reason is that once a sub-population at the root of the network is lost it cannot be recovered. Self-renewal is a mandatory prerequisite for a system that is structurally robust against repeated damage and extensive stress. We, therefore, predict that *self-renewal* is an essential property of stem-cell systems, but it may be a very slow and selective process and, therefore, difficult to detect.

Our reasoning has emphasized the role of cell–cell and cell–microenvironment interactions. This implies that specific attention needs to be paid to the role of the microenvironment, which is a complex subject itself. GEs encompass an element of spatial neighborhood to other stem cells and matrix cells, ways to adhere to them and ways to

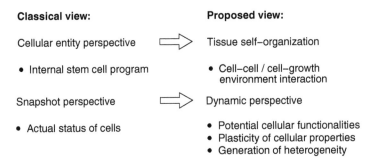

Figure 7   Classical versus proposed view on tissue stem-cell systems.

receive signals (growth factors, direct cell contacts, gap junctions, and pseudopods). GEs may home a cell for a certain while and can then be called a niche. However, such niches may have limited life times, and currently little is known about the dynamic changes of GEs. Any kinetic changes present will, however, increase the fluctuations in the stem-cell population. Our approach to include biomechanical properties of cells and, therefore, to include a spatial component into the control of cellular fates is one possible way to get more insight into the underlying mechanisms of cellular interaction.

In summary, our modeling approaches prove that one can conceive regenerative tissue systems fully consistent with the functional definition of stem cells, without assumptions on unidirectional hierarchies, preprogramed asymmetric divisions, or other assumptions implying a priori, the entity of predetermined tissue stem cells. It has been shown by our modeling that functional, self-organizing systems with stochastic components (sources for generation and for elimination of variance) are powerful, alternative concepts to explain tissue stem-cell organization consistently. We, therefore, propose a revised conceptual view on tissue stem-cell organization, replacing the classical perspective of a pre-defined stem-cell entity by considering stem-cell potential as a system property resulting from dynamically controlled cell–cell and cell–microenvironment interactions (Fig. 7).

Concluding from these conceptual insights, the major experimental challenge is, in our opinion, to explore the potential repertoire of cell populations containing tissue stem cells, that is, to focus on the scope of skills rather than on selected individual abilities. Also, modeling approaches need to be extended in several regards. First, more simulation studies are required to demonstrate that the concepts proposed comply with a broad spectrum of data. Furthermore, it will be important to show that the same general model principles hold for tissue stem cells as diverse as the blood-forming stem cells, epithelial stem cells, and other systems. The major challenge in the field of theoretical modeling, however, is the design of predictive models, which can bridge the different levels of description (i.e., tissue, cells, and molecules) and, hence, link a molecular description of tissue stem cells to the functional definition. It is evident that modeling, besides the new bioinformatic methods in data analysis, will be important to link data from all these three description levels into one comprehensive framework.

ACKNOWLEDGMENT

This work has been supported by the Deutsche Forschungsgemeinschaft (DFG) grants LO 942/9-1,2 and BIZ-6 1/1.

REFERENCES

1. Potten CS, Loeffler M. Stem cells: attributes, cycles, spirals, pitfalls and uncertainties. Lessons for and from the crypt. Development 1990; 110:1001–1020.
2. Loeffler M, Roeder I. Tissue stem cells: definition, plasticity, heterogeneity, self-organization and models—a conceptual approach. Cells Tissues Organs 2002; 171:8–26.
3. Frimberger AE, Stering AI, Quesenberry PJ. An in vitro model of hematopoietic stem cell homing (abstr). Blood 1997; 90.
4. Rolink AG, Nutt SL, Melchers F, Busslinger M. Long-term in vivo reconstitution of T-cell development by Pax5-deficient B-cell progenitors see comments. Nature 1999; 401:603–606.
5. Sato T, Laver JH, Ogawa M. Reversible expression of CD34 by murine hematopoietic stem cells. Blood 1999; 94:2548–2554.
6. Quesenberry P, Habibian H, Dooner M, et al. Physical and physiological plasticity of hemato-poietic stem cells. Blood Cell Mol Dis 2001; 27:934–937.
7. Wagers AJ, Christensen JL, Weissman IL. Cell fate determination from stem cells. Gene Ther 2002; 9:606–612.
8. Goodell MA. Stem-cell "plasticity": befuddled by the muddle. Curr Opin Hematol 2003; 10:208–213.
9. Colvin GA, Lambert JF, Abedi M, et al. Differentiation hotspots: the deterioration of hierarchy and stochasm. Blood Cell Mol Dis 2004; 32:34–41.
10. Quesenberry PJ, Abedi M, Aliotta J, et al. Stem cell plasticity: an overview. Blood Cell Mol Dis 2004; 32:1–4.
11. Zhang CC, Lodish HF. Murine hematopoietic stem cells change their surface phenotype during ex vivo expansion. Blood 2005; 105(11):4314–4320.
12. Quesenberry PJ, Dooner G, Colvin G, Abedi M. Stem cell biology and the plasticity polemic. Exp Hematol 2005; 33:389–394.
13. Bradford GB, Williams B, Rossi R, Bertoncello I. Quiescence, cycling, and turnover in the primitive hematopoietic stem cell compartment. Exp Hematol 1997; 25:445–453.
14. Cheshier SH, Morrison SJ, Liao X, Weissman IL. In vivo proliferation and cell cycle kinetics of long-term self-renewing hematopoietic stem cells. Proc Natl Acad Sci USA 1999; 96:3120–3125.
15. Habibian HK, Peters SO, Hsieh CC, et al. The fluctuating phenotype of the lymphohematopoie-tic stem cell with cell cycle transit. J Exp Med 1998; 188:393–398.
16. Goodell MA. CD34+ or CD34-: Does it really matter? Blood 1999; 94:2545–2547.
17. Geiger H, Sick S, Bonifer C, Muller AM. Globin gene expression is reprogrammed in chimeras generated by injecting adult hematopoietic stem cells into mouse blastocysts. Cell 1998; 93:1055–1065.
18. Van Zant G, Scott-Micus K, Thompson BP, Fleischman RA, Perkins S. Stem cell quiescence/ activation is reversible by serial transplantation and is independent of stromal cell genotype in mouse aggregation chimeras. Exp Hematol 1992; 20:470–475.
19. Abkowitz JL, Catlin SN, Guttorp P. Evidence that hematopoiesis may be a stochastic process in vivo. Nat Med 1996; 2:190–197.
20. Abkowitz JL, Golinelli D, Harrison DE, Guttorp P. In vivo kinetics of murine hemopoietic stem cells. Blood 2000; 96:3399–3405.
21. Kamminga LM, Akkerman I, Weersing E, et al. Autonomous behavior of hematopoietic stem cells. Exp Hematol 2000; 28:1451–1459.
22. Roeder I, Kamminga LM, Braesel K, Dontje B, Haan Gd, Loeffler M. Competitive clonal hematopoiesis in mouse chimeras explained by a stochastic model of stem cell organization. Blood 2005; 105:609–616.
23. Winton DJ, Ponder BA. Stem-cell organization in mouse small intestine. Proc R Soc Lond B Biol Sci 1990; 241:13–18.
24. Loeffler M, Birke A, Winton D, Potten C. Somatic mutation, monoclonality and stochastic models of stem cell organization in the intestinal crypt. J Theor Biol 1993; 160:471–491.

25. Jordan CT, Lemischka IR. Clonal and systemic analysis of long-term hematopoiesis in the mouse. Genes Dev 1990; 4:220–232.

26. Van Zant G, Chen JJ, Scott-Micus K. Developmental potential of hematopoietic stem cells determined using retrovirally marked allophenic marrow. Blood 1991; 77:756–763.

27. Kim HJ, Tisdale JF, Wu T, et al. Many multipotential gene-marked progenitor or stem cell clones contribute to hematopoiesis in nonhuman primates. Blood 2000; 96:1–8.

28. Drize NJ, Olshanskaya YV, Gerasimova LP, et al. Lifelong hematopoiesis in both reconstituted and sublethally irradiated mice is provided by multiple sequentially recruited stem cells. Exp Hematol 2001; 29:786–794.

29. Kuramoto K, Follman D, Hematti P, et al. The impact of low-dose busulfan on clonal dynamics in nonhuman primates. Blood 2004; 104:1273–1280.

30. McIvor ZJ, Heyworth CM, Johnson BA, et al. A transient assay for regulatory gene function in haemopoietic progenitor cells. Br J Haematol 2000; 110:674–681.

31. Zhang P, Behre G, Pan J, et al. Negative cross-talk between hematopoietic regulators: GATA proteins repress PU.1. Proc Natl Acad Sci USA 1999; 96:8705–8710.

32. Orkin SH. Diversification of haematopoietic stem cells to specific lineages. Nat Rev Genet 2000; 1:57–64.

33. Zhang P, Zhang X, Iwama A, et al. PU.1 inhibits GATA-1 function and erythroid differentiation by blocking GATA-1 DNA binding. Blood 2000; 96:2641–2648.

34. Nerlov C, Querfurth E, Kulessa H, Graf T. GATA-1 interacts with the myeloid PU.1 transcription factor and represses PU.1-dependent transcription. Blood 2000; 95:2543–2551.

35. Heyworth C, Pearson S, May G, Enver T. Transcription factor-mediated lineage switching reveals plasticity in primary committed progenitor cells. Embo J 2002; 21:3770–3781.

36. Back J, Dierich A, Bronn C, Kastner P, Chan S. PU.1 determines the self-renewal capacity of erythroid progenitor cells. Blood 2004; 103:3615–3623.

37. Letting DL, Chen YY, Rakowski C, Reedy S, Blobel GA. Context-dependent regulation of GATA-1 by friend of GATA-1. Proc Natl Acad Sci USA 2004; 101:476–481.

38. Cross MA, Enver T. The lineage commitment of haemopoietic progenitor cells. Curr Opin Genet Dev 1997; 7:609–613.

39. Bjornson CR, Rietze RL, Reynolds BA, Magli MC, Vescovi AL. Turning brain into blood: a hematopoietic fate adopted by adult neural stem cells in vivo see comments. Science 1999; 283:534–537.

40. Brazelton TR, Rossi FM, Keshet GI, Blau HM. From marrow to brain: expression of neuronal phenotypes in adult mice. Science 2000; 290:1775–1779.

41. Seale P, Rudnicki MA. A new look at the origin, function, and "stem-cell" status of muscle satellite cells. Dev Biol 2000; 218:115–124.

42. Herzog EL, Chai L, Krause DS. Plasticity of marrow-derived stem cells. Blood 2003; 102:3483–3493.

43. Filip S, English D, Mokry J. Issues in stem cell plasticity. J Cell Mol Med 2004; 8:572–577.

44. Wei G, Schubiger G, Harder F, Muller AM. Stem cell plasticity in mammals and transdetermination in drosophila: common themes? Stem Cells 2000; 18:409–414.

45. Anderson DJ, Gage FH, Weissman IL. Can stem cells cross lineage boundaries? Nat Med 2001; 7:393–395.

46. Blau HM, Brazelton TR, Weimann JM. The evolving concept of a stem cell: entity or function? Cell 2001; 105:829–841.

47. Goodell MA, Jackson KA, Majka SM, et al. Stem cell plasticity in muscle and bone marrow. Ann NY Acad Sci 2001; 938:208–218.

48. Lemischka I. Rethinking somatic stem cell plasticity. Nat Biotechnol 2002; 20:425.

49. Theise ND. Blood to liver and back again: seeds of understanding. Haematologica 2003; 88:361–362.

50. Theise ND, Wilmut I. Cell plasticity: flexible arrangement. Nature 2003; 425:21.

51. Rothenberg EV. T-lineage specification and commitment: a gene regulation perspective. Semin Immunol 2002; 14:431–440.

52. Ohishi K, Katayama N, Shiku H, Varnum-Finney B, Bernstein ID. Notch signalling in hematopoiesis. Semin. Cell Dev Biol 2003; 14:143–150.
53. Heasley LE, Petersen BE. Signalling in stem cells. EMBO Rep 2004; 5:241–244.
54. Bonifer C. Long-distance chromatin mechanisms controlling tissue-specific gene locus activation. Gene 1999; 238:277–289.
55. Tagoh H, Melnik S, Lefevre P, Chong S, Riggs AD, Bonfier C. Dynamic reorganization of chromatin structure and selective DNA demethylation prior to stable enhancer complex formation during differentiation of primary hematopoietic cells in vitro. Blood 2004; 103:2950–2955.
56. Bonifer C. Epigenetic plasticity of hematopoietic cells. Cell Cycle 2005; 4.
57. Rosmarin AG, Yang Z, Resendes KK. Transcriptional regulation in myelopoiesis: hematopoietic fate choice, myeloid differentiation, and leukemogenesis. Exp Hematol 2005; 33:131–143.
58. Roeder I, Loeffler M. A novel dynamic model of hematopoietic stem cell organization based on the concept of within-tissue plasticity. Exp Hematol 2002; 30:853–861.
59. Roeder I, Loeffler M, Quesenberry PJ, Colvin GA, Lambert JF. Quantitative tissue stem cell modeling. Blood 2003; 102:1143–1144; author reply 1144–1145.
60. Teller IC, Beaulieu JF. Interactions between laminin and epithelial cells in intestinal health and disease. Expert Rev Mol Med 2001; 2001:1–18.
61. Batlle E, Henderson JT, Beghtel H, et al. Beta-catenin and TCF mediate cell positioning in the intestinal epithelium by controlling the expression of EphB/ephrinB. Cell 2002; 111:251–263.
62. Zhu J, Emerson SG. A new bone to pick: osteoblasts and the haematopoietic stem-cell niche. Bioessays 2004; 26:595–599.
63. Taichman RS. Blood and bone: two tissues whose fates are intertwined to create the hematopoietic stem-cell niche. Blood 2005; 105:2631–2639.
64. Charras GT, Horton MA. Determination of cellular strains by combined atomic force microscopy and finite element modeling. Biophys J 2002; 83:858–879.
65. Schwarz US, Balaban NQ, Riveline D, Bershadsky A, Geiger B, Safran SA. Calculation of forces at focal adhesions from elastic substrate data: the effect of localized force and the need for regularization. Biophys J 2002; 83:1380–1394.
66. Lincoln B, Erickson HM, Schinkinger S, et al. Deformability-based flow cytometry. Cytometry A 2004; 59:203–209.
67. Galle J, Loeffler M, Drasdo D. Modeling the effect of deregulated proliferation and apoptosis on the growth dynamics of epithelial cell populations in vitro. Biophys J 2005; 88:62–75.
68. Hogeweg P. Evolving mechanisms of morphogenesis: on the interplay between differential adhesion and cell differentiation. J Theor Biol 2000; 203:317–333.
69. Meineke FA, Potten CS, Loeffler M. Cell migration and organization in the intestinal crypt using a lattice-free model. Cell Prolif 2001; 34:253–266.
70. Drasdo D. On selected individual-based approaches to the dynamics in multicellular systems. In: Alt W, Chaplain M, eds. Polymer and Cell Dynamics. Basel: Birkhäuser, 2003:169–203.
71. Webb A, Clark P, Skepper J, Compston A, Wood A. Guidance of oligodendrocytes and their progenitors by substratum topography. J Cell Sci 1995; 108:2747–2760.
72. Teixeira AI, Abrams GA, Bertics PJ, Murphy CJ, Nealey PF. Epithelial contact guidance on well-defined micro- and nanostructured substrates. J Cell Sci 2003; 116:1881–1892.
73. Engler AJ, Griffin MA, Sen S, Bonnemann CG, Sweeney HL, Discher DE. Myotubes differentiate optimally on substrates with tissue-like stiffness: pathological implications for soft or stiff microenvironments. J Cell Biol 2004; 166:877–887.
74. Yeung T, Georges PC, Flanagan LA, et al. Effects of substrate stiffness on cell morphology, cytoskeletal structure, and adhesion. Cell Motil Cytoskeleton 2005; 60:24–34.
75. El-Sabban ME, Sfeir AJ, Daher MH, Kalaany NY, Bassam RA, Talhouk RS. ECM-induced gap junctional communication enhances mammary epithelial cell differentiation. J Cell Sci 2003; 116:3531–3541.
76. Phillips RL, Ernst RE, Brunk B, et al. The genetic program of hematopoietic stem cells. Science 2000; 288:1635–1640.

77. Ivanova NB, Dimos JT, Schaniel C, Hackney JA, Moore KA, Lemischka IR. A stem cell molecular signature. Science 2002; 298:601–604.
78. deHaan G, Bystrykh LV, Weersing E, et al. A genetic and genomic analysis identifies a cluster of genes associated with hematopoietic cell turnover. Blood 2002; 100:2056–2062.
79. Venezia TA, Merchant AA, Ramos CA, et al. Molecular signatures of proliferation and quiescence in hematopoietic stem cells. PLoS Biol 2004; 2:301.
80. Jeong JA, Hong SH, Gang EJ, et al. Differential gene expression profiling of human umbilical cord blood-derived mesenchymal stem cells by DNA microarray. Stem Cell 2005; 23:584–593.
81. Zhong JF, Zhao Y, Sutton S, et al. Gene expression profile of murine long-term reconstituting vs. short-term reconstituting hematopoietic stem cells. Proc Natl Acad Sci USA 2005; 102:2448–2453.

3

Mechanisms of Genetic Fidelity in Mammalian Adult Stem Cells

James L. Sherley
Division of Biological Engineering, Center for Environmental Health Sciences, Biotechnology Process Engineering Center, Center for Cancer Research, Massachusetts Institute of Technology, Cambridge, Massachusetts, U.S.A.

EVOLUTION OF MECHANISMS OF MAMMALIAN TISSUE CELL GENETIC FIDELITY: THE NEEDS OF A FEW

Although DNA is thought of as the most stable known biological molecule for storage and retrieval of phenotypic information required for the production and functional integration of mammalian tissue cells, it is still quite mutable on the timescale of mammalian life-spans. It is estimated that one of every 1500 of the 3 billion base pairs of DNA in a human cell undergoes either a chemical conversion or a replicative mismatch each day (1–3). Highly efficient DNA repair mechanisms revert most of these changes, but those that escape are the engines of creation in the DNA code. On the evolutionary timescale, the mutable nature of DNA is thought beneficial to species by giving rise to variants that are more fit for survival. However, for individuals, the same mutability leads to chronic diseases such as cancer, adversely affects offspring, and may contribute to tissue aging.

Most adult tissue cells contain a copy of the genome of their zygotic precursor. The importance of the fidelity of a given cell's copy to the health of a mammal depends on that cell's position in the cell kinetics architecture of its tissue of residence. The majority of mammalian tissues are formed by arrays of repeating micro-anatomical tissue units (e.g., pits, crypts, columns, follicles, papillae, and tubules) that undergo reiterative development throughout the adult lifespan. This process has been referred to as cell renewal or tissue turnover (4–7). As a result of cell renewal, genetic fidelity may not matter much at all in most adult tissue cells, because their lifetimes in tissues are short compared with the mammalian lifespan (6–10).

Although the specific rates vary, most mammalian tissues undergo continuous cell turnover. This property is best described for epithelia (4–7,9,10), but it has even been observed in tissues like those of the brain that, until very recently, were considered by many to be devoid of cell renewal (11). Invariably, cell renewal division in mammalian tissues is compartmentalized. It is limited to relatively undifferentiated cells found exclusively in discrete segments of repeating tissue units. Cell production by these cells

37

replenishes associated larger segments of the unit that contain differentiating cells and mature terminally arrested cells. As expired or damaged terminal cells are lost from differentiated segments of tissue units because of natural tissue processes such as apoptosis and wear, they are replaced by the immigration of differentiating progeny from divisions in the associated proliferative segments of tissue units. One of the best-studied examples of this cell kinetics architecture is the small intestinal epithelium. The repeating crypt-villus unit of this epithelium exemplifies the micro-anatomical insulation of generative cells from their adjoining maturing differentiating descendants (7,9,10).

Two types of cells are postulated to co-exist in the proliferative compartments of renewing tissue units, adult stem cells (ASCs) and their proximate dividing progeny, variably called progenitor cells or early transit cells. The non-stem-cell progeny proceed along a developmental path that involves overlapping programs of division, cell-cycle arrest, maturation, and terminal differentiation. In contrast, ASCs persist in a state characterized by phenotypic immaturity and long-term division capacity (12). The founding motivation for invoking two cell types, instead of only one, was primarily a mathematical logic, with little scientific verification (12,13). The proliferative compartments of tissue units have no known entry point for an exogenous source of new cells and there is a continual exit of differentiating progeny. Thus, to maintain its relatively undifferentiated phenotype, the proliferative zone must possess a mechanism for *asymmetric self-renewal* (13). It must continuously produce cells that become mature differentiated cells, while preserving sufficient immature generative cells to maintain the zone.

Both stochastic and deterministic mathematical models have been advanced to account for ASC asymmetric self-renewal, but the answer to this vexing cell kinetics riddle remains elusive (12,14). Mathematically, the two models are equivalent, but biologically they have fundamentally different consequences. In stochastic constructs, individual stem cells are permitted to undergo direct differentiation to a non-stem cell phenotype. ASC differentiation, which alone would result in extinctions of tissue units, is balanced by symmetric ASC divisions that produce two ASCs (12,15). In deterministic constructs, ASCs do not differentiate into non-stem cells (12,13,16,17). Their programmed state is individual asymmetric self-renewal in which each cell division produces a new ASC and a non-stem cell daughter. Tissue requirements for multiplication of tissue units are met by regulated shifts of ASCs to transient symmetric self-renewal. Although intense effort has been brought to bear on the question of the exact mathematical form of ASC asymmetric self-renewal in tissues over the past two-and-a-half decades, it has proven to be a single challenging problem in mammalian cell biology, and to this day it remains unresolved.

The discussion of the nature of cell identities in the proliferative zones of tissue units has been a major driving force for ideas on the role of gene mutation in tissue disease etiology, especially carcinogenesis. The basic concept that is gaining wider recognition and acceptance is that non-stem cells (i.e., progenitor cells, transit cells, and terminally differentiated cells) do not have sufficient lifetimes in tissues to be frequent sources of cancer cell development. Mutations would have to disrupt their programed march through differentiation, maturation, and terminal arrest to final death or loss from the tissue before they could be effective in tumorigenesis. Although not impossible, the alignment of mutations and tissue alterations required for such an effect is considered too improbable to account for most cancers (6–8,10).

It is more likely that the cells in which carcinogenic mutations matter most are ASCs. Mutations in ASCs are inherited by all of their progeny and their descendants. For cancers that may require multiple genetic changes, long-lived ASCs provide incubator genomes for accumulation of mutations that are passed on to progeny cells. Even though transit cells and terminal cells might inherit sufficient alterations from their ancestral ASCs to initiate tumor formation, their turnover kinetics is still postulated to be sufficiently rapid to flush them from the tissue before they can do so. In contrast, the originating mutations in ASCs, being in long-lived retained cells, are able to initiate cancers (6–8,10).

The ASC-subtended tissue unit, or tissue turnover unit (8), may be the fundamental biological unit for cancer and other pathological processes that develop in the post-embryonic tissues of diverse mammalian species and potentially other vertebrate species as well (18). This paradigm predicts that the key determinants of cancer development will not be simply mutation rate and total body cell number. Instead, they will be the number of long-lived ASCs, the relative number of non-stem cells in their turnover units (6), cell transit rates through turnover units relative to lifespan, and the rates of mutation fixation by the stem cells. Although there are still many intriguing facets of the ASC turnover unit that remain to be elucidated, this basic concept resolves perplexing problems in cancer cell biology such as Peto's paradox (18–20) and the cellular basis for cancer stem cells (21). It also focuses the discussion of the evolution of mechanisms for tissue cell genetic fidelity sharply on the need for ASCs to remain error-free until mammals reach reproductive maturity.

AN EXACT DEFINITION FOR THE LONG-LIVED NATURE OF ASCs

In light of the historical basis for the above discussion (6,7,10,13) and the writings of others on the topic as well (22–24), the recent excitement over the revival of the concept of tumor stem cells (21) is both an amusing and welcomed development. In a similar vein, there is one aspect of the previous characterization of normal ASCs, and similarly cancer stem cells, that is so imprecise that its wide acceptance and usage is rather surprising. This is the convention that ASCs are "long-lived." What does it mean for a continuously dividing cell to be "long-lived?" As division occurs with renewal of nearly all cellular constituents (semi-conservatively replicated molecules such as DNA and centrosomes being exceptions), with each ASC division, a newly made ASC is born. Therefore, ASCs are not long-lived at all. As individuals, they are, in fact, rather *short-lived*.

Now this point is not a frivolous semantic. Recognizing it and refining the scientific language used to discuss it is critical for the formulation of applicable ideas about the nature of ASC genetic fidelity. A more exact characterization of ASCs is that their phenotypic program is long-lived. The adult "stemness" (13) program has at least two identifiable components: the ASC's genome sequence and its phenotypic expression. Alterations in either of these can have dire effects on the tissue units that stem cells subtend, and, accordingly, the tissues and mammals in which they reside. Although the stemness phenotype is ultimately derivative of the ASC's genomic DNA sequence, the epigenetic, post-transcriptional, post-translational, and extra-cellular determinants, which are not explicitly encoded in an ASC's genome, may also play a role in the maintenance of the stemness phenotype. The main goal of this chapter is to review published evidence for mechanisms that function to preserve the DNA sequence of ASCs and contemplate their possible consequences for mammalian tissue function.

Before specifically addressing mechanisms of mutation avoidance, a further step of refinement of the precept of "long-lived" is needed. If DNA, which is continuously synthesized in cycling ASCs, were randomly segregated at mitosis to non-stem cell daughters, then the long-lived entity responsible for stemness would be the *information* encoded by the DNA and not the DNA molecules per se. This is because both newly synthesized DNA strands and hybridized older template strands would be randomly segregated to non-stem progeny. With a geometric rate of dilution, similar to the ASCs in which they reside, old DNA molecules would continuously be replaced by new molecules from semi-conservative DNA replication. In this scheme, when a mutation occurred in an ASC, its fixation in the genetic code of the ASC compartment would be the event that eroded stemness. This would occur even though its actual DNA molecules were not long-lived at all. Thus, the only long-lived feature of ASCs would be the unique information they held, the complete program for constructing and maintaining their tissue units. Mutations that became fixed in their short-lived DNA molecules would become a fixed part of their information code. We turn now to how such mutations in ASCs may arise.

MUTAGENESIS MECHANISMS IN ASCs

A discussion of the nature of mutagenesis mechanisms in ASCs must begin with acknowl-edgement that very little is known about the exact origin of mutations within mammalian somatic tissue cells in general. Unlike germline mutations that can be deciphered directly by analyses of gene mutations in offspring (25,26), the nature of somatic cell mutations has been largely inferred from observations made with cultured cells (20), transgenic engin-eered reporter mice (27), and analyses of diseased tissues such as tumors (20). Each of these approaches has contributed to important advances in understanding mammalian cell DNA mutagenesis, culminating in an exhaustive catalogue of the many ways (1–3) in which mutations may arise. However, each has shortcomings because of their indirect nature (20,27). For example, in the case of mutations detected in tumors, it is difficult to determine whether they are responsible for the observed tissue pathology or secondary to it. In some specific cases, the mutation signature is sufficiently specific to support infer-ence of the responsible mechanism with a high degree of confidence. An example of this is C to T transitions that occur as a result of the higher rate of spontaneous deamination of C to T at methyl-CpG sites (28). For most somatic mutations, such a specific contextual feature is not available. Thus, although there are a large number of possible mechanisms of mutation in mammalian tissue cells in vivo, which actually occur and their relative frequencies are unknown. This being the case for mammalian tissue cells in general, it follows, of course, that the answers are even more elusive for more rare ASCs.

The calculated potential mutagenic events in a single cell per day are an astounding number. Janion (1) puts it at 2×10^6 affected base pairs. If each of these were productive for mutation, the human genome would sustain a mutation rate of approximately 0.1% per base pair per day, which would soon lead to error catastrophe and cell death. Of course, this does not occur because of the remarkable network of DNA repair machines that function in mammalian cells (2,3). In a recent accounting, the number of known human DNA repair genes was noted as 130 (29). These molecular systems work to keep spontaneous and damage-induced mutation rates exceedingly low in mammalian tissue cells (2,3). In fact, it is so low as to be a formidable technical challenge to direct detection, quantification, and characterization of mutations in mammalian tissues. By using sensitive mutational spectrometry approaches, a handful of investigators have attempted this feat (reviewed in Ref. 30). However, the majority of these studies have evaluated mutations in mitochon-drial DNA, because it has higher per cell gene copy number than nuclear DNA, and the fewer nuclear DNA analyses are limited to target regions of only a few chosen genes.

Despite the limitations of mutational analyses performed with cultured cells, esti-mates from them for spontaneous mutation rates in a few specific genes ($1-2 \times 10^{-10}/$ base pair/cell generation; 20) agree very well with estimates based on the more direct determinations of germ cell mutation rates ($2-4 \times 10^{-10}/$base pair/cell generation; 26).

Germ cells are specialized ASCs responsible for gametogenesis in fetal (oogenesis) and adult (spermatogenesis) mammals. Therefore, although inherited mutations are limited to a subset of genes that can be evaluated, the mechanisms responsible for them may be very relevant to mechanisms that determine mutations in ASCs of somatic tissues. In a similar fashion, as both tumor-derived cell lines and spontaneously immortalized cultured cell lines are likely to be derived from mutated ASCs (17,31,32), some aspects of their mutation rates and mutation spectra may be relevant to mutagenesis in ASCs in vivo.

Mutations in mammalian cells occur as a result of at least three events: DNA damage, repair of DNA damage, and replicative DNA synthesis. In theory, DNA damage could be due to several mechanisms, including environmental agents such as ultraviolet light, ionizing irradiation, and genotoxic chemicals; oxidative damage from reactive products of cellular metabolism; or spontaneous chemical disintegration (1,2). However, with the exception of ultraviolet light and skin cell mutagenesis, the available data indicate that human tissue cell mutations are ultimately produced by the action of DNA polymerases (30,33), either as a result of unrepaired mis-incorporated bases during replicative DNA synthesis (6) or error-prone DNA repair synthesis (34–36). In fact, half of the estimated 2×10^6 daily base pair exchanges in mammalian cells are calculated to be due to mismatch repair (1). Consistent with this idea, in mutational spectra analyses of inherited mutations, the average rate of single nucleotide substitutions was about 25 times greater than that of any other type of mutation (25). Single base substitutions also predominate the spectra of spontaneous mutations recovered for three different adult somatic tissues in transgenic reporter mice (27).

The foregoing discussion of observations and concepts suggests the inference that the predominant basis for mutations in ASCs is mis-incorporation of nucleotides by either replicative DNA polymerases or repair polymerases. Allowing this interpretation, there is still one remaining essential question. What is the relative contribution of these two types of DNA synthesis to fixation of mutations by ASCs? A mutation is fixed when it becomes a stable change in the DNA sequence that is no longer recognized by DNA repair mechanisms. The number of mutations fixed per cell generation, as a result of the action of a given DNA polymerase, will be a function of the number of base incorporations by the polymerase and its mis-incorporation rate. It is now well appreciated that although many previously described error-prone repair polymerases have high error rates on undamaged DNA, their fidelity at sites of DNA damage is comparable to that of replicative polymerases (3,35). Considering that the 6 billion bases incorporated by replicative polymerases during each human cell-cycle dwarfs, the estimated 1 million sites of damage (1) available to repair polymerases, an excellent case can be made that the origin of most mutations in ASCs will be the result of mis-repaired errors of replicative DNA synthesis. A quantitative treatment for this conclusion is given a later section.

IMMORTAL DNA STRAND CO-SEGREGATION AND A CARPENTER'S RULE FOR GENETIC FIDELITY IN ASCs

The first formal conceptualization of the idea that ASCs must have evolved additional mechanisms beyond basic DNA repair processes to reduce their fixation of mutations was presented by Cairns nearly 30 years ago (6). Cairns observed that estimates of tissue gene mutation rates predicted significantly higher rates of cancers in adult human somatic tissues than were observed. Given the place of ASCs in tissue cell kinetics architecture, this discrepancy led him to consider how ASCs might avoid mutations that occur as a result of replication errors, which he proposed were a major source of carcinogenic mutations. On the basis of the earlier studies by Lark et al. (37), indicating non-random mitotic chromosome segregation in embryonic mouse cells, Cairns proposed a remarkable hypothesis for a unique mechanism of chromosome segregation that could ensure that ASCs did not acquire mutations that resulted from "unrepaired" errors of replicative DNA polymerases (6). In the present refinement of the hypothesis, these errors are referred to as "mis-repaired" to recognize the high efficiency of DNA repair systems (1–3). As will be discussed in detail in "Estimation of the mutation-avoidance effect of an immortal DNA strand mechanism in ASCs," it is more likely that replication errors are mis-repaired by mismatch repair DNA polymerases than that they go unrepaired at all.

Cairns suggested that ASCs, which divide continuously with asymmetric cell kinetics throughout life, would quickly accumulate transforming mutations before the reproductive age of humans, unless the cells possessed a unique mitotic chromosome segregation mechanism. The traditional view of mitotic chromosome segregation is based on the characteristics of meiotic chromosome segregation. After semi-conservative DNA replication, paired sister chromatids, each made of an old template DNA strand and a newly synthesized DNA strand, segregate at mitosis, one to each new daughter cell. A fundamental aspect of the segregation event is that it is random and independent, meaning that there is an equal probability for either sister chromatid to go to either new daughter cell; and the manner in which one chromatid pair segregates does not affect another.

Cairns proposed that mitotic chromosome segregation in ASCs lacked these fundamental properties. Instead, it was non-random and dependent in nature. To fully understand Cairns' idea, it is important to recognize that the four DNA strands in paired sister chromatids differ in age. At mitosis, DNA strands of three different ages are present in each set of paired sister chromatids. Because replication is semi-conservative, both paired sister chromatids contain a newly synthesized DNA strand. However, their parental DNA strands are older and unequal in age. Because of the inheritance pattern of semi-conservative DNA replication, one of the parental strands was made in the previous cell generation, but the other is ≥2-cell generations old, depending on whether it was synthesized in its grandparent cell or an even earlier ancestor. In symmetrically cycling mitotic cell populations, random segregation of chromosomes bearing these oldest DNA strands leads to their dilution among chromosomes with younger DNA strands. Thus, the inherent "age asymmetry" of the genome is lost by randomization at each mitotic metaphase.

Cairns proposed that asymmetrically cycling ASCs continuously co-segregate to themselves the chromosomes that have the oldest DNA strands, thereby preserving the inherent age asymmetry of their genome. This mechanism would effectively allow them to repeatedly use the same DNA template for replication and segregate all mis-repaired replication errors made in newly synthesized DNA copies to their transient progeny cells. This hypothesis is referred to as the "immortal DNA strand hypothesis" (6), with reference to the predicted longevity of the oldest template DNA strands in ASCs.

The mutation-avoidance mechanism proposed by Cairns operates by a "carpenter's rule." When building, good carpenters do not use sequentially cut boards to measure subsequent boards, because of the well-known problems of copy drift and error amplification. Thus, all measurements are made with either a ruler or a single carefully measured first template. The immortal strand hypothesis predicts that ASCs follow the same rule. At ASCs' inception, an original DNA strand for each chromosome is selected as a template. These template strands are thereafter retained by ASCs through repeated cycles of asymmetric cell division by co-segregation of the sister chromatids that contain them. Therefore, all copied DNA with mis-repaired bases is passed on to transit cell progeny and eventually lost from the tissue as a result of cell turnover.

IMPLICATIONS OF THE CARPENTER'S RULE FOR ASC AGING, CELL KINETICS, AND DNA REPAIR FUNCTIONS

The carpenter's rule imposed on ASCs by the immortal DNA strand hypothesis has several noteworthy implications for ideas on ASC function. First of all, it provides a physical basis for ASC longevity. The long-lived nature of the stemness program in the ASC compartment is predicted to be physically embodied in immortal DNA strands. By preserving the sequence and integrity of immortal DNA strands, ASCs are predicted to achieve the main evolutionary goal of preserving the function of their subtended tissue units well beyond the time of reproductive maturity. However, the selection to protect the fidelity of the stemness code may have come at the cost of later cumulative defects in the chemical integrity of immortal DNA strands (38). Any poorly repaired chemical damage (from reactions with exogenous or endogenous agents or chemical disintegration) to immortal DNA strands will accrue in stem cells, leading to their eventual malfunction and/or death. (The consequences of actively repaired lesions will be addressed at the end of this section.) Such events in ASCs may contribute to reductions in tissue cellularity and proliferative capacity associated with chronological age (38).

If the carpenter's rule proves to be a fundamental biological law for the ASC function, then the issue of stochastic versus deterministic asymmetric self-renewal in ASC compartments will finally be resolved. Cairns' original formulation of the immortal DNA strand hypothesis was predicated on the idea of deterministic asymmetric cell kinetics by ASCs (6). An implicit precept for the hypothesis is that immortal DNA co-segregation only occurs in cells dividing with deterministic asymmetric cell kinetics that require deterministic asymmetric self-renewal. Each cell division produces a new ASC and a non-stem cell daughter. The latter of the two, thereafter, divides with symmetric cell kinetics to produce an expanded lineage of differentiating progeny cells. By retaining the immortal DNA strand complement, each successive new ASC preserves the stemness program for its tissue unit.

Now, on the other hand, if ASCs renewed as prescribed by stochastic asymmetric models, then all the cells in the stem cell compartment would have an equal probability of exiting as a result of differentiation events. With time, immortal DNA strands in any given stem cell and, accordingly, stemness identity would be lost due to a combination of differentiation and dilution. When an ASC with a set of immortal DNA strands underwent differentiation by chance, any immortal template strands that it contained would be lost from the compartment. Moreover, symmetric divisions by ASCs would result in randomization of immortal DNA strands among newly synthesized DNA strands in one of the two ways, depending on whether the mechanism functioned during symmetric mitoses that produced two ASCs.

In the first case, if the implicit ideas of the immortal DNA strand hypothesis held, then symmetrically dividing stem cells would simply not co-segregate chromosomes with their oldest template DNA strands. In the second, even if the co-segregation mechanism still functioned, it would no longer have the same effect. Each one of the two newly born ASCs from symmetric divisions would get a set of DNA strand copies that would become immortal DNA strands at its next mitosis. One cell would get its parent's immortal DNA strands, whereas the other would have to specify a new set. Thus, in either case, a stochastic self-renewal program is predicted to cause rapid dilution of the original ASC DNA templates among mutation-bearing copied DNA strands due to either mitotic randomization or continual replacement in new ASCs produced by symmetric cell divisions. This inherent dilution of immortal DNA strands combined with their continuous removal for the ASC compartment by chance stem-cell differentiation events would quickly erode any advantage they might provide.

A common objection to the immortal DNA strand hypothesis is that well-known DNA modification mechanisms would wreak havoc on such a mechanism. To address this caveat, it has been suggested that ASCs may suppress error-prone DNA repair and DNA recombination activities. The well-known greater sensitivity of cells in adult stem compartments to exogenous agents that damage DNA supports this idea (10,39). However, the formidable challenge of accurately quantifying the activities of suspected DNA repair systems in rare stem cells in adult tissues has, thus far, precluded further progress in evaluating these hypotheses experimentally.

Although it has not been possible to measure DNA repair and recombination in ASCs directly, recent developments in ideas in the DNA repair field suggest that reduction of error-prone DNA repair and DNA recombination efficiency may not be necessary for ASCs to realize an advantage from immortal DNA strands. There are three main categories of DNA modification activities to consider for their impact on the predicted effectiveness of the carpenter's rule in maintaining the genetic fidelity of ASCs: repair of DNA replication mismatches, repair of damage in immortal DNA strands, and mitotic recombination in the form of sister chromatid exchanges (SCEs). Repair of mismatches that occur during copying of immortal DNA strands poses no problems. If unrepaired, or more likely mis-repaired, the DNA copy with the error is segregated to a non-stem-cell daughter at the next mitosis. Given the speed and efficiency of DNA repair mechanisms, most mismatches will be repaired; and, as discussed in "Estimation of the mutation-avoidance effect of an immortal DNA strand mechanism in ASCs," a small fraction will be mis-repaired and therefore persist as mutations. The likelihood of the immortal DNA strand template being altered during this event is very small, because mismatch repair is strand-specific, targeting only mis-paired bases in the newly synthesized DNA strand (2,40).

There is nothing in the conceptualization of the immortal DNA strand hypothesis that imbues immortal DNA templates with immunity from the many different types of damage encountered by DNA (1–3). If such damages were repaired by the previously envisioned error-prone DNA polymerases, the ASC genetic code would be eroded in short order. However, the recent discovery of translesion polymerases, which faithfully replicate across a variety of forms of DNA damage (2,3,35), provides a solution by which ASCs might tolerate many forms of DNA damage while safeguarding their stemness code. Although translesion polymerases have lower fidelity when replicating undamaged DNA, their error rate at sites of their targeted damage matches that of replicative polymerases (3,35). Their key advantage, for the purpose of this discussion, is that they obviate the suggested necessity for repair of damaged immortal DNA strands. Even if translesion polymerases do occasionally introduce errors into newly synthesized DNA strands (2,36), the immortal strand mechanism would ensure their segregation to non-stem cell daughters. As better tools for identification and isolation of ASCs become available, it will be of interest to know whether translesion DNA polymerases are more highly expressed in ASCs when compared with their non-stem cell progeny.

On initial consideration, SCEs may appear to pose an obvious serious complication for an ASC carpenter's rule. Like many sought after properties of ASCs, the actual spontaneous rate of SCE in these cells is unknown. If spontaneous SCE rates determined from tissue cell preparations and cultured cells apply, in the absence of a specific suppression mechanism, the expected number of SCEs in ASCs would range from one to 10 at each mitosis (41,42). Given estimates of several thousand divisions in the lifetime of some ASCs (10), this number of SCE rates could be viewed to neutralize the advantages of an immortal strand mechanism. However, such a conclusion misses the purpose for which the carpenter's rule was postulated. Its purpose is to reduce the risk of ASC malfunction, disease, and death prior to the attainment of reproductive success. The ideal of an ASC that never acquires a mutation is incompatible with the biological reality that many cancers do occur; and they occur most probably as a result of cumulative mutations in ASCs. However, many cancers occur late in life after reproductive maturity at a time when their evolutionary impact on species survival is minimal.

Thus, it is important to recognize that, implicitly, the immortal DNA strand mechanism was never envisioned to be perfect. Because of its presence, ASCs will accrue detrimental gene mutations at a reduced rate. Because of its imperfections, ASCs will eventually fix sufficient mutations to precipitate their malfunction, death, or neoplastic transformation. Both SCEs and repair of certain types of DNA damage may cause mutations in immortal DNA strands. The balance among replication errors retro-fixed into immortal strands by SCEs, DNA repair-induced mutations, and stable unrepaired damage in immortal strands will be an important determinant of ASC function. As alluded to earlier in this chapter on the subject of ASC aging mechanisms, accumulated stable damage in immortal DNA strands could compromise DNA replication and gene transcription, leading to malfunction and death. DNA repair and SCE would renew immortal DNA strands at the cost of introducing mutations that could ultimately lead to aberrant functions such as neoplastic transformation.

In the special context of ASCs undergoing immortal DNA strand co-segregation, SCEs might have another untoward effect on cell viability that would not occur in non-stem cells. If immortal DNA strands have stable, dominant, distributive marking (e.g., at a minimum of two widely separated sites) for co-segregation to the ASC, then, after an SCE, both the two new hybrid sister chromatids might be recognized for segregation to the ASC. Such an event would effectively be a non-disjunction, resulting in aneuploidy. The ASC would acquire an extra chromosome, and the non-stem-cell daughter would lose one chromosome. Such gene dosage imbalances are often lethal, especially if more than one chromosome is involved. This idea suggests another potential explanation for the greater sensitivity of ASC compartments to DNA damaging agents that may also induce SCEs (6,10).

ESTIMATION OF THE MUTATION-AVOIDANCE EFFECT OF AN IMMORTAL DNA STRAND MECHANISM IN ASCs

Mathematical modeling has been undertaken as a means to evaluate predictions of the immortal DNA strand hypothesis with respect to observed cancer rates in some human populations (6). However, quantitative mathematical modeling to estimate the magnitude of the effect of an immortal DNA strand mechanism on ASC mutagenesis has not been reported. In an elementary fashion, this question can be addressed by considering the estimated number of replication errors relative to the number of repair synthesis errors in the absence of an immortal DNA strand mechanism. Mismatch errors from both sources will be repaired by the high-fidelity mismatch repair system with equal efficiency. Therefore, the final relative rate of mutation by the two mechanisms will be directly related to their relative number of respective errors of each cell generation. The number of errors per cell generation for a given type of DNA synthesis will be related to the number of base incorporations and the error rate of the responsible DNA polymerase. On the basis of these ideas, the magnitude of the effect of an immortal DNA strand mechanism on ASC mutagenesis can be estimated.

An estimate can be developed from the following calculations. One million DNA base pair exchanges are estimated to occur during each 24-hour human cell generation as a consequence of repair of modified or damaged bases (1). Depending on the kind of damage, either base excision repair (BER) or nucleotide excision repair (NER) is responsible. Three different DNA polymerases perform repair synthesis for BER and NER. They are pol-beta, pol-delta, and pol-epsilon (43). pol-Beta has an in vitro determined average substitution error rate of approximately 7×10^{-4} (44). This rate is a log higher than that of pol-delta and pol-epsilon (44) and is, therefore, the limiting rate that determines the number of DNA repair synthesis errors per cell generation. Multiplication of the pol-beta error rate (7×10^{-4} per base incorporated) by the estimated number of bases incorporated by BER and NER repair synthesis per cell generation (1×10^{6}) yields ~700 errors per cell generation. This value is a maximal expectation for the number of errors that would occur as a consequence of repair synthesis before the action of the mismatch repair system. On average, half of these errors, ~350, would be expected in the oldest DNA strands in the cell. If every one of these errors was subsequently acted on by the high-fidelity mismatch repair system, which is thought to use pol-delta and pol-epsilon (2,43; average substitution error rate equals 8×10^{-6} per base incorporated, 44), then the final number of mutations is estimated to be 0.003/cell/generation (Fig. 1). For a human cell with 3×10^{9} mutable base pairs, this corresponds to an estimated damage-dependent mutation rate in template DNA strands of 1×10^{-12}/base pair/cell/generation as a result of DNA damage and repair.

$$MR = MR_d + MR_r$$

$$TC\ MR = 0.5\ [(BE_d)(ER_{dp})(ER_{mp})] + 0.5[(BI)(ER_{rp})(ER_{mp})]$$

$$ASC\ MR = 0.5\ [(BE_d)(ER_{dp})(ER_{mp})] + 0$$

MR = cell mutation rate (number of fixed errors/cell/day; human)
MR_d = cell mutation rate due to DNA damage
MR_r = cell mutation rate due to replication
BE_d = number of damaged bases exchanged/cell/day ($\sim 1 \times 10^6$)
ER_{dp} = error rate of damage repair polymerases
     (pol-beta $\sim 7 \times 10^{-4}$ base errors/base)
ER_{mp} = error rate for mismatch repair polymerases
     (pol-delta, pol-epsilon $\sim 8 \times 10^{-6}$ fixed errors/base error)
BI = number of bases incorporated during replication
     ($\sim 6 \times 10^9$ bases/cell/day)
ER_{rp} = error rate of replication polymerases
     (pol-alpha $\sim 1.6 \times 10^{-4}$ base errors/base)
ASC, adult stem cell; TC, transit cell

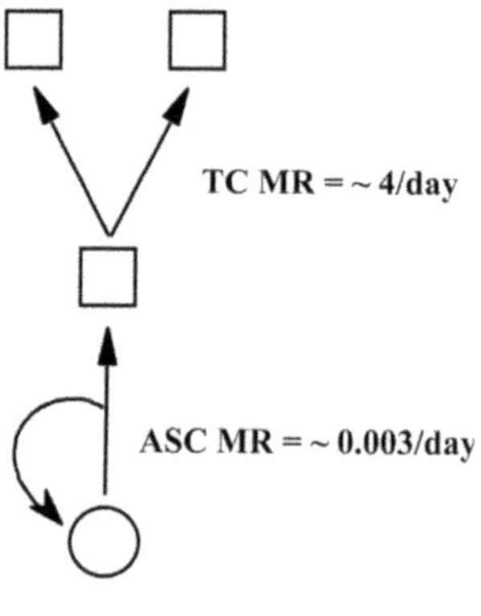

TC MR = ~ 4/day

ASC MR = ~ 0.003/day

Figure 1 Estimate of the expected mutation avoidance effect of an immortal DNA strand mechanism in ASCs. ASCs divide asymmetrically to replace themselves (*circle*) while simultaneously producing transit cell progeny (TC; *squares*). Asymmetric self-renewal proceeds with a non-random immortal DNA strand co-segregation mechanism that is modeled to yield an ASC mutation rate of 0.003/day as a consequence of DNA damage events. This low mutation rate is a consequence of the 0 term for mutations due to replication errors. In contrast, transit cells that divide symmetrically with random chromosome segregation are estimated to acquire an additional approximately 4 mutations per day as a result of replication errors. This nearly 1000-fold greater mutation rate, not experienced by ASCs, is the magnitude of the protection from mutations that is predicted to be afforded by an immortal DNA strand mechanism.

When considering the number of replicative polymerase errors, for the purposes of this analysis, only those occurring in newly replicated strands need to be considered. It is estimated that 1×10^6 replication errors occur per 24-hour cell generation (1,44) and these are repaired by the mismatch repair system. Multiplying by the mismatch repair synthesis substitution error rate of 8×10^{-6} (approximate for pol-delta and pol-epsilon; 44) gives a value of ~ 8 errors due to replication per cell generation. In the absence of an immortal strand mechanism, half of these, ~ 4, would be fixed by stem cells (Fig. 1). For a human cell with 3×10^9 mutable base pairs, this corresponds to an estimated replication-dependent mutation rate of 1×10^{-9}/base pair/cell/generation as a result of DNA replication errors. This rate is in good agreement with mutation rates determined for inherited mutations and cultured cell mutations ($1-4 \times 10^{-10}$/base pair/cell/generation, 20,26; "Mutagenesis mechanisms in ASCs"). Thus, this analysis provides quantitative support for the earlier proposal ("Mutagenesis mechanisms in ASCs") that the main mechanism for mutation in mammalian tissues is mis-repaired replication errors.

When the carpenter's rule of an immortal DNA strand mechanism is active, the ASC mutation rate is predicted to be equivalent to the damage-dependent rate (1×10^{-12}). In its absence, ASCs are predicted to experience the dramatically higher replication-dependent mutation rate (1×10^{-9}). Therefore, an immortal strand mechanism is predicted to afford ASCs a 1000-fold reduction in mutation rate compared with rates in their non-stem-cell progeny. Given estimates of 5000 cell divisions for human intestinal stem cells during the human lifespan (10), about 15 mutations are predicted. So, a small number of mutations are predicted to affect critical genes only rarely. Therefore, it may be that particular forms of DNA repair in immortal strands, that have not been considered in this treatment and which have a higher error rate, and SCEs, which would continuously retro-fix replication-dependent mutations in immortal DNA strands at a low rate, may be responsible for conversion of ASCs to cancer cells.

EVIDENCE FOR IMMORTAL DNA STRAND CO-SEGREGATION IN ASCs

The earliest evidence for non-random mitotic chromosome segregation is found in the elegant experiments of Lark et al. (37,45–48) with cultured mammalian cells and plant root tips. Lark was motivated to examine the symmetry of chromosome segregation in eukaryotic cells because of his earlier ideas on mechanisms of bacterial chromosome segregation (47,48). Lark proposed that the essential units of genetic segregation in bacteria were the individual template DNA strands. He postulated that both had to be attached to the bacterial cell wall, on either side of the future septum, before DNA replication could be initiated. This control point was proposed as a mechanism to ensure that each new bacterial daughter cell received a complete copy of the genome. Many of Lark's ideas on this topic have subsequently been confirmed experimentally (49).

From the single bacterial chromosomes, Lark advanced the idea of stable attachment of template DNA strands to a parent cell structure for the more complex segregation of numerous eukaryotic chromosomes. He looked for evidence of non-random mitotic segregation by cohorts of sister chromatids that contained one unlabeled DNA strand that existed before the introduction of ^{3}H–thymidine. This DNA inheritance tracer is incorporated into all DNA strands synthesized after its addition. In these studies, the presence of co-segregating chromosomes with an unlabeled DNA strand was detected by quantifying the amount of radioactivity in daughter cells after two generations of continuous labeling with ^{3}H–thymidine. If chromosome segregation were random, the distribution of radioactivity per cell would be predicted to be uniform. If non-random chromosome co-segregation occurred as envisioned by Lark, then the distribution would be bimodal, with two equal-size populations of daughter cells that differ by a factor of 2 in their ^{3}H–DNA content. According to the modeling, the population with the greater amount of radioactivity would contain chromosomes with two labeled DNA strands and the population with the lesser amount would contain chromosomes with one labeled and one unlabeled DNA strands. Consistent with the proposal for non-random chromosome segregation, bimodal distributions were observed for cells in plant root tips, primary cultures of embryo fibroblast, and a hamster cell strain (37,45–48).

In complementary experiments, it was also shown that if these cells were cultured with ^{3}H–thymidine for a one-generation period followed by a one-generation period of culture with the ^{3}H–thymidine removed, then about half of the cells released their entire previously incorporated label. This finding was predicted if non-random chromosome segregation occurred. It was consistent with a co-segregation of chromosomes that contained a previously unlabeled oldest template DNA strand. After the first generation of labeling, these strands would become paired with newly synthesized ^{3}H–thymidine containing DNA strands and co-segregate together. After the next period of labeling without ^{3}H–thymidine, they would become hybridized again with newly synthesized DNA, but it would be unlabeled. Their previously hybridized labeled DNA complement would now reside in their sister chromatids. Thus, at mitosis, cells co-segregating the chromosomes with the oldest DNA templates would effectively release their entire label. In studies with plant root tips, these label-releasing segregations were visualized by autoradiography of cells in anaphase and telophase. These images show a highly asymmetric localization of radioactivity to one set of segregating chromosomes (46,47). The relatively label-free complement of chromosomes would contain the oldest complement of template DNA strands that would later become Cairns' immortal DNA strands (6).

Lark's ideas met with much resistance from geneticists who then and now adhere fervently to the paradigm of random chromosome segregation (47). However, this long-standing precept is based entirely on experimentation for meiotic chromosome segregation. Results from meiotic chromosome segregation studies have been applied to mitotic chromosome segregation by analogy, and there have been very few studies that have addressed this issue for mitosis by direct experimentation. The main unavoidable shortcoming of Lark's proposal was that he suggested it for all eukaryotic cells. Although not noted at the time of his studies, it can now be appreciated, retrospectively, that all the cells that Lark observed to exhibit non-random chromosome segregation were likely to share the same special property, deterministic asymmetric cell kinetics ("Evolution of mechanisms of mammalian tissue cell genetic fidelity: the needs of a few"). It is now recognized that early passage murine embryo fibroblasts (31) and somatic stem cells in the root tip (50) divide with deterministic asymmetric cell kinetics. Asymmetrically dividing ASCs in cultures of primary adult tissue cells and pre-senescent cell strains become progressively diluted by their symmetrically dividing progeny (31,32), which are predicted to exhibit random chromosome segregation; and all cells in cultures of symmetrically dividing tumor-derived cell lines are expected to display random segregation. Lark's observations are well explained by these new concepts. The best distinction was observed for non-random chromosome segregation in primary cell cultures of mouse and plant cells. With a cultured hamster cell strain, the distinction was less dramatic and with tumor-derived HeLa cells it was not evident (47,48).

The earliest attempts to demonstrate the existence of immortal DNA strands in adult mammalian stem cells in vivo applied the label-release strategy of Lark, and also introduced a new strategy, called label retention (51). In concept, the label-retention strategy required that DNA inheritance tracers such as ^{3}H–thymidine be incorporated into immortal DNA strands prior to their selection for co-segregation. Potten et al. (52) developed two different strategies for this purpose. They either labeled juvenile mice with ^{3}H–thymidine prior to completion of gut development or labeled adult mice after gamma irradiation to induce crypt regeneration in the small intestinal epithelium. In both procedures, autoradiography detects rare cells in the stem-cell compartment of intestinal crypts that retain ^{3}H radioactivity for extended periods after other crypt cells are label-free. Subsequent labeling of mice with another DNA inheritance tracer, bromodeoxyuridine (BrdU) that can be detected with specific antibodies, was used to demonstrate that about 90% of detected label-retaining cells (LRCs) incorporate BrdU. This result indicates that these cells continue to cycle actively, as expected for ASCs. Moreover, after the BrdU labeling period, although they continue to retain ^{3}H radioactivity, they rapidly lose their BrdU label at a rate consistent with release as a result of non-random chromosome segregation. It is also noteworthy that in this remarkable demonstration, no LRCs with BrdU are produced, further validating the need for the two special strategies for labeling immortal DNA strands.

In a more recent study, Smith (53) applied the approach of Potten to demonstrate that LRCs in the mammary epithelium of the mouse also exhibit label retention/release kinetics indicative of immortal DNA strand co-segregation. Cells with this property were detected in a mammary epithelium ASC transplantation model. The transplanted mammary epithelium tissue fragments were shown to contain adult mammary stem cells based on their ability to confer serial repopulation.

In Smith's studies, it was possible to introduce label into label-retaining breast epithelium cells during allometric expansion in response to estradiol administration after tissue transplantation. This feature suggests that hormone-induced allometric growth of the mammary epithelium proceeds by symmetric divisions of adult mammary stem cells followed by their initiation of asymmetric cell kinetics and immortal DNA strand co-segregation. Smith proposes that the well-known effect of parity to reduce breast cancer incidence in rodents and humans may reflect parity-induced ASCs that cycle asymmetrically with a mutation-protective immortal DNA strand mechanism (53).

The recent work of Potten et al. (52) is by far the best-reported evidence for immortal DNA strand co-segregation in ASCs in vivo. Although the mammary epithelium studies of Smith were performed in transplanted tissues (53), it seems very likely that the cell processes defined will also be found in normal mammary epithelium. However, these studies together address only two ASC compartments in one species, leaving open the question of how general the mechanism is for ASCs in different tissues and in different mammalian species. A strong teleological argument can be made that, given the importance of ASC genetic fidelity in mammalian evolution, if an immortal DNA strand mechanism occurs in one tissue, it will occur in all tissues that possess ASCs that cycle throughout the mammalian lifespan. In support of this proposal, there is a very common experimental observation reported for many other renewing tissues that may indicate immortal strand co-segregation in their ASC compartments. LRCs have been reported in a diverse collection of ASC compartments in several different mammalian species, including: mouse oral mucosae (54,55), epidermis (54–56), and hair follicles (57,58); hamster oral mucosae and epidermis (59); rat pancreas (60), kidney (61), and colonic epithelium (62); and human embryonic and fetal epidermis in organ culture (63). In many of these studies, the introduction of DNA base analogues occurred at times in fetal and neonatal development (54,55,57–60,63) or tissue regeneration (56) when immortal DNA strands are predicted to undergo establishment (52,53). In some cases, LRCs were observed to persist for approximately as long as half an animal's lifespan, despite several rounds of cell division (58).

Surprisingly, although in all cases LRCs have been regarded as ASCs, the basis for their label retention has not been uniformly attributed to immortal DNA strands retained as a result of non-random chromosome segregation in ASCs. In fact, although many reports have discussed this basis as a possibility (54–56,58,59,61), many have not considered it at all (57,60,62,63). In at least one case, an attempt was made to show that LRCs in the epidermis of the mouse had cell kinetics that ruled out an immortal DNA strand mechanism (56). However, the authors of this report did not consider that immortal DNA strands might be re-established during tissue regeneration. Therefore, their conclusions are equivocal at best.

Much of the hesitation to interpret LRCs as evidence of non-random chromosome segregation is due to the pervasive idea that ASCs divide rarely during the adult mammalian lifespan. The origin of this idea can be traced to Lajtha (13,54,63), who put forth the hypothesis that ASCs might have low division frequencies compared with their transit cell progeny. Over the years, hypothesis has become dogma. Thus, LRCs have been primarily interpreted to be infrequently cycling ASCs that incorporated label because of a rare cycle during a period of labeling. Thereafter, because of their infrequent cycling, they are expected to retain the label for long periods. Given the complex nature of tissues' cell kinetics architecture, there are likely to be different classes of "LRCs," not all of which are ASCs. In fact, this feature has been noted. Depending on the time of labeling (e.g., neonatal vs. adult; 57) and the time of assessment after the labeling period (54,55,59), the basis for detected LRCs may differ. Even under the same conditions of detection, LRCs differ quantitatively in the amount of label they retain (54,55,58,59). However, in all reported cases, rare cells are detected that maintain close to their initial level of label after very long periods. In these cases, if the cells have divided more than five times after their incorporation of label, then the retention of label is consistent with non-random co-segregation of immortal DNA strands. This is because five generations of random chromosome segregation would reduce a chromosomal label to about 3% of its starting level, which is undetectable in typical label-retention analyses.

Despite this straightforward approach to clarifying the basis for LRC, very few groups have independently evaluated the cell kinetics of LRCs detected in their studies. Only Potten's group and Smith performed this evaluation in the ideal manner, simultaneously in situ without additional experimental manipulations. Their independent confirmation of active cycling by rare LRCs in intestinal crypts and mammary epithelium was essential to the conclusion that these cells retain immortal DNA strands (52,53). Another group has shown that LRCs detected after labeling colonic pit cells in adult rats continue to cycle at a low rate (62). However, for two reasons, the significance of their findings is difficult to decipher. First, because they labeled adult animals, the disposition of the label is less certain. Immortal DNA strands are not predicted to be elaborated, unless as a result of new pit maintenance formation or as a result of label-induced (BrdU) tissue injury and regeneration. Second, the independent measurement of proliferation was not performed for individual LRCs. So, a small subpopulation of more actively cycling LRCs would have been overlooked. Several groups have also observed evidence that growth promoters (54) and tissue injury (58,61) can induce the active proliferation of some LRCs, and in one study LRCs were shown to be quiescent prior to tissue injury (61).

Thus, few studies have evaluated the cell kinetics of LRCs in the ideal manner required to discern whether some are in fact the manifestation of immortal DNA strand co-segregation in ASCs. The present availability of specific antibodies that detect independent markers of cycling cells (Ki67 antigen, proliferating cell nuclear antigen, cyclins, histone H2b) makes this issue quite straightforward to address. However, the ideal experiment is that of Potten and Smith (52,53), in which LRCs are shown to incorporate a second DNA base analog (BrdU) and then rapidly release it while still retaining their original label.

On a final note, it is quite instructive to realize that only recently have there been direct experimental assessments of Lajtha's hypothesis that ASCs divide infrequently. These evaluations were performed with cell populations enriched for murine hematopoietic stem cells (HSCs). Although, as predicted by Lajtha, isolable HSCs were found to cycle at lower rates than their progeny, to the surprise of many, quantitatively, they cycled frequently compared with the scale of the murine lifespan (64,65). If the estimated cycling rate for HSCs were shared by ASCs in other tissues, it would be more than sufficient to allow for interpretation of LRCs as cells that harbor immortal DNA strands. As luck would have it, LRC analyses have not been reported for the HSC compartment, which poses a greater experimental challenge because of the absence of well-defined anatomical landmarks. However, it might be possible to combine flow cytometric enrichment procedures with a label-retention strategy to look for LRCs in this ASC compartment as well. With the a priori evidence that ASCs in this compartment cycle sufficiently, detecting LRCs would further strengthen the case that ASCs safeguard their genetic fidelity with an immortal DNA strand co-segregation mechanism. In addition, isolated populations enriched for HSCs might be more accessible than ASCs in other tissues for investigation of the molecular basis of non-random chromosome segregation.

Recent studies in cell culture promise more accessible experimental models for the investigation of molecular mechanisms responsible for non-random chromosome segregation. Merok et al. (38) demonstrated that genetically engineered cultured cell lines with conditional asymmetric cell kinetics exhibit immortal DNA strand co-segregation. The main cells used for these studies were p53-null murine embryo fibroblasts engineered to conditionally express normal levels of the wild-type p53 protein (31). When the conditional p53 gene is off, the cells divide with symmetric cell kinetics. However, under culture conditions that induce normal levels of p53 protein, the cells switch to asymmetric cell kinetics (31). The kinetics are characterized by cycling adult stem-like cells that continuously produce a non-cycling, non-stem-cell daughter every 20–24-hour cell cycle.

Two different strategies were used to demonstrate that immortal DNA strand co-segregation occurred only in asymmetrically cycling adult stem-like daughter cells. The first was label retention as described for in vivo studies. BrdU was introduced into cells under conditions of symmetric cell kinetics, when non-random chromosome segregation did not occur. Thereafter, cells were induced to cycle asymmetrically in BrdU-free medium. In this experiment, cycling adult stem-like cells were shown to co-segregate a set of chromosomes that contained the same BrdU-labeled DNA strands for at least seven generations, the longest period that was evaluated (38). The second strategy was continuous labeling as first described by Lark et al. (37). Symmetrically cycling and asymmetrically cycling cells were compared after introduction into BrdU-containing medium for several generations of division. Although the chromosomes of symmetrically cycling cells became uniformly labeled with BrdU, asymmetrically cycling adult stem-like cells maintained chromosomes that had one unlabeled DNA strand, corresponding to their non-randomly co-segregated immortal DNA strands. It is of note here, with respect to the earlier described mammary epithelium studies of Smith (53), that a less studied cell line, which also showed evidence of non-random chromosome segregation by the continuous-labeling method, was a mouse mammary epithelial cell line (C127; 38).

These studies with cultured ASC models provided for the first time direct visualization of immortal DNA strands' co-segregation and confirmation of their predicted chemical topology in large numbers of cells. Consistent with the critical role of ASCs in tumorigenesis, the p53 cancer gene was implicated as a key determinant of immortal DNA strand co-segregation mechanisms in vivo. There is a large body of scientific literature that considers the primary function of p53 in tissues to be regulation of cellular responses to DNA damage. The work of Merok et al. (38) raises the hypothesis that p53 may also function as a carpenter to ensure the genetic fidelity of asymmetrically self-renewing ASCs. The relationship between p53 genotype and responses to DNA damage may reflect these basic ASC functions. The demonstration of immortal DNA strand co-segregation in a cultured cell model provides for the first time opportunities to evaluate predictions of the impact of immortal DNA strand co-segregation on cell mutation rates and to elucidate the responsible cellular mechanisms. In 1969, Lark (47) considered that perhaps "all new templates were attached to a common segregation apparatus which was distinct from the one to which all old templates were already attached." Now, 36 years later, the tools are in hand to test this hypothesis directly. The outcomes of these evaluations are predicted to inform many longstanding problems in mammalian biology, among them discovery of the nature of ASCs in tissue youth and age, health, and disease.

ACKNOWLEDGMENTS

I am grateful to J. Cheng for identification of several key references. Many thanks to Dr. J.-F. Paré, Dr. J.A. Lansita, A.M. Nichols, S. Ram-Mohan, and R. Taghizadeh for review of the manuscript.

REFERENCES

1. Janion C. Some provocative thoughts on damage and repair of DNA. J Biomed Biotechnol 2001; 1:50–51.
2. Hoeijmakers JHJ. Genome maintenance mechanisms for preventing cancer. Nature 2001; 411:366–374.
3. Friedberg EC. DNA damage and repair. Nature 2003; 421:436–440.
4. Leblond CP, Walker BE. Renewal of cell populations. Physiol Rev 1956; 36:255–276.
5. Thrasher JD. Analysis of renewing epithelial cell populations. In: Prescott DM, ed. Methods in Cell Physiology. New York: Academic Press, 1966:323–357.
6. Cairns J. Mutation selection and the natural history of cancer. Nature 1975; 255:197–200.
7. Potten CS, Morris RJ. Epithelial stem cells in vivo. J Cell Sci Suppl 1988; 10:45–62.
8. Herrero-Jimenez P, Thilly G, Southam PJ, Tomita-Mitchell A, Morgenthaler S, Furth EE, Thilly WG. Mutation, cell kinetics, and subpopulations at risk for colon cancer in the United States. Mutat Res 1998; 400:553–578.
9. Marshman E, Booth C, Potten CS. The intestinal epithelial stem cell. Bioessays 2002; 24:91–98.
10. Potten CS, Booth C, Hargreaves D. The small intestine as a model for evaluation of adult tissue stem cell drug targets. Cell Prolif 2003; 36:115–129.
11. Temple S. The development of neural stem cells. Nature 2001; 414:112–117.
12. Loeffler M, Potten CS. Stem cells and cellular pedigrees—a conceptual introduction. In: Potten CS, ed. Stem Cells. London: Academic Press, 1997:1–27.
13. Lajtha LG. Stem cell concepts. Differentiation 1979; 14:23–34.
14. Blackett N, Gordon M. "Stochastic"—40 years of use and abuse. Blood 1999; 93:3148–3149.
15. Matioli G, Niewisch H, Vogel H. Stochastic stem cell renewal. Rev Eur Ètudes Clin Biol 1970; XV:20–22.
16. Sherley JL. Stem cell differentiation: what does it mean? Proc Second Joint EMBS-BMES Conf 2002; 1:741–742.
17. Sherley JL. Asymmetric cell kinetics genes: the key to expansion of adult stem cells in culture. Stem Cells 2002; 20:561–572.

18. Leroi AM, Koufopanou V, Burt A. Cancer selection. Nat Rev 2003; 3:226–231.
19. Peto R, Roe FJ, Lee PN, Levy L, Clack J. Cancer and aging in mice and men. Br J Cancer 1975; 32:411–442.
20. Loeb LA. Mutator phenotype may be required for multistage carcinogenesis. Cancer Res 1991; 51:3075–3079.
21. Reya T, Morrison SJ, Clarke MF, Weissman IL. Stem cells, cancer, and cancer stem cells. Nature 2001; 414:105–111.
22. Knudson AG. Stem cell regulation, tissue ontogeny, and oncogenic events. Sem Cancer Biol 1992; 3:99–106.
23. Sherley JL, Stadler PB, Johnson DR. Expression of the wild-type p53 antioncogene induces guanine nucleotide-dependent stem cell division kinetics. Proc Natl Acad Sci 1995; 92:136–140.
24. Sherley JL. The p53 tumor suppressor gene as regulator of somatic stem cell renewal division. Cope 1996; 12:9–10.
25. Kondrashov AS. Direct estimates of human per nucleotide mutation rates at 20 loci causing Mendelian diseases. Hum Mutat 2002; 21:12–27.
26. Crow J. Spontaneous mutation as a risk factor. Exp Clin Immunogenet 1995; 12:121–128.
27. Stuart GR, Oda Y, de Boer JG, Glickman BW. Mutation frequency and specificity with age in liver, bladder, and brain of *lac*I transgenic mice. Genetics 2000; 154:1291–1300.
28. Rodin SN, Rodin AS. Strand asymmetry of CpG transitions as indicator of G1 phase-dependent origin of multiple tumorigenic p53 mutations in stem cells. Proc Natl Acad Sci USA 1998; 95:11927–11932.
29. Wood RD, Mitchell M, Sgouros J, Lindahl T. Human DNA repair genes. Science 2001; 291:1284–1289.
30. Thilly WG. Have environmental mutagens caused oncomutations in people? Nat Genet 2003; 34:255–259.
31. Rambhatla L, Bohn SA, Stadler PB, Boyd JT, Coss RA, Sherley JL. Cellular senescence: ex vivo p53-dependent asymmetric cell kinetics. J Biomed Biotechnol 2001; 1:27–36.
32. Merok JL, Sherley JL. Breaching the kinetic barrier to in vitro somatic stem cell propagation. J Biomed Biotechnol 2001; 1:24–26.
33. Muniappan BP, Thilly WG. Polymerase beta creates APC mutations found in human tumors. Cancer Res 2002; 62:3271–3275.
34. Branum ME, Reardon JT, Sancar A. DNA repair excision nuclease attacks undamaged DNA. J Biol Chem 2001; 276:25421–25426.
35. Friedberg EC, Wagner R, Radman M. Specialized DNA polymerases, cellular survival, and the genesis of mutations. Science 2002; 296:1627–1630.
36. Goodman MF. Error-prone repair DNA polymerases in prokaryotes and eukaryotes. Annu Rev Biochem 2002; 71:17–50.
37. Lark KG, Consigli RA, Minocha HC. Segregation of sister chromatids in mammalian cells. Science 1966; 154:1202–1205.
38. Merok JR, Lansita JA, Tunstead JR, Sherley JL. Cosegregation of chromosomes containing immortal DNA strands in cells that cycle with asymmetric stem cell kinetics. Cancer Res 2002; 62:6791–6795.
39. Cairns J. Somatic stem cells and the kinetics of mutagenesis and carcinogenesis. Proc Natl Acad Sci USA 2002; 99:10567–10570.
40. Yang W. Structure and function of mismatch repair proteins. Mutat Res 2000; 460:245–256.
41. Natarajan AT. Chromosome aberrations: past, present, and future. Mutat Res 2002; 504:3–16.
42. Helleday T. Pathways for mitotic homologous recombination in mammalian cells. Mutat Res 2003; 532:103–115.
43. Friedberg EC, Wood RD. DNA excision repair pathways. In: DePamphilis ML, ed. Concepts in Eukaryotic DNA Replication. Cold Spring Harbor: Cold Spring Harbor Laboratory Press, 1999:249–269.
44. Roberts JD, Kunkel TA. Fidelity of DNA replication. In: DePamphilis ML, ed. Concepts in Eukaryotic DNA Replication. Cold Spring Harbor: Cold Spring Harbor Laboratory Press, 1999:217–247.

45. Lark KG. Nonrandom segregation of sister chromatids in *Vicia faba* and *Triticum boeoticum*. Proc Natl Acad Sci USA 1967; 58:352–359.

46. Lark KG. Sister chromatid segregation during mitosis in polyploidy wheat. Genetics 1969; 62:289–305.

47. Lark KG. Nonrandom segregation of sister chromatids. In: Teas HJ, ed. Genetics and Developmental Biology. Lexington: University of Kentucky Press, 1969:8–24.

48. Lark KG, Eberle H, Consigli RA, Minocha HC, Chai N, Lark C. Chromosome segregation and the regulation of DNA replication. In: Vogel HJ, Lampen JO, Bryson V, eds. Organizational Biosynthesis: A Symposium. New York: Academic Press, 1967:63–89.

49. Lewis PJ. Bacterial chromosome segregation. Microbiology 2001; 147:519–526.

50. Laurenzio LD, Wysocka-Diller J, Malamy JE, Pysh L, Helariutta Y, Freshour G, Hahn MG, Feldmann KA, Benfey PN. The SCARECROW gene regulates an asymmetric cell division that is essential for generating the radial organization of the Arabidopsis root. Cell 1996; 86:423–433.

51. Potten CS, Hume WJ, Reid P, Cairns J. The segregation of DNA in epithelial stem cells. Cell 1978; 15:899–906.

52. Potten CS, Owen G, Booth D. Intestinal stem cells protect their genome by selective segregation of template DNA strands. J Cell Sci 2002; 115:2381–2388.

53. Smith G. Label-retaining epithelial cells in mouse mammary gland divide asymmetrically and retain their template DNA strands. Development 2005; 132:681–687.

54. Mackenzie IC, Bickenbach JR. Patterns of epidermal cell proliferation. Carcinogenesis 1982; 7:311–317.

55. Bickenbach JR, McCutecheon J, Mackenzie IC. Rate of loss of tritiated thymidine label in basal cells in mouse epithelial tissues. Cell Tissue Kinet 1986; 19:325–333.

56. Kuroki T, Murakami Y. Random segregation of DNA strands in epidermal basal cells. Jpn J Cancer Res 1989; 80:637–642.

57. Cotsarelis G, Sun T-T, Lavker RM. Label-retaining cells reside in the bulge area of pilosebaceous unit: implications for follicular stem cells, hair cycle, and skin carcinogenesis. Cell 1990; 61:1329–1337.

58. Morris RJ, Potten CS. Highly persistent label-retaining cells in the hair follicles of mice and their fate following induction of anagen. J Invest Dermatol 1999; 112:470–475.

59. Bickenbach JR, Mackenzie IC. Identification and localization of label-retaining cells in hamster epithelia. J Invest Dermatol 1984; 82:618–622.

60. Duvillié B, Attali M, Aiello V, Quemeneur E, Scharfmann R. Label-retaining cells in the rat pancreas: location and differentiation potential in vitro. Diabetes 2003; 52:2035–2042.

61. Maeshima A, Yamashita S, Nojima Y. Identification of renal progenitor-like tubular cells that participate in the regeneration processes of the kidney. J Am Soc Nephrol 2003; 14:3138–3146.

62. Kim SJ, Cheung S, Hellerstein MK. Isolation of nuclei from label-retaining cells and measurement of their turnover rates in rat colon. Am J Physiol Cell Physiol 2004; 286:C1464–C1473.

63. Bickenbach JR, Holbrook KA. Label-retaining cells in human embryonic and fetal epidermis. J Invest Dermatol 1987; 88:42–46.

64. Bradford GB, Williams B, Rossi R, Bertoncello I. Quiescence, cycling and turnover in the primitive hematopoietics stem cell compartment. Exp Hematol 1997; 25:445–453.

65. Cheshier SH, Morrison SJ, Liao X, Weissman IL. In vivo proliferation and cell cycle kinetics of long-term self-renewing hematopoietic stem cells. Proc Natl Acad Sci USA 1999; 96:3120–3125.

4

Neural Stem Cells: Isolation and Self-Renewal

Hideyuki Okano, Jun Kohyama, Hiroyuki Ohba, Masanori Sakaguchi, Akinori Tokunaga, Takuya Shimazaki, and Hirotaka James Okano
Department of Physiology, Keio University School of Medicine, Tokyo and Core Research for Evolutional Science and Technology (CREST), Japan Science and Technology Agency, Saitama, Japan

INTRODUCTION

The human brain is composed of more than 100 billion neurons and more than 10 times that many glia, and in spite of having a wide variety of functions and morphology depending on the individual site, they function superbly as a single community. Neural stem cells (NSCs) can be described as the source of this wide variety of cells. Stem cells are generally defined as cells that fulfill four conditions (1). They are capable of (1) proliferation, (2) self-renewal, (3) multipotency, and (4) tissue-repair ability (discussed subsequently), and NSCs are likely to fulfill those conditions. In mice, NSCs are known to be maintained by self-renewal from the time they first appear around embryo, day 8.5, until adulthood. A lineage relationship between embryonic and adult NSCs, however, has not been demonstrated. Experiments have shown that it is possible to selectively culture NSCs in the presence of growth factors by monolayer culture on an adhesive substrate (2) and by suspension culture (3), which is called the "neurosphere method" (Fig. 1). As they differentiate into the neurons, astrocytes, and oligodendrocytes that comprise the central nervous system (CNS) when the growth factors are removed, they can be said to possess multipotency. In adult mammalian brains in vivo, NSCs or NSC-like cells have been shown to be involved in neurogenesis under physiological conditions at particular sites, that is, such as the subventricular zone (SVZ) of the lateral ventricle and the subgranular zone (SGZ) of the hippocampal formation (4–6). Furthermore, recent reports have suggested that NSCs also have the ability to partly repair the damaged CNS (7,8).

IN VIVO LOCALIZATION OF NSCs

NSCs are present from the developmental period until adulthood, and they are maintained throughout life (9) (Fig. 2). Depending on the stage of development, the mitotic cycle starts at 7 to 10 hours, is 18 hours in the late fetal period, and on the order of several

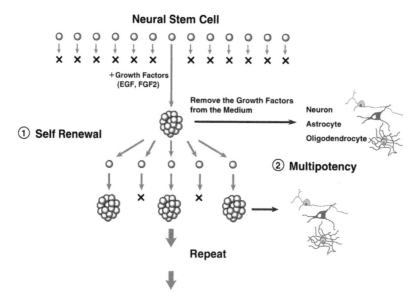

Figure 1 Neurosphere culture. NSCs can be selectively grown in the presence of the growth factor EGF or FGF-2, and they form cell aggregates called neurospheres. When the cell population composing a neurosphere is broken apart, similar neurospheres form again (self-renewal); and when the growth factor is removed and they are cultured on an adhesive substrate, neurons, astrocytes, and oligodendrocytes are produced. *Abbreviation*: NSCs, neural stem cells. *Source*: From Ref. 12. (*See color insert.*)

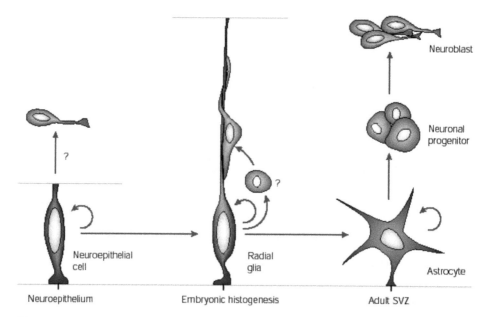

Figure 2 NSCs are maintained in the perventricular area all through the animals. NSCs are present as "purple cells," throughout the animals' lives. These cells have been given several different names (neuroepithelical cells, radial glia, SVZ astrocytes). *Abbreviations*: NSCs, neural stem cells; SVZ, subventricular zone. *Source*: From Ref. 9. (*See color insert.*)

days in adulthood (10). NSCs can be extracted in test tubes by the selective culture method, but can they be prospectively identified and isolated without going through that procedure? And how do they behave in vivo?

What Are NSCs?

NSCs could be defined as cells that have the ability to generate the multiple cell types of CNS (multipotency) and are capable of self-renewing (11). However, this definition is fairly conceptual. How should NSCs be defined empirically? One of the breakthroughs in NSC research has been the development of a selective culture method for NSCs by Samuel Weiss and co-workers in 1992 called the "neurosphere method" (3), in which floating clonal colonies of cells are formed. A cell group that contains NSCs is cultured in serum-free liquid culture medium containing the mitogen epidermal growth factor (EGF) and/or fibroblast growth factor-2 (FGF-2) in insulin, transferrin, serine, and progesterone. The viability of the cells, other than stem cells, is impaired in serum-free culture medium, which makes it possible to start growing only NSCs, capable of surviving in serum-free medium. When this is done, the proliferating NSCs form neurospheres, and the neurospheres float in nonadhesive culture conditions. Moreover, when the neurospheres that have been produced are divided into individual cells and cultured in serum-free culture medium as described earlier, new neurospheres are produced in the same way. Removal of the mitogens EGF and FGF-2 from the above serum-free liquid culture medium induces them to differentiate and three types of cells are generated: neurons, astrocytes, and oligodendrocytes. Thus, the neurosphere method makes it possible to amplify NSCs that possess multipotency and are capable of self-renewal (Fig. 1) (12).

The advantages of the neurosphere method are: (*i*) enabling NSCs to be grown in an undifferentiated state [which is still difficult to achieve with hematopoietic stem cells (HSCs)] and (*ii*) enabling the multipotency and self-renewal capacity of NSCs to be evaluated numerically, and at the same time it makes it possible to define NSCs empirically as "cells that have the ability to form neurospheres when cultured in vitro." The "cells that initiate neurosphere formation when cultured in vitro" are called "neurosphere-initiating cells (NS-ICs)." Their neurosphere-forming efficiency is referred to as NS-IC activity and the numerical data make it possible to compare cell populations whose relative NSC content differs. In the fetal CNS, the NS-ICs are highly enriched in proliferating cells located in the ventricular zone, where NSCs are likely to be present in vivo (13).

As previously described, NSCs or NSC-like cells have been shown to be involved in the production of new neuronal cells under physiological and pathological conditions in the adult mammalian CNS in vivo. However, the identity of NS-ICs and NSCs in the adult mammalian forebrain has been debated (6,14). In mammalian forebrain, adult NSCs are likely to be relatively quiescent (15), GFAP expressing subpopulation of cells (type B cells) located in the SVZ of the lateral ventricle (14) that can give rise to new neurons migrating into the olfactory bulb. Type B cells (NSCs) give rise to Dlx-2 positive, transit-amplifying cells (type C cells) that in turn give rise to migrating neuroblasts (type A cells) (Fig. 3). Recent report by Doetsch et al. (16) showed that, in addition to NSCs (type B cells), Dlx-2-positive transit amplifying cells (type C cells), which do not have strong self-renewal capacity but have high mitotic activity, have the ability to form neurospheres (16).

Identification and Isolation of NSCs

In the hematopoietic system, the prospective identification of stem cells (HSCs) has been achieved either by using a method to combine antibodies that recognize cell surface

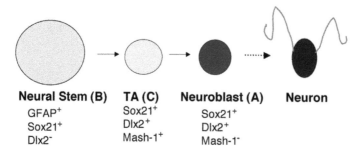

Neural Stem (B)   **TA (C)**   **Neuroblast (A)**   **Neuron**

GFAP$^+$ Sox21$^+$ Sox21$^+$
Sox21$^+$ Dlx2$^+$ Dlx2$^+$
Dlx2$^-$ Mash-1$^+$ Mash-1$^-$

Figure 3 NSCs in the adult SVZ. Slowly dividing GFAP-positive cells (type B cells) in the sub-ventricular zone SVZ are said to be NSCs in the adult. Since under conditions that would selectively kill fast-dividing cells (type C cells) by exposing animals to the anti-cancer drug AraC, the slowly dividing type B cells are AraC-resistant, and the type C cells subsequently reappear. Neurogenesis is thought to proceed by the cell lineage as illustrated in the figure. Representative markers expressed in type B, C, and A cells are shown in the figure. *Abbreviations*: NSCs, neural stem cells; SVZ, sub-ventricular zone. (*See color insert.*)

antigens (17,18) or by using that method in combination with other methods (19), which greatly contributed to our understanding of HSCs. On the other hand, our understanding of the biology of NSCs has lagged far behind that of HSCs. This has been due in part to the lack of available methodologies for the prospective identification or purification of NSCs, as well as to the lack of in vivo repopulation assays, capabilities that have proven of seminal value to studies of the HSC (20). In recent years, however, methods for the prospective identification of stem cells have been developed for the CNS as well. First, there is a method in which a gene for a fluorescent protein, such as enhanced green fluorescent protein (EGFP) or enhanced yellow fluorescent protein (EYFP), is introduced and expressed downstream of the gene expression control region of a marker gene that is selectively expressed in NSCs (20). Genes that are known to be selectively expressed in NSCs are the genes encoding the intermediate protein Nestin (21,22), the RNA-binding protein Musashi-1 (23–25), and the transcription factor Sox family (Fig. 4) (26,27). We have produced transgenic mice that express the fluorescent protein EGFP under the control of the NSC-selective second intronic enhancer element of the nestin gene and succeeded in concentrating NSCs without using the neurosphere method (12,28). Many similar techniques have been reported, but actually there is the problem of having to introduce genes from outside. Second, there are techniques in which NSCs are collected by using antibodies to surface antigens, as in the HSCs and neural crest stem cells (NCSCs), prospectively identified as p75$^+$P0$^-$ (29). It was recently reported that it is possible to isolate cells at different stages of differentiation in the nervous system, such as NSCs, neuronal progenitor cells, glia progenitor cells, etc., by using combinations of antigens, such as the cholera toxin B subunit (ChTx), tetanus toxin fragment C (TnTx), and A2B5 (30). A recent study indicated that the cells collected from the telencephalon of E13 rats in the ChTx$^-$TnTx$^-$A2B5$^-$ Jones fraction were NSCs, that those in the ChTx$^+$TnTx$^+$ fraction were neurons or neuronal progenitors, and that those in the A2B5$^+$Jones$^+$ fraction were either neurons or glia. The group of cells in the adult mouse brain that has a diameter of 12 μm or more weakly express CD24 and bind weakly with peanut agglutinin has been identified as a cell group that possesses high neurosphere-forming ability in vitro (31). Third, there are techniques (32,33) that do not use exogenously introduced genes or surface antigens (34,35). These techniques involve the use of a cell sorter and identify stem cells based on parameters that indicate

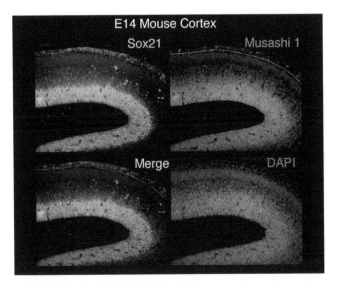

Figure 4 Expression of Musashi-1 and Sox21, which are strongly expressed in NSCs/progenitor cells, in the telencephalon of fetal mice. Immunohistochemistry of E14 mouse cerebral cortex with antibodies to Sox21 and Musashi-1. *Source*: From Refs. 24, 27. (*See color insert.*)

the size and internal structural complexity of the cell, the degree of cell uptake of Hoechst 33342, etc. We used a cell sorter to fractionate mouse corpus-striatum-derived cells according to two parameters: forward scattering (FSC), which indicates cell size; and side scattering (SSC), which indicates the complexity of the internal structure of cells. We evaluated them by means of the neurosphere method (33). The results showed that this technique makes it possible to concentrate NSCs by fractionating extremely large cells, 20 μm or more in diameter (GATE8), at any developmental stage. Although this technique separates stem cells according to degree of uptake of Hoechst 33342, as that may be a property common to all stem cells, this cell population is referred to as "side population (SP)" cells. A cell population in the hematopoietic system that has strong Hoechst 33342 efflux capacity due to strong expression of ABC transporters, functions as stem cells (36) and they were given that name because of their characteristic fluorescence-activated cell sorting (FACS) pattern. A great deal of significance was attached to this because of the possibility that it might be a common property of stem cells and allow stem cells from other organs to be concentrated as well. By combining the characteristics of possessing strong ability to efflux Hoechst 33342 and having characteristic cell surface antigens, we succeeded in almost completely purifying hematopoietic stem cells in terms of "Tip"-SP $CD34^-$ $c\text{-}Kit^+$ $Sca\text{-}1^+$ Lin^- cells (19). We then demonstrated the presence of SP cells in the CNS as well and conducted an analysis to define the cell population by comparing it with the *nestin*-EGFP transgenic mice, which had already been analyzed (33). Although a Notch1-positive undifferentiated cell population was very frequently enriched in the SP cell group during the embryonic period, it was found to contain many other cells besides NSCs. No SP cells were detected in the top 10% of the cell population in terms of EGFP fluorescence intensity, where NSCs are thought to be enriched based on the analysis of *nestin*-EGFP transgenic mice. On the other hand, the high frequency of matching the NSC fraction in adults (>12 μm, CD24 low/PNA low) suggested that it might correspond to the stem-cell population in adults, as was reported previously (31).

In Vivo Localization of NSCs

In addition to isolation of NSCs with a cell sorter, vigorous research is being conducted on the localization and dynamics of NSCs in vivo. Cells called "radial glia" have been reported to function as NSCs at least in the fetal period (37–39). That discovery is said to have been made by conducting serial observations of the neurosphere-forming ability of cells isolated with GFP under the control of the *gfap* gene promoter as the marker (37) and of radial glia labeled with retrovirus vector (38) and Dil (39), with the results showing that the radial glia, which had been thought to be the support cells of neurons that migrate vertically through the cerebral wall, give rise to astrocytes and neurons as well.

Studies to identify NSCs in the adult have been attracting a great deal of attention in recent years. As already described, in the adult mammalian brains, NSCs or NSC-like cells are thought to be present in the SGZ of the hippocampal dentate gyrus (40) and the SVZ facing the lateral ventricle in the forebrain, and neurogenesis by GFAP-positive astrocytes is said to occur at these sites (14,41). On the other hand, there is also a report of the presence of NSCs in the ependymal cell layer by Frisen's group (42). According to that report, Notch1-positive cells are present in the ependymal layer, and when they were collected by using the Notch1 antibody, predominantly neurospheres were formed. When Dil, a pigment coexisting in the lipid bilayer of membranes, was injected into a lateral ventricle, ependymal cells were labeled with Dil. Some time later neurons labeled with Dil were detected in the pathway to the olfactory bulb; thus, investigators claimed that the ependymal cells were NSCs. However, questions about the experiment remain, including whether labeling with Dil really occurs only in the ependymal cells. Actually, it has been reported that when ependymal cells and SVZ cells were directly harvested from mouse brain and their NS-IC ability and differentiating ability were compared, while the ependymal cells divided, they only formed small neurospheres and could not be subcultured, and the ependymal cell-derived neurospheres were incapable of producing neurons (41). There has also been a report of neurosphere-initiating ability in a cell population that was Nestin-positive and glia-marker- and ependymal-cell-marker-negative (31). The notion that NSCs are present in the ependymal cell layer has not received much support from the above findings. The finding that Lex/SSEA-1, which is expressed by embryonic stem (ES) cells, etc., is also expressed in the adult SVZ and that neurosphere formation occurs predominantly in the Lex/SSEA-1-positive cell group has been reported as the basis for the claim that NSCs are present in the SVZ in the adult (34). That report stated that as Lex was not simultaneously expressed in the ependymal cell group and that no neurospheres were formed by it, there is little possibility of ependymal cells being NSCs.

CONTROL OF THE SELF-RENEWAL AND DIFFERENTIATION OF NSCs

Differentiation into Neurons and Differentiation into Astrocytes

It is known that mainly neurons are produced in the CNS in the early embryonic period and that glia, including astrocytes and oligodendrocytes, are produced later. When NSCs obtained from E10 to E17 brain tissue were actually compared in a monolayer culture system, mainly neurons were produced in the E10 system, but the ability to produce neurons had declined in the E17 system and gliogenesis occurred (43). All through the development, the NSCs are present for a long time, from the early embryonic period until adulthood, and they appear to play a role in tissue formation by exquisitely controlling growth and differentiation according to the time and the environment. How are the maintenance and differentiation of NSCs controlled?

Differentiation toward neurons is positively controlled by basic helix–loop–helix (bHLH)-type transcription factors, such as Mash1 and Neurogenin, and they are referred to as proneural bHLH factors (44). They form heterodimers with E proteins, such as E47/E12, which are expressed ubiquitously in a wide variety of tissues, and activate transcription by binding to a sequence called the E box. They promote the process of differentiation from NSCs toward neurons and are thought to be involved in neuron production. The bHLH-type transcription factors are also involved in the production of specific neuron types as they mutually repress each other. With respect to astrocytic differentiation, on the other hand, progress is being made in analyzing the control of the *gfap* gene transcription by diffusible factors, including cytokines. Astrocytic differentiation is induced synergistically by leukemic inhibitory factor (LIF), a member of the IL-6 superfamily, and by BMP2, a member of the TGF-β superfamily. It is now known that during the process, their respective downstream factors, STAT3 and Smad, form complexes with the co-activator protein, p300/CBP and activate transcription of the *gfap* gene (45). On the other hand, differentiation into astrocytes is known to be inhibited by repression of their signals by the proneural bHLH factor Neurogenin1 (Ngn1) (46). Astrocytic differentiation is prematurely induced earlier during the development of Ngn/Mash1 double knock-out mice, which suggests that the expression level of proneural bHLH factors not only promotes differentiation toward neurons in the early embryonic period, but also controls the timing of the differentiation toward astrocytes that occurs in the late embryonic period (47).

Another reason why differentiation toward astrocytes does not occur in the early embryonic period is that the CpG sequence of the STAT3-binding region on the *gfap* gene promoter is highly methylated during that period, and it has been shown to be demethylated on mouse E14 (48). This methylation interferes with the binding of phosphorylated STAT3 to the *gfap* gene promoter, suggesting that expression of GFAP no longer occurs in the presence of LIF or the gp130-STAT3 signaling in neurons in which division has ended as a result of inactivation of *gfap* gene transcription. This type of epigenetic modification of chromatin is also suspected of having become an important control mechanism of cell differentiation switching.

Role of Notch Signaling in Deciding the Fate of NSCs

Notch signaling in the control of NSC differentiation has also been attracting interest. Notch is known to be a single-transmembrane-domain-type protein containing 36 EGF repeats and to be activated by stimulation by ligands in adjacent cells. *Notch* gene was originally discovered in *Drosophila*, but its structure is also conserved in the mouse, and it is thought to have an important function in relation to the development of various organs including CNS (49). The Notch signal transmission mechanism is well conserved from invertebrates to mammals. It is activated by binding to the same membrane protein ligands, Delta and Serrate (Jagged), and transmits the signal into the nucleus. Thus, the signal is activated by ligands expressed in neighboring cells. The activation essentially occurs after ligand binding because the intracellular ligand is cleaved by the protease presenilin, which possesses γ-secretase activity (50), and the signal is transmitted by direct translocation of the intracellular domain into the nucleus. There it forms a complex with the transcription control factor RBP-J/CSL and activates expression of its target gene *E(spl)*, in *Drosophila*, or *Hes1* or *Hes5*, in mammals, which encodes a bHLH-type transcription control factor. This signal is known to constantly act in an inhibitory manner on neurons being produced in the process of development of the CNS and

PNS in *Drosophila*. Notch receptors are strongly expressed in the periventricular area of the mouse CNS from the embryonic period until adulthood and an analysis in relation to Hes1 and Hes5 has suggested a role of Notch signaling in maintaining the undifferentiated state of NSCs and inhibiting their differentiation into neurons (51,52). Hes1 is the mammalian homolog of the *Drosophila* Notch-signaling effector molecule E(spl) and is a bHLH-type transcriptional repressor. As stated earlier, Hes1 has a functionally redundant relationship with the Hes family member Hes5 and has been shown to function as a downstream effector molecule of Notch1 (52). We analyzed the role of Notch/Hes1 signaling in maintaining NSCs and deciding their fate by analyzing the NSCs of *Hes1* knock-out mice (51). The results of analyses by the neurosphere method, low-density monolayer culture method, etc., showed that the self-renewal ability of the NSCs was reduced in the *Hes1* knock-out mice and that at the same time the fate decision of NSCs toward the neuron cell lineage was promoted and differentiation by NSCs into neurons had increased. When these results are considered together with the fact that the *Hes1* gene product is the downstream target of Notch signaling, the Notch signal would appear to inhibit the fate decision of NSCs toward the neuronal lineage as well as to positively inhibit the self-renewal ability of NSCs.

Recently, however, there have been reports that Notch signaling does not simply inhibit neuron differentiation and maintain undifferentiation, but may vigorously promote differentiation into glia cells as well (53–56). The retina is composed of six types of neurons and of glia cells called, Müller glia, and expression of Müller glia markers was confirmed in rat retina cells after active-type Notch was introduced into them (54). Moreover, as they differentiated into Müller glia even when Hes1 was introduced and they did not express glia markers when the Hes1 dominant negative type was introduced, Notch signaling has been reported to play an important role in the production of Müller glia. Although the mechanism of Notch signaling in glia cell production is unclear, based on these reports, a model in which Hes1 and Hes5 activate genes that promote glia differentiation is possible. On the other hand, the RBP-J/CSL-binding sequence, which is present in the promoter region of the *Hes1* and *Hes5* genes, is also present in the promoter region of the *gfap* gene and there is also a report that Notch signaling decides differentiation (or activation of transcription of the *gfap* gene) toward astrocytes directly, without any mediation by Hes1, Hes5, etc. (57). It has been shown that the transcriptional co-repressor NcoR inhibits astrocytic differentiation by NSCs and neural progenitor cells and represses transcription of the *gfap* gene by physically interacting with RBP-J/CSL, which binds directly to the repressor region of the *gfap* gene promoter (57,58). Upon Notch1 activation, N1-ICD translocates into the nucleus to form a complex with RBP-J/CSL (59), thereby converting RBP-J/CSL from a transcriptional repressor into an activator and stimulating transcription of its target genes, including the *gfap* gene. Thus, it is tempting to hypothesize that transient activation of Notch1 de-represses *gfap* gene by recruiting N1-ICD into the CSL complex through exclusion of NcoR from the CSL complex bound to the *gfap* gene promoter. Consistent with this hypothesis, removal of NCoR is indeed sufficient to induce GFAP expression both in vivo and in vitro (58). However, detailed molecular studies are needed to determine whether the transcriptional activator complex containing N1-ICD and CSL is maintained on the *gfap* gene promoter during astrocytic maturation.

On the other hand, there is also a report that although Notch signaling is essential to maintain NSCs in an undifferentiated state, it has no effect on astrocytogenesis based on the loss-of-function studies of presenilin, which possesses γ-secretase activity (60). NSCs induced from ES cells in which the RBP-J, related to Notch signaling, had been knocked out; NSCs derived from mice lacking presenilin1 were used in that study and the use of the

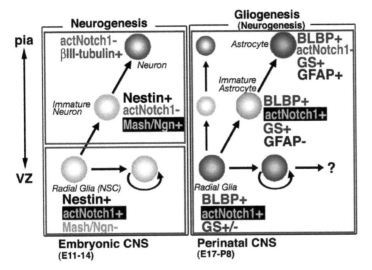

Figure 5 Activation of Notch1 signal during CNS development. The Notch1-activation pattern, determined by anti-activated Notch1, is likely to be associated with a self-renewal of NSCs, the inhibition of neurogenesis, and astrocytic differentiation. Note that Notch1 is transiently activated in the astrocytic differentiation of radial glia, not in the fully matured astrocytes. *Abbreviations*: CNS, central nervous system; NSCs, neural stem cells. *Source*: From Ref. 61.

neurosphere method showed reduced neurosphere formation by NSCs derived from these mutants. Both RBP-J-deficient NSCs and presenilin-deficient NSCs showed a greater tendency to differentiate, notch type compared to the wild-type. When active-type Notch was introduced into wild-type NSCs, they were maintained in the undifferentiated state. The investigators also reported that they had not obtained any results that would indicate a role of Notch signaling in promoting their differentiation into astrocytes. Thus, though some of the reports on the relation between Notch signaling and gliogenesis are contradictory, the discrepancies are thought to be attributable to differences in the timing of the introduction of active-type Notch and the role of Notch signaling in the developmental stage.

More recently, we immunohistochemically investigated the state of activation of the Notch1 signal in the development process in the mouse forebrain region in situ by using an antibody that specifically recognizes active-type Notch protein (Notch1 proteolytic fragment cleaved by γ-secretase). We found that the Notch1 signal was activated in the growth process of NSCs/precursor cells in the embryonic period and in the early stage of the process of astrocyte differentiation from radial glia and that it was repressed in the neuronal cell lineage (61) (Fig. 5). Interestingly, the active form of Notch1 was below the threshold of immunohistochemical detection in the astrocytes in the SVZ, where NSCs are thought to be maintained in the adult brain (61). This indicates that some mechanism other than Notch1 signaling may be involved in the maintenance of adult NSCs; this point is discussed in the following section.

Self-Renewal and Long-Term Maintenance Mechanism of NSCs

How is the self-renewal and maintenance of the undifferentiated state of NSCs regulated by signaling? The most likely possibility seems to be that they are regulated by extrinsic factors such as the microenvironment, including cell adhesion and cell interactions, at the site where the cells are located. The first extrinsic factor candidates that can be cited are the

NSC mitogens EGF and FGF-2. They promote the self-renewal of NSCs and make it possible to subculture them long-term in vitro (62,63). Moreover, it has recently been shown in vitro that activation of the IGF-I receptor by IGF-I is essential for promotion of stem-cell division by EGF and FGF-2. However, as the cycling time of adult NSCs is very slow (average in the corpus striatum: 15 days) (64) or mitosis has almost stopped, it cannot be completely explained by these mitogens alone and thus other factors appear to be necessary to maintain adult NSCs in an undifferentiated state. Shimazaki et al. (65) recently discovered that one of them is a signal mediated by gp130. gp130 is a receptor subunit common to the members of the Class I cytokine family (CLC/CLF, CNTF, CT-1, IL-6, IL-11, LIF, and Oncostatin M) and it is known to activate transcription factor STAT1/3 via JAK kinase and repress expression of the target gene, as well as to activate the signal transmission pathway of the RAS-MAPKinase system and P13Kinase system via the docking protein Gab1. The signal mediated by gp130 has been shown to have a variety of biological actions in addition to maintaining mouse ES cells undifferentiated; and it is known to be required in the CNS development for the survival of specific neurons, such as motor neurons, the survival of oligodendrocytes, and differentiation into astrocytes. As stated above, the signal mediated by gp130 has attracted particular attention as possibly causing NSCs to differentiate into astrocytes by coupling with the BMP signal (45). Nevertheless, in an analysis of knock-out mice for the LIF receptor (LIFR), which is a receptor subunit required for activation of gp130 signal transmission by CNTF, LIF, etc. Shimazaki et al. (65) demonstrated that the gp130 signal promotes maintenance of NSCs in the undifferentiated state, the same as in ES cells. When NSCs are suspension cultured in the presence of EGF or FGF-2, they form single-cell-derived aggregates called neurospheres and can be made to grow for long periods, but they never grow in response to CNTF or LIF. In contrast, human NSCs are more difficult to subculture for long periods than rodent NSCs, but subculture has been reported to be easier when LIF is added to culture medium containing EGF and FGF-2 (66). Shimazaki et al. (65), therefore, first investigated the dynamics of NSCs in mice lacking LIFR, but they did not detect any difference from the wild-type in number of NSCs that they were able to confirm by neurosphere formation in the corpus striatum of E14 mice lacking LIFR ($LIFR^{-/-}$). However, when the $LIFR^{-/-}$ NSCs were subcultured seven times or more in vitro in the presence of EGF, cell growth became impossible in low-density cultures. Subsequent subculture by high-density culture was possible, but neurosphere-forming ability had been lost, and the cells assumed a fibroblast-like form. Even when these fibroblast-like cells were adhesion cultured in the absence of EGF, which are ordinary differentiation conditions, they failed to differentiate and died. These findings indicate that LIFR is essential for long-term maintenance of NSCs in vitro. Well then, is LIFR actually also involved in the long-term maintenance of NSCs in vivo? This issue was addressed by analyzing the heterozygotes of LIFR knock-out mice (67). Shimazaki et al. (65) determined the number of NSCs in the adult striatum of heterozygotes ($LIFR^{+/-}$) by the neurosphere method and the number of progenitor cells they generated by labeling by BrdU uptake, and they compared them with the wild-type. The results suggested that LIFR is necessary in NSCs to maintain them in vivo at least from birth until adulthood. However, it was unclear from these results alone whether LIFR promotes the self-renewal of NSCs or is necessary for survival. Then, Shimazaki et al. (65) intraventricularly injected adult mice with CNTF or EGF or both for six days and monitored changes in the numbers of NSCs in the corpus striatum and concluded that CNTF promotes self-renewal of NSCs rather than their survival. In other words, the LIFR/gp130 signal appears to promote self-renewal during NSC division. However, it is still unclear at which stage in brain development it is prominent, especially after birth, that is, whether it is only in the early postnatal stage when large numbers of glia are produced or whether it is similar in the adult as well. With regard to the molecular mechanisms, whether STAT3 is the chief control factor for self-renewal, as in ES cells, and how the RAS-MAPKinase pathway and the PI3Kinase pathways are involved also remain to be elucidated. Interestingly, the activation of gp130 in NSCs was shown to rapidly increase Notch1 expression, indicating a link between gp130 signaling and Notch1 in regulating NSC self-renewal (68).

Needless to say, in addition to extrinsic factors, cell autonomous intrinsic factors are involved in the self-renewal of NSCs. What are the candidates for such intrinsic factors involved in the self-renewal of NSCs? Stem cells derived from various tissues (e.g., HSCs and NSCs) are thought to share several parts of their self-renewal mechanism (69) and the fact that NSCs (23,24,70), intestinal epithelial stem cells (71,72) and other epithelial stem cells (or stem-like cells) (73,74) all share expression of the RNA-binding protein Musashi1, may be one basis for this. Musashi1 binds to the 3'UTR of *m-Numb* mRNA and activates the Notch1 signal by repressing translation of *m-Numb* mRNA (25,75), and it has been postulated to increase the self-renewal of the stem cells among these cells and to be responsible for maintaining them in the undifferentiated state (25,76) (Fig. 6). By analyzing the upstream signals involved in the regulation of Musashi1 expression in the future, we hope to elucidate the entire signal mechanism that leads to the self-renewal of these stem cells.

Recently, the polycomb group transcriptional repressor Bmi-1 was shown to be required for the postnatal maintenance of HSCs as well as for the self-renewal of NSCs and NCSCs (77,78), suggesting that a common mechanism regulates the self-renewal and postnatal persistence of diverse types of stem cells. Furthermore, the detailed clonal analyses showed that Bmi-1 is required for the self-renewal of NSCs and NCSCs, but not for proliferation of restricted neural progenitors from the gut and forebrain, suggesting that Bmi-1 dependence distinguishes stem-cell self-renewal from restricted progenitor proliferation, at least in the CNS and PNS. Determining the integrative interactions of

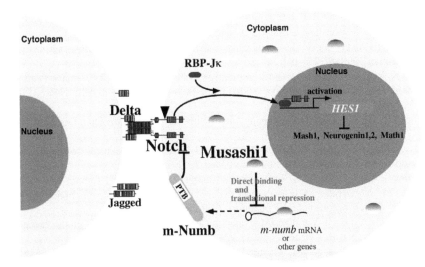

Figure 6 Function of Neural RNA-binding protein Musashi 1 in the Notch1-activation and self-renewal of NSCs. An RNA-binding protein, Musashi1 inhibits the translation of m-Numb, a Notch1 antagonist, thereby inducing the Notch1 signaling and the self-renewal of NSCs. *Abbreviation:* NSCs, neural stem cells. *Source:* From Ref. 25. (*See color insert.*)

extrinsic and intrinsic factors involved in the self-renewal of NSCs and other stem cells will be of considerable interest.

CONCLUSION AND PERSPECTIVES

The term "regenerative medicine" has come into general use. The word "regeneration" seems to mean returning a lost function or part to its original state after it has been lost, and it is often used coupled with "stem-cell system." Treatment of one degenerative disease of the CNS, Parkinson's disease, by transplantation of brain cells derived from fetuses has already been tried, and obvious efficacy has been confirmed (79,80). However it should be noted that neural transplantation is still at an experimental stage in the treatment of Parkinson's disease due to the use of fetal tissue, including lack of sufficient amounts of tissue for transplantation in a large number of patients, variation of functional outcome due to poor standardization, and ethical issues (80). Against this background, research on stem-cell technology has flourished in recent years and naturally their biological aspects are often taken up by the mass media precisely because of the expectation of clinical applications. Of course, if stem-cell technology can be used to treat degenerative diseases of the CNS, with degenerative efficacy and safety there could be no better news for patients. However, there is no doubt about the need for much greater basic medical analysis in the future (11,12).

REFERENCES

1. Loeffler M, Potten CS. Stem Cells and Cellular Pedigrees—a Conceptual Introduction. London: Academic Press, 1997:1–27.
2. Davis AA, Temple S. A self-renewing multipotential stem cell in embryonic rat cerebral cortex. Nature 1994; 372:263–266.
3. Reynolds BA, Tetzlaff W, Weiss S. A multipotent EGF-responsive striatal embryonic progenitor cell produces neurons and astrocytes. J Neurosci 1992; 12:4565–4574.
4. Alvarez-Buylla A, Temple S. Stem cells in the adult mammalian central nervous system. Curr Opin Neurobiol 1999; 9:135–141.
5. Doetsch F, Petreanu L, Caille I, Garcia-Verdugo JM, Alvarez-Buylla A. EGF converts transit-amplifying neurogenic precursors in the adult brain into multipotent stem cells. Neuron 2002; 36:1021–1034.
6. Morshead C, van der Kooy D. Disguising adult neural stem cells. Curr Opin Gen Dev 2004; 14:1–7.
7. Arvidsson A, Collin T, Kirik D, Kokaia Z, Lindvall O. Neuronal replacement from endogenous precursors in the adult brain after stroke. Nat Med 2002; 8:963–970.
8. Nakatomi H, Kuriu T, Okabe S, Yamamoto S, Hatano O, Kawahara N, Tamura A, Kirino T, Nakafuku M. Regeneration of hippocampal pyramidal neurons after ischemic brain injury by recruitment of endogenous neural progenitors. Cell 2002; 110:429–441.
9. Alvarez-Buylla A, Garcia-Verdugo JM, Tramontin AD. A unified hypothesis on the lineage of neural stem cells. Nat Rev Neurosci 2001; 2:287–293.
10. Doetsch F, Garcia-Verdugo JM, Alvarez-Buylla A. Regeneration of a germinal layer in the adult mammalian brain. Proc Natl Acad Sci USA 1999; 96:11619–11624.
11. Okano H. The stem cell biology of the central nervous system. J Neurosci Res 2002; 69:698–707.
12. Okano H. Neural stem cells: progression of basic research and perspective for clinical application. Keio J Med 2002; 51:115–128.

13. Kawaguchi A, Miyata T, Sawamoto K, Takashita N, Murayama A, Akamatsu W, Ogawa M, Okabe M, Tano Y, Goldman SA, Okano H. Nestin-EGFP transgenic mice: visualization of the self-renewal and multipotency of CNS stem cells. Mol Cell Neurosci 2001; 17:259–273.

14. Doetsch F, Caille I, Lim DA, Garcia-Verdugo JM. Alvarez-Buylla A. Subventricular zone astrocytes are neural stem cells in the adult mammalian brain. Cell 1999; 97:703–716.

15. Morshead CM, Reynolds BA, Craig CG, McBurney MW, Staines WA, Morassutti D, Weiss S, van der Kooy D. Neural stem cells in the adult mammalian forebrain: a relatively quiescent subpopulation of subependymal cells. Neuron 1994; 13:1071–1082.

16. Doetsch F, Petreanu L, Caille I, Garcia-Verdugo JM, Alvarez-Buylla A. EGF converts transit-amplifying neurogenic precursors in the adult brain into multipotent stem cells. Neuron 2002; 36:1021–1034.

17. Uchida N, Jerabek L, Weissman IL. Searching for hematopoietic stem cells. II. The heterogeneity of Thy-1.1(lo) Lin(-/lo)Sca-1+ mouse hematopoietic stem cells separated by counterflow centrifugal elutriation. Exp Hematol 1996; 24:649–659.

18. Osawa M, Hanada K, Hamada H, Nakauchi H. Long-term lymphohematopoietic reconstitution by a single CD34-low/negative hematopoietic stem cell. Science 1996; 273:242–245.

19. Matsuzaki Y, Kinjo K, Mulligan RC, Okano H. Unexpectedly efficient homing capacity of purified murine hematopoietic stem cell. Immunity 2004; 20:87–93.

20. Okano H, Goldman SA. Identification and selection of neural progenitor cells. NeuroScience News 2000; 3:27–31.

21. Hockfield S, McKay RD. Identification of major cell classes in the developing mammalian nervous system. J Neurosci 1985; 5:3310–3328.

22. Lendahl U, Zimmerman LB, McKay RD. CNS stem cells express a new class of intermediate filament protein. Cell 1990; 23(60):585–595.

23. Sakakibara S, Imai T, Aruga J, Nakajima K, Yasutomi D, Nagata T, Kurihara Y, Uesugi S, Miyata T, Ogawa M, Mikoshiba K, Okano H. Mouse-Musashi-1, a neural RNA-binding protein highly enriched in the mammalian CNS stem cell. Dev Biol 1996; 176:230–242.

24. Kaneko Y, Sakakibara S, Imai T, Suzuki A, Nakamura Y, Sawamoto K, Ogawa Y, Toyama Y, Miyata T, Okano H. Musashi-1: an evolutionarily conserved marker for CNS progenitor cells including neural stem cells. Dev Neurosci 2000; 22:139–153.

25. Okano H, Imai T, Okabe M. Musashi: a translational regulator of cell fates. J Cell Sci 2002; 115:1355–1359.

26. Pevny LH, Sockanathan S, Placzek M, Lovell-Badge R. A role for SOX1 in neural determination. Development 1998; 125:1967–1978.

27. Ohba H, Chiyoda T, Endo E, Yano M, Hayakawa Y, Sakaguchi M, Darnell RB, Okano HJ, Okano H. Sox21 is a repressor of neuronal differentiation and is antagonized by YB-1. Neurosci Lett 2004; 358:157–160.

28. Sawamoto K, Nakao N, Kakishita K, Ogawa Y, Toyama Y, Yamamoto A, Yamaguchi M, Mori K, Goldman SA, Itakura T, Okano H. Generation of dopaminergic neurons in the adult brain from mesencephalic precursor cells labeled with a nestin-GFP transgene. J Neurosci 2001; 21:3895–3903.

29. Morrison SJ, White PM, Zock C, Anderson DJ. Prospective identification, isolation by flow cytometry, and in vivo self-renewal of multipotent mammalian neural crest stem cells. Cell 1999; 96:737–749.

30. Maric D, Maric I, Chang YH, Barker JL. Prospective cell sorting of embryonic rat neural stem cells and neuronal and glial progenitors reveal selective effects of basic fibroblast growth factor and epidermal growth factor on self-renewal and differentiation. J Neurosci 2003; 23:240–251.

31. Rietze RL, Valcanis H, Brooker GF, Thomas T, Voss AK, Bartlett PF. Purification of a pluripotent neural stem cell from the adult mouse brain. Nature 2001; 412:736–739.

32. Cai J, Wu Y, Mirua T, Pierce JL, Lucero MT, Albertine KH, Spangrude GJ, Rao MS. Properties of a fetal multipotent neural stem cell (NEP cell). Dev Biol 2002; 251:221–240.

33. Murayama A, Matsuzaki Y, Kawaguchi A, Shimazaki T, Okano H. Flow cytometric analysis of neural stem cells in the developing and adult mouse brain. J Neurosci Res 2002; 69:837–847.

34. Capela A, Temple S. LeX/ssea-1 is expressed by adult mouse CNS stem cells, identifying them as nonependymal. Neuron 2002; 35:865–875.

35. Uchida N, Buck DW, He D, Reitsma MJ, Masek M, Phan TV, Tsukamoto AS, Gage FH, Weissman IL. Direct isolation of human central nervous system stem cells. Proc Natl Acad Sci USA 2000; 97:14720–14725.

36. Goodell MA, Brose K, Paradis G, Conner AS, Mulligan RC. Isolation and functional properties of murine hematopoietic stem cells that are replicating in vivo. J Exp Med 1996; 183:1797–1806.

37. Malatesta P, Hartfuss E, Gotz M. Isolation of radial glial cells by fluorescent-activated cell sorting reveals a neuronal lineage. Development 2000; 127:5253–5263.

38. Noctor SC, Flint AC, Weissman TA, Dammerman RS, Kriegstein AR. Neurons derived from radial glial cells establish radial units in neocortex. Nature 2001; 409:714–720.

39. Miyata T, Kawaguchi A, Okano H, Ogawa M. Asymmetric inheritance of radial glial fibers by cortical neurons. Neuron 2001; 31:727–741.

40. Seri B, Garcia-Verdugo JM, McEwen BS, Alvarez-Buylla A. Astrocytes give rise to new neurons in the adult mammalian hippocampus. J Neurosci 2001; 21:7153–7160.

41. Chiasson BJ, Tropepe V, Morshead CM, Van der Kooy D. Adult mammalian forebrain ependymal and subependymal cells demonstrate proliferative potential, but only subependymal cells have neural stem cell characteristics. J Neurosci 1999; 19:4462–4471.

42. Johansson CB, Momma S, Clarke DL, Risling M, Lendahl U, Frisen J. Identification of a neural stem cell in the adult mammalian central nervous system. Cell 1999; 96:25–34.

43. Qian X, Shen Q, Goderie SK, He W, Capela A, Davis AA, Temple S. Timing of CNS cell generation: a programmed sequence of neuron and glial cell production from isolated murine cortical stem cells. Neuron 2000; 28:69–80.

44. Ross SE, Greenberg ME, Stiles CD. Basic helix–loop–helix factors in cortical development. Neuron 2003; 39:13–25.

45. Nakashima K, Yanagisawa M, Arakawa H, Kimura N, Hisatsune T, Kawabata M, Miyazono K, Taga T. Synergistic signaling in fetal brain by STAT3–Smad1 complex bridged by p300. Science 1999; 284:479–482.

46. Sun Y, Nadal-Vicens M, Misono S, Lin MZ, Zubiaga A, Hua X, Fan G, Greenberg ME. Neurogenin promotes neurogenesis and inhibits glial differentiation by independent mechanisms. Cell 2001; 104:365–376.

47. Nieto M, Schuurmans C, Britz O, Guillemot F. Neural bHLH genes control the neuronal versus glial fate decision in cortical progenitors. Neuron 2001; 29:401–413.

48. Takizawa T, Nakashima K, Namihira M, Ochiai W, Uemura A, Yanagisawa M, Fujita N, Nakao M, Taga T. DNA methylation is a critical cell-intrinsic determinant of astrocyte differentiation in the fetal brain. Dev Cell 2001; 1:749–758.

49. Artavanis-Tsakonas SR, MD Lake RJ. Notch signaling: cell fate control and signal integration in development. Science 1999; 284:770–776.

50. Selkoe D, Kopan R. Notch and presenilin: regulated intramembrane proteolysis links development and degeneration. Annu Rev Neurosci 2003; 26:565–597.

51. Nakamura Y, Sakakibara S, Miyata T, Ogawa M, Shimazaki T, Weiss S, Kageyama R, Okano H. The bHLH gene hes1 as a repressor of the neuronal commitment of CNS stem cells. J Neurosci 2000; 20:283–293.

52. Ohtsuka T, Ishibashi M, Gradwohl G, Nakanishi S, Guillemot F, Kageyama R. EMBO J 1999; 18:2196–2207.

53. Gaiano N, Fishell G. The role of notch in promoting glial and neural stem cell fates. Annu Rev Neurosci 2002; 25:471–490.

54. Furukawa T, Mukherjee S, Bao ZZ, Morrow EM, Cepko CL. rax, Hes1, and notch1 promote the formation of Muller glia by postnatal retinal progenitor cells. Neuron 2000; 26:383–394.

55. Tanigaki K, Nogaki F, Takahashi J, Tashiro K, Kurooka H, Honjo T. Notch1 and Notch3 instructively restrict bFGF-responsive multipotent neural progenitor cells to an astroglial fate. Neuron 2001; 29:45–55.

56. Wang S, Barres BA. Up a notch: instructing gliogenesis. Neuron 2000; 27:197–200.

57. Ge W, Martinowich K, Wu X, He F, Miyamoto A, Fan G, Weinmaster G, Sun YE. Notch signaling promotes astrogliogenesis via direct CSL-mediated glial gene activation. J Neurosci Res 2002; 69:848–860.

58. Hermanson O, Jepsen K, Rosenfeld MG. N-CoR controls differentiation of neural stem cells into astrocytes. Nature 2002; 419:934–939.

59. Kato H, Taniguchi Y, Kurooka H, Minoguchi S, Sakai T, Nomura-Okazaki S, Tamura K, Honjo T. Involvement of RBP-J in biological functions of mouse Notch1 and its derivatives. Development 1997; 124:4133–4141.

60. Hitoshi S, Alexson T, Tropepe V, Donoviel D, Elia AJ, Nye JS, Conlon RA, Mak TW, Bernstein A, van der Kooy D. Notch pathway molecules are essential for the maintenance, but not the generation, of mammalian neural stem cells. Genes Dev 2002; 16:846–858.

61. Tokunaga A, Kohyama J, Yoshida T, Nakao K, Sawamoto K, Okano H. Mapping spatiotemporal activation of Notch signaling during neurogenesis and gliogenesis in the developing mouse brain. J Neurochem 2004; 90:142–154.

62. Palmer TD, Takahashi J, Gage FH. The adult rat hippocampus contains primordial neural stem cells. Mol Cell Neurosci 1997; 8:389–404.

63. Reynolds BA, Weiss S. Clonal and population analyses demonstrate that an EGF-responsive mammalian embryonic CNS precursor is a stem cell. Dev Biol 1996; 175:1–13.

64. Tropepe V, Craig CG, Morshead CM, Van der Kooy D. Transforming growth factor-alpha null and senescent mice show decreased neural progenitor cell proliferation in the forebrain subependyma. J Neurosci 1997; 17:7850–7859.

65. Shimazaki T, Shingo T, Weiss S. The ciliary neurotrophic factor/leukemia inhibitory factor/gp130 receptor complex operates in the maintenance of mammalian forebrain neural stem cells. J Neurosci 2001; 21:7642–7653.

66. Carpenter MK, Cui X, Hu ZY, Jackson J, Sherman S, Seiger A, Wahlberg LU. In vitro expansion of a multipotent population of human neural progenitor cells. Exp Neurol 1999; 158:265–278.

67. Ware CB, Horowitz MC, Renshaw BR, Hunt JS, Liggit D, Koblar SA, Gliniak BC, McKenna HJ, Papayannopoulou T, Thoma B, et al. Targeted disruption of the low-affinity leukemia inhibitory factor receptor gene causes placental, skeletal, neural and metabolic defects and results in perinatal death. Development 1995; 121:1283–1299.

68. Chojnacki A, Shimazaki T, Gregg C, Weinmaster G, Weiss S. Glycoprotein 130 signaling regulates Notch1 expression and activation in the self-renewal of mammalian forebrain neural stem cells. J Neurosci 2003; 23:1730–1741.

69. Reya T, Morrison SJ, Clarke MF, Weissman IL. Stem cells, cancer, and cancer stem cells. Nature 2001; 414:105–111.

70. Sakakibara S, Nakamura Y, Yoshida T, Shibata S, Koike M, Takano H, Ueda S, Uchiyama Y, Noda T, Okano H. RNA-binding protein Mushashi family: roles for CNS stem cells and a subpopulation of ependymal cells revealed by targeted disruption and antisense ablation. Proc Natl Acad Sci USA 2002; 99:15194–15199.

71. Potten CS, Booth C, Tudor GL, Booth D, Brady G, Hurley P, Ashton G, Clarke R, Sakakibara S, Okano H. Identification of a putative intestinal stem cell and early lineage marker; Musashi-1. Differentiation 2003; 71:28–41.

72. Sakatani T, Kaneda A, Iacobuzio-donahue CA, Carter MG, Witzel SB, Okano H, Ko MSH, Ohlsson R, Longo DL, Feinberg AP. Loss of imprinting of Igfll alter intestinal maturation and tumorigenesis in mice. Science 2005; 307:1976–1978.

73. Clarke RB, Spence K, Anderson E, Howell A, Okano H, Potten CS. A putative human breast stem cell population is enriched for steroid receptor-positive cells. Dev Biol 2005; 277:443–456.

74. Akasaka Y, Saikawa Y, Fujita K, Kubota T, Ishii T, Okano H, Kitajima M. Expression of a candidate marker for progenitor cells, Musashi-1, in the proliferative regions of human antrum and its decreased expression in intestinal metaplasia. Histopathology 2005; 47:348–356.

75. Imai T, Tokunaga A, Yoshida T, Hashimoto M, Mikoshiba K, Weinmaster G, Nakafuku M, Okano H. The neural RNA-binding protein Musashi1 translationally regulates mammalian numb gene expression by interacting with its mRNA. Mol Cell Biol 2001; 21:3888–3900.

76. Okano H, Kawahara H, Toriya M, Nakao K, Shibata S, Takao I. Function of RNA binding protein Musashi-1 in stem cells. Exp Cell Res 2005; 306:349–356.

77. Molofsky AV, Pardal R, Iwashita T, Park IK, Clarke MF, Morrison SJ. Bmi-1 dependence distinguishes neural stem cell self-renewal from progenitor proliferatin. Nature 2003; 425:962–967.

78. Park IK, Qian D, Kiel M, Becker MW, Pihalja M, Weissman IL, Morrison SJ, Clarke MF. Bmi-1 is required for maintenance of adult self-renewing haematopoietic stem cells. Nature 2003; 423:302–305.

79. Winkler C, Kirik D, Björklund A. Cell transplantation in Parkinsion's disease: how can we make it work? Trends Neurosci 2005; 28:86–92.

80. Lindvall O, Björklund A. Cell Therapy in Parkinson's disease. Neurorx 2004; 1:382–393.

5
Stem Cells in Mammary Epithelium

Gilbert H. Smith
Mammary Biology and Tumorigenesis Laboratory, Center for Cancer Research,
National Cancer Institute, National Institutes of Health, Bethesda, Maryland, U.S.A.

Robert B. Clarke
Division of Cancer Studies, University of Manchester, Paterson Institute,
Manchester, U.K.

INTRODUCTION

A long history of scientific interest is associated with the mammary gland because of its seminal role in infant nutrition and well being, and because it is often afflicted by cancer development. In fact, before the beginning of the twentieth century, there were already more than 10,000 scientific references to published articles relating to mammary biology (1). It was an interest in cancer and cancer development in the breast that brought about the first series of experiments that led to our current concept of tissue-specific mammary epithelial stem cells. The occurrence of what appeared to be premalignant lesions of the glandular epithelium led DeOme et al. (2) to develop a biologic system to recognize, characterize, and study hyperplastic nodules in the mammary glands of mouse mammary tumor virus (MMTV)-infected mice. These investigators developed a surgical method for removing the endogenous mammary epithelium from the fourth mammary fat pad. Subsequently, the "cleared" pad was used as a site of implantation where suspected premalignant lesions could be placed and their subsequent growth and development could be observed. Using this approach, they were able to show that both premalignant and normal mammary implants could grow and fill the empty fat pad within several weeks. During this growth period, the premalignant implants recapitulated their hyperplastic phenotype, whereas normal implants produced normal branching mammary ducts. Serial transplantation of normal and premalignant outgrowths demonstrated that while normal implants invariably showed growth senescence after several generations, hyperplastic outgrowths did not. It soon became apparent that any portion of the normal mammary parenchyma could regenerate a complete mammary tree over several transplant generations, suggesting the existence of cells capable of reproducing new mammary epithelium through several rounds of self-renewal. However, it was some time later before this property was recognized as representative of the presence of mammary epithelial stem cells (3).

AGING AND REPRODUCTIVE SENESCENCE

The discovery that all portions of the mouse mammary gland appeared competent to regenerate an entire new gland upon transplantation triggered a series of papers relating to the reproductive lifetime of mammary cells (4–7). It was determined that no difference existed in the regenerative ability of mammary tissue taken from very old mice versus that taken from very young mice during serial transplantation. In addition, neither reproductive history nor developmental state had a significant impact on the reproductive longevity of mammary tissue implants. The ability of grafts from old donors to proliferate equivalently to those from young in young hosts suggested to these authors that the lifespan of mammary cells is primarily affected by the number of mitotic divisions rather than by the passage of chronological or metabolic time. The authors in a series of experiments tested this where mammary implants were serially transplanted. In one series, fragments were taken from the periphery of the outgrowth for subsequent transplantation. In the other, the fragments for transplant were removed from the center. The supposition was that the cells at the periphery had undergone more mitotic events than those in the center and therefore peripheral tissue would show growth senescence more quickly than tissue near the center. Outgrowths from fragments taken from the periphery repeatedly showed senescent growth in earlier passages when compared to those generated from implants from the centers of outgrowths (5). The authors concluded that the growth senescence in transplanted mammary epithelium was related primarily to the number of cell divisions. In contrast, mouse mammary epithelial cells could be transformed to unlimited division potential either spontaneously, by MMTV infection, or by treatment with carcinogens (4,8). At the time this observation was taken to signify that "immortalization," that is, attainment of unlimited division potential, was an important early step in malignant transformation. More recently, Medina et al. (9) have shown that mammary epithelium from p53−/− mice also exhibits an "immortal" phenotype upon serial transplantation. This is a striking discovery because of the essential role that p53 signaling plays in the maintenance and genomic stability of the stem cells within the crypts of the small intestine. For example, radiation sensitivity is absent in the intestinal crypt stem cells in p53 null mice (10).

With respect to transplantation of mammary fragments to epithelium-free fat, extensive studies indicate that rat mammary epithelium shows similar clonogenic activity to that of the mouse. In fact, rat mammary implants grow extensively to complete glandular structures within "cleared" mouse mammary fat pads (11). In addition, there is a similar indication that all parts of the rat gland have regenerative capacities. Little is known regarding the regenerative ability of human breast upon transplantation. Human mammary fragments were maintained and could be stimulated to functional differentiation in mouse mammary fat pads, but did not grow extensively (12). Xenografts of human breast in immuno-compromised Nu/Nu mice have been shown to exhibit a mitogenic response upon exposure to increased levels of estrogen and progesterone (13). Because of the lack of a functional transplantation assay for human breast epithelium, virtually nothing is known about its growth, longevity, or capacity to self-renew, although emerging techniques for transplantation of human breast cells into the sub-renal capsules or cleared and humanized mammary fat pads of recipient mice should enable this to be tested in the near future (14,15).

IN VITRO STUDIES

Dispersed mouse mammary epithelial cells have been shown to be able to recombine and grow to form a new gland within the epithelium-free mammary fat pad (16–19). In these experiments, both normal and transformed mammary outgrowths were developed, indicating that both normal and abnormal mammary cells could exist within any given apparently normal glandular population. More recently, irradiated feeder cells have been employed to propagate primary cultures of mouse mammary epithelium. Under these conditions, the cells were maintained for nine passages and produced normal mammary outgrowths upon introduction into cleared mammary fat pads (20). The number of dispersed mammary cells required to produce a positive take, that is, form a glandular structure within the fat pad increased with increasing passage number. This observation applies to all mouse mammary epithelial cell lines that have been developed in vitro and maintained through serial passages. Eventually, with passage, as with the fragment implants, either no growth is attained or neoplastic development is achieved when the cells are placed into cleared fat pads (21). Some mouse mammary cell lines that were grown for various periods in culture demonstrated an extended reproductive lifespan when reintroduced into cleared mammary fat pads and transplanted serially. The resulting outgrowths appeared in every way to be normal and did not exhibit hyperplastic or tumorigenic growth (22). The authors concluded that the immortalization phenotype could be dissociated from the preneoplastic phenotype and suggested that these mammary cell lines may represent an early stage, perhaps the earliest, in progression to mammary tumorigenesis. Human breast epithelium in culture endures at least two growth senescent periods before progressing to an immortalized population. The molecular events accompanying these conversions have been studied very extensively (23). Nothing from these in vitro studies has shed any light on either the biology or characterization of human mammary epithelial stem cells.

During the last decade, a number of authors have investigated the endpoint of the clonogenic capacity of dispersed rodent mammary epithelial cells in limiting dilution transplantation experiments (24–27). Both in the mouse and in the rat, 1000–2000 mammary epithelial cells represent the smallest number required for the establishment of an epithelial growth in a fat pad. Earlier, it was shown that genes could be introduced into primary mammary epithelial cell cultures with retroviral vectors. Subsequently, the genetically modified epithelial cells were reintroduced into cleared mammary fat pads for evaluation in vivo (28). Although stable transduction of gene expression could be achieved in a high percentage of mammary cells in culture, recovery of these retroviral-marked cells in regenerated glandular structures was only possible when virtually 100% of the implanted cells were stably modified. It was determined that this resulted from the fact that only a very small proportion of the primary epithelial cells inoculated were capable of contributing to tissue renewal in vivo. This was the first indication that only a subset of the mammary epithelial population possessed the capacity to regenerate mammary tissue upon transplantation. From this followed the possibility that this cellular subset represented the mammary epithelial stem-cell compartment.

For two entirely different purposes, dispersed rat and mouse mammary cells were tested for their ability to form epithelial structures in empty fat pads at limiting dilution. The possibility that lobule and ductal lineage-limited cells existed among the mouse mammary epithelial population was investigated based upon the common observation that lobular development could be suppressed in transgenic mouse models when ductal branching morphogenesis was unaffected. The results of this study provided evidence for distinct lobue-limited and ductal-limited progenitors in the mouse mammary gland (24). Figure 1 depicts a growing implant in an impregnated host, and both ductal branching morphogenesis and lobulogenesis occur simultaneously under these circumstances. In Figure 1A, an arrowhead indicates the growing terminal end bud of a duct, while the

(A)

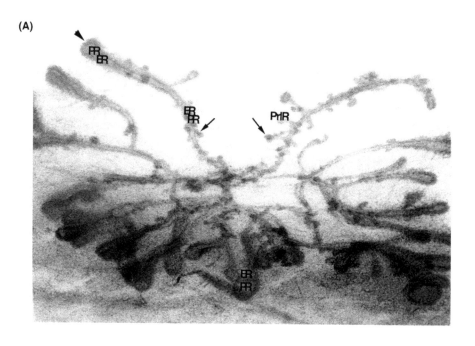

(B)   **Mammary Stem Cells and Progeny**

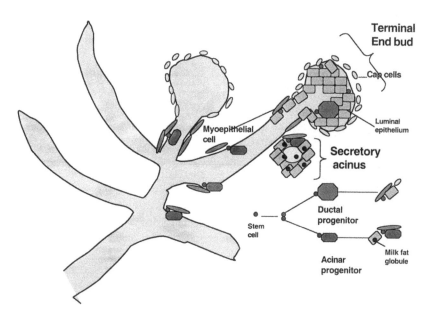

Figure 1   A growing mammary implant is shown in top panel (**A**). The outgrowth is 11 days old and is in the cleared mammary fat pad of a four-day pregnant host. Active ductal growth and elongation is present with the extension of ducts occurring radially from the implant. Terminal end buds (*arrowhead*) are enlarged and actively growing, and along the subtending ducts, small secretory acini (*arrows*) are developing. Estrogen, progesterone, and prolactin signaling through their cognate receptors (ER, PR, PrlR) are essential in this activity. An illustration in bottom panel (**B**) indicates the type and location of pluripotent mammary epithelial cells and their respective progeny.

small arrows point out developing secretory acini on the subtending duct. Figure 1B represents our current understanding of the location and type of mammary epithelial progenitors. In an effort to establish the total number of clonogenic cells in the rat mammary gland as a measure of radiogenic susceptibility to cancer induction, Kamiya et al. (25,27) conducted similar experiments. These authors found that like the mouse, rat mammary glands possessed distinct lobule-committed and duct-committed progenitors. In the mouse, it was shown in clonal-dominant mammary populations that both of these progenitors arose from a common antecedent, that is, a primary mammary epithelial stem cell (26).

As described earlier, efforts to propagate mammary epithelial cells in continuous culture and subsequently demonstrate their ability to reconstitute the mammary gland in vivo have met with limited success. A different approach to understanding mammary epithelial cell lineage was applied by using cell surface markers to distinguish basal (myoepithelial) from luminal (secretory) epithelial cells. With fluorescence-activated cell sorting (FACS), human mammary epithelial cells were separated into myoepithelial (CALLA-positive) and luminal (MUC-positive) populations and evaluated for their respective capacity to produce mixed colonies in cloning assays (29,30). These authors reported that individual epithelial cells bearing luminal markers alone or both luminal and myoepithelial surface markers could give rise to colonies with a mixed lineage phenotype. Cells bearing only the CALLA marker (basal/myoepithelial) were only able to produce like epithelial progeny. Using a similar approach, another group (31) demonstrated that CALLA-positive (myoepithelial) and MUC-1 positive (luminal) mammary epithelial cells could be purified to essentially homogeneous populations and maintained as such under certain specific culture conditions in vitro. Expression of distinctive keratin gene patterns and other genetic markers also characterized these disparate cellular populations. It was further demonstrated that only the luminal epithelial cell population was able to produce both luminal and myoepithelial cell progeny in vitro, providing further evidence that the multipotent cellular subset in mammary epithelial tissue resided among the luminal rather than the myoepithelial lineage. More recently, this same group has shown that unlike myoepithelial cells from normal glands, tumor-derived myoepithelial cells were unable to support three-dimensional growth when combined with normal luminal cells in vitro (32). This deficiency was shown to be due to the inability of the tumor myoepithelial cells to express a specific laminin gene (LAM1) product.

Mouse mammary epithelial cells have been FACs separated according to their luminal or myoepithelial surface markers. Subsequent study of these different populations in vitro gave results that agree with those reported for human cells. The cells capable of giving rise to mixed colonies in cloning studies were only found among the cells bearing luminal epithelial cell markers (33).

Another approach using FACS-purified cells was reported by Clayton et al. (34) who predicted that stem/progenitor cells in the human breast may be either double positive (DP) for CALLA and MUC-1 or possibly double negative (DN). When they analyzed colony formation on either mouse embryonic or human mammary fibroblast feeder layers, they found that DN cells gave mostly luminal only, some myo-epithelial only, and a few mixed colonies. In contrast, DP cells gave rise to approximately equal numbers of either luminal or myoepithelial only colonies, some mixed lineage colonies, but also some DP and DN cell colonies. This suggests that DP cells may be capable of self-renewal, an important defining characteristic of a stem cell.

A tissue culture approach previously applied to brain stem cells has been neurosphere suspension cultures in which the capacity of a stem cell for self-renewal can be measured. This culture method prevents adherence of cells to the tissue culture plastic and induces cell death by anoikis due to lack of attachment. Since differentiated mammary epithelial cells require attachment for survival (35), this method has been applied to human mammary epithelial cells (36). Mammospheres (MS) were demonstrated to be clonal and produced by approximately 1/250 cells. MS contain both luminal and myoepithelial cell markers and can form mixed colonies when dispersed and plated on feeder layers or into three-dimensional culture in matrigel. On passage of dispersed MS cells, a similar number form MS (1/250), which suggests that symmetric self-renewal divisions occur under these conditions. However, the addition of Notch receptor agonists increases the number of MS by 10× suggesting the Notch signaling pathway can stimulate mammary stem-cell self-renewal (37).

MAMMARY STEM-CELL MARKERS

Several recent studies have demonstrated that the multipotent cells in mammary epithelium reside within the luminal cell population in humans and mice (31,33). However, no specific molecular signature for mammary epithelial stem cells was revealed. Smith and Medina (3) presented an earlier marker that held promise for identifying mammary stem cells in the ultrastructural description of mitotic cells in mammary epithelial explants. These investigators noticed that mouse mammary explants, such as mammary epithelium in situ, contained pale or light-staining cells and that it was only these cells that entered mitosis when mammary explants were cultured.

Chepko and Smith (38) analyzed light cells in the electron microscope utilizing their ultrastructural features to distinguish them from other mammary epithelial cells. The following basic features expected of stem cells were applied in the ultrastructural evaluation: division-competence (presence of mitotic chromosomes) and an undifferentiated cytology (Fig. 2). Figure 2 shows the side-by-side appearance of an undifferentiated large light cell (ULLC) and a small, undifferentiated light cell (SLC) in a secretory acinus of a lactating rat mammary gland. The pale-staining (stem) cells are of distinctive morphology; therefore, their appearance in side-by-side pairs or in one-above-the-other pairs (relative to the basement membrane) was interpreted as the result of a recent symmetric mitosis. In addition to pairs, other informative images would be of juxtaposed cells that were morphologically intermediate between a primitive and differentiated morphology based on the number, type, and development of cytoplasmic organelles. Cells were evaluated for cytological differentiation with respect to their organelle content and distribution, that is, cells differentiated toward a secretory function might contain specific secretory products, such as milk protein granules or micelles, which have been ultrastructurally and immunologically defined (39). In addition, the presence and number of intracellular lipid droplets, the extent and distribution of Golgi vesicles, and rough endoplasmic reticulum (RER) attest to the degree of functional secretory differentiation of a mammary epithelial cell. These features are characteristically well developed in the luminal cells of active lactating mammary gland. Myoepithelial cells are flattened, elongated cells located at the basal surface of the epithelium, and their prominent cytoplasmic feature is the presence of many myofibrils and the absence of RER or lipid droplets.

In a retrospective analysis of light and electron micrographs, a careful and detailed scrutiny of mammary tissue was performed to determine the range of morphological features among the cell types that had previously been reported. The samples evaluated included mouse mammary explants, pregnant and lactating mouse mammary glands, and rat mammary glands from 17 stages of development, beginning with nulliparous through pregnancy, lactation, and involution (38,40–42). From this analysis, we were

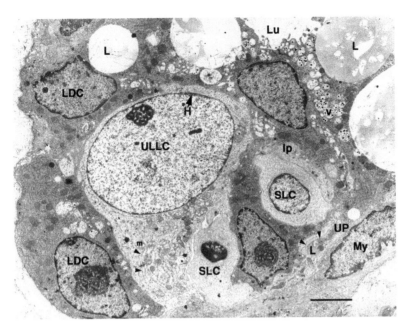

Figure 2 In a secretory acinus from a lactating rat mammary gland, SLC and ULLC appear juxtaposed, suggesting they result from a single mitotic event. To the right, a second pair is present where only the SLC is completely within the plane of section. Portions of its undifferentiated neighbor (Ip) and (UP) are seen beside it. Differentiated secretory mammary epithelial cells (LDC) lie on either side in an adjacent acinus. Milk fat globules (L) and casein micelles in secretory vesicles (v) are present within the DSC and in the lumen (Lu). A portion of a myoepithelial cell cytoplasm (My) also appears near the SLC. The bar equals 4.0 micrometer. *Abbreviations*: ULLC, undifferentiated large light cell; SLC, small light cell; DSC, dark secretory cell.

able to expand the number of cell types in the epithelium from two secretory (or luminal) and myoepithetlal cells to five distinguishable structural phenotypes or morphotypes. Our observations strengthened the conclusion that the undifferentiated (light) cells are the only cell type to enter mitosis. The undifferentiated cells were found in two easily recognized forms: small (~8 microns) and large (15–20 microns). Mitotic chromosomes were never found within the differentiated cells, namely, secretory and myoepithelial cells, suggesting that they were terminally differentiated and out of the cell cycle. Using all of the above features, we were able to develop a more detailed description of the epithelial subtypes that comprise the mammary epithelium.

The characteristics used to develop a standardized description of five mammary epithelial cellular morphotypes were: staining of nuclear and cytoplasmic matrix, cell size, cell shape, nuclear morphology, amount and size of cytoplasmic organelles, location within the epithelium, cell number, and grouping relative to each other and to other morphotypes. These characteristics were used to perform differential cell counts and morphometric analysis of the cell populations in rat mammary epithelium (38). Figure 3 presents an illustration of each mammary cell type and can be used on both the light and electron levels to help form a search image for recognizing them in situ. The five morphotypes we recognize in rodent mammary epithelium are a primitive small light cell (SLC), a ULLC, a very differentiated large light cell (DLLC), the classic cytologically differentiated luminal cell (LDC), and the myoepithelial cell. We described three sets of division-competent cells in rodent mammary epithelium and demonstrated that mammary epithelial stem cells and

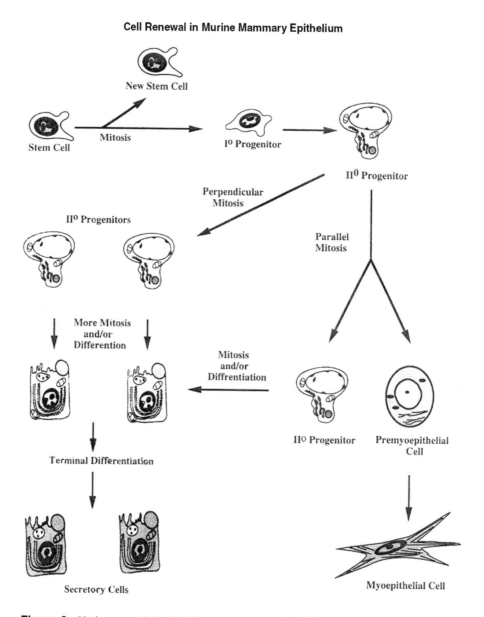

Cell Renewal in Murine Mammary Epithelium

Figure 3   Various morphological forms are portrayed, which make up the fully differentiated murine mammary epithelium. It is not drawn to scale; rather, it indicates our interpretation of the lineal relationships between the mammary stem cell at the top left, the lineage-limited I^0 and II^0 progenitors, and the fully differentiated secretory and myoepithelial cells. *Source*: Adapted from Ref. 38.

their downstream progenitors are morphologically much less differentiated than either the secretory or the myoepithelial cells. We counted a total of 3552 cells through 17 stages of rat mammary gland development and calculated the percent of each morphotype. This analysis showed that the population density (number of cells/mm^2) of SLC among mammary epithelium did not change from puberty through post-lactation involution.

The proportion of SLC remained at 3%. This means that although the number of mammary epithelial cells increased by 27-fold during pregnancy in the mouse (26,43), the percent of SLC in the population does not change. Therefore, SLC increases and decreases in absolute number at the same relative rate as the differentiating epithelial cells.

If these undifferentiated epithelial cells represent structures essential for self-renewal and stem-cell function, they would be rare or absent in growth and regeneration senescent populations. In support of this conclusion, neither small light cells (SLC) nor ULLCs was observed in an extensive study of growth senescent mouse mammary transplants. Examination of growth-competent implants in the same host reveals easily detectable SLC and ULLC (44). Further, "immortal" premalignant mammary outgrowths, which never show a growth senescent phenotype upon serial transplantation persistently contain both SLCs and ULLCs. These observations lend additional support to the conclusion that both SLC and ULLC represent important components in the mechanism for mammary epithelial stem-cell maintenance and self-renewal in situ.

Gudjonsson et al. (45) predicted that if human mammary epithelium contained cells similar to the SLC and ULLC described in rodents, then these cells would be low or negative for the luminal surface marker, sialomucin (MUC-1), as they do not commonly contact the luminal surface. Coincidentally, such cells would be positive for epithelial specific antigen (ESA) but negative for the basal myoepithelial cell marker, smooth muscle actin (SMA). Using this approach, they isolated two luminal epithelial cell populations. One, the major population, co-expressed MUC-1 and ESA. The other, a minor population, was found in a suprabasal location in vivo and expressed ESA but not MUC-1 or SMA. These latter cells were multipotent and formed elaborate branching structures composed of both luminal and myoepithelial lineages, both in vitro and in vivo. The outgrowths produced and resembled terminal duct lobular units both by morphology and marker expression. These data provide strong evidence for the presence of mammary epithelial stem cells in the human breast with characteristics similar, if not identical, to those described earlier for rodent mammary gland.

Several specific cellular markers that identify the "stemness" of any particular mammary epithelial cell have been reported but none of them infallibly identifies the actual stem-cell pool. In many cases, the markers identify a sub-set of cells enriched for stem-cell-like behavior or only identify a proportion of the stem cells. Several features known to define stem cells in other organs have been applied to the mammary gland, for example, the property of retaining DNA synthesis incorporated label over a long chase following labeling, that is, long label retaining cells (LRCs) in ^{3}H-thymidine or 5-bromodeoxyuridine (5BrdU) pulsed mammary tissue. Mammary cells in the mouse with this property have been identified and were found scattered along the mammary ducts. Estrogen receptor immuno-staining suggests that these cells are often estrogen receptor (ER) positive (46). These authors made no further characterization of these cells.

In a recent attempt to further characterize label-retaining cells in the mouse mammary gland, specific cellular markers were applied to mammary cells pulsed for 14 days in vivo with BrdU and chased for nine weeks (47). In situ, LRCs represent 3% to 5% of the population after nine weeks in good agreement with the number of SLCs (38). Some LRCs were found to be negative for both luminal and myoepithelial markers (CK14 and 18), suggesting that they were undifferentiated. In contrast to the above study, they were also mainly steroid receptor-negative. Two characteristics, efficient efflux of Hoechst dye side population (SP) and the presence of stem-cell antigen-1 (Sca-1) known to associate with stem cells in other organ systems, were used to enrich for putative mammary epithelial stem cells by FACS. LRCs were enriched for Sca-1 expression and SP dye-effluxing properties. In addition, the SP mammary cells possessed a frequency and size distribution that was very similar to SLCs and were highly enriched for Sca-1 expression. The Sca-1 positive mammary cells showed a greater regenerative potential in cleared fat pads than similar numbers of Sca-1 negative cells. This study represents a first important step toward prospectively isolating mammary epithelial progenitors and may permit the identification of additional markers useful in determining the biological potential of mammary stem cells.

The presence of an SP in mouse mammary epithelial preparations was also reported by Alvi et al. (48). In this study, the SP formed $0.45 \pm 11\%$ ($n = 17$) of mouse mammary epithelial cells. Mouse mammary SP cells had high levels of Bcrp1 (Hoechst-effluxing protein) expression but were depleted for cells that expressed cytokeratins and a mouse luminal epithelial cells surface marker, suggesting that the SP cells were undifferentiated. Interestingly, the SP was enriched for cells that expressed the telomerase catalytic subunit and α_6-integrin, both potential stem-cell markers, but no difference was found in expression of the oestrogen receptor between SP and nonSP cells (48). SP percentage has been used as a surrogate marker for stem-cell numbers in the mouse mammary epithelium in studies of Wnt signaling (49). Hyperplastic glands of MMTV-Wnt-1 and MMTV-ΔN-catenin mice had SP percentages increased by three- and ninefold respectively, compared to wild type mice. When MMTV-Wnt-1 or MMTV-ΔN-catenin mice were crossed with syndecan-1 null mice, the hyperplastic response was reduced, the glands were hypomorphic, and, furthermore, the SP fraction was reduced by at least 50%. Addition of soluble Wnt-3a or epidermal growth factor (EGF) to primary mammary epithelial cell cultures increased SP percentages, indicating that growth factors can indeed have a direct effect on the percentage of cells in the SP (49).

A similar SP to that observed in the mouse mammary gland has also been identified by several groups in normal human breast tissue obtained from reduction mammoplasty and other noncancer breast surgery (34,36,48,50,51). In the three groups who have performed human breast tissue SP analyses, the proportion of breast SP cells varied from $\sim$0.2% (34,48) to $\sim$1% (36) to $\sim$5% (50,51). Age, parity, day of menstrual cycle, and contraceptive use were not correlated with %SP in an analysis of the nine women from whom breast tissue was obtained (48). Although different proportions of isolated human breast SP cells were reported in the above studies, their stem-cell nature has been analyzed and compared to the nonSP cells using various in vitro cell culture methods. The growth of SP and nonSP at clonal densities in monolayer culture in vitro either on feeder layers or on collagen produced three types of colonies: those consisting of myoepithelial or luminal epithelial cells alone and mixed colonies of both cell types. However, depending on the substratum, SP cells produced two to seven times more colonies than nonSP cells. In support of their putative stem-cell nature, only the SP cells possessed the ability to produce colonies with both myoepithelial and luminal epithelial cell types (34,36). Another published method for the culture of undifferentiated tissue-specific stem cells is the growth of colonies from single cells in nonadherent suspension culture such as neurospheres from brain tissue, which are enriched in neural stem cells (52). Where this has been applied to human breast cells grown as "mammospheres," SP cells made up 27% of the total population of sphere cells. Conversely, only SP, and not nonSP cells, from fresh breast cell digests were capable of forming mammospheres in nonadherent suspension culture (36). Finally, in three-dimensional (3D) cultures in basement membrane preparations such as matrigel, breast cells can differentiate to form acini (small hollowed out or solid colonies) or large branching structures reminiscent of lobular structures in vivo. Human breast epithelial (BER$-$EP4$+$) SP cells were demonstrated to produce branching type structures, while nonSP cells produced only acinus-like structures. Only the SP cells structures contained differentiated cells expressing cytokeratins (CK) of both myoepithelial (CK14) and luminal epithelial (CK18) type (50).

Putative stem-cell markers and differentiation markers have been analyzed in the human SP and compared to the nonSP cells by two research groups (34,50,51). Using antibodies to the cell surface markers of differentiated myoepithelial and luminal epithelial cells, CALLA and MUC-1 respectively, it was demonstrated by both groups that ~70% of epithelial SP cells expressed neither protein, whereas most nonSP cells expressed one or the other of these differentiated cell markers (34,50,51), strongly suggesting that SP cells include an undifferentiated population of cells. Since the breast is a steroid hormone-responsive tissue, both groups analyzed ER expression. Clarke et al. (50) found a sixfold increased ER-α mRNA and protein expression in SP compared to nonSP cells, whereas Clayton et al. (34) found no SP cells expressing either the ER-α or ER-β mRNA. In agreement with Clarke et al. (50), Alvi et al. (48) had reported that up to half of mouse mammary SP cells expressed ER-α protein. The expression of other putative stem-cell markers such as p21$^{\text{CIP1/WAF1}}$ (twofold) and Musashi-1 (sixfold) mRNA was demonstrated to be increased in SP compared to nonSP cells (50). Interestingly, these proteins were co-expressed with ER-α in breast epithelial cells examined by dual label immuno-fluorescence, suggesting that SP cells may express all three proteins (50). The proliferation marker Ki67 was absent in SP cells by QRT-PCR (34), which would fit with the established fact that cells expressing ER-α do not proliferate in breast epithelium in vivo (38) and the long-recognized quiescence of tissue-specific stem cells.

In the small intestine, the interfollicular integument and in hair follicles, evidence has accumulated that strongly supports the existence of an "immortal strand" in somatic stem cells, that is, during asymmetric division, the stem cell retains its template DNA in a semi-conservative manner. This feature protects the stem cell from genetic errors arising from DNA replication. Direct evidence demonstrates that this phenomenon occurs in the ultimate stem cells of the crypts in the small intestine (53) and in tissue culture lines modified to undergo asymmetric divisions under specified culture conditions (54). The stability of the pattern of proviral insertions in serial transplants of retroviral-infected, clonal dominant mammary epithelial outgrowths argues that this may be the case. New proviral insertions occur during DNA synthesis in the cell cycle of chronically infected cells. Therefore, in cells replicating exponentially as opposed to asymmetrically, new proviral insertions should be common in the renewing population but they are not (55). Evidence for active MMTV replication in these mammary populations is provided by the demonstration of easily detectable unintegrated proviral DNA by Southern analysis (26). In a very recent study, self-renewing mammary epithelial stem cells that were originated during allometric growth of the mammary ducts in puberfal females were labeled using [^{3}H]-thymidine (^{3}H-TdR). After a prolonged chase, during which much of the branching duct morphogenesis was completed, ^{3}H-TdR-label retaining epithelial cells (LRECs) were detected among the epithelium of the maturing glands. Labeling newly synthesized DNA in these glands with a different marker, 5BrdU, resulted in the appearance of doubly labeled nuclei in a large percentage of the LRECs (Fig. 4). In contrast, label-retaining cells within the stroma did not incorporate 5BrdU during the pulse, indicating that they were not traversing the cell cycle. Upon chase, the second label (5BrdU) was distributed from the double-labeled LREC to unlabeled mammary cells while ^{3}H-TdR was retained (56). These results demonstrate that mammary LREC selectively retain their ^{3}H-TdR-labeled template DNA strands and pass newly synthesized 5BrdU-labeled DNA to their progeny during asymmetric divisions. Similar results were obtained in mammary transplants containing self-renewing, pluripotent, LacZ-positive epithelial cells (57), suggesting that cells capable of expansive self-renewal may repopulate new mammary stem-cell niches during the allometric growth of new mammary ducts (56). These studies imply that during epithelial morphogenesis, mammary stem cells, newly formed by symmetric self-renewal, enter a stem-cell niche and undertake asymmetric

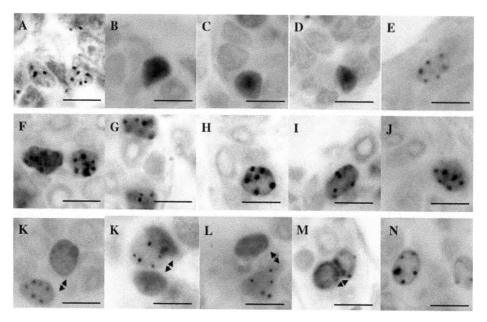

Figure 4   Singly labeled nuclei are shown in **A**–**E**, either positive for ^{3}H-TdR alone, grains (**A** and **E**), or 5BrdU alone, *dark gray color* (**B**–**D**). Doubly labeled 5BrdU/^{3}H-TdR cell nuclei are shown in **F**–**J**, following the 5BrdU incorporation. After a five-day chase of the 5BrdU label, the frequency of doubly labeled 5BrdU/^{3}H-TdR nuclei decreased, and the number of singly labeled ^{3}H-TdR-positive nuclei and 5BrdU-labeled nuclei increased. These nuclei were often juxtaposed, suggesting that they resulted from a recent mitotic event (*double arrows* in **K**–**M**). **E** and **N** are examples of singly labeled ^{3}H-TdR-labeled nuclei in 5BrdU-labeled mammary tissues for comparison with those shown in **K**–**M**. Bar = 10 microns. *Abbreviations*: 5BrdU, 5-bromodeoyuridine; H-TdR, [H]-thymidine.

cell division kinetics, traversing the cell cycle, retaining their template DNA strands, and giving rise to differentiating epithelial progeny, indefinitely maintaining tissue homeostasis.

MAMMARY STEM CELLS IN CARCINOGENESIS

Contiguous portions of the human mammary gland possess the identical pattern of X chromosome inactivation. Thus, local portions of the gland are derived from a single antecedent (58). In a further study of human mammary tissue, this same group (59) showed that mammary cancer in situ and the apparently normal tissue surrounding the lesion shared similar genetic alterations. This was interpreted to indicate that mammary lesions arise as a result of the clonal expansion of previously affected epithelium subsequent to further genetic change. The results imply that local genetically damaged mammary stem cells may give rise to premalignant lesions, which may progress to frank malignancy. Studies by several other laboratories (60–62) have confirmed and extended these observations, supporting the concept of clonal progression in the development of breast cancer in humans. Therefore, it is conceivable that mammary hyperplasia and tumors develop locally from damaged clonogenic epithelial progenitors (stem cells). Using an immunological, rather than a genetic approach, Boecker et al. (63) reported a bipotent progenitor cell in normal breast tissue capable of giving rise to glandular and myoepithelial cell lineages, characterized by its expression of cytokeratin 5/6 (CK5/6). Subsequent analysis of benign usual ductal hyperplasia, atypical hyperplasia, and ductal cell carcinoma in situ by these authors led them to speculate that there was no required biological continuum in the development of these three types of intraductal lesions of the breast. Instead, they suggested that all three could arise independently and directly from the progeny of a committed stem (progenitor) mammary cell.

Experimental evidence from MMTV-induced mouse mammary hyperplasia and tumorigenesis (64) provides strong genetic support for the concept of clonal progression from normal through premalignant to malignant epithelium in the rodent mammary gland. In an effort to provide a proof of principle, that is, mammary stem cells may contribute to mammary tumor development, mice exhibiting a mammary growth senescent phenotype in transplant experiments were challenged with the oncogenic retrovirus MMTV (65). Only one tumor was induced by MMTV in these mice. On the other hand, more than half of their MMTV-infected wild type female littermates developed mammary tumors. The result indicates that premature regenerative senescence in mammary epithelial stem cells can reduce the subsequent risk for mammary tumorigenesis in MMTV challenged mice.

Previous experimentation with retrovirus-marked (MMTV) clonal-dominant mammary populations demonstrated that an entire functional mammary glandular outgrowth might comprise the progeny of a single antecedent (26). These populations have been serially transplanted to study the properties of aging, self-renewing mammary clonogens derived from the original progenitor. Premalignant, malignant, and metastatic clones arose from these transplants during passage. All of these bore a lineal relationship with the original antecedent, because all of the original proviral insertions were represented in each of these lesions (55). While this does not prove that mammary stem cells may directly give rise to cancerous lesions within the mammary gland, it demonstrates that normal, premalignant, and malignant progeny are all within the "repertoire" of an individual mammary cell.

It has been proposed that tumors may contain a small population of cancer stem cells (CSCs). These may be either mutated stem cells or alternatively mutated differentiated or lineage-restricted progenitor cells that acquired mutations, granting them the stem-cell-like capacity for self-renewal (66,67). There is some limited but intriguing evidence for a tumorigenic human breast population that may constitute the CSCs. In the report, $CD44^+/CD24^{low}$ cells, mainly from breast cancer cells removed in patients' pleural effusions (fluid in the thoracic cavity), were shown to readily generate solid tumors in mice, whereas other tumor cells did not (68). The presence of CSCs may explain the phenotypic heterogeneity seen within solid tumors, which are composed of a mixture of differentiated tumor cell types with limited proliferative capacity and a small population of proliferative, undifferentiated stem cells. The possible existence of a CSC has important implications for cancer therapy. The current chemotherapeutic endpoint is a reduction in tumor size, using drugs which target actively proliferating cells. CSCs, however, may divide infrequently and be refractory to the chemotherapeutic hit. Additionally, if stem cells synthesize proteins such as Bcrp1, which is responsible for the SP phenomenon, then this may serve to efflux toxic drugs (69), effectively selecting for a population of cells resistant to chemotherapy. Therefore, for the successful development of new anti-cancer therapies, it will be necessary to target cancer stem cells.

PREGNANCY AND BREAST CANCER RISK

In mice, rats, and humans, a single early pregnancy provides a significant lifelong reduction in mammary cancer risk. In rats and mice, the protective effect of pregnancy can be mimicked through hormonal application in the absence of pregnancy. This refractoriness to chemical induction of mammary tumorigenesis has recently been linked to the absence of a proliferative response in the parous epithelium when confronted with the carcinogen as compared with the nulliparous gland (70,71). Concomitant with the reduction in proliferative response is the appearance of stable activation of p53 in epithelial cell nuclei. This suggests that in response to the hormonal stimulation of pregnancy that a new cellular population is created with an altered response to carcinogen exposure. A new parity-induced mammary epithelial cell population was discovered (72), employing a conditionally activated Cre/lox recombinase/LacZ system to identify mammary cells in situ, which had differentiated during pregnancy and survived post-lactation involution. Transplantation studies indicate that the surviving, LacZ-positive, parity-specific epithelial cells have the capacity for self-renewal and contribute extensively to regeneration of mammary glands in cleared fat pads. This population accumulates in parous females upon successive pregnancies. In situ, these cells are committed to secretory cell fate and proliferate extensively during the formation of secretory lobule development upon successive pregnancies. In this process, both secretory and myoepithelial cell lineages arise from the LacZ-positive survivors, as well as ER-positive and progesterone receptor (PR) positive epithelial progeny (57). Transplantation of dispersed cells indicates that this population is preferentially included in growth-competent mammary cell reassembly and has an individual capacity to undergo at least eight cell doublings. Studies are in progress to isolate and characterize these cells and to determine their contribution to the refractoriness of parous mammary tissue to cancer development.

FUTURE PROSPECTS

The existence of epithelial stem cells in the mammary glands of rodents and humans has been established. Much remains to be learned about the mechanism(s) involved in the maintenance of these cells in situ and the signals governing their behavior. A number of candidate genes, which may play a role in mammary stem-cell biology, have appeared during the study of mammary gland growth and development in transgenic and gene deletion models. However, none of these genes has been fully assessed under conditions where mammary stem-cell function is required, namely, during regeneration of the glandular epithelium. The MMTV-induced Notch4/Int3 mutation results in the unregulated constitutive signaling of the Notch intracellular domain in the affected epithelium, invariably leading to the development of mammary cancer. The presence of this mutation in mammary epithelium prevents the development of the secretory cell fate (73). Transplantation of mammary epithelium containing MMTV-Notch4/Int3 into cleared fat pads routinely fails to result in growth. Hormonal stimulation with estrogen and progesterone rescues ductal growth and development in these implants but not secretory cell fate. These results imply that Notch signaling is essential in regulating mammary stem-cell function. Expression of a Notch4/Int3 transgene lacking the CBF-1 (mammalian homolog of suppressor of hairless) binding domain and the ability to affect the cascade of genes effected by Hairy Enhancer of Split (HES) in mammary gland does not block secretory development or ductal growth in transplants (74). This result implicates Notch signaling through HES in mammary cell fate decisions.

The vast array of genetic models and manipulations developed in the mouse has yet to be fully employed in the dissection of stem-cell biology in the mammary gland, or for that matter, in a number of other organ systems. This will change with the increased awareness of multipotent cells in adult organs and mounting evidence for the importance of somatic cell signaling upon stem-cell behavior in tissue-specific stem-cell niches (75). The application of conditional gene deletion or expression in stem-cell populations in the epidermis provides an excellent example of this approach (76). Here, conditional activation of the proto-oncogene myc, even transiently, in epidermal stem cells commits them to the production of sebaceous epithelial progeny at the expense of hair follicle progeny. In the mammary gland, only indirect evidence supports the possible role of somatic cell control of stem-cell behavior for mammary tumor induction by MMTV (65). Modulation of stem-cell behavior holds exceptional promise of a new prophylactic approach for controlling mammary cancer risk. An important step toward the achievement of this control will be the characterization of the stem-cell niche in the rodent mammary gland and ultimately in human mammary glands.

REFERENCES

1. Lyons WR. Hormonal synergism in mammary growth. Proc Royal Soc London 1958; 149:303–325.
2. DeOme KB, Fauklin LJ, Bern HA, Blair PB. Development of mammary tumors from hyperplastic alveolar nodules transplanted into glan-free mammary fat pads of female C3H mice. J Natl Cancer Inst 1959; 78:751–757.
3. Smith GH, Medina D. A morphologically distinct candidate for an epithelial stem cell in mouse mammary gland. J Cell Sci 1988; 90:173–183.
4. Daniel C, DeOme K, Young L, Blair P, Faulkin L. The in vivo life span of normal and preneoplastic mouse mammary glands: a serial transplantation study. Proc Natl Acad Sci USA, 1968; 61:53–60.
5. Daniel CW, Young LJ. Influence of cell division on an aging process. Life span of mouse mammary epithelium during serial propagation in vivo. Exp Cell Res 1971; 65:27–32.
6. Daniel CW, Young LJ, Medina D, DeOme KB. The influence of mammogenic hormones on serially transplanted mouse mammary gland. Exp Gerontol 1971; 6:95–101.
7. Young LJ, Medina D, DeOme KB, Daniel CW. The influence of host and tissue age on life span and growth rate of serially transplanted mouse mammary gland. Exp Gerontol 1971; 6:49–56.
8. Daniel CW, Aidells BD, Medina D, Faulkin LJ Jr. Unlimited division potential of precancerous mouse mammary cells after spontaneous or carcinogen-induced transformation. Fed Proc 1975; 34:64–67.
9. Medina D, Kittrell FS, Shepard A, Stephens LC, Jiang C, Lu J, Allred DC, McCarthy M, Ullrich RL. Biological and genetic properties of the p53 null preneoplastic mammary epithelium. FASEB J 2002; 16:881–883.
10. Merritt AJ, Potten CS, Kemp CJ, Hickman JA, Balmain A, Lane, DP, Hall, PA. The role of p53 in spontaneous and radiation-induced apoptosis in the gastrointestinal tract of normal and p53-deficient mice. Cancer Res 1994; 54:614–617.
11. Welsch CW, O'Connor DH, Aylsworth, CF, Sheffield LG. Normal but not carcinomatous primary rat mammary epithelium: readily transplanted to and maintained in the athymic nude mouse. J Natl Cancer Inst 1987; 78:557–565.
12. Sheffield LG, Welsch, CW. Transplantation of human breast epithelia to mammary-gland-free fat-pads of athymic nude mice: influence of mammotrophic hormones on growth of breast epithelia. Int J Cancer 1988; 41:713–714.
13. Anderson E, Clarke RB, Howell A. Estrogen responsiveness and control of normal breast proliferation. J Mammary Gland Biol Neoplasia 1998; 3:23–35.

14. Kuperwasser C, Chavarria T, Wu M, Magrane G, Gray JW, Carey L, Richardson A, Weinberg RA. Reconstruction of functionally normal and malignant human breast tissues in mice. Proc Natl Acad Sci USA 2004; 101:4966–4971.

15. Parmar H, Young P, Emerman JT, Neve RM, Dairkee S, Cunha GR. A novel method for growing human breast epithelium in vivo using mouse and human mammary fibroblasts. Endocrinology 2002; 143:4886–4896.

16. Daniel CW, DeOme KB. Growth of mouse mammary gland in vivo after monolayer culture. Science 1965; 149:634–636.

17. DeOme KB, Miyamoto MJ, Osborn RC, Guzman RC, Lum K. Effect of parity on recovery of inapparent nodule-transformed mammary gland cells in vivo. Cancer Res 1978; 38:4050–4053.

18. DeOme KB, Miyamoto MJ, Osborn RC, Guzman RC, Lum K. Detection of inapparent nodule-transformed cells in the mammary gland tissues of virgin female BALB/cfC3H mice. Cancer Res 1978; 38:2103–2113.

19. Medina D, Oborn CJ, Kittrell FS, Ullrich RL. Properties of mouse mammary epithelial cell lines characterized by in vivo transplantation and in vitro immunocytochemical methods. J Natl Cancer Inst 1986; 76:1143–1156.

20. Ehmann UK, Guzman RC, Osborn RC, Young JT, Cardiff RD, Nandi S. Cultured mouse mammary epithelial cells: normal phenotype after implantation. J Natl Cancer Inst 1987; 78:751–757.

21. Kittrell FS, Oborn CJ, Medina D. Development of mammary preneoplasias in vivo from mouse mammary epithelial cell lines in vitro. Cancer Res 1992; 52:1924–1932.

22. Medina D, Kittrell FS. Immortalization phenotype dissociated from the preneoplastic phenotype in mouse mammary epithelial outgrowths in vivo. Carcinogenesis 1993; 14:25–28.

23. Tlsty TD, Romanov SR, Kozakiewicz BK, Holst CR, Haupt LM, Crawford YG. Loss of chromosomal integrity in human mammary epithelial cells subsequent to escape from senescence. J Mammary Gland Biol Neoplasia 2001; 6:235–243.

24. Smith GH. Experimental mammary epithelial morphogenesis in an in vivo model: evidence for distinct cellular progenitors of the ductal and lobular phenotype. Breast Cancer Res Treat 1996; 39:21–31.

25. Kamiya K, Gould MN, Clifton KH. Quantitative studies of ductal versus alveolar differentiation from rat mammary clonogens. Proc Soc Exp Biol Med 1998; 219:217–225.

26. Kordon EC, Smith GH. An entire functional mammary gland may comprise the progeny from a single cell. Development 1998; 125:1921–1930.

27. Kamiya K, Higgins PD, Tanner MA, Gould MN, Clifton KH. Kinetics of mammary clonogenic cells and rat mammary cancer induction by X-rays or fission neutrons. J Radial Res Tokyo 1999; 40(suppl):128–137.

28. Smith G, Gallahan D, Zweibel J, Freeman S, Bassin R, Callahan R. Long-term in vivo expression of genes introduced by retrovirus-mediated transfer into mammary epithelial cells. J Virol 1991; 65:6365–6370.

29. Stingl J, Eaves CJ, Kuusk U, Emerman JT. Phenotypic and functional characterization in vitro of a multipotent epithelial cell present in the normal adult human breast. Differentiation, 1998; 63:201–213.

30. Stingl J, Eaves CJ, Zandich I, Emerman JT. Characterization of bipotent mammary epithelial progenitor cells in normal adult human breast tissue. Breast Cancer Res Treat 2001; 67:93–109.

31. Pechoux C, Gudjonsson T, Ronnov-Jessen L, Bissell MJ, Petersen OW. Human mammary luminal epithelial cells contain progenitors to myoepithelial cells. Dev Biol 1999; 206:88–99.

32. Gudjonsson T, Ronnov-Jessen L, Villadsen R, Rank F, Bissell MJ, Petersen OW. Normal and tumor-derived myoepithelial cells differ in their ability to interact with luminal breast epithelial cells for polarity and basement membrane deposition. J Cell Sci 2002; 115:39–50.

33. Smalley MJ, Titley J, Paterson H, Perusinghe N, Clarke C, O'Hare MJ. Differentiation of separated mouse mammary luminal epithelial and myoepithelial cells cultured on EHS matrix analyzed by indirect immunofluorescence of cytoskeletal antigens. J Histochem Cytochem 1999; 47:1513–1524.

34. Clayton H, Titley I, Vivanco M. Growth and differentiation of progenitor/stem cells derived from the human mammary gland. Exp Cell Res 2004; 297:444–460.

35. Pullan S, Wilson J, Metcalfe A, Edwards GM, Goberdhan N, Tilly J, Hickman JA, Dive C, Streuli CH. Requirement of basement membrane for the suppression of programmed cell death in mammary epithelium. J Cell Sci 1996; 109(Pt 3):631–642.

36. Dontu G, Abdallah WM, Foley JM, Jackson KW, Clarke MF, Kawamura MJ, Wicha MS. In vitro propagation and transcriptional profiling of human mammary stem/progenitor cells. Genes Dev 2003; 17:1253–1270.

37. Dontu G, Jackson KW, McNicholas E, Kawamura MJ, Abdallah, WM, Wicha MS. Role of Notch signaling in cell-fate determination of human mammary stem/progenitor cells. Breast Cancer Res 2004; 6:R605–R615.

38. Chepko G, Smith GH. Three division-competent, structurally-distinct cell populations contribute to murine mammary epithelial renewal. Tissue Cell 1997; 29:239–253.

39. Hogan DL, Smith GH. Unconventional application of standard light and electron immunocytochemical analysis to aldehyde-fixed, araldite-embedded tissues. J Histochem Cytochem 1982; 30:1301–1306.

40. Smith GH, Vonderhaar BK. Functional differentiation in mouse mammary gland epithelium is attained through DNA synthesis, inconsequent of mitosis. Dev Biol 1981; 88:167–179.

41. Smith GH, Vonderhaar BK, Graham DE, Medina D. Expression of pregnancy-specific genes in preneoplastic mouse mammary tissues from virgin mice. Cancer Res 1984; 44:3426–3437.

42. Vonderhaar BK, Smith GH. Dissociation of cytological and functional differential in virgin mouse mammary gland during inhibition of DNA synthesis. J Cell Sci 1982; 53:97–114.

43. Nicoll CS, Tucker HA. Estimates of parenchymal, stromal, and lymph node deoxyribonucleic acid in mammary glands of C3H/Crgl-2 mice. Life Sci 1965; 4:993–1001.

44. Smith GH, Strickland P, Daniel CW. Putative epithelial stem cell loss corresponds with mammary growth senescence. Cell Tiss Res 2002; 310:313–320.

45. Gudjonsson T, Villadsen R, Nielse HL, Ronnov-Jessen L, Bissell MJ, Petersen OW. Isolation, immortalization, and characterization of a human breast epithelial cell line with stem cell properties. Genes Dev 2002; 16:693–706.

46. Zeps N, Bentel JM, Papadimitriou JM, D'Antuono MF, Dawkins, HJ. Estrogen receptor-negative epithelial cells in mouse mammary gland development and growth. Differentiation 1998; 62:221–226.

47. Welm BE, Tepera SB, Venezia T, Graubert TA, Rosen JM, Goodell MA. Sca-1[pos] cells in the mouse mammary gland represent an enriched progenitor cell population. Dev Biol 2002; 245:42–56.

48. Alvi AJ, Clayton H, Joshi C, Enver T, Ashworth A, Vivanco MM, Dale TC, Smalley MJ. Functional and molecular characterisation of mammary side population cells. Breast Cancer Res 2003; 5:R1–R8.

49. Liu BY, McDermott SP, Khwaja SS, Alexander CM. The transforming activity of Wnt effectors correlates with their ability to induce the accumulation of mammary progenitor cells. Proc Natl Acad Sci USA 2004; 101:4158–4163.

50. Clarke RB, Spence K, Anderson E, Howell A, Okano H, Potten CS. A putative human breast stem cell population is enriched for steroid receptor-positive cells. Dev Biol 2005; 277:443–456.

51. Clarke RB, Anderson E, Howell A, Potten CS. Regulation of human breast epithelial stem cells. Cell Prolif 2006; 36(suppl 1):45–58.

52. Reynolds BA, Weiss S. Clonal and population analyses demonstrate that an EGF-responsive mammalian embryonic CNS precursor is a stem cell. Dev Biol 1996; 175:1–13.

53. Potten CS, Owen G, Booth D. Intestinal stem cells protect their genome by selective segregation of template DNA strands. J Cell Sci 2002; 115:2381–2388.

54. Merok JR, Lansita JA, Tunstead JR, Sherley JL. Cosegregation of chromosomes containing immortal DNA strands in cells that cycle with asymmetric stem cell kinetics. Cancer Research 2002; 62:6791–6795.

55. Smith GH, Boulanger CA. Mammary stem cell repertoire: new insights in aging epithelial populations. Mech Aging Dev 2002; 123:1505–1519.

56. Smith GH. Label-retaining epithelial cells in mouse mammary gland divide asymmetrically and retain their template DNA strands. Development 2005; 132:681–687.

57. Boulanger CA, Wagner KU, Smith GH. Parity-induced mammary epithelial cells are pluripotent, self-renewing and sensitive to TGF-β1 expression. Oncogene 2005; 24:552–560.

58. Tsai YC, Lu Y, Nichols PW, Zlotnikov G, Jones PA, Smith H. Contiguous patches of normal human epithelium derived from a single stem cell: Implications for breast carcinogenesis. Cancer Res 1996; 56:402–404.

59. Deng G, Lu Y, Zlotnikov G, Thor AD, Smith HS. Loss of heterozygosity in normal tissue adjacent to breast carcinomas. Science 1996; 274:2057–2059.

60. Lakhani SR, Slack DN, Hamoudi RA, Collins N, Stratton MR, Sloane JP. Detection of allelic imbalance indicates that a proportion of mammary hyperplasia of usual type are clonal, neoplastic proliferations. Lab Invest 1996; 74:129–135.

61. Lakhani SR, Chaggar R, Davies S, Jones C, Collins N, Odel C, Stratton MR, O'Hare MJ. Genetic alterations in 'normal' luminal and myoepithelial cells of the breast. J Pathol 1999; 189:496–503.

62. Rosenberg CL, Larson PS, Romo JD, De Las Morenas A, Faller DV. Microsatellite alterations indicating monoclonality in atypical hyperplasias associated with breast cancer. Hum Pathol 1997; 28:214–219.

63. Boecker W, Moll R, Dervan P, Buerger H, Poremba C, Diallo R, Herbst H, Schmidt A, Lerch MM, Buchalow IB. Usual ductal hyperplasia of the breast is a committed stem (progenitor) cell lesion distinct from atypical ductal hyperplasia and ductal carcinoma in situ. J Pathol 2002; 198:458–467.

64. Callahan R, Smith GH. MMTV-induced mammary tumorigenesis: gene discovery, progression to malignancy and cellular pathways. Oncogene 2000; 19:992–1001.

65. Boulanger CA, Smith GH. Reducing mammary cancer risk through premature stem cell senescence. Oncogene 2001; 20:2264–2272.

66. Dontu G, Al-Hajj M, Abdallah WM, Clarke MF, Wicha MS. Stem cells in normal breast development and breast cancer. Cell Prolif 2003; 36(suppl 1):59–72.

67. Reya T, Morrison SJ, Clarke MF, Weissman IL. Stem cells, cancer, and cancer stem cells. Nature 2001; 414:105–111.

68. Al-Hajj M, Wicha MS, Benito-Hernandez A, Morrison SJ, Clarke MF. Prospective identification of tumorigenic breast cancer cells. Proc Natl Acad Sci USA 2003; 100:3983–3988.

69. Doyle LA, Ross DD. Multidrug resistance mediated by the breast cancer resistance protein BCRP (ABCG2). Oncogene 2003; 22:7340–7358.

70. Sivaraman L, Stephens LC, Markaverich BM, Clark JA, Krnacik S, Conneely OM, O'Malley BW, Medina D. Hormone-induced refractoriness to mammary carcinogenesis in Wistar-Furth rats. Carcinogenesis 1998; 19:1573–1581.

71. Sivaraman L, Conneely OM, Medina D, O'Malley BW. p53 is a potential mediator of pregnancy and hormone-induced resistance to mammary carcinogenesis. Proc Natl Acad Sci USA 2001; 98:12379–12384.

72. Wagner K-U, Boulanger CA, Henry MD, Sagagias M, Hennighausen L, Smith GH. An adjunct mammary epithelial cell population in parous females: its role in functional adaptation and tissue renewal. Development 2002; 129:1377–1386.

73. Smith GH, Gallahan D, Diella F, Jhappan C, Merlino G, Callahan R. Constitutive expression of a truncated INT3 gene in mouse mammary epithelium impairs differentiation and functional development. Cell Growth Differ 1995; 6:563–577.

74. Raafat A, Callahan R. Unpublished data.

75. Spradling A, Drummond-Barbosa D, Kai T. Stem cells find their niche. Nature, 2001; 414: 98–104.

76. Arnold I, Watt FM. c-Myc activation in transgenic mouse epidermis results in mobilization of stem cells and differentiation of their progeny. Curr Biol 2001; 11:558–568.

6

Lineage Tracking, Regulation, and Behaviors of Intestinal Stem Cells

Melissa H. Wong
Department of Dermatology, Cell and Developmental Biology, Oregon Health and Science University, Portland, Oregon, U.S.A.

Adnan Z. Rizvi
Department of Surgery, Oregon Health and Science University, Portland, Oregon, U.S.A.

INTRODUCTION

Stem cells hold the promise of the development of novel therapies for treating diseases. Unfortunately, the use and study of embryonic stem cells are currently clouded by ethical controversy. Adult stem cells offer a unique alternative in that they may be isolated, studied, or manipulated without harming the donor. However, the adult stem-cell field is still in its infancy. Several obstacles for manipulation of adult stem cells exist. First, the ability to identify most adult stem cells is impeded by lack of stem-cell-exclusive markers. Second, in vitro systems for manipulating adult stem-cell populations are not well defined for all tissues. Third, the ability to reconstitute stem-cell function in vivo has not been demonstrated for most organs. Finally, our understanding of how adult stem cells are regulated within their niche is just beginning to be elucidated. Next to the hematopoietic stem cell, epithelial stem cells are one of the most widely studied adult stem-cell population. Even so, the diversity between epithelial functions in different organs makes it difficult to determine if common themes exist in regulating these related stem cells. In the intestine, insights into the stem-cell behavior have been primarily inferred by lineage tracking experiments. These studies have been invaluable in establishing the foundation for our understanding of intestinal stem cells. This chapter reviews the historical use of lineage tracking of intestinal epithelial cells and presents recent findings in our understanding of regulation of stem cells in order to anticipate where the intestinal stem-cell field is heading in the future.

INTESTINAL EPITHELIUM

The adult small intestinal epithelium is a rapidly renewing epithelium that completely turns over approximately every three to five days in the mouse (reviewed in 1). To support the perpetual epithelial renewal, while concurrently maintaining the intestinal function, the adult small intestine is composed of well-defined, functionally active and proliferative units—the crypt-villus units (Fig. 1). The functionally active region is represented by villi, that is, finger-like projections that extend into the intestinal lumen perpendicular to the intestinal floor. The proliferative region, the crypts of Lieberkühn, lines the floor of the intestine, surrounding and populating adjacent villi. The epithelium covers both the crypt and the villus, representing a continuum of proliferating and differentiating cells along this axis. The villus epithelium is composed of differentiated cells that convey four primary functions: (*i*) absorption of nutrients and fluids, (*ii*) secretion of protective mucins to prevent damage to the epithelium, (*iii*) maintenance of a barrier between lumenal contents and the organism, and (*iv*) secretion of hormones that aid in digestion. The villi are populated with terminally differentiated cells, and the crypts of Lieberkühn are primarily populated with undifferentiated, differentiating, and proliferating cells (with the exception of the differentiated Paneth cells that reside at the base of the crypt). In the normal state, all proliferation of the epithelium is confined to the crypts.

The adult large intestine is composed similarly to the small intestine (Fig. 1). However, although it is composed of two functionally distinct regions, the regions are not physically well defined. The large intestine lacks villi. Therefore, the functional portion of the large intestine is represented by the colon cuff cells that surround the opening of the crypts. The stem cell resides in the colonic crypts. Both

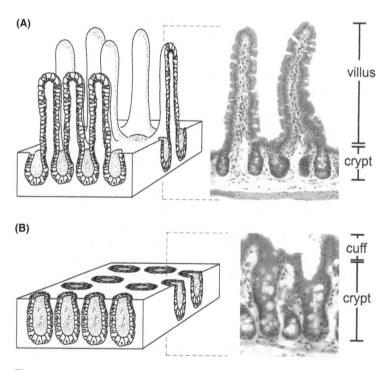

Figure 1   Intestinal structure. (**A**) The small intestine is composed of a functional region and a proliferative region. The epithelium lining the villi represents the functional portion of the intestine, whereas the epithelium lining the intestinal crypts represents the proliferative compartment. (**B**) The large intestine is also composed of functional and proliferative regions. The large intestine lacks villi. The surface cuff epithelium that lines the crypt openings represents the functional portion of the large intestine. Colonic crypts contain both proliferative and differentiated cells. (*See color insert.*)

differentiated and undifferentiated cells reside in the colonic crypts. Goblet cells are the primary epithelial lineage.

Intestinal Stem Cell

By definition, the epithelial stem cell retains the ability to (*i*) give rise to multiple cell lineages, (*ii*) be anchored within its niche, and (*iii*) undergo asymmetric cell division (self-renew and produce a daughter cell population) (Fig. 2). The intestinal epithelial stem cell resides in the crypt of Lieberkühn at approximately the fourth cell strata from the crypt base (reviewed in 2,3). This location was established through labeling experiments using ^{3}H-thymidine or bromodeoxyuridine (BrdU). It is believed that stem cells seldom divide and therefore retain the incorporated label. Daughter cells, on the other hand, rapidly divide, diluting the label beyond detectible limits, and also migrate away from the stem-cell environment. Label retention within the stem-cell population was thought to reflect the lack of proliferative activity. However, it is now thought that retention of label is not indicative of stem-cell division, but that stem cells are capable of asymmetrically segregating their DNA upon cell division. The original DNA (labeled DNA) would be allocated to the stem cell replacing the dividing parent stem cell, whereas the newly synthesized, unlabeled DNA would be segregated to its daughter cell (4).

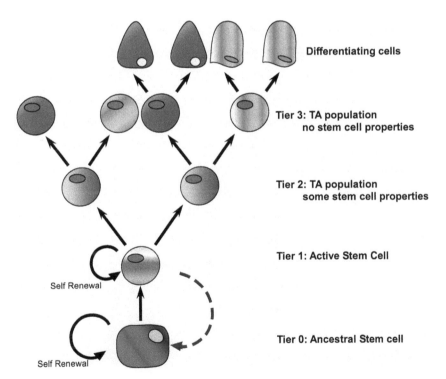

Figure 2   Stem-cell hierarchy. An ancestral stem cell is selected during developmental formation of crypt and populates each mature crypt (tier 0). Before becoming quiescent, the ancestral stem cell undergoes an asymmetric cell division, self-renewing while giving rise to an active stem-cell population (tier 1). The active stem-cell population retains stem-cell properties and is responsible for actively populating the crypt and villus epithelium. Tier 1 stem cells give rise to the TA population. The first layer of TA cells (tier 2) retains some stem-cell properties. Tier 3 cells no longer retain stem-cell properties and are lineage committed. *Abbreviation*: TA, transient amplifying. (*See color insert.*)

Differentiated Intestinal Epithelial Lineages

In the adult small intestinal crypt, multipotent epithelial stem cells give rise to the four principal epithelial lineages of the intestine (Fig. 3). Three of the lineages, the absorptive enterocyte, the mucin-secreting goblet cell, and the peptide hormone-secreting enteroendocrine cell, differentiate as they migrate up and out of the crypt onto adjacent villi. The epithelial cells journey up the villus, which takes approximately three to five days. As these cells near the villus tip, they undergo apoptosis or are exfoliated into the lumen of the intestine. In this manner, the epithelial barrier is maintained. The fourth lineage, the Paneth cell, differentiates as it undergoes a downward migration to reside at the crypt's

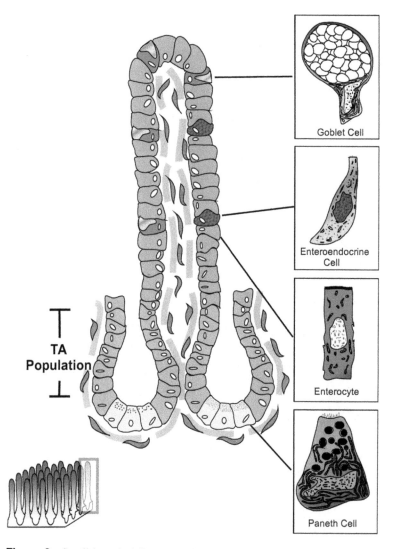

Figure 3 Small intestinal lineages. The multipotent stem-cell (*blue cell*) intestine of the small intestine gives rise to two types of epithelial cells: absorptive and secretory. The absorptive enterocyte, along with the secretory enteroendocrine and goblet cells, differentiates as they migrate up and out of the crypt. The Paneth cell, which shares a common lineage precursor with the goblet cell, undergoes a downward migration to reside at the crypt base. (*See color insert.*)

base. Paneth cells are involved in mucosal immunity and secrete proteins including tumor necrosis factor, lysozyme, and cryptins (5). Paneth cells are longer lived than cells that populate the villus, surviving 18 to 23 days before they are phagocytosed by surrounding epithelial cells as macrophages (6).

Stem-Cell Hierarchy

The rapid renewal of the intestinal epithelium necessitates the need for a physically well-defined stem-cell niche that promotes an ordered stem-cell hierarchy. Epithelial turnover in the intestine is rapid and, therefore, a constant supply of newly formed cells is required to accommodate for the daily loss of epithelium. The multipotent epithelial stem cell is the cell source. The actual number of active stem cells within each crypt is debated. Two schools of thought exist. One suggests that a single multipotent stem cell is selected from the proliferative region of the developing intestine, the intervillus region (IVR), during crypt morphogenesis to populate each adult crypt. This hypothesis is primarily supported by the observation that the IVR is composed of multiple stem cells (polyclonal) in chimeric animals, whereas the adult crypt appears to be monoclonal or derived from a single clone (monoclonal; 7,8). The second school of thought suggests that crypts are populated by multiple (four to six or as many as 60) dividing stem cells. This hypothesis is based upon two different studies of intestinal stem cells. One approach evaluated the stem-cell numbers using a combination of cell proliferation studies and mathematical modeling (2,3,9). The second approach suggests multiple active stem cells based upon the presence of different DNA methylation patterns of cells within the crypt (10). Ultimately, this debate cannot be resolved until reliable stem-cell markers are identified for this population.

It is likely that a combination of these two models actually occurs. During intestinal morphogenesis (Fig. 4) and formation of mature crypts, a single ancestral stem cell is

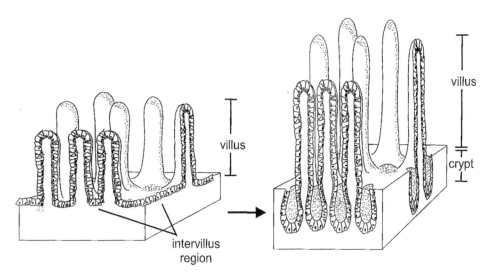

Figure 4 Intestinal crypt morphogenesis. The proliferative region of the developing intestine is represented by the linear row of cells that is situated between villi (IVR). Multiple stem cells reside within the IVR. During the course of crypt formation, a single ancestral stem cell is selected to populate the adult crypt. *Abbreviation*: IVR, intervillus region.

selected from the developing intestinal IVR to populate a single adult crypt. This ancestral stem cell undergoes asymmetric divisions to renew itself and give rise to a subset of active stem cells before becoming quiescent (Fig. 2). The immediate daughter cells of the ancestral stem cell comprise the active stem cells that are responsible for populating the crypt, and the active stem cells number between four and six (11). These cells, in turn, undergo asymmetric cell division to self-renew and give rise to a rapidly dividing cell population, termed the transient amplifying (TA) population. These cells are responsible for amplifying the cellular census in the crypt and proliferate more frequently than the anchored stem cell. Cells of the TA population are mostly undifferentiated but become committed to one of the four epithelial lineages as they migrate away from the stem cell. In the small intestinal crypt, all these different cells populate the stem-cell niche, presenting the challenge of imparting functional cellular diversity to these cells within a confined physical region.

Cells that reside at different cell strata within the crypt possess unique characteristics. Potten and co-workers (12) propose a three-tier hypothesis for stem-cell hierarchy based upon a cell's response to variable levels of gamma irradiation dependent upon its location within the crypt. Using a microcolony clonogenic stem-cell assay to assess the number of stem cells that survive variable levels of gamma irradiation, they found that low levels of irradiation resulted in survival of approximately six clonogenic cells per crypt. This observation supports the mathematical model of four to six active stem cells per crypt. These cells make up the first tier of stem cells. They undergo apoptosis in response to gamma irradiation rather than attempt to undergo DNA repair. A second tier of cells is less susceptible to a higher level of radiation-induced apoptosis and is again composed of six cells. These cells retain stem-cell characteristics and can be recruited to repopulate the crypt after radiation-induced death of the tier 1 stem cells. Finally, at a much higher dose of radiation a third tier of cells was identified that is comprised of approximately 24 cells. These cells are radio-resistant and therefore possess repair capabilities (13,14). These data suggest a scenario where the active stem cells are the most susceptible to apoptosis upon DNA damage and are replaced by second-tier cells that are capable of de-differentiating into tier 1 cells. This intriguing hypothesis brings up a number of interesting issues. First, is the process of de-differentiation context-dependent (e.g., tier 2 cells migrate into and are influenced by a tier 1 environment) or is it cell autonomous (e.g., tier 2 cells express stem-cell markers that allow them to fill into the tier 1 cellular void)? Second, does this mechanism ultimately preserve the genetic code? It is not intuitive to replace a damaged active stem cell with a progeny that has equal or greater potential for genetic error. Finally, if a stem-cell hierarchy such as this makes up the crypt, we should expect to see clonal differences within the cell population with aging. Interestingly, this appears to be the case when tracking changes in methylation patterns within the crypt (10).

Asymmetric Division

Stem cells of the small intestine have devised mechanisms to protect their original DNA content in the presence of damaging agents. As previously mentioned, Potten et al. (4) suggest that the asymmetric division of stem cells results in preferential segregation of the original DNA to the self-renewed stem cell and the newly synthesized DNA to the daughter cell. Thus, replication-induced errors are segregated to the daughter cells, effectively protecting the stem cell from retaining genetic errors (15). In addition, DNA damage to the stem cell induces a p53-dependent apoptosis, which would allow a stem cell to sacrifice itself in order to prevent retention of genetic errors (4,16).

Definition of the Stem-Cell Niche

The stem-cell niche is composed of both an epithelial and a mesenchymal compartment. It is structured to promote signaling to occur between the stem cell and neighboring cells of both epithelial and mesenchymal origin. Although it is difficult to discern which signaling pathways critically impact the stem cell's behavior and which are important for influencing the TA population, it is clear that the stem cell either receives different extrinsic signals than the TA population or responds differently to similar signals. Although the stem cell remains anchored within the niche and divides at a slower rate, the TA population rapidly proliferates and differentiates along one of several terminal differentiation cell fates. In the adult small intestine, the crypts of Lieberkühn and the surrounding pericryptal mesenchyme (Fig. 5) compose the intestinal epithelial stem-cell niche. The niche is a specialized environment that not only acts to protect the stem-cell population from externally induced damage, but also supports an atmosphere that nurtures divergent cellular

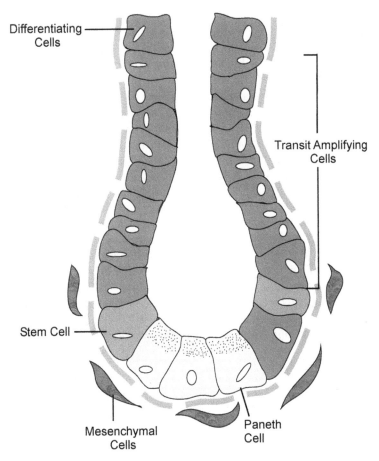

Differentiating Cells

Transit Amplifying Cells

Stem Cell

Mesenchymal Cells

Paneth Cell

Figure 5 The small intestinal stem-cell niche. The crypts of Lieberkühn provide the niche for the small intestinal epithelial stem cell. The stem cell resides at approximately the fourth cell strata in the crypt. The stem-cell niche is composed of differentiated Paneth cells that reside at the base of the niche, as well as undifferentiated cells that make up the TA population. Cells of the TA population are at different stages of terminal differentiation. *Abbreviation*: TA, transient amplifying. (*See color insert.*)

states (quiescent, proliferating, and differentiating), infrequent asymmetric division of the stem cell, and rapid proliferation of the stem cell's daughter cell population.

Epithelial Component of the Stem-Cell Niche

The epithelial compartment is populated with differentiated Paneth cells, the quiescent epithelial stem cell, active stem cells, a TA cell population, and undifferentiated, committed epithelial cells (Fig. 5). Communication between these cell populations is integral in defining a favorable stem-cell environment. Cell adhesion status is likely to play a major role in anchoring the active stem cells within the niche while allowing differentiating cells to migrate out of the niche.

The only terminally differentiated cell population that resides within the epithelial portion of this niche is the Paneth cell. Because of its close proximity to the stem cell, it was thought that Paneth cells might secrete factors that influence stem-cell survival. Elegant lineage ablation studies in transgenic mice demonstrated that Paneth cell ablation had no impact on the viability of the epithelial stem cell (6).

Mesenchymal Component of the Stem-Cell Niche

The mesenchymal component of the stem-cell niche directly surrounds the crypt epithelium and is composed of extracellular matrix, enteric neurons, blood vessels, intraepithelial lymphocytes, and pericryptal fibroblasts. The cells in the mesenchyme secrete factors that directly instruct the overlying epithelium, setting up epithelial–mesenchymal cross-talk. Although a number of secreted factors and corresponding receptors have been identified, it is by no means a comprehensive list. Platelet-derived growth factor (PDGF) is one such secreted factor. Mice deficient for intestinal expression of PDGF-α or its receptor PDGFR-α developed the abnormal intestinal epithelium and depletion of the pericryptal mesenchyme (17). In addition, Sonic hedgehog (Shh), bone morphogenic protein (BMP), Forkhead-6 (Fkh6), Wnt, Notch, and the nuclear transcription factor Nkx2–3 are among other factors that have been shown to influence the intestinal epithelium (18–22). In most cases, the exact cellular origin of these factors has not yet been defined.

Pericryptal fibroblasts play a major role in influencing the overlying epithelium of the stem-cell niche. Not only do they secrete a number of growth factors including hepatocyte growth factor, tissue growth factor-β (TGF-β), and keratinocyte growth factor (23), but also they influence epithelial migration. Experiments using ^{3}H-thymidine labeling indicate that the pericryptal fibroblasts migrate up along the crypt-villus axis at a similar rate as the differentiating epithelium (24). These observations all strongly support the existence of an intimate relationship between the epithelium and the mesenchyme.

LINEAGE TRACKING AS AN APPROACH TO UNDERSTANDING THE STEM-CELL BEHAVIOR

Can lineage tracking reveal information regarding the stem-cell behavior within the proliferative crypt? Tracking cell lineages or cell markers within the crypt have the potential to reveal information on the cellular behavior with the stem-cell niche. Traditionally, morphological characteristics were used to trace the different lineages within the stem-cell niche. These studies laid the foundation for our current understanding of how lineages within the intestinal epithelium are related to each other. However, defining the behavior of the stem-cell population through understanding the behavior of their descendents leaves us wondering if the readout is an accurate readout. As previously stated, the identification of reliable stem-cell markers, in vitro systems to determine lineage relationships, and an in vivo reconstitution assay are required before we can gain further insight into the stem cell.

Unitarian Theory of Epithelial Cell Formation

The concept that undifferentiated cells exist in the intestinal crypt and possess the ability to give rise to different epithelial cell types originated in the early 1960s (25). Several groups identified a cell type that was morphologically intermediate to the stem cell, yet a committed precursor to the differentiated lineages. From these studies, the hypothesis was formed that all epithelial cells are derived from a single precursor cell or stem cell (26,27). Thus, the Unitarian theory of epithelial cell formation was born, stating that all differentiated lineages within the intestinal epithelium were derived from a single common precursor.

Tracking Intestinal Lineages by BrdU or ^{3}H-thymidine

The evidence for a common precursor for all epithelial cells that populate the intestine was first reported in the stomach. Experiments grafting newborn mouse stomachs initially resulted in total ablation of all cells except mucous cells, but eventually repopulation of all lineages. Slowly, parietal, enteroendocrine, and chief cell lineages reappeared, suggesting that these cell lineages are all derived from a common mucous progenitor (28). Likewise, Chang and Leblond (29) reported similar findings using radioautography in the mouse colon. However, it was a series of five tandemly reported studies that clearly illustrated the relationship between a common epithelial stem cell and each of the four differentiated epithelial lineages of the intestine (30–34). ^{3}H-thymidine injected into mice resulted in radio-damage to the intestinal epithelia residing in the stem-cell niche. Dying ^{3}H-thymidine-labeled stem cells were phagocytosed by undamaged neighboring epithelial cells. Phagosomes were then used as markers to follow the evolution of the crypt-based columnar cells, tracking their cellular fate. A common crypt-based columnar cell gave rise to all four intestinal lineages. These observations definitively supported the Unitarian theory of epithelial cell origin, representing the first such large-scale lineage tracking study in the small intestine.

Using an Epigenetic Event to Track Lineages Derived from the Stem Cell

Tracking DNA methylation patterns in the intestine provides a unique approach toward tracking cellular hierarchies within the intestinal crypt in order to determine the behavior of stem cells within the niche. This concept is based upon the idea that epigenetic variants with different patterns of methylation at CpG sites arise during stem-cell division. The distribution of methylation variants among and within tissue regions conveys information about stem-cell population dynamics (10). Heterogeneous methylation patterns have been observed in human colonic crypts (35). To address the debate of whether crypts are populated by a single ancestral stem cell or by multiple stem cells, Yatabe et al. (35) used methylation tags to fate-map human colonic crypts and to study the dynamics of stem cells. They reasoned that because methylation patterns are somatically inherited, drift within a crypt's lifetime would reveal relationships between cells populating that colonic crypt. In human colonic tissues, they isolated individual colonic crypts and used a polymerase chain reaction (PCR)-based method to clone and sequence methylation patterns at three independent loci. Using genes that are not expressed in intestinal cells to prevent selective methylation, they reasoned that differences were likely due to the random process of methylation associated with cellular aging. If there were little difference in methylation patterns within the crypts, this would support the notion that a single ancestral stem cell might populate the entire crypt. However, if there were diverse methylation "tags" within the crypt, this would support the notion that crypts are stochastically populated by multiple stem cells. Their data revealed a number of diverse methylation "tags" within the crypt, indicating that the random changes in methylation patterns reflected lineage propagation from multiple stem cells within the crypt. They went on to use mathematical modeling to suggest that as many as 64 actively dividing stem cells populate the human colonic crypt.

Kim and Shibata (10) extended their study to examine ancestry among crypts located in close proximity of each other to determine if adult crypts share more recent common ancestors they frequently divide by crypt fission to form clonal patches of crypts. Methylation patterns among crypts that were in close proximity had the same amount of variation as crypts that were located far away. This observation suggested that the human colonic crypt is a long-lived structure.

Although the epigenetic tagging approach to studying the stem-cell behavior within a crypt is an intriguing approach to gaining information, several issues must first be resolved (reviewed in 36). First, the experimental approach must be absolutely accurate. If errors predict a higher number of methylation patterns, this would skew the results toward favoring the multiple stem cells/crypt hypothesis. Second, predicted methylation patterns may not be similar among all types of somatic cells. Among differentiated cells, genomic methylation patterns are generally stable and are inherited, however, among germ cells, methylation patterns are variable reflecting a broad developmental potential (37). The question arises whether methylation patterns are stable and inherited between adult stem cells and their immediate daughter cells (Potten's tier 2 cells). Alternatively, if adult stem cells and their immediate daughter cells act more like germ cells, it would impart variability in predicting inherited methylation patterns and lend to misinterpretation.

Using Histological Markers to Track Epithelial Lineage

Use of histochemical markers to track epithelial lineages to gain insight into the intestinal epithelial stem-cell behavior was illustrated in female mice that were mosaic for the X-linked alleles Pgk-1a and Pgk-1b. These studies took advantage of X-inactivation in conjunction with the ability to follow inheritance of a cell autonomous marker in stem cells. By tracking the behavior of daughter cells, the behavior of the stem cell might be inferred. In these studies, it was observed that adult intestinal epithelial crypts were derived from either all Pgk-1a or Pgk-1b expressing cells. Therefore, it was inferred that all cells within the crypt were derived from a single parent cell (38).

To further explore these observations, Winton et al. (7) used a mutation-induced marker in mice heterozygous at the locus, which determines the expression of binding sites for an intestinal epithelial lectin, *Dolichos bifluorus* agglutinin (DBA), to study the relationship between the stem cell and its progeny. The Dlb-1 gene resides on chromosome 11. Inbred mouse strains are either Dlb-1b homozygotes, which bind the conjugate on epithelial surfaces of the intestine, or Dlb-1a homozygotes, which do not bind. Mutatgens such as *N*-nitroso-*N*-ethylurea or dimethylhydrazine are used to randomly induce mutations in the genome. Random cells within the intestinal epithelium are mutated to change their DBA-binding affinity and can be used to track lineages (39).

In these studies, adult animals displayed patches of Dlb-1a and Dlb-1b expressing crypts. The boundaries between these patches represented regions that were populated by cells of both genotypes. However, even though these regions encompassed coexisting genotypes, it was observed that the crypts were completely monoclonal (39,40).

Winton et al. (7) went on to confirm that adult mouse intestinal crypts were derived from single clones of cells (i.e., monoclonal) using the DBA tracking approach. They extended their studies to the developing intestine to determine if the proliferative compartments of neonatal intestines are composed of multiple stem cells (polyclonal; 41). During the course of crypt morphogenesis, which spans both a neonatal and postnatal time frame, polyclonal proliferative regions become monoclonal in nature. At approximately 14 to 21 days postnatal (P), the mature crypt becomes populated by a single genotype (40). These studies using mosaic mice presented the basis for the hypothesis that adult crypts are populated by a single active stem cell.

Bjerknes and Cheng (42) used the Dlb-1 mutagenesis approach to characterize cells within the crypts that are neither stem cells nor differentiated cells, that is, cells that are the intermediate progenitor or the early lineage progenitor. They randomly mutagenized intestinal epithelial cells in adult mice, then analyzed crypts that possessed cells that had undergone somatic mutation at the Dlb-1 locus, and tracked their behavior in intact isolated crypt and villus preparations. Using a time course to track the longevity of mutated epithelial cells, three distinct groups of clones were identified: short-lived progenitor cells, long-lived progenitor cells, and a population of pluripotent stem cells. Short-lived clones lived only 10 to 14 days and presumably represented relatively differentiated cells. Long-lived progenitors gave rise to either only columnar or mucosal cells (although there was also a group of "mixed" long-lived progenitors). These clones lived for >154 days and by mathematical modeling were thought to have divided two or three times. Pluripotent stem cells gave rise to all lineages.

These studies also addressed the issue of crypt replication. A number of crypt-villus isolates were identified that displayed branched crypts. These branched structures were thought to be crypts that were undergoing division. In a subset of these branched crypts, one crypt was completely populated by a single genotype, whereas the other crypt was completely populated by the opposite genotype. This is an interesting observation because it supports the notion that multiple stem cells exist within a crypt. Crypts undergo asymmetric division, segregating stem cells and all descendants of one genotype to one crypt. However, because this system does not lend itself to a real-time analysis, it is difficult to determine if these branching crypts are indeed the result of crypt fission or of crypt fusion. Currently, we know little about the mechanism regulating crypt numbers. Analysis of the developing intestine where crypt numbers are rapidly expanding should result in a greater number of these branched crypts and resolve this issue.

The mutagenesis studies performed by Bjerknes and Cheng (42) offer support for the hypothesis that four to five stem cells exist within each adult crypt. They reason that three out of 1000 crypts contain a long-lived progenitor-type (stem cell) mutant clone, in a background where they estimate a mutation rate of one out of 1500 crypts. Therefore, an average crypt contains four to five of these long-lived progenitor-type cells (stem cell; 0.003/0.00066; 42).

Transgenic Markers for Tracking Lineages

Studies using DBA as a marker for studying clonal organization within intestinal crypts led to the use of mosaically expressed transgenic markers. Saam and Gordon (43) established an inducible gene expression system in transgenic mice using the bacterial gene Cre recombinase. Their system mosaically expressed Cre recombinase in a subset of epithelial stem cells (8). Cre recombinase excises DNA sequences that are flanked by 34 bp loxP sites and allows for induction of a traceable marker in the intestinal stem-cell population to evaluate its behavior and the behavior of its descendents. The expression of the marker can be induced in adulthood and allows for assessment of the adult stem cell at steady state. Wong et al. (8) induced expression of Cre recombinase in the adult to stimulate expression of β-galactosidase (LacZ). In these animals, adult crypts of the small intestine and the cecum and colon were predominantly monoclonal in nature, supporting the view established by Winton et al. (7,39).

Interestingly, a small subset of crypts was not monoclonal and displayed both genotypes. A portion of these mixed crypts displayed LacZ expression in the bottom portion of the crypts, suggesting that Cre recombinase was activated in Paneth cell precursors. However, a portion of mixed crypts displayed LacZ expression on the right- or left-hand side of the crypt. Interpretation of these crypts presents a challenge. These crypts could represent a crypt that is undergoing changes in its lifecycle. Although this phenomenon is difficult to assess using a static system, it is clear that the inducible gene expression system in the intestinal epithelium can be used to express genes that regulate the stem-cell behavior in the adult or developing intestine. This type of lineage tracking has also been performed in the ovarian follicular stem cells of genetically mosaic flies (44).

Genetic mosaic analysis of chimeric–transgenic mouse intestines offers a powerful approach for studying the importance of various factors in the regulation of the stem cell. If a particular molecule has a deleterious effect on stem-cell propagation, a transgenic mouse or knockout mouse may display a lethal phenotype. Mosaic expression of molecules allows survival of the intestinal tissue because only a subset of cells will harbor the transgene, or will have a gene deletion. For example, the role of GATA-4 in specification of the definitive gastric endoderm was explored by introducing Gata-4^{null} ES cells into ROSA26 blastulae (45). Resulting mice had patches of normal epithelium juxtaposed to patches of Gata-4^{null} epithelium. The Gata-4 null epithelium displayed a squamous morphology and lacked expression of gastric differentiation markers, suggesting that Gata-4 is involved in the transition from proliferation to differentiation of gastric epithelia within the stem-cell niche. In addition, Jacobsen et al. (45) illustrated that Gata-4 null epithelia had perturbed expression of Shh. Studies such as these allow the dissection of molecules that influence stem-cell proliferation or differentiation of the stem-cell progeny and will ultimately allow us to understand the dynamic interactions of the signaling pathways that play a role in maintaining the stem-cell niche.

Mosaic analysis of the intestine has also led to the clarification of cell lineage distribution and the stem-cell behavior during development. Shiojiri and Mori (46) generated mice chimeric for the spf^{ash} mutation, which is located on the X chromosome, and causes ornithine transcarbamylase (Otc) deficiency. The small intestine of female heterozygotes had small aggregates of Otc-positive cells. This study looked in-depth at the mosaism that occurs during the intestinal development and confirmed the results that were presented in previous studies by Schmidt and coworkers.

Tracking Changes in Mitochondrial DNA

The stem-cell behavior can be inferred by tracking changes in mitochondria. Taylor et al. (47) identified inherited changes in mitochondria DNA of colonic stem cells. Mitochondria are semi-autonomous organelles that are ubiquitously present in all cells (48).

Mutations in mitochondria DNA are somatically inherited and also accumulate with age (49). Using an enzyme assay for respiratory chain deficiency in colonic crypts and crypt stem cells, Taylor et al. (47) were able to show that mutations in mitochondrial DNA induced defects in cytochrome C oxidase activity that were trackable. Therefore, like methylation patterns, mutations in mitochondrial DNA can infer information about stem-cell division and their progeny (49).

TRACKING STEM-CELL FATE THROUGH UNDERSTANDING WHAT REGULATES THEIR PROLIFERATION AND DIFFERENTIATION

In order to understand the stem-cell behavior, knowledge of what regulates its behavior is invaluable. This knowledge will allow us to anticipate the stem-cell behavior within the context of development, homeostasis, or disease states. In addition, identification of the signaling pathways or molecules that ultimately impart stemness is the critical step toward gaining the ability to manipulate stem cells for therapeutic purposes.

Currently, all the major developmental signaling pathways have been implicated in regulation of stem cells (reviewed in Ref. 23). The challenge is to understand how to integrate each of these influences into a coherent regulatory network capable of modulating different behaviors in the active stem cell and simultaneously in the TA cell population. Unfortunately, coordinating regulation of these signaling pathways is complex. The more we learn about how these signaling pathways interact, the more complex the scenario becomes. However, great strides have been made in the initial elucidation of how proliferation and differentiation within the stem-cell niche are achieved.

Wnt Signaling

The canonical Wnt signaling pathway plays a key role in development, cellular homeostasis, and disease. Ablation of key components of the Wnt signaling pathway in the mouse results in early embryonic lethality, thereby highlighting the importance of the pathway in general developmental themes (50,51). Studies in chimeric–transgenic mice as well as the generation of tissue-specific gene ablation and inducible gene ablation systems have allowed further study of this pathway in its role in regulation of stem cells (52,53).

Wnt signaling is transduced intracellularly when secreted Wnt proteins bind to frizzled and low-density lipoprotein-related receptor protein (Lrp) receptors (Fig. 6). Receptor activation acts to inhibit the phosphorylation activity of glycogen-synthase kinase-3β (Gsk-3β) through a mechanism involving the protein Disheveled. The absence of phosphorylation activity allows β-catenin to escape degradation and translocate to the nucleus. In the nucleus, β-catenin can interact with Lef/Tcf HMG-box transcription factors to drive expression of target genes. In the absence of a Wnt signal, Gsk-3β phosphorylates β-catenin/Apc/Axin complexes to promote degradation of β-catenin through a ubiquitin-mediated proteosome pathway (reviewed in Ref. 54). Many of the Wnt target genes participate in cell proliferation, cell polarity, and cell fate decisions (55).

Wnt Signaling in the Intestine

Wnt signaling has been implicated in maintenance of epithelial homeostasis. Disruption in the Wnt signal through mutations in the Apc gene results in stabilization of β-catenin and increased transactivation of Wnt target gene expression. In both humans and in mice,

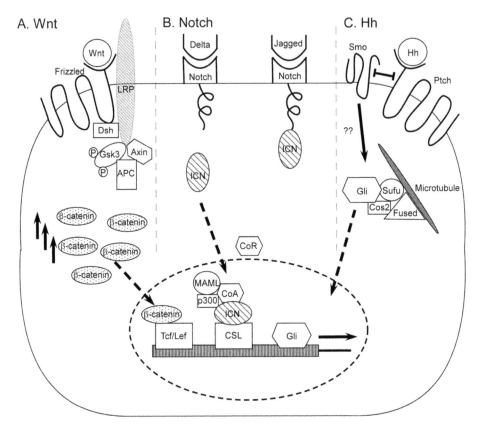

Figure 6 Signaling pathways that are involved in stem-cell regulation. The Wnt, Notch, Hh, and TGF-β3/BMP pathways have all been implicated in regulation of intestinal stem cells or regulation of the intestinal stem-cell niche. *Abbreviations*: Hh, hedgehog; TGF, tissue growth factor; BMP, bone morphogenic protein.

increased β-catenin signaling results in intestinal adenomatous polyp formation (56–58). Unregulated activation of Wnt target genes leads to unregulated epithelial proliferation and overgrowth of the epithelium. The result is the formation of benign adenomatous polyps, which is a precursor or risk factor for colorectal cancer. As defects in the Wnt signaling pathway are associated with disruption of epithelial homeostasis, it is implied that Wnt signaling impacts the status of the epithelial stem cell.

Wnt Signaling During Intestinal Development. Wnt signaling has been implicated in regulation of the intestinal stem cell. During intestinal morphogenesis, Wnt signaling is critical for maintaining the proliferative pressure within the stem-cell niche. The intestines from mice deficient for the HMG box transcription factor, Tcf-4, developed normally until embryonic day (E) 16.5 when crypt structures begin to form (59). At this developmental time point, the normally proliferative stem-cell niche became devoid of all proliferating cells. Electron microscopy of cells within this region revealed the presence of an apical brush border signifying that normally undifferentiated cells had become inappropriately differentiated. Therefore, it was concluded that Tcf-4 or Wnt signaling, in particular, was critical for maintaining the proliferative stem-cell niche during development. Studies in transgenic mice overexpressing the Wnt inhibitor, Dickoff-1 (Dkk-1), resulted in suppression of Wnt signaling and suppression of proliferation in the stem-cell niche (22). Interestingly, although Dkk-1 was expressed in the intestine at the same time Tcf-4^{null} was ablated in the previous study, the Dkk-1 mice survived to adulthood. This discrepancy may highlight an additional role for Tcf-4 independent of the Wnt signal, or the ability of Dkk-1 to completely ablate the Wnt signal. Regardless, these experiments suggest that Wnt is a potent growth factor in the intestine.

Although Wnt signaling is critical for sustaining proliferation in the intestinal stem-cell niche, overexpression of Wnt signaling during intestinal morphogenesis results in perturbation of stem-cell selection during crypt morphogenesis. A fusion molecule between β-catenin and the HMG box transcription factor Lef-1 represents a constitutively active Wnt signaling molecule. This fusion molecule was overexpressed in the intestinal epithelium in chimeric mice (53). Interestingly, intestinal stem cells that expressed the fusion protein underwent apoptosis and were not selected to populate the adult crypts during crypt morphogenesis and stem-cell selection. These results suggest that Wnt signaling is a critical factor in designating which stem cells will be anchored in each adult stem crypt during crypt morphogenesis. Moreover, these observations, taken within the context of the Tcf-4 knockout experiments, suggest that levels of Wnt signaling are important in defining the balance among cell death, cell proliferation, and cell differentiation. In addition, the act of anchoring a stem cell within a developing crypt may require an absolute level of the Wnt signal to maintain the ancestral or populating stem cell.

Wnt Signaling During Adulthood: Impact on Intestinal Homeostasis. The role of Wnt signaling in the adult intestinal epithelium is less clear. Clearly, stem cells and the TA population within the adult crypts must proliferate, but whether or not Wnt signaling plays a role in this is uncertain. One line of indirect evidence suggests that Wnt may play a minor role in the TA population proliferation. First, transgenic mice expressing a reporter for Wnt signaling do not express overt reporter activity in adult crypts (60). Mice expressing a LacZ reporter composed of seven tandem Lef/Tcf binding sites upstream of the *siamois* gene promoter and the LacZ gene were analyzed for reporter expression in adult mouse crypts. When this mouse was crossed to a mouse model that has characterized defects in intestinal proliferation (the Min mouse harbors a mutation in the Apc gene, which results in formation of intestinal adenomas; 57), LacZ expression was readily detectable. We have determined that Wnt signaling is present in the adult crypt at low levels using a similar reporter mouse designed by Elaine Fuchs' laboratory, the TopGal mouse (61). Comparison of LacZ expression levels between the adult and developing intestines revealed high levels of Wnt reporter activity during development, but dramatically lower levels during adulthood (Fig. 7). Although preliminary, these results suggest that Wnt signaling may play very different roles in stem-cell regulation during development as compared to adult. This notion may not be too difficult to fathom, as during development, there is rapid cellular expansion, perhaps dependent upon a strong Wnt signal, whereas during adulthood the epithelium is being maintained at a steady-state level.

Role of Wnt Signaling in Cellular Differentiation. Within the stem-cell niche, there is a delicate balance between proliferation and differentiation. Cells near the stem-cell zone are more proliferative, whereas cells near the crypt-villus junction are more differentiated. The Wnt pathway likely plays a role in directing cell differentiation. Whether or not this role is an active role or a passive role (absence of Wnt and the proliferative influence results in differentiation) remains to be elucidated. However, there is evidence that suppression of Wnt signaling results in up-regulation of cellular differentiation markers. Using Caco-2 cells in culture, Mariadason et al. (62) showed that down-regulation of Wnt/β-catenin signaling resulted in an increase in promoter activities of

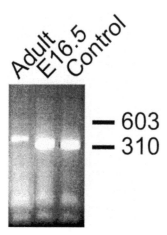

Figure 7 Wnt signaling in the intestine. Wnt reporter mice were assayed for reporter expression at different developmental time points by RT–PCR. Primer to the 3′-end of the β-galactosidase gene amplified 320 bp product and were used to assay for expression. Adult, P28, small intestines express low levels of β-galactosidase, whereas the embryonic (E) 16.5 intestine has higher levels of Wnt reporter expression. *Abbreviation*: RT–PCR, reverse transcription–polymerase chain reaction.

alkaline phosphatase and intestinal fatty-acid-binding protein, two markers of epithelial cell differentiation. In addition, Clevers' group (63) used DNA microarrays and a colon carcinoma cell line to identify genes that respond to Wnt signaling. Most of the genes identified were localized to proliferative crypts. Identification of c-MYC (myelocytomatosis oncogene) supported the previous report that this gene is a Wnt target (64). Further, they went on to show that expression of c-MYC disrupted expression of the cell-cycle inhibitor p21$^{CIP1/WAF1}$. Because p21$^{CIP1/WAF1}$ was previously reported to be expressed in differentiated colon epithelium (65), van de Wetering and co-workers propose that Wnt signaling results in increased c-MYC expression to allow cell proliferation and concomitantly inhibits p21$^{CIP1/WAF1}$ to suppress epithelial differentiation. Therefore, in the absence of high levels of Wnt signaling, epithelial differentiation persists. Additional support for a role of Wnt signaling in epithelial differentiation comes from the transgenic mice expressing Dkk-1. Intestines from these mice were devoid of secretory lineages, suggesting that Wnt expression may play an active role in promoting or directing differentiation of the three secretory lineages (22).

Wnt Signaling in Modulating Adhesive Properties Within the Stem-Cell Niche. Adhesive status within the stem-cell niche is an important area of stem-cell-related research. The active stem cell must be able to undergo asymmetric division but remain anchored within the niche. The daughter cell, however, has different adhesive properties, providing the challenge to modulate differences in adhesiveness within a tight localized region. Wnt signaling has been implicated in directing cellular migration of Paneth cells (63). Paneth cells are the only differentiated epithelial cell lineage in the small intestine to undergo a downward migration within the stem-cell niche. A recent report identifying Wnt target genes in a colorectal cell line identified EphB2 and EphB3 as targets (63). Eph receptors are part of the tyrosine kinase receptor family and are involved in vascular development, tissue-border formation, regulation of cell shape, and migration (66). EphB2 and EphB3 are normally expressed in the crypts of wild-type adult mice, whereas EphB3 expression is restricted to the Paneth cell population. Interestingly, EphB3$^{+/-}$ and EphB3$^{-/-}$ mice developed normal appearing villi, but display abnormal distribution of Paneth cells (67). Paneth cells in EphB3$^{+/-}$ and EphB3$^{-/-}$ mice do not migrate to the base of the crypt as they do in wild-type mice; instead, they undergo a disorganized migration with a final destination of both the lower-third and the upper-third of the intestinal crypt. This observation suggests that Ephrins and Wnt signaling are involved in restricting intermingling of proliferative and differentiated cell populations.

Wnt Signaling Influence on Stem Cells in Other Organs

Wnt signaling plays an active role in maintaining the stem-cell niche in the intestine, however, the full extent of how it impacts active stem cells is still not known. Studying the impact of Wnt signaling on stem cells in other tissues will help direct future research for understanding the regulation of the intestinal stem cell.

Epidermis. Wnt signaling plays a major role in regulating the epidermal stem-cell niche (68). Briefly, the epidermal epithelium is similar to the intestinal epithelium in that it represents a rapidly renewing population, turning over every 10 to 14 days in the mouse (69). The stem-cell niche in the epidermis is located in the bulge region of the hair follicle. In mammals, it is thought that skin is maintained by stem cells whose daughters differentiate along the lineages of the hair follicle, interfollicular epidermis, and sebaceous gland (70–74). Wnt signaling is clearly important in the developmental process of the epidermis as well as in the cycling of the hair follicle. Mice expressing the Wnt reporter displayed developmental expression of the LacZ reporter as well as cyclic expression in the adult parallel with hair follicle cycling (61). In addition, an elegant microarray analysis of epidermal stem cells (detailed subsequently) determined that the stem-cell niche expressed Wnt inhibitors sFRP1, DKK3, and WIF1 (68). Inhibition of Wnt signaling in the stem-cell niche makes sense, as the epidermal stem cell seldom divides. Furthermore, unlike the intestine where the stem cell and daughter cells are closely situated within the niche, the epidermal TA cells and the committed lineage precursors migrate away from the epidermal stem-cell niche.

In the epidermis, Wnt signaling may also impact cellular adhesion and migration from the stem-cell niche. Forced expression of the downstream Wnt target c-Myc resulted in depletion of the stem cells within the epidermal stem-cell niche as the animal aged (75,76). Frye et al. (77) suggest that c-Myc acts to stimulate cells to exit the stem-cell compartment by modulating the adhesiveness of the stem-cell niche. They go on to suggest that a cell's failure to differentiate may reflect its failure to migrate from the niche (77).

Hematopoietic Stem Cells. Wnt signaling has been implicated in the survival and proliferation of hematopoietic stem cells, both in vitro and in vivo (78,79). In vitro analysis of Wnt genes on CD34$^+$Lin hematopoietic progenitors determined that the number of progenitor cells in the presence of soluble Wnts was increased relative to the number of cells present in the absence of Wnt. Therefore, it was concluded that Wnt acts as a hematopoietic growth factor, perhaps exhibiting a higher specificity for the earlier progenitor cells (80). More recently, Reya et al. (79,81,82) illustrated that overexpression of Wnt inhibitors, Axin, or a frizzled ligand-binding domain led to inhibition of hematopoietic stem-cell growth in vitro and also reduced reconstitution of this population in an in vivo assay. Further, overexpression of β-catenin in these hematopoietic stem cells resulted in the ability to sustain the cultures long term. This observation, in conjunction with the observation that the Wnt reporter was activated in the normal hematopoietic stem-cell niche, strongly suggests a role for Wnt signaling in hematopoietic stem-cell homeostasis.

Notch Signaling

Notch and Wnt signaling are intimately regulated in the intestine. Intestines of transgenic mice overexpressing the Wnt inhibitor, Dkk-1, displayed reduced expression of the Notch pathway molecule, Math-1 (22). Notch proteins are involved in various aspects of vertebrate cell fate determination including lateral inhibition of adjacent cells to direct cell fate (83,84). Briefly, Notch signaling functions through interaction of the transmembrane receptor, Notch, and two cell surface ligands, Delta and Jagged, that reside on neighboring cells (Fig. 6). Upon ligand binding, the intracellular domain of Notch is cleaved and translocates to the nucleus, activating the transcription factor Supressor of Hairless (SuH) and up-regulating target genes [such as Hairy/Enhancer of Split (Hes)] (84). Hes proteins inhibit the activity of various basic helix–loop–helix transcriptional activators including Math-1 and Neurogenin-3.

Immunohistochemistry of the mouse intestine reveals expression of the four Notch receptors (Notch 1–4), five ligands (Delta 1, 3, 4 and Jagged 1, 2), and four Hes genes (Hes 1, 5, 6, 7) at both embryonic and adult time points (85).

Notch signaling actively designates the intestinal secretory cell lineage. Intestinal phenotypes described from a series of knockout mice support the role of Notch signaling in defining the intestinal stem-cell hierarchy. Hes1[null] mice revealed precocious development of endocrine cells in the stomach and small intestine at embryonic time points, as well as an increased number of goblet cells and fewer enterocytes (86). Because expression of Hes-1 is normally restricted to nonproliferating cells (villus epithelium) in wild-type mice during development, the changes in cell fate allocation was thought to be independent of a proliferative influence.

Additional evidence implicating Notch signaling in cellular differentiation decision was observed in the Math-1 null mice (21). The loss of Math-1 in embryonic mouse intestines resulted in complete depletion of all the three secretory lineages: Paneth, goblet, and enteroendocrine cells. The proliferative regions of these mice also exhibited an increase in the number of cycling cells, which might reflect a proliferative, compensatory mechanism to maintain villus cell census. Further, neurogenin-3 null mice also failed to develop enteroendocrine cells within their intestines (87). Interestingly, Paneth and goblet cells were detected, suggesting that perhaps additional factors are important in defining these three secretory cell lineages. For example, activation of Rac1, a member of the Rho GTPase family of GTP-binding proteins, in the mouse intestinal epithelium resulted in Paneth and goblet cell depletion, suggesting that Rac1 may be involved in differentiation of a Paneth/goblet precursor cell (88).

The discrepancy in lineage allocation in the various Notch factor knockout mice suggests that perhaps levels of Notch signaling are critical for defining the various secretory cell fates. Variable levels of expression of Delta are critical in defining Notch responsiveness in the epidermis and may be similar for the intestinal epithelium. Although it is unclear what supports the proliferation to differentiation gradient within the crypt, it may in part be influenced by Notch-induced lateral inhibition similar to what is seen during *Drosophila* eye development (89).

Hedgehog Signaling

Hedgehog (Hh) signaling is involved in various aspects of embryonic development such as left–right asymmetry, anterior–posterior patterning of the limb bud, and neural tube formation (90–92). In vertebrates, there are three Hh genes that share similar homology: Shh, Indian hedgehog (Ihh), and Desert hedgehog. Shh is a protein secreted by endodermal epithelium and induces the expression of its receptor, Patched (Ptch), in the surrounding mesenchyme (Fig. 6). Hh proteins bind to a transmembrane receptor Ptch, which normally inhibits downstream signaling through a second transmembrane protein, Smoothened (Smo) (93). Uninhibited Smo acts upon downstream transcription factors Gli and HRK4 through unknown mechanisms to transduce the signal (94,95).

Shh and Ihh expressions during gastrointestinal development mediate anterior–posterior patterning, radial patterning, and epithelial stem-cell proliferation and differentiation. Mice null for Shh or Ihh died before birth, but exhibited interesting intestinal phenotypes that ranged from intestinal transformation of the stomach, duodenal stenosis, aganglionic colon, and imperforate anus (20). Interestingly, only Ihh[null] mice displayed a reduction in villus size and a repression of cell proliferation within the stem-cell niche. Expression of the Wnt signaling mediator, Tcf-4, was normal in these intestines, suggesting that suppression of proliferation was not due to loss of Tcf-4 expression. In contrast, experiments suppressing Hh signaling by systemic injection of an anti-Hh antibody resulted in "disorganized" intestines that contained vacuolated epithelium and defective lipid processing, but no effect on proliferation within the stem-cell compartment (96). Perhaps this difference in phenotype is the result of a critical temporal requirement for Hh signaling during intestinal morphogenesis that can only be appreciated if the pathway is perturbed during embryogenesis. Alternatively, systemic suppression of Hh signaling may elicit a different phenotype by indirectly affecting the intestine.

Although Hh signaling impacts overall intestinal morphogenesis, it also plays a role in epithelial differentiation. Inhibition of Ihh signaling, by injection of the Hh inhibitor, cyclopamine, in the colon epithelium resulted in abnormal villin expression and loss of carbonic anhydrase IV expression, two enterocyte differentiation products. This suggests that Hh signaling may directly impact enterocyte differentiation (97). In addition, Ihh may coordinately regulate Wnt expression within the colonic stem-cell niche, as downstream Wnt target genes, engrailed-1, Cyclin D1, and BMP-4 were up-regulated and mislocalized. These studies suggest that Ihh may act to restrict Wnt-responsive cells to the stem-cell compartment. Interestingly, a recent report suggests that Ihh acts to restrict Wnt-responsive epithelium to the proliferative zone of the colon crypt and these signaling pathways act to reciprocally inhibit one another (97).

TGF-β/BMP Signaling

BMPs are members of the TGF-β superfamily of secreted signaling molecules. BMPs have important functions in many biological contexts including those important during embryogenesis. BMPs bind to specific serine/threonine kinase receptors, which transduce the signal to the cell nucleus through Smad proteins (Fig. 6). Although not much is known about BMP signaling within the intestinal stem-cell niche, a role for this pathway is clear as mutations in BMP-4 are associated with the polyp forming disease juvenile polyposis in humans (98). Mice deficient in the BMP molecule, Smad4, also form polyps in their intestinal epithelium but lacked some of the hallmarks of the human disease (99). Most recently, intestinal expression of the BMP inhibitor, Noggin, resulted in de novo, ectopic crypt formation perpendicular to the villus axis in the mouse small intestine (100). Except for their inappropriate location, these crypts appear normal, expressing factors such as Wnt target genes c-Myc and EphB3 that are found in wild-type crypts. Because BMP-4 is expressed in the mesenchymal compartment of the intestine, this suggests an instructive role for BMP signaling in crypt morphogenesis.

The winged helix transcription factor *Fkh6* is a regulatory protein expressed only in the pericryptal mesenchyme. Fkh6[null] mice developed proliferation of cells in the IVR as well as along the villi (19). The villi of the small intestine contained all four cell lineages; however, there was an increase in goblet cells, suggesting a role of Fkh6 in epithelial differentiation. Possible downstream mediators involved may include BMP-2 and BMP-4, as these were both reduced in the mutant mice. Interestingly, the morphological changes seen in the Fkh6^{null} mice were restricted to the stomach and proximal small intestine, even though Fkh6 is expressed along the entire intestinal tract in a wild-type mouse.

Two Requirements for Defining the Stem-Cell Niche

Understanding how stem cells are regulated within their epithelial stem-cell niche is not an easy task. In vitro manipulation of signaling factors in cell culture provides insight into this process; however, these systems lack the critical epithelial–mesenchymal complexity to appreciate interplay between various signaling pathways. In vivo studies in the intestine provide the complexity of biological context, but are often difficult to interpret due to the complex inter-regulatory nature of the signaling network. Given the data presented within this chapter, it is clear that cell signaling pathways function for two basic needs. The first is to physically define the stem-cell niche and the second is to define the proliferation to differentiation gradient within the niche.

Defining the Physical Stem-Cell Niche

Cell signaling pathways such as TGF-β/BMPs may act to physically restrict the Wnt expressing mesenchymal cells to the base of the crypt in the pericryptal mesenchyme, thus participating in defining the physical niche. Wnt signaling may also act to regulate its own expression, as BMP-4 is a Wnt target gene. In addition, Ihh expression is restricted to the differentiated villus epithelium by a (yet to be identified) Wnt-responsive factor within the crypt. Further, Ihh may also modulate the Wnt expression domain indirectly through its regulation of BMP-4. These factors act together to define a border between the proliferative stem-cell niche and the differentiated villus epithelium. They may act to restrict selected mesenchymal cells to the crypt base where they can influence the epithelial cells of the stem-cell niche.

Defining the Gradient of Proliferation to Differentiation Within the Stem-Cell Niche

A gradient of factors set up a microenvironment within the stem-cell niche to promote adhesiveness of the stem cell, proliferation of the TA population, and differentiation of the epithelium. Wnt signaling is a likely candidate for creating morphogen-dependent differential responses within the microenvironment. High levels of Wnt participate in promoting apoptosis of stem cells that are not selected to be anchored within the crypt during development. In the adult crypt, Wnt or another factor may be responsible for maintaining proliferation of the TA population. As the epithelium receives less Wnt signal, differentiation ensues. The differentiation of enterocytes may occur passively, whereas differentiation of the secretory lineage occurs actively through Notch signaling. Wnt signaling within the crypt stimulates Notch signaling factors, and Notch acts to suppress the Wnt signal. Complicating this view in the intestinal crypt microenvironment is the downward migration, accompanied by differentiation, of the Paneth cell lineage. It is possible that Wnt signaling is suppressed to low levels within the adult crypt by the presence of Notch signaling and that some other unidentified factor is responsible for maintaining proliferation of the TA population. Supporting this notion, Wnt reporter mice do not display reporter expression in adult crypts (60).

It is clear that the signaling pathways that define the stem-cell niche are intertwined. The maintenance and propagation of the stem cell critically depend upon the tight regulatory relationships within this signaling network. A systematic approach to manipulating these signaling pathways during development and an inducible approach for manipulation in the adult is greatly needed. Additionally, comparison of epithelial stem-cell niches from different organs will continue to broaden our understanding of how these regions are defined and regulated.

IDENTIFYING MARKERS FOR INTESTINAL STEM CELLS

It is the hope that studies performed to track intestinal epithelial lineages will shed insight into the stem-cell behavior. However, all these studies can only infer the stem-cell behavior. Markers of the stem-cell population are needed to definitively identify and track it behavior. Although some stem-cell populations, such as hematopoietic stem cells, have been thoroughly characterized, the molecular profile of the intestinal epithelial stem cell has yet to be elucidated.

A Candidate "Market" Approach

The Musashi-1 gene encodes an RNA-binding protein that is required for asymmetric divisions in sensory organ precursor cells in the *Drosophila* (101). In mammals, the Musashi-1 homolog is expressed in neural stem cells and is down-regulated in differentiated progeny (102). Three groups present the possibility that Musashi-1 is also a stem-cell marker for intestinal stem cells (103–105). Potten et al. assayed Musashi-1 expression during intestinal morphogenesis and determined that it was ubiquitously expressed throughout the proliferative IVR. In the adult, Musashi-1 expression was confined to the lower cellular strata of the crypt. These findings are consistent with Musashi-1 marking a stem-cell population. During development, multiple stem cells reside in the polyclonal proliferative IVR, whereas stem cells in adult crypts are thought to be restricted to the lower portion of the crypt. However, the broad expression pattern of Musashi-1 in the adult crypts is inconsistent with the notion that Musashi-1 exclusively marks active stem cells. It may mark active stem cells as well as their immediate daughter cells. There is the possibility that the estimation of active stem cells is low and that the broad Musashi-1 expression pattern reflects this. Alternatively, Musashi-1 may be expressed in all cells with stem-cell properties, which would include the Potten tier two cells that have the potential to replace damaged stem cells. These two populations of cells may express different levels of Musashi-1 that are beyond the limit of detection using immunohistochemistry.

Although several other factors have been shown to be restricted to proliferating cells in the intestinal crypts, including Tcf-4 and Cdx-1, their expression patterns have been too broad to constitute an exclusive stem-cell marker. Rather, their expression pattern appears to reflect their cellular proliferative status.

Taking Clues from Other Organ Systems

A number of genes have been identified in the elegant stem-cell molecular profiling study performed in the skin (106). These molecules include Wnt inhibitors: sFrp1, Dkk2, Wif1;

cell-cycle inhibitors: Gas1, Ak1, Inhbb; and TGF-β signaling components: TGFβ-2, Ltbp, Igfbp. Although these potential candidate markers are intriguing candidates for markers in the intestinal stem cell, their expression pattern has not yet been defined in the intestine (discussed in detail subsequently).

Transcriptional Profiling of the Stem Cell

Although the morphologic features of the intestinal stem-cell niche have been well characterized using a histological approach, the precise molecular definition of the active stem cell remains a mystery. Recent studies by Stappenbeck et al. (107) took a global approach toward identifying gene expression patterns in the undifferentiated region of the stem-cell niche. Using laser capture microdissection (LCM) to isolate cells near the crypt base, DNA microarray analysis was performed. A clever scheme was used to compare crypts that were enriched with stem cells. Mice lacking Paneth cells presumably have expansion of stem cells/progenitor cells. The epithelial cells at the base of the crypts in these Paneth-cell-ablated mice were compared with regions from conventionally raised wild-type mice. The two regions were isolated by LCM into two populations: (*i*) cell strata 1–3, which may represent the stem cell in the ablated Paneth cell crypts, and (*ii*) cell strata 4–6 in which the stem cell resides in the wild-type crypts. The 163 transcripts that were identified as up-regulated in the progenitor-enriched population (1–3 cell layer, Paneth cell ablated mice), were functionally categorized into seven broad groups: cell cycle, protein folding, protein processing, chromatin, intracellular signaling, RNA-binding proteins, and protein synthesis. It is not surprising that the stem-cell and TA cell population had an increase in expression of genes that participate in cellular functions required for maintaining proliferation.

The results from other genome anatomy projects including those from neural stem cell and hematopoietic stem cell were compared with the intestinal progenitor cell results (107–109). Interestingly, these tissue-specific cell populations shared ~20% and ~8.5% of identified genes with the neural stem cell and hematopoietic stem cell, respectively. In addition, other such studies have found similar gene expression patterns between keratinocytes and intestinal epithelium at various stages of differentiation (110). Although it is intriguing to believe that all adult stem cells share a complement of common genes, especially in light of the reports that hematopoietic stem cells can circulate and give rise to cells in other organs within an animal (stem-cell plasticity; 111), the exact relationship between adult stem cells from discrete organs has yet to be elucidated. Although this list of genes characterizing small intestinal epithelial progenitors presents an intriguing starting point for identification of markers of the intestinal stem cell, it will be interesting to determine if it indeed reflects the expression profile of active intestinal stem cells. Isolation of stem cells from the intestine has yet to be accomplished.

An elegant approach to isolate the epidermal stem-cell population in the absence of stem-cell markers to allow for their molecular characterization was utilized by Tumbar et al. (106). They developed a transgenic mouse expressing Histone H2B-green fluorescent protein (GFP) controlled by a tetracycline-responsive regulatory element to activate GFP expression by a keratin-specific promoter. Mice expressed GFP in all stem cells and descendants but were then "chased" with doxycycline to inactivate GFP expression. All long-lived stem cells in the bulge remained labeled with GFP. Then, GFP expressing stem cells were isolated and characterized using microarray analysis. Clever schemes to mark and isolate stem cells can be used in other systems such as the intestine.

Together, transcriptional profiling of stem cells from multiple organs such as the intestine, epidermis, blood, and brain provide invaluable resources for defining the behavior of the stem cell within its niche (106–109). Comparison between these populations will yield important similarities and differences that will begin to help shape our understanding of how stem cells behave.

CONCLUSION

There has been a tremendous amount of work accomplished in tracking lineages in the adult intestine. Much to the credit of the scientists that have forged the path to ask questions regarding the relationship between the intestinal stem cell and their lineages, a strong foundation for future studies exploring regulation of adult stem cells has been built. However, several obstacles for manipulation of adult stem cells still remain. First, the ability to identify intestinal stem cells requires identification of stem-cell-exclusive markers. Second, an in vitro system for manipulating intestinal stem cells is required for studying the regulatory factors that designate lineage differentiation. Finally, an in vivo reconstitution assay must be established. Future experimentation within these areas will lead to a greater understanding of how the intestinal stem cell is regulated within its niche, leading to the ability to manipulate adult stem cells for the development of novel therapies for treating diseases.

REFERENCES

1. Simon TC, Gordon JI. Intestinal epithelial cell differentiation: new insights from mice, flies and nematodes. Curr Opin Genet Dev 1995; 5:577–586.
2. Potten CS, Loeffler M. A comprehensive model of the crypts of the small intestine of the mouse provides insight into the mechanisms of cell migration and the proliferation hierarchy. J Theor Biol 1987; 127:381–391.
3. Potten GS, Hendry JH, Moore JV. Estimates of the number of clonogenic cells in crypts of murine small intestine. Virchows Arch B Cell Pathol Incl Mol Pathol 1987; 53:227–234.
4. Potten CS, Owen G, Booth D. Intestinal stem cells protect their genome by selective segregation of template DNA strands. J Cell Sci 2002; 115:2381–2388.
5. Wright NA. Epithelial stem cell repertoire in the gut: clues to the origin of cell lineages, proliferative units and cancer. Int J Exp Pathol 2000; 81:117–143.
6. Garabedian EM, Roberts LJ, McNevin MS, Gordon JI. Examining the role of Paneth cells in the small intestine by lineage ablation in transgenic mice. J Biol Chem 1997; 272:23729–23740.
7. Winton DJ, Blount MA, Ponder BA. A clonal marker induced by mutation in mouse intestinal epithelium. Nature 1988; 333:463–466.
8. Wong MH, Saam JR, Stappenbeck TS, Rexer CH, Gordon JI. Genetic mosaic analysis based on Cre recombinase and navigated laser capture microdissection. Proc Natl Acad Sci USA 2000; 97:12601–12606.
9. Loeffler M, Birke A, Winton D, Potten C. Somatic mutation, monoclonality and stochastic models of stem cell organization in the intestinal crypt. J Theor Biol 1993; 160:471–491.
10. Kim KM, Shibata D. Methylation reveals a niche: stem cell succession in human colon crypts. Oncogene 2002; 21:5441–5449.
11. Potten CS, Owen G, Roberts SA. The temporal and spatial changes in cell proliferation within the irradiated crypts of the murine small intestine. Int J Radiat Biol 1990; 57:185–199.
12. Bach SP, Renehan AG, Potten CS. Stem cells: the intestinal stem cell as a paradigm. Carcinogenesis 2000; 21:469–476.
13. Roberts SA, Hendry JH, Potten CS. Deduction of the clonogen content of intestinal crypts: a direct comparison of two-dose and multiple-dose methodologies. Radiat Res 1995; 141:303–308.

14. Roberts SA, Potten CS. Clonogen content of intestinal crypts: its deduction using a microcolony assay on whole mount preparations and its dependence on radiation dose. Int J Radiat Biol 1994; 65:477–481.

15. Marshman E, Booth C, Polten CS. The intestinal epithelial stem cell. Bioessays 2002; 24:91–98.

16. Merritt AJ, Polten CS, Kemp CJ et al. The role of p53 in spontaneous and radiation-induced apoptosis in the gastrointestinal tract of normal and p53-deficient mice. Cancer Res 1994; 54:614–617.

17. Karlsson L, Lindahl P, Heath JK, Betsholtz C. Abnormal gastrointestinal development in PDGF-A and PDGFR-(alpha) deficient mice implicates a novel mesenchymal structure with putative instructive properties in villus morphogenesis. Development 2000; 127:3457–3466.

18. Pabst O, Zweigerdt R, Arnold HH. Targeted disruption of the homeobox transcription factor Nkx2–3 in mice results in postnatal lethality and abnormal development of small intestine and spleen. Development 1999; 126:2215–2225.

19. Kaestner KH, Silberg DG, Traber PG, Schutz G. The mesenchymal winged helix transcription factor Fkh6 is required for the control of gastrointestinal proliferation and differentiation. Genes Rev 1997; 11:1583–1595.

20. Ramalho-Santos M, Melton DA, McMahon AP. Hedgehog signals regulate multiple aspects of gastrointestinal development. Development 2000; 127:2763–2772.

21. Yang Q, Bermingham NA, Finegold MJ, Zoghbi HY. Requirement of Math1 for secretory cell lineage commitment in the mouse intestine. Science 2001; 294:2155–2158.

22. Pinto D, Gregorieff A, Begthel H, Clevers H. Canonical Wnt signals are essential for homeostasis of the intestinal epithelium. Genes Dev 2003; 17:1709–1713.

23. Brittan M, Wright NA. Gastrointestinal stem cells. J Pathol 2002; 197:492–509.

24. Marsh MN, Trier JS. Morphology and cell proliferation of subepithelial fibroblasts in adult mouse jejunum. II. Radioautographic studies. Gastroenterology 1974; 67:636–645.

25. Behnke O, Moe H. An electron microscope study of mature and differentiating Paneth cells in the rat, especially of their endoplasmic reticulum and lysosomes. J Cell Biol 1964; 22:633–652.

26. Corpron RE. The ultrastructure of the gastric mucosa in normal and hypophysectomized rats. Am J Anat 1966; 118:53–90.

27. Rubin W, Ross LL, Sleisenger MH, Jefries GH. The normal human gastric epithelia: A fine structural study. Lab invest 1968; 19:598–626.

28. Matsuyama M, Suzuki H. Differentiation of immature mucous cells into parietal, argyrophil, and chief cells in stomach grafts. Science 1970; 169:385–387.

29. Chang WW, Leblond CP. Renewal of the epithelium in the descending colon of the mouse. I. Presence of three cell populations: vacuolated-columnar, mucous and argentaffin. Am J Anat 1971; 131:73–99.

30. Cheng H, Leblond CP. Origin, differentiation and renewal of the four main epithelial cell types in the mouse small intestine. V. Unitarian theory of the origin of the four epithelial cell types. Am J Anat 1974; 141:537–561.

31. Cheng H, Leblond CP. Origin, differentiation and renewal of the four main epithelial cell types in the mouse small intestine. I. Columnar cell. Am J Anat 1974; 141:461–479.

32. Cheng H, Leblond GP. Origin, differentiation and renewal of the four main epithelial cell types in the mouse small intestine. III. Entero-endocrine cells. Am J Anat 1974; 141:503–519.

33. Cheng H, Leblond CP. Origin, differentiation and renewal of the four main epithelial cell types in the mouse small intestine. II. Goblet cells. Am J Anat 1974; 141:480–502.

34. Cheng H, Leblond CP. Origin, differentiation and renewal of the four main epithelial cell types in the mouse small intestine. IV. Paneth cell. Am J Anat 1974; 141:520–536.

35. Yatabe Y, Tavare S, Shibata D. Investigating stem cells in human colon by using methylation patterns. Proc Natl Acad Sci USA 2001; 98:10839–10844.

36. Ro S, Rannala B. Methylation patterns and mathematical models reveal dynamics of stem cell turnover in the human colon. Proc Natl Acad Sci USA 2001; 98:10519–10521.

37. Reik W, Dean W, Walter J. Epigenetic reprogramming in mammalian development. Science 2001; 293:1089–1093.

38. Ponder BA, Schmidt GH, Wilkinson MM, Wood MJ, Monk M, Reid A. Derivation of mouse intestinal crypts from single progenitor cells. Nature 1985; 313:689–691.

39. Wintan DJ. Mutation induced clonal markers from polymorphic loci: application to stem cell organisation in the mouse intestine. Semin Dev Biol 1993; 4:293–302.

40. Schmidt GH, Wilkinson MM, Ponder BA. Detection and characterization of spatial pattern in chimaeric tissue. J Embryol Exp Morphol 1985; 88:219– 230.

41. Schmidt GH, Winton DJ, Ponder BA. Development of the pattern of cell renewal in the crypt-villus unit of chimaeric mouse small intestine. Development 1988; 103:785–790.

42. Bjerknes M, Cheng H. Clonal analysis of mouse intestinal epithelial progenitors. Gastroenterology 1999; 116:7–14.

43. Saam JR, Gordon JI. Inducible gene knockouts in the small intestinal and colonic epithelium. J Biol Chem 1999; 274:38071–38082.

44. Margolis J, Spradling A. Identification and behavior of epithelial stem cells in the *Drosophila* ovary. Development 1995; 121:3797–3807.

45. Jacobsen CM, Narita N, Bielinska M, Syder AJ, Gordon JI, Wilson DB. Genetic mosaic analysis reveals that GATA-4 is required for proper differentiation of mouse gastric epithelium. Dev Biol 2002; 241:34–46.

46. Shiojiri N, Mori M. Mosaic analysis of small intestinal development using the spf(ash)-heterozygous female mouse. Histochem Cell Biol 2003; 119:199–210.

47. Taylor RW, Barron MJ, Borthwick GM, et al. Mitochondrial DNA mutations in human colonic crypt stem cells. J Clin Invest 2003; 112:1351–1360.

48. Schon EA. Tales from the crypt. J Clin Invest 2003; 112:1312–1316.

49. Khrapko K, Nekhaeva E, Kraytsberg Y, Kunz W. Clonal expansions of mitochondrial genomes: implications for in vivo mutational spectra. Mutat Res 2003; 522:13–19.

50. Huelsken J, Vogel R, Brinkmann V, Erdmann B, Birchmeier C, Birchmeier W. Requirement for beta-catenin in anterior–posterior axis formation in mice. J Cell Biol 2000; 148:567–578.

51. Haegel H, Larue L, Ohsugi M, Fedorov L, Herrenknecht K, Kemler R. Lack of beta-catenin affects mouse development at gastrulation. Development 1995; 121:3529–3537.

52. Huelsken J, Vogel R, Erdmann B, Cotsarelis G, Birchmeier W. beta-Catenin controls hair follicle morphogenesis and stem cell differentiation in the skin. Cell 2001; 105:533–545.

53. Wong MH, Huelsken J, Birchmeier W, Gordon JI. Selection of multipotent stem cells during morphogenesis of small intestinal crypts of Lieberkühn is perturbed by stimulation of Lef-1/beta-catenin signaling. J Biol Chem 2002; 277:15843–15850.

54. Polakis P. Wnt signaling and cancer. Genes Dev 2000; 1:1837–1851.

55. Wodaz A, Nusse R. Mechanisms of Wnt signaling in development. Annu Rev Cell Dev Biol 1998; 14:59–88.

56. Kinzler KW, Vogelstein B. Lessons from hereditary colorectal cancer. Cell 1996; 87:159–170.

57. Moser AR, Dove WF, Roth KA, Gordon JI. The Min (multiple intestinal neoplasia) mutation: its effect on gut epithelial cell differentiation and interaction with a modifier system. J Cell Biol 1992; 116:1517–1526.

58. Harada N, Tamai Y, Ishikawa T, et al. Intestinal polyposis in mice with a dominant stable mutation of the beta-catenin gene. EMBO J 1999; 18:5931–5942.

59. Korinek V, Barker N, Moerer P, et al. Depletion of epithelial stem-cell compartments in the small intestine of mice lacking Tcf-4. Nat Genet 1998; 19:379–383.

60. Maretto S, Cordenonsi M, Dupont S, et al. Mapping Wnt/beta-catenin signaling during mouse development and in colorectal tumors. Proc Natl Acad Sci USA 2003; 100:3299–3304.

61. DasGupta R, Fuchs E. Multiple roles for activated LEF/TCF transcription complexes during hair follicle development and differentiation. Development 1999; 126:4557–4568.

62. Mariadason JM, Bordonaro M, Aslam F, et al. Down-regulation of beta-catenin TCF signaling is linked to colonic epithelial cell differentiation. Cancer Res 2001; 61:3465–3471.

63. van de Wetering M, Sancho E, Verweij C, et al. The beta-catenin/TCf-4 complex imposes a crypt progenitor phenotype on colorectal cancer cells. Cell 2002; 111:241–250.
64. He TC, Sparks AB, Rago C, et al. Identification of c-MYC as a target of the APC pathway. Science 1998; 281:1509–1512.
65. el-Deiry WS, Tokino T, Waldman T, et al. Topological control of p21WAF1/CIP2 expression in normal and neoplastic tissues. Cancer Res 1995; 55:2910–2919.
66. Kullander K, Klein R. Mechanisms and functions of Eph and ephrin signalling. Nat Rev Mol Cell Biol 2002; 3:475–486.
67. Batlle E, Henderson JT, Beghtel H, et al. beta-Catenin and TCF mediate cell positioning in the intestinal epithelium by controlling the expression of EphB/ephrinB. Cell 2002; 111: 251–263.
68. Fuchs E, Tumbar T, Guasch G. Socializing with the neighbors: stem cells and their niche. Cell 2004; 116:769–778.
69. Potten CS. Epidermal transit times. Br J Dermatol 1975; 93:649–658.
70. Morris RJ, Potten CS. Highly persistent label-retaining cells in the hair follicles of mice and their fate following induction of anagen. J Invest Dermatol 1999; 112:470–475.
71. Cotsarelis G, Sun TT, Lavker RM. Label-retaining cells reside in the bulge area of pilosebaceous unit: implications for follicular stem cells, hair cycle, and skin carcinogenesis. Cell 1990; 61:1329–1337.
72. Lavker RM, Sun TT. Epidermal stem cells: properties, markers, and location. Proc Natl Acad Sci USA 2000; 97:13473–13475.
73. Lavker RM, Sun TT, Oshima H, et al. Hair follicle stem cells. J Invest Dermatol Symp Proc 2003; 8:28–38.
74. Oshima H, Rochat A, Kedzia C, Kobayashi K, Barrandon Y. Morphogenesis and renewal of hair follicles from adult multipotent stem cells. Cell 2001; 104:233–245.
75. Waikel RL, Kawachi Y, Waikel PA, Wang XJ, Roop DR. Deregulated expression of c-Myc depletes epidermal stem cells. Nat Genet 2001; 28:165–168.
76. Arnold I, Watt FM. c-Myc activation in transgenic mouse epidermis results in mobilization of stem cells and differentiation of their progeny. Curr Biol 2001; 11:558–568.
77. Frye M, Gardner G, Li ER, Arnold I, Watt FM. Evidence that Myc activation depletes the epidermal stem cell compartment by modulating adhesive interactions with the local microenvironment. Development 2003; 130:2793–2808.
78. Austin TW, Solar GP, Ziegler FC, Liem L, Matthews W. A role for the Wnt gene family in hematopoiesis: expansion of multilineage progenitor cells. Blood 1997; 89:3624–3635.
79. Reya T, Duncan AW, Ailles L, et al. A role for Wnt signalling in self-renewal of haematopoietic stem cells. Nature 2003; 423:409–414.
80. Van Den Berg DJ, Sharma AK, Bruno E, Hoffman R. Role of members of the Wnt gene family in human hematopoiesis. Blood 1998; 92:3189–3202.
81. Willert K, Brown JD, Danenberg E, et al. Wnt proteins are lipid-modified and can act as stem cell growth factors. Nature 2003; 423:448–452.
82. Reya T. Regulation of hematopoietic stem cell self-renewal. Recent Prog Horm Res 2003; 58:283–295.
83. Gaiano N, Fishell G. The role of notch in promoting glial and neural stem cell fates. Annu Rev Neurosci 2002; 25:471–490.
84. Artavanis-Tsakonas S, Rand MD, Lake RJ. Notch signaling: cell fate control and signal integration in development. Science 1999; 284:770–776.
85. Schroder N, Gossler A. Expression of Notch pathway components in fetal and adult mouse small intestine. Gene Expr Patterns 2002; 2:247–250.
86. Jensen J, Pedersen EE, Galante P, et al. Control of endodermal endocrine development by Hes-1. Nat Genet 2000; 24:36–44.
87. Jenny M, Uhl C, Roche C, et al. Neurogenin3 is differentially required for endocrine cell fate specification in the intestinal and gastric epithelium. EMBO J 2002; 21:6338–6347.
88. Stappenbeck TS, Gordon JI. Rac1 mutations produce aberrant epithelial differentiation in the developing and adult mouse small intestine. Development 2000; 127:2629–2642.

89. Stewart BA. Membrane trafficking in *Drosophila* wing and eye development. Semin Cell Dev Biol 2002; 13:91–97.
90. Tsukui T, Capdevila J, Tamura K, et al. Multiple left–right asymmetry defects in Shh($-/-$) mutant mice unveil a convergence of the shh and retinoic acid pathways in the control of Lefty-1. Proc Natl Acad Sci USA 1999; 96:11376–11381.
91. Riddle RD, Johnson RL, Laufer E, Tabin C. Sonic hedgehog mediates the polarizing activity of the ZPA. Cell 1993; 75:1401–1416.
92. Zhu G, Mehler MF, Zhao J, Yu Yung S, Kessler JA. Sonic hedgehog and BMP2 exert opposing actions on proliferation and differentiation of embryonic neural progenitor cells. Dev Biol 1999; 215:118–129.
93. Stone DM, Hynes M, Armanini M, et al. The tumour-suppressor gene patched encodes a candidate receptor for Sonic hedgehog. Nature 1996; 384:129–134.
94. Villavicencio EH, Walterhouse DO, Iannaccone PM. The sonic hedgehog-patched-gli pathway in human development and disease. Am J Hum Genet 2000; 67:1047–1054.
95. Murone M, Rosenthal A, de Sauvage FJ. Sonic hedgehog signaling by the patched–smoothened receptor complex. Curr Biol 1999; 9:76–84.
96. Wang LC, Nassir F, Liu ZY, et al. Disruption of hedgehog signaling reveals a novel role in intestinal morphogenesis and intestinal-specific lipid metabolism in mice. Gastroenterology 2002; 122:469–482.
97. van den Brink GR, Bleuming SA, Hardwick JC, et al. Indian hedgehog is an antagonist of Wnt signaling in colonic epithelial cell differentiation. Nat Genet 2004; 36:277–282.
98. Howe JR, Bair JL, Sayed MG, et al. Germline mutations of the gene encoding bone morphogenetic protein receptor 1A in juvenile polyposis. Nat Genet 2001; 28:184–187.
99. Takaku K, Miyoshi H, Matsunaga A, Oshima M, Sasaki N, Taketo MM. Gastric and duodenal polyps in Smad4 (Dpc4) knockout mice. Cancer Res 1999; 59:6113–6117.
100. Haramis AP, Begthel H, van den Born M, et al. De novo crypt formation and juvenile polyposis on BMP inhibition in mouse intestine. Science 2004; 303:1684–1686.
101. Nakamura M, Okano H, Blendy JA, Montell C. Musashi, a neural RNA-binding protein required for *Drosophila* adult external sensory organ development. Neuron 1994; 13:67–81.
102. Kaneko Y, Sakakibara S, Imai T, et al. Musashi 1: an evolutionarily conserved marker for CNS progenitor cells including neural stem cells. Dev Neurosci 2000; 22:139–153.
103. Kayahara T, Sawada M, Takaishi S, et al. Candidate markers for stem and early progenitor cells, Musashi-1 and Hes 1, are expressed in crypt base columnar cells of mouse small intestine. FEBS Lett 2003; 535:131–135.
104. Nishimura S, Wakabayashi N, Toyoda K, Kashima K, Mitsufuji S. Expression of Musashi-1 in human normal colon crypt cells: a possible stem cell marker of human colon epithelium. Dig Dis Sci 2003; 48:1523–1529.
105. Potten CS, Booth C, Tudor GL, et al. Identification of a putative intestinal stem cell and early lineage marker; musashi-1. Differentiation 2003; 71:28–41.
106. Tumbar T, Guasch G, Greco V, et al. Defining the epithelial stem cell niche in skin. Science 2004; 303:359–363.
107. Stappenbeck TS, Mills JG, Gordon JI. Molecular features of adult mouse small intestinal epithelial progenitors. Proc Natl Acad Sci USA 2003; 100:1004–1009.
108. Ramalho-Santos M, Yoon S, Matsuzaki Y, Mulligan RC, Melton DA. "Stemness": transcriptional profiling of embryonic and adult stem cells. Science 2002; 298:597–600.
109. Ivanova NB, Dimos JT, Schaniel C, Hackney JA, Moore KA, Lemischka IR. A stem cell molecular signature. Science 2002; 298:601–604.
110. Dabelsteen S, Telsen JT, Olsen J. Identification of keratinocyte proteins that mark subsets of cells in the epidermal stratum basale: comparisons with the intestinal epithelium. Oncol Res 2003; 13:393–398.
111. Krause DS, Theise ND, Collector MI, et al. Multi-organ, multi-lineage engraftment by a single bone marrow-derived stem cell. Cell 2001; 105:369–377.

7

Stem Cell Populations in Skin

Richard P. Redvers and Pritinder Kaur
Epithelial Stem Cell Biology Laboratory, Peter MacCallum Cancer Centre, Melbourne, Victoria, Australia

INTRODUCTION

It has been evident for some time that cell replacement in the epidermis of the skin is a highly ordered process with a central role for keratinocyte stem and progenitor cells. In recent years, many investigators have sought to distinguish keratinocyte stem cells (KSCs) from their immediate progeny using molecular markers, both in situ and ex vivo, and a number of molecular regulators that can perturb ordered cell renewal in skin epithelium have also been identified. Although we are far from having a clear understanding of the precise mechanisms that regulate ordered epidermal tissue morphogenesis and cell renewal, significant progress has been made that has begun to shed light on these processes. Unequivocal identification and isolation of viable keratinocyte stem and progenitors are now possible; this combined with the advent of molecular technologies, such as high-throughput genome-wide scanning and the ability to generate mice with designer skin, and the development of assays for these cells, albeit at an early stage, places us at an exciting time of experimental investigation and discovery, poised to capitalize on the collective efforts expended by many laboratories across the world.

Emergence of Stem Cell Concepts in Skin Biology

The skin provides a protective barrier and sensory interface that represents the largest organ system in the body (1), functioning in thermoregulation, electrolyte, and fluid balance; immune, nervous, and endocrine systems; psycho-social communication; and the synthesis, processing, and metabolism of an assortment of structural proteins, glycans, lipids, and signaling molecules (2). The epidermis forms the outermost layer, consisting of a pluristratified keratinizing epithelium, resting upon a basement membrane apposed to the underlying dermis (3). In glabrous or interfollicular epidermis, cells of the lowest stratum proliferate laterally and progressively differentiate as they migrate suprabasally, terminating in flat, tightly packed, cornified enucleated squames, enmeshed within a lipid matrix to create an impermeable barrier (4–6). The entire process, from the birth of a basal cell to surface corneocyte and desquamation, lasts 8 to 14 days in mice (7–10) and

14 to 75 days in humans (11–14), requiring continuous cell proliferation in the basal layer to maintain the tissue (15,16). The situation is more complex in regions of skin where appendages undergo alternative differentiation programs in the form of a pilosebaceous or sweat gland apparatus (17).

The first indication of the existence of epidermal stem cells can be traced back to 1949 when Berenblum and Shubik (18) observed that a delay between initiation and promotion had no effect on tumor yields, suggesting the presence of long-lived cells. Over the next 20 years, biologists described the regenerative and pluripotent properties of epidermal cells (19–21) without invoking the stem cell paradigm. The earliest description of stem cell activity as an assayable quantity came from in vivo studies of epidermal tissue regeneration following radiation damage, that is, epidermal micro-colony formation derived from single cells (22), providing the first functional means to identify "stemness" in the epidermis. The basis of this assay lies in the experimental approaches adopted to study the hemopoietic system, one of the best characterized adult stem cell systems to date (23). Indeed, definitions of hemopoietic stem cells (HSCs) have provided a conceptual framework to begin to define epidermal stem cells. A literal adoption of these definitions of HSCs to all other stem cell populations is perhaps inappropriate and does not allow for variations based upon the structural organization and turnover rates of particular tissues. Perhaps the most relevant functional definition applicable to all stem cells is that provided by Lajtha in 1979 (24), that is, the ability to regenerate the tissue of origin for the lifespan of an organism, which implies long-term self-renewal of both stem cells and tissue. Although one might reasonably expect all stem cells to be relatively quiescent, unspecialized blast-like cells with the capacity to renew their tissue indefinitely, stipulations such as infinitesimally low incidence, confinement to a definable niche, ability to give rise to many lineages, etc. (25–30), are not always applicable to all tissues.

Definitions of stem cells are further complicated by the behavior of cells in homeostatic (normal) versus damaged tissue; for instance, early lineage bone marrow cells appear to retain the flexibility to function as stem cells in exceptional circumstances (e.g., severe trauma) (31) although they represent a "short-term subset that self-renews for a defined interval" (26) as it gradually differentiates while losing its stemness under steady-state conditions (14). Thus, a very real caveat in characterizing/defining stem cell populations, recognized by Potten and Loeffler (31) early on, is that perturbing the tissue in any way is likely to alter cell behavior, and the conclusions drawn have to take this into account. Nowhere is this a greater issue than when studying epithelial cells after removing them from the tissue—after all, epithelial renewal occurs in vivo in a physically constrained environment with strong adhesion to neighboring epithelial cells and to their extracellular matrix, and intimate association with the dermal environment. With the exception of in situ analyses of stem cell behavior performed largely in murine epidermis, all experiments place KSCs and their progeny into unnatural circumstances thereby activating these cells. Although this is a rather self-evident concept, it has nevertheless been under-appreciated by skin stem cell biologists until recently (32,33). Thus, it is important to remember that much of the current literature is interpreted with the assumption that stem cells are solely responsible for cell replacement during homeostasis and in injury. As very little work has been done with *prospectively* defined populations of stem cells, we have no knowledge at present about which class of keratinocytes actually heal wounds or contribute to cancer. Given that the epidermis has an overriding function to cover wounds as rapidly as possible, it is critical to assess current data in light of how experimental design is likely to influence the behavior of keratinocyte stem or progenitor cells under specific experimental regimes.

Proliferative Hierarchical Organization of the Epidermis

Prior to the emergence of the epidermal stem cell field, some investigators asserted that all cells of the basal epithelial layer had uniform proliferative potential (34). Mitotically active cells were restricted to the *stratum basale* (35–38), dividing randomly and migrating suprabasally due to "population pressure" (39–41). Iversen et al. (42) hinted at an age structure and hierarchical organization, suggesting that some basal cells were postmitotic differentiating cells and that migration was restricted to the oldest G_1 cell in the vicinity of a mitosis. Heterogeneity and hierarchy in the basal layer were recognized due to local morphologic variations (43), the presence of various cell types (44), early differentiating cells (15), and the first suggestions of a rare subpopulation of clonogenic stem cells (45,46). Ordered structure was first elucidated by Mackenzie in 1969 (4). Christophers (47) and Menton and Eisen (48) demonstrated vertical columnar stacking in the *stratum corneum* of nonvolar skin, and subsequently Potten (46) visualized hexagonal units within the surface view of epidermal sheets. A theoretical structure–function relationship emerged, wherein the basal cells directly underlying a squame column divided and migrated suprabasally at the periphery (49,50), giving rise to all cells of their column (47,51), termed the epidermal proliferative unit (EPU) (46); a central stem cell among a subunit of 10 basal cells was ultimately responsible for maintenance of the EPU (16,50). In support of autonomous units, Kam et al. (52) demonstrated that fluorescent dyes spread in columnar EPU-like patterns when injected into excised neonatal murine skin, suggesting intimate connectivity and communication within discrete units and physical compartmentalization of the tissue. In addition to autonomous EPU controls, a coordinated inter-EPU behavior was proposed to account for the complex morphological network underlying homeostasis (53).

Thus, cell replacement in the epidermis involves a slow-cycling subpopulation of stem cells generating "a hierarchical series of progressively 'aging' cell cycles" (30). Mathematical modeling based on kinetic data predicted heterogeneity with respect to cycle time, comprising slow-cycling stem cells, up to three "transit proliferative" populations and postmitotic cells (54). Morphologic and kinetic data were correlated to demonstrate the existence of these populations in monkey palm epidermis, with the "transient amplifying" cohort responsible for populating the bulk of the tissue (55,56).

Estimating Epidermal Stem Cell Frequency

Estimates of epidermal stem cell frequency vary widely (0.01% to 40% of basal cells) depending upon species, anatomical sites, and, particularly, the methodologies employed (Table 1). For example, estimates of 1% to 8% come from radiation studies that may impair or destroy some stem cells, 1% to 2% from DNA label retention studies with somewhat arbitrary chase periods, 6% from a follicle-specific repository that excludes other reservoirs, and 0.01% from mathematical modeling of a competitive assay that may not account for inherent technical limitations—all with the potential to underestimate the frequency of KSCs in steady-state conditions. One explanation put forth for the disparity and range of frequencies asserts that there is no clear delineation between stem and nonstem entities, but rather a "diminishing stemness spiral" reflecting a spectrum of capabilities, with an inverse relationship between stemness and differentiation/maturity (31). Other discrepancies in the reported incidence of "stem cells" can be attributed to the definitions employed to ascribe "stemness" such as rapid adhesion to particular substrates or short-term tissue reconstitution that may or may not reflect stem cell properties. A major flaw in any in vitro assay used to estimate stem cell frequencies is the fact that a very small

Table 1   Estimates of Epidermal Stem Cell Frequency in the Basal Layer

% Incidence	Methodology	References
0.01	Competitive repopulation of GFP-marked cells + mathematical modeling	(169)
<1	Radiation response	(16,46,90)
1	^{3}H-Tdr LRC; unit gravity sedimentation, ^{3}H-Tdr LRC, cell cycle, size, RNA, N:C ratio	(124,149)
1–2	^{3}H-Tdr LRCs, GFP-LRCs	(163,208,209)
2–7	Radiation response, mathematical modeling, ^{3}H-Tdr LRCs	(16,30,210)
2–8	Re-analysis of previously published radiation responses	(45)
3	Radiation response	(211)
4–8$^+$	Lit review	(14)
6	K15-EGFP expression at base of telogen follicle	(65)
8	α_6^{bri}/CD71^{dim} phenotype + ^{3}H-Tdr LRCs, cell cycle, size, N:C ratio	(150)
9–10	Ultrastructure; resistance to pulse labeling; EPU model	(46,50,212)
10	α_6^{bri}/CD71^{dim} phenotype + cell cycle, keratins, total proliferative output	(155)
10–12	In vitro retroviral labeling + in vivo reconstitution	(125)
10–30	β_1^{bri} phenotype + rapid adhesion to ColIV (20 min) or FN (5 min) or KC-ECM (10 min) *not* LN (30 min) + CFU ($\geq$32 cells)	(154)
9.5–40	β_1^{bri} (α_2 or α_3)/K19$^+$ phenotype + 5 min adhesion to ColIV + CFU ($\geq$32 cells)	(130)

Abbreviations: LRC, label-retaining cell; GFP, green fluorescent protein; EPU, epidermal proliferative unit.

proportion of primary cells plated in culture are recruited to proliferate (routinely <1% in most laboratories). The identification of factors that can promote the attachment *and* subsequent proliferation of *all* keratinocytes in vitro would greatly facilitate quantification and characterization of epidermal stem cells. An important question that remains unanswered to date is whether the tissue culture media used to propagate keratinocytes are capable of recruiting both stem cells and transit amplifying cells into cycle, given their natural selection for cells that are actively growing. It is plausible that only those cells in specific phases of the cell cycle are selected in vitro and that perhaps the most deeply quiescent stem cells never proliferate. As cultured keratinocytes can reconstitute grafts on severely burned patients for decades following transplantation, long-term tissue-reconstituting keratinocytes are clearly not lost. Whether this reconstitution is being obtained from cells lower in the proliferative hierarchy that retain stem cell properties or from "actual" stem cells remains to be determined.

Do Stem Cells Segregate Their Template DNA Strand—Supporting Evidence from Intestinal Epithelium?

Perhaps the only feature that discriminates an ancestral stem cell from an early "potential" stem cell daughter is the retention of its template DNA. In 1975, Cairns (57) hypothesized that selective segregation of template DNA in a rare subset of immortal stem cells was an evolutionary strategy to minimize mutation and tumorigenesis. The latest refinement of the hierarchical cell replacement scheme suggests that ancestral stem cells residing in their niche give rise to "potential" stem cell progenies that retain the flexibility to re-occupy the ancestral niche and assume the requisite responsibilities if necessary (14). This is compatible with the original proposal that the immortal stem cells were a subset within a population of cells that could all qualify as stem cells by virtue of their ability to re-epithelialize radiation-damaged epidermis (57). It is reasonable to suspect that the ancestor stem cell and the immortal stem cell are one and the same.

Experimental proof of the concept of template DNA strand segregation is technically difficult to obtain given that the ancestral stem cell must be labeled within a narrow timeframe, at the precise division where tissue stem cells are laid down. Elegant experimental evidence in support of Cairns hypothesis has only recently been provided by Potten et al. (58), utilizing a double-labeling technique in the small intestinal epithelium. Stem cells were labeled with tritiated thymidine during stem cell expansion—and template DNA synthesis—in neonates, followed by BrdU after the expansionary phase. Double-labeled label-retaining cells (LRCs) continued to retain tritiated thymidine despite subsequent depletion of BrdU label, providing irrefutable evidence of a DNA label that persists in cells undergoing multiple rounds of division. Whether similar DNA segregation occurs in other epithelial stem cell populations remains to be determined.

Stem Cell Lineages and Locations

Although the structure and cell lineage diversity within the *stratum basale* of the epidermis has been well documented, the precise nature and location of the various stem or "stem-like" precursor cells in this perpetually renewing tissue is the subject of intense investigation and vigorous debate. Differences may be attributed to comparisons between various species or anatomical sites, the varied experimental approaches and manipulations, steady-state versus perturbed epidermis, the developmental stage of the host, and the complexity of hair follicle lineage composition and cyclic remodeling. It is generally agreed that the hair follicle bulge is a repository for KSCs in murine adnexal epidermis (59–62) that is capable of contributing to the regeneration of follicles, sebaceous glands, and interfollicular epidermis (63–65). These findings have convinced many that bulge stem cells represent the ultimate stem cells of this tissue (61,66,67). However, some seemingly incongruous observations (65,68–75) necessitate more complex hypotheses to corral disparate opinions on the locations of epidermal stem cell reservoirs and their hierarchical relationship—if any—to bulge stem cells, as discussed below.

Hair Follicle KSCs

It has long been suspected that hair follicles harbor cells capable of regenerating new follicles after damage, and such cells were believed to originate from the upper permanent portion (19) of the outer root sheath (21). Many ensuing studies demonstrated continued hair growth after removal of a significant portion of the lower follicle (21,76–83). Remarkably, early indications also suggested that follicle cells could contribute to re-epithelialization of damaged epidermis (20,84–89). In a radiation–response assay that permits an approximation of clonogenic cell frequency in the epidermis, apparent migration of surviving clonogenic cells from follicles into the interfollicular epidermis further complicated calculations (90,91), identifying the hair follicle as a potential source of cells capable of repopulating the epidermis. Subsequently, dermabrasion studies corroborated the existence of stem cells in the upper hair follicle (92), and dissected follicles were able to regenerate fully differentiated interfollicular epidermis in an in vitro organotypic model (93). The hair follicles are believed to harbor the majority of the clonogenic cells, estimated at 3000 to 6000 mm^{-2} in human scalp versus 1000 to 2000 mm^{-2} in glabrous epidermis (94). It is, therefore, not surprising that epidermal regeneration is proportional to the number of residual hair follicles that remain (95).

Bulge KSCs

One of the most universally accepted stem cell attributes that can be readily demonstrated is long-term retention of a DNA label (96). Hence, the localization of a cluster of LRCs to the bulge region provides compelling evidence that this well-protected structure at the lower end of the permanent portion of murine hair follicles is a stem cell repository (59,62,63,97). Stem cells permanently affixed to this "well-nourished" region would be ideally placed to participate in hair follicle cycling and regeneration, while surviving degeneration of the lower portion during catagen remodeling—a scenario that inspired the "bulge-activation hypothesis" (59,98). This model stipulates that dermal papilla cells are brought into close proximity to the bulge during late catagen, whereupon instructive signals stimulate the normally quiescent bulge cells to transiently proliferate at the onset of anagen (99), giving rise to transit amplifying matrix cells that generate new hair growth (59,98,100). The extremely long telogen (35 to 70 days) during the second cycle suggests that mere apposition of the dermal papilla and bulge is insufficient to initiate anagen by the third cycle; an as-yet-unknown factor may be an additional requirement (99). The susceptibility of skin to carcinogen initiation during early anagen suggests that stem cells in the bulge that proliferate at that time are selectively targeted (101). However, reports that selective killing of highly proliferative cells during early anagen I had no impact on tumor yield would appear to contradict this, although quiescent long-lived stem cells would be implicated in tumorigenesis (102). Pathogenesis of a genetic form of alopecia involves disconnection between the matrix and underlying dermal sheath, which leaves the dermal papilla deep in the dermis, thereby arresting any potential communication with the bulge (103,104). Consequently, no further hair growth ensues as would be predicted by the bulge-activation hypothesis (96,99,105).

In an effort to localize the functionally superior cells within follicles, Kobayashi et al. (60) microdissected and subdivided rat vibrissa follicles to demonstrate that the bulge was indeed the region most highly enriched for colony-forming cells, although such cells were not exclusively bulge-derived. This was corroborated by a similar study utilizing human hair that also found enrichment for colony-forming cells in the presumptive bulge region and demonstrated that bulge keratinocytes had superior in vitro clonogenicity to unfractionated interfollicular keratinocytes (61). However, subsequent studies in human follicles variously reported the major clonogenic cell enrichment to be in the sub-bulge region (94,106) or upper central outer root sheath (107). Once again, all showed that the principal repository of quiescent stem cells was not an exclusive locale for colony-forming cells. The apparent lack of consensus on a discrete stem cell repository in human follicles is not surprising, as the presumptive bulge region falls within a morphologically indistinct area that is virtually indistinguishable from its surrounds in adult human hair follicles (108,109). Further complications arise from the assumption that colony-forming ability is a surrogate assay for stem cells only—presumably the immediate progeny of stem cells that are in fact the largest actively proliferating pool of epidermal cells in situ are also capable of forming colonies in vitro.

Is the Hair Follicle Bulge Stem Cell Population the Source of All Epidermal Tissue Renewal?

That bulge stem cells were able to contribute to multiple tissues, including the matrix, sebaceous gland, and interfollicular epidermis was suggested by histologic data (103,110) in response to injury (20,84,88–92). Indeed, Lavker et al. (96) argued for a hierarchical organization in the follicles with bulge cells giving rise to "germ" and matrix cells in the proximal direction and to isthmus, sebaceous, infundibulum, and interfollicular cells in the distal direction. This proposition remained theoretical until Taylor et al. (63) utilized DNA double-labeling to demonstrate migration of bulge-derived cells into the lower and upper follicles and showed emigration of upper follicle cells into the epidermis. Subsequent studies utilizing lineage-marked cells in tissue recombination and regeneration assays reached similar conclusions, demonstrating full follicular contribution, and adding sebaceous gland development to the bulge cell repertoire (64,65).

Not surprisingly, this impressive body of work has led many to suggest that bulge stem cells represent the "ultimate" stem cells of the epidermis (61,66,67). However, it is important to note that the contributions to interfollicular epidermis have been observed only in tissue expansionary (neonatal) or regenerative phases (following wounding), after complex manipulations or from admixtures of many cells—not from foci of single cells under steady-state conditions. Given the location of bulge cells within deep recesses, it is highly unlikely that these participate in routine maintenance of the interfollicular epidermis (94). This view was vindicated by exquisite long-term lineage marking studies by Ghazizadeh and Taichman (71) that showed lineage restriction, with follicular cells contributing a mere "rim of epidermis" (Fig. 1A), venturing no further than the margin of the follicle in the absence of wounding. Importantly, self-sustaining units of epidermal cells not associated with hair follicles were consistently observed (Fig. 1B), providing elegant proof of Potten's EPU model. In addition, histological and immunohistochemical examination of human follicles suggested that differentiation proceeds horizontally inward from the outer root sheath (95,111,112). Intuitively, it would seem to be a more favorable evolutionary strategy to have as many equipotent stem cell reservoirs as possible in disparate locations, to call upon if necessary. As shown by the study of Ghazizadeh and Taichman (71), distinct stem cell populations giving rise to clonal growth reside in the hair follicle, sebaceous gland, and interfollicular epidermis. Hence, it is more likely that the bulge does not participate in routine epidermal maintenance, rather it serves as a backup reservoir capable of impressive multilineage contribution in extraordinary circumstances, even in very hairy skin. Interestingly, Miller et al. (70) have shown that sweat gland cells can also contribute to wound healing in a porcine model. By excising a circular wound down to the muscle fascia and leaving a denuded central region harboring only sweat glands, they were able to remove lateral keratinocyte migration from the equation to demonstrate, for the first time, re-epithelialization from the sweat apparatus and re-establishment of rete ridges (70). Interestingly, although the sweat gland keratinocytes exhibited extensive proliferation and tissue regeneration, they were unable to fully recapitulate the appendages or keratinization of unwounded skin.

Bulb/Matrix/Germinative Epidermal Stem Cells

The cells within the matrix of follicles exhibit considerable proliferative and differentiative potential. Proliferation during anagen is so rapid that the growth fraction of matrix cells approaches 1.0 (113), making them among the fastest dividing cells in any adult tissue (114). Hair matrix cells are able to divide continuously for around 1000 days in humans, giving rise to several distinct hair follicle lineages (95). Their close proximity to the base of the follicle suggests that matrix cells may communicate with the dermal papilla (77), on a more regular basis than bulge cells. Hence, it was initially believed that the matrix of the bulb region was the source of follicle renewal and regeneration (48,113,115,116). In the ensuing years, proponents of bulb stem cells were faced with mounting conundrums and apparent contradictions, particularly when long-term label retention was localized to the bulge region in a number of studies (59,62,63,97).

(A)

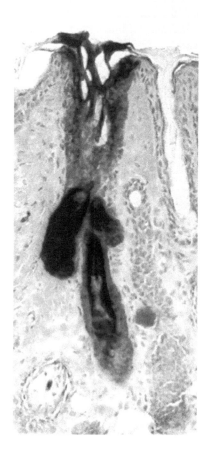

(B)

Figure 1   Evidence for distinct self-renewing stem cell populations in the hair follicle bulge and interfollicular epidermis. (**A**) Lineage analysis of hair follicles marked with β-galactosidase showing lack of contribution to interfollicular epidermis (*non black*) from hair-follicle-derived stem cells. (**B**) Lineage analysis revealing the presence of self-maintaining interfollicular EPUs. *Abbreviation*: EPU, epidermal proliferative unit. *Source*: From Ref. 71. (*See color insert.*)

Once labeled, stem cells that are quiescent or selectively segregate labeled DNA strands retain that label for long periods, whereas non–stem cells do not selectively segregate DNA and proliferate vigorously, leading to depletion of label. However, the length of the chase period is important in that examination with short chase periods will reveal many more labeled cells than with long chase periods. Indeed, chase periods of 10 weeks reveal LRCs in the inter- and intrafollicular epidermis, perisebaceous region, external root sheath, and the bulge (97), whereas a 14-month chase preserved only those highly persistent LRCs in the bulge (62). Although many have gone to great lengths to exclude the matrix and bulb region from label retention, others have demonstrated their presence in the medium term (111,117) and recent reports demonstrate label retention in the hair germ after eight (73) and 10 weeks (75). Although DNA label retention or quiescence is a defining characteristic of stem cells, it is important to remember that this is relative to other more rapidly cycling cells within the same hierarchy. Thus, it is erroneous to compare the persistence of LRCs across different stem cell populations over the same time interval after initial labeling as an indicator of stemness given that their rates of tissue replacement are not identical. In other words, the demands for cell proliferation (and therefore loss of labeled cells) on specific stem cell populations are not identical. Other factors that influence DNA label retention include the efficiency of labeling (i.e., where all stem cells labeled), and whether the template strand was labeled, at the beginning of the experiment. We suggest that sustained tissue renewal has to take precedence over label retention as a stem cell characteristic. For instance, despite the disappearance of LRCs due to hyperproliferation in transgenic mice expressing ΔNLef1, the interfollicular epidermis remained viable for over two years, thereby demonstrating robust stem cell maintenance in the absence of quiescence (118). In addition, angora rabbits, poodle dogs, and merino sheep are believed to grow follicles continuously without pause (119). Clearly, the existence of stem cells in the bulb region cannot be excluded merely because of their proliferative status or relative lack of label retention at this site.

Many studies have shown that follicular regeneration can still ensue after removal of the lower portion, provided a dermal papilla is in close proximity (21,76–83,120). Although this apparent dispensability of the bulb region has been cited repeatedly as powerful evidence of an upper follicular stem cell reservoir, it does not exclude the existence of a bulb reservoir that shoulders significant responsibility in normal circumstances. Interestingly, there has been occasion to doubt the contribution of bulge cells to follicular growth at the onset of anagen, as they appeared not to divide amidst a flurry of proliferative activity leading up to hair growth (62). In contrast to experimental removal of the lower follicle, removal or tearing out of hair fibers may be the most common injury to follicles (121). In the latter more physiologically relevant injury, germinative epidermal cells are retained after plucking (68). Interestingly, Ito et al. (73) have demonstrated that label-retaining bulge cells undergo apoptosis upon plucking and that consequent hair follicle regeneration occurs from residual label-retaining hair germ cells that are protected from this type of injury.

Reynolds and Jahoda (68) utilized microdissection after plucking to liberate a population of morphologically distinct and highly fastidious germinative epidermal cells. These cells displayed characteristics of stem cells, having small size, few organelles, abundant free ribosomes, and firm attachment to a well-vascularized niche that protects them from injury (68). Importantly, it was found that the immense proliferative capacity of germinative epidermal cells was unleashed only in the presence of dermal papilla cells (68). It is noteworthy that a number of studies localized enrichment of colony-forming keratinocytes to the upper regions, yet still found some colony-forming cells albeit reduced in number in the bulb region (60,94,106,107). Given that in some studies cell growth was assessed on irradiated human fibroblasts (106) or from tissue explants (107), it is reasonable to speculate that the latent proliferative potential in the lower follicle was grossly underestimated.

Arguments for bulb/matrix/germinal stem cells versus bulge stem cells need not be pitted against each other if a model incorporating both hair follicle stem cell reservoirs is accepted. The hair follicle predetermination model asserts that two stem cell populations are present, each with a distinct fate. Although they remain separate, the populations interact to coordinate the follicular growth and differentiation program, with anagen activation originating in the hair germ leading to activation of the bulge (72). The "split-fuse hypothesis" reconciles some of the differences and apparent contradictions in the two camps by proposing that the two follicle stem cell populations coalesce during the catagen–telogen transition and individualize again during anagen (122). This co-mingling of bulge and bulb/germinal populations would make them indistinguishable during telogen and arguments on their origins and position in the hierarchy immaterial. Interestingly, although Cotsarelis and co-workers (59,96,98) have gone to great lengths to distinguish label-retaining bulge cells from transit amplifying matrix cells, they have subsequently included the secondary hair germ within their "operational definition of the bulge" (123) and have localized lineage-marked stem cells to the bulge during telogen, precisely when bulge and bulb regions are most intimately fused (65).

KSCs of Interfollicular Epidermis

The proposal that interfollicular and glabrous skin harbor stem cells at the center of EPUs (46,91) has been vindicated by numerous studies (71,124–126). Indeed, the mere existence of appendage-free regions of self-renewing skin offers irrefutable testimony to that assertion. However, although the ability of these presumptive stem cells to exhibit foci of clonal regeneration is undisputed, their autonomy or position in the hierarchy within hairy epidermis has been contentious, due in no small part to examples of interfollicular regeneration emanating from the bulge as described above.

From the outset, stem cells were believed to reside within the center of EPUs (16,46,53), as kinetic data demonstrated that mitotic cells were invariably found at the periphery (49,124), whereas 2% of basal cells retained label after 28 days, and 90% of these were within one nuclear diameter of the central cell (124). LRCs have been detected in the interfollicular epidermis up to 20 weeks post-labeling (75). Numerous studies utilizing in vitro retroviral lineage marking have demonstrated foci of clonal growth giving rise to columnar units in vivo that persist from 12 to 40 weeks (125,127). Importantly, in situ lineage marking removed any possible in vitro artifacts to show that columnar EPU-like foci of clonal growth persisted after 37 epidermal turnovers and five hair growth cycles after depilation (71). Similarly, EPU-like columns emanating from a clonogenic cell were also evident in the footpad of a transgenic mouse (126,128). Taken together, these data confirm the existence of long-lived interfollicular stem cell residents with considerable proliferative potential for routine maintenance of the epidermis.

The EPU-like organization is most evident in mouse interfollicular epidermis but is also apparent in the *stratum corneum* of humans in thin epidermal regions such as abdomen, forearm, thigh, and buttocks (14). However, alternative models have been invoked to account for the dissimilar organization of volar (palm, sole) epidermis. Lavker and Sun (55,56) addressed this conundrum by correlating morphological and structural observations in primate volar skin with a functional model wherein "nonserrated" stem cells residing in the deep pockets of rete ridges give rise to progeny that migrate upward and laterally to the tips of dermal papillae. Dsg3 has been identified as a negative stem cell marker of the deep rete ridge region (129). A completely opposing model has also been proposed, suggesting that stem cells occur as clusters located at the tips of dermal papillae (130–136). The latter model has been controversial and difficult to reconcile, with the preferred location of stem cells in protected sites in the deeper rete ridges and mounting data to support the presence of single stem cells in the interfollicular epidermis (75).

The persistence and multipotency of bulge stem cells has been cited as proof of their supremacy and ancestral place in the cellular hierarchy (67). However, some of the longest-lived (stem cell) targets of carcinogens reside within the interfollicular epidermis (137), and many studies have demonstrated that interfollicular keratinocytes are capable of generating pilosebaceous and sweat gland structures (65,69,74,75,138,139). Therefore, it may be a reasonable supposition that equipotent ancestor stem cells are seeded throughout the nascent epidermis during ontogenesis and that the divergence between stem cells in terms of folliculogenesis at discrete locations is contextual, depending upon connective tissue and microenvironmental influences. In this context, the role of the wnt signaling pathway in lineage specification is highly relevant. Overexpression of active β-catenin, an activator of the wnt signaling pathway, can cause ectopic hair formation specifying interfollicular epidermal cells down the folliculogenesis pathway (140). The converse is also true, that is, blocking the wnt signaling pathway by targeted deletion of β-catenin (141) or Dkk-1 (142) led to inhibition of hair follicle formation. Lef1, a co-activator of the wnt pathway, is also important for hair follicle development as demonstrated by the loss of these appendages in Lef1 knockout mice (143). Thus, it is clear that multipotency is neither an intrinsic nor an immutable property of hair follicle stem cells, but can be conferred on interfollicular epidermal cells by tinkering with the molecular regulation controlling the fate specification of epidermal progenitors. Whether this is an exclusive property of stem cells or any basal keratinocyte remains to be determined because the promoters utilized to date, target the entire basal layer rather than stem cells.

Enrichment and Isolation of KSCs

Over the years, a number of approaches have been used to identify and isolate viable epidermal stem cells for biological characterization. The validity of all experimental approaches and purported stem cell markers is directly linked to the kind of assays used to define the isolated cells as stem cells. The behavior of epidermal stem cells in situ is understood well enough that correlation of isolated populations with these properties without extensive experimental manipulation is a valid approach. As we have yet to determine exactly how epidermal stem cells behave in culture or in different biological assays, it is difficult to know whether the criteria used to assign "stemness" are appropriate or not. However, with increasing experimentation and exploration in this area, significant progress is being made to permit further refinement of stem cell-purification strategies.

It has been reported that stem cells have a smaller size (55) and consequently a higher density (144). These attributes have been exploited with unit gravity and density gradient sedimentation to enrich for colony-forming cells. Small size has been correlated with proliferative capacity, low RNA content, quiescence, and label retention (68,144–149) in blast-like cells with a high nuclear to cytoplasmic ratio (150). However, cell size selection alone is not sufficient to allow resolution of stem cells from their immediate progeny.

An early enrichment strategy termed "panning" involved selective adherence of keratinocytes labeled with antibodies to a basal cell marker onto a surface coated with anti-mouse IgG antibodies, resulting in 2.5-fold enrichment for basal keratinocytes (151). Given that specific adhesion reactions may facilitate attachment in less than a second (152) and that gradients in cell–extracellular matrix adhesiveness (153) and differences in integrin expression have been observed in the basal layer (130,132,150,153–155), panning would seem to be a promising strategy to employ if stem cell-specific extracellular matrices and their receptors can be found. Although it has been claimed that rapid adherence to various extracellular matrix-coated surfaces enhances stem cell enrichment (129,130,154,156,157), other data demonstrate that this supposition does not stand up to close scrutiny given that rapidly adhering cells from both murine and human epidermis comprise the majority of basal keratinocytes (158).

In efforts to more specifically isolate epidermal stem cells, investigators have adapted fluorescence-activated cell sorting (FACS) techniques utilized by HSC biologists to separate viable populations for functional analyses (159–161). Indeed, the search for markers that permit isolation of viable epidermal stem cells has been one of the more controversial aspects of the field (32). Early efforts targeted integrin β_1^{bri} populations to enrich for human epidermal cells with higher colony-forming efficiency (130,154). However, this marker is expressed at high levels up to 30% to 40% of basal cells, a rather high incidence for a stem cell population. Moreover, subsequent work has demonstrated that integrin α_6 is a more specific marker for basal keratinocytes and when used in conjunction with CD71 (specifically cells expressing low levels of CD71 α_6^{bri} CD71^{dim}) facilitating greater enrichment for stem cells than the β_1^{bri} CD71^{dim} phenotype (162). Keratinocytes with the phenotype α_6^{bri} CD71^{dim} have been demonstrated to fulfill many stem cell criteria: in murine epidermis, cells with this phenotype are small, and blast-like cells enriched for slow-cycling LRCs found in the interfollicular epidermis and hair follicle bulge region (150). The observation that murine hair follicles exhibit undetectable levels of CD71 protein in the bulge region compared with the actively growing hair bulb regions as shown in Figure 2 (150) has recently been confirmed by molecular profiling analysis of green fluorescent protein (GFP)-marked hair follicle stem cells derived from transgenic animals (163). The α_6^{bri} CD71^{dim} fraction of *human* epidermis is also enriched for stem cells given their low incidence, blast-like morphology, slow-cycling nature, and extensive cell regeneration capacity in long-term culture (155).

Unequivocal Identification of Markers for the Murine Hair Follicle Bulge Region

The ability to identify stem cells of the murine hair follicle as slow-cycling DNA LRCs localized to a morphologically identifiable niche in situ has been instrumental in devising and validating techniques for their viable isolation using cell surface markers and flow cytometry. These combined techniques provided validation for the strategy to use the surface markers α_6 and CD71 to enrich epidermal stem cells (150) and subsequently CD34 (164). Both CD71 and CD34 have also been utilized for FACS isolation of HSCs; notably, CD34 is expressed on both stem and progenitor cells of the bone marrow, whereas low levels of CD71 distinguish stem cells from their immediate progeny in both the bone marrow and epidermis (165). Whether all cells of the bulge region represent stem cells or a hierarchy within the follicular stem cell compartment remains to be elucidated.

The recent development of genetic strains of mice bearing GFP-positive LRCs in the hair follicle bulge region (65,163) has been the culmination of many decades of work permitting in situ visualization of bulge stem cells and their viable isolation for further biological characterization. Specifically, the Fuchs laboratory generated mice expressing GFP-tagged histone under the regulation of a K14 promoter rendering all basal cells green. The use of a tetracycline-regulatable construct permitted them to extinguish its expression in neonates, permitting a subsequent loss of GFP label from rapidly cycling cells and its retention in slowly cycling cells, particularly the bulge region (163). In

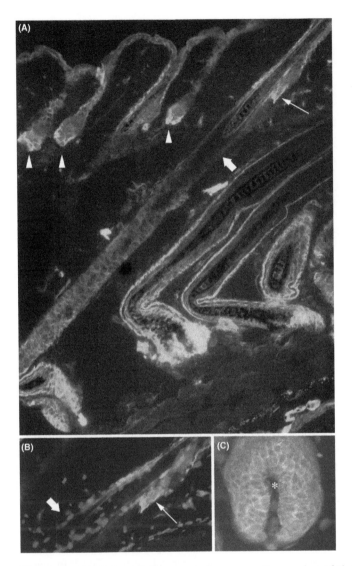

Figure 2 CD71 (transferrin receptor) as a negative marker of the hair follicle bulge region. (**A**) Staining for CD71 in early anagen hair follicles is restricted to the base of early anagen follicles (*arrowheads*). Note guard hair follicle in mid-anagen showing strong CD71 staining on either side of the unstained bulge region (marked with *block arrows*) directly below the sebaceous gland (*arrow*). (**B**) Dual staining for CD71 (*light gray*) and nuclei with propidium iodide (*dark gray*) illustrating the presence of nuclei in the CD71^{dim} bulge region. (**C**) CD71^{bri} cells in the bulb region. Asterisk denotes dermal papilla region. *Source*: From Ref. 150. (*See color insert.*)

contrast, Cotsarelis and co-workers (65) used the K15 promoter, active only in the bulge region, to drive GFP expression thus generating green bulge region cells (Fig. 3). These strains of mice will undoubtedly provide an elegant means of furthering our biological understanding of murine KSCs and have already been used to isolate viable bulge region cells for transcriptional profiling using gene arrays. In addition to providing a bulge KSC "molecular signature" at least at the mRNA level, this should prove valuable in identifying new markers that could be used for further refinement of stem cell-purification strategies. These data will also permit the validity of several reported markers for KSCs (Table 2) and have already confirmed the use of CD71 and CD34 to resolve

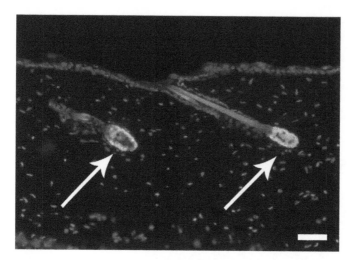

Figure 3 Generation of GFP-labeled hair follicle bulge stem cells in K15-EGFP transgenic mice. *Gray* indicates marked bulge cells that can be isolated from dorsal skin by FACS following enzymatic dispersion of the skin. *Abbreviations*: GFP, green fluorescent protein; EGFP, enhanced green fluorescent protein; FACS, fluorescence-activated cell sorting. *Source*: From Ref. 65. (*See color insert.*)

epidermal stem cells. The identification of signaling pathways that work to promote "stemness" or indeed inhibit transit amplifying cell activities (e.g., proliferation) is also feasible [see Ref. (166) for review]. Perhaps the most exciting information that can be gleaned from the elucidation of the KSC "transcriptome" is the identification of promoters uniquely active in bulge stem cells or conserved genes expressed in well-defined stem cell populations from different tissues. The former will allow investigators to target the expression of genes of their choice exclusively to the bulge stem cells in transgenic mice; the latter may permit the identification of conserved mechanisms of stem cell maintenance. However, the identification of unique proteins expressed in stem cells versus their progeny is also required to elucidate the control of important biological processes regulating stem cell maintenance, proliferation, and differentiation. Identification of unique proteins on the surface of stem cells will assist in understanding how these cells interact with their environment as well as providing markers for viable cell sorting. This information is not too far from being generated given the rapid development of proteomic technologies. Careful distinctions need to be made about the unique mRNAs versus proteins expressed in stem cells. Thus, although keratin 15 is widely expressed in the basal layer of the follicular and intrafollicular epidermis at the protein level, and thus not a suitable marker for identifying stem cells, its promoter is active only in stem cells making it a valuable tool for transgenics. CD34 is a very useful marker for cell separation strategies given that its expression at the protein level is restricted to the bulge region, although staining just outside the bulge region has been detected (163). The applicability of CD34 as a stem cell marker of interfollicular epidermis remains to be determined.

Enrichment and Identification of Human Epidermal Stem Cells via Surrogate Assays

Unequivocal identification of human epidermal or KSCs has been hampered given that for ethical reasons, one cannot generate LRCs in humans. Importantly, the development of culture techniques for keratinocytes (167) has led to the establishment of a variety of

Table 2 Putative KSC Markers

Marker	References
^{3}H-Tdr-LRC	(59,62,97,124,208,209,213–215)
K19	(95,122,216–218)
CD71^{dim,a,b}	(150,155,163,165)
$\beta_1^{bri,c}$	(129,130,133,154,219–223)
Bcl-2 (2X↑)^{a,d}	(163,224)
EGF-R, EGF, TGFα, PDGF α and β chains	(225)
p75NTR^{a,e}	(223,225)
DCC (deleted in colon carcinoma)	(226)
E-cadherin^{lo}/β-catenin^{lo}/(plakoglobin)γ-catenin^{high}	(132,153)
Basonuclin (3X↑)^{a,f}	(65,227,228)
BrdU-LRC	(63,75,156,229)
K15^g	(65,75,219,230–232)
TRAF-4	(233)
α_6^{bri}/CD71^{dim,h}	(75,150,155,163,165,218,234)
BDNF (8X↑, 5.6X↑)^{a,e}	(65,163,235)
c-Myb	(236)
p63^{a,i}	(237–239)
Tcf3 (3X↑)^a	(163,240–242)
Barx-2 (2X↑)	(163,243)
Delta1^{bri}	(244)
c-myc	(3,222,245)
Hoechst 33342 efflux^{a,j}	(205,246)
S100A4 (35X↑, 144X↑; 5X↑)	(65,73,163,247,248)
S100A6 (3X↑; 3X↑)^{a,k}	(65,163,247)
AC133-2	(249)
Connexin43$^-$(3X↓)	(65,250,251)
Adh^{3+}/EGF-R^{lo}	(157)
CD34^a (9X↑, 34X↑, 43X↑, 189X↑)	(65,163,164,205)
Dsg3^{lo}/β_1^{bri}	(129)
MCSP^l	(136,218,252)
Nestin^a	(253)
Thioredoxin^m	(65)
GFP^{high}-LRC	(163)

[a]Also reported as stem cell markers in nonepidermal lineages.

[b]Reported as 10G7 (155), later identified as CD71 (150).

[c]Rapid adherence to ColIV and FN; keratinocyte ECM (but not laminin) was also employed. Also reportedly nonspecific (162,163,218,254).

[d]Also reportedly a TA/non–stem cell marker (255,256). Anti-apoptotic, although bulge cells are reportedly apoptotic after plucking (73).

[e]p75NTR and BDNF are receptor–ligand partners; both are reportedly stem cell markers in epidermis and other tissues in mice and humans.

[f]Widespread transcription factor in cytoplasm of cells throughout basal layer, but nuclear in germinal region of telogen follicles (228), targeting rRNA (257); nuclear location associated with more rapid proliferation (258)—perhaps a germinative epidermal stem cell marker that translocates to nucleus in preparation for immense proliferation at anagen onset, though Tumbar et al. (163) report that GFP-LRCs are BSN^{low}.

[g]Also reported to be nonspecific (163,218,259–261) (our unpublished results, 2001).

[h]Targets integrin of hemidesmosomes on basal noncycling keratinocytes. It has been claimed that α_6 does not facilitate enrichment for KSCs (154) and that the bulge is α_6^{dim} (2.46-fold lower expression) (65).

[i]Also reportedly a TA/non–stem cell marker (3).

[j]Gating strategy and profile of Dunnwald et al. differs significantly from standard HSC protocol of Goodell et al. (174). Montanaro et al. did not separate dermal from epidermal cells in in vivo plasticity assay. Terunuma et al. (206) report that effluxing cells are not LRCs.

[k]Reported in hair germ and bulge—may be evidence of the "split-fuse" hypothesis of Commo et al. (122).

[l]Couchman et al. (252) noted chondroitin sulphate proteoglycans in bulge and matrix [though Legg et al. (136) and Ghali et al. (218) did not cite this study], suggesting stem and TA expression patterns.

[m]KSCs are reportedly slightly hypoxic (27,262) and thioredoxin is a stress-sensing protein induced by hypoxia that is believed to increase cell growth, in part by increasing sensitivity of cells to cytokines and growth factors (263).

surrogate in vitro assays, that is, clonogenicity, colony-forming efficiency, long-term proliferative output, believed to reflect the extensive capacity for self-renewal, and superior
proliferative potential expected of KSCs in vivo. In clonogenic assays, the status of
stem or transit amplifying (TA) cells was initially assigned retrospectively based on
expected behavior of stem cells versus TA cells (168). The use of relative short-term
colony-forming efficiency of subpopulations of prospectively isolated keratinocytes
using differential expression of β_1 integrin (130,154) has been subsequently discredited
as discussed before, based on multiparameter analysis (including long-term proliferative
output) of fractions of primary basal keratinocytes separated on the basis of α_6 integrin
and CD71 (155). Indeed, it is becoming increasingly clear that virtually all basal keratinocytes of neonatal skin retain extensive proliferative potential in vitro with equivalent life-
spans obtained from KSCs, transit amplifying cells, and even early differentiating
keratinocytes derived from neonatal foreskin epidermis (155). Thus, although there is a
strong case in support of the high clonogenic and replicative potential of KSC in vitro,
the functional properties of transit amplifying cells may be greater than previously
suspected and difficult to distinguish from that of KSCs with respect to short-term
clonogenicity in vitro.

It should now be possible to experimentally address the in vitro behavior of murine
hair follicle bulge versus nonbulge keratinocytes given recent developments in the field
and indeed some work has begun to take place. However, limitations that remain in this
type of work are that (*i*) murine keratinocytes are notoriously difficult to propagate and
long-term culture analysis can be complicated by the high rate of spontaneous transformation in these cells, and (*ii*) whether the markers available to date are good enough to truly
provide a stem versus transit amplifying population. Although it is possible to purify the
quiescent bulge cells with CD34 or from custom GFP-marked transgenics, the population
used for comparison is a mixture of interfollicular stem, progenitor and maturing cells, hair
follicle progenitors, and differentiating cells; and presumably sebaceous gland stem, progenitor, and differentiating cells. Thus, claims of greater colony-forming efficiency by
bulge region cells compared to undefined so-called progeny are fraught with misinterpretation, given the possibilities that (*i*) the latter are disadvantageous due to dilution, (*ii*) the
readouts may represent the clonogenic capacity of interfollicular or sebaceous gland stem
cells, or (*iii*) a combination of the two. Identification of further markers for true transit
amplifying cells only and negative selection for nonbulge epithelial stem cell populations
as well as lineage differentiation markers are required (analogous to CD38-negative
selection in bone marrow stem cell-purification strategies).

Long-Term Epidermal Tissue Reconstitution as an
Assay for KSC Activity

As stem cells are responsible for the lifelong production of epidermal keratinocytes of the
skin in vivo, the most important functional validation for any candidate KSC population
must be its capacity to exhibit sustained epidermal tissue regeneration in long-term repopulation assays. Morris and colleagues have utilized an in vivo transplant model to reconstitute hair follicles from FACS-isolated murine bulge keratinocytes adding a vital
technological advance to the complete characterization of what is surely the best-characterized cutaneous stem cell population to date. Another transplant assay was recently
described whereby GFP-marked unfractionated primary keratinocytes derived from
murine interfollicular epidermis were placed in the hat chamber model together with
unmarked keratinocytes to assess their competitive regenerative capacity (169). Long-
term reconstitution (five to nine weeks) was estimated to be achieved from 1/35,000

basal epidermal cells based on mathematical modeling of data from inoculation of decreasing numbers of GFP-positive cells. This assay provides an excellent means to test the relative tissue-regenerative capacity of candidate epidermal stem cell populations when competed with unenriched cells, provided that one population is genetically tagged. The estimates of stem cell frequency (0.01%) obtained, however, are difficult to reconcile with in situ analyses of murine epidermis placing the number of basal cells capable of sustaining an EPU at 10%. It is very likely that all cells capable of tissue regeneration are not recruited by this assay due to sub-optimal conditions or other technical reasons as is the case with virtually all experimental approaches for assaying keratinocytes, and further optimization is required.

Very few investigators have used in vitro and in vivo tissue regeneration to define or characterize *human* KSC populations. Further, in the absence of a comparison with tissue-regenerative ability of unfractionated keratinocytes or better still, non–stem cell populations, it is difficult to assess the validity of short-term reconstitution as a measure of stem cell activity which by analogy with HSCs may also be a property of committed progenitor populations rather than an exclusive characteristic of stem cells. The ability of autologous grafts of cultured epidermal cells to rescue patients with extensive full thickness burns for over a decade (170,171) suggests that stem cell activity is maintained in culture. However, whether this is an exclusive property of stem cells is not clear given that experimental long-term epidermal tissue reconstitution studies (up to 40 weeks) have been performed with transduced bulk cultures of human keratinocytes (127). Studies with prospectively isolated KSCs and their progeny have demonstrated that significant short-term (two weeks) and relatively long-term (6 to 10 weeks) tissue-regenerative ability can be elicited from *all* classes of basal keratinocytes in vitro and in vivo following transplantation (172). Consequently, there is a need to re-evaluate purported markers of human epidermal stem cells in the literature, as it is becoming increasingly clear that many parameters thought to measure stem cell behavior in various assays may not be attributed solely to stem cells. Interestingly, Morris et al. (65), who compared the hair-follicle-regenerative capacity of murine bulge versus nonbulge follicular keratinocytes, reported that the latter non–stem cell population was capable of giving rise to hair follicle morphogenesis albeit at a decreased frequency compared with bulge region cells. Thus, even in murine studies, there is little information available on the comparative tissue-regenerative ability of stem cells versus their progeny. Hopefully, this will be an area of extensive investigation over the coming years so that the skin stem cell field can evolve to the enviable stage of HSC biology with a plethora of assays for stem and progenitor cells.

Do Keratinocytes Capable of Effluxing Hoechst 33342 Represent a Candidate Stem Cell Population?

Many investigators have expended a considerable amount of effort to determine whether the ability to exclude the vital DNA-staining dye Hoechst 33342 is a common feature of stem cells from various tissues. The underlying notion is that stem cells should be able to actively pump out drugs or other toxins to prevent damage to these long-lived residents of rapidly renewing tissues. Originally, Hoechst 33342 was used by Baines and Visser (173) to enrich for hematopoietic progenitors in bone marrow, by sorting a subset of cells with low Hoechst fluorescence as detected in a single-emission wavelength. When Goodell et al. (174) displayed the Hoechst fluorescence of bone marrow cells in red versus blue emission wavelengths, a complex profile emerged, allowing resolution of a rare Hoechst[low] subpopulation of cells with superior dye-efflux ability—termed the side population (SP). It was shown that the bone marrow SP was enriched at least 1000-fold for hematopoietic reconstituting activity (174), with the subset capable of highest efflux possessing the greatest HSC activity (175) and enrichment for primitive cells (176). In the ensuing years, an SP resembling that in bone marrow has been resolved in many other tissues, including brain (177–181), heart (181–183), liver (181,184–186), lung (181,187–189), mammary gland (190–195), and muscle (181,196–203). The mounting reports of SPs in diverse tissues led to the concept that Hoechst efflux represented a universal stem cell trait (191,204) and motivated the search for this population in many tissues, including the epidermis.

Recent reports have established that human and murine epidermis harbor an SP-like population (205–207). On the basis of the data from the bone marrow, it would be reasonable to adopt the hypothesis that the epidermal SP population is the most potent of keratinocyte progenitors. Although many laboratories have attempted to study this intriguing population, this has proved difficult due in large part to their low incidence in the epidermis, making it difficult to get enough cells to place in various assays. Murine tissue has been used to circumvent this problem, but this is problematic given that mouse keratinocytes are difficult to propagate in vitro. The clonogenicity of Hoechst-treated cells also appears to be compromised, suggesting that the drug may be toxic to keratinocytes.

Terunuma et al. (206) examined SPs in human epidermis in an attempt to determine these resembled KSCs. The investigators were successfully able to generate LRCs in human neonatal foreskin by grafting the human tissue onto mice and subjecting the mice to BrdU labeling albeit with an unorthodox approach (using topical application of *O*-tetradecanoylphorbol-13-acetate to stimulate cell proliferation). On the basis of the differential expression of cell surface integrin levels on SP cells ($\alpha_6^{low}/\beta_1^{low}$) and BrdU LRCs ($\alpha_6^{bri}/\beta_1^{bri}$), these investigators concluded that the epidermal SP fraction (K14-positive) was different from "traditional" KSCs. Interestingly, it was not possible to directly analyze the SP population for enrichment of LRCs given that BrdU appeared to quench Hoechst 33342 generating Hoechst^{low} cells artificially. In contrast, Triel et al. (207) have reported that 80% of BrdU LRCs from murine epidermis are co-isolated in the nonSP fraction allaying concerns about this quenching effect. Importantly, these data suggest that the SP population is not enriched for quiescent stem cells, although a caveat to this interpretation is that should this epidermal subset represent a deeply quiescent subpopulation; it may have eluded detection by failing to acquire any BrdU during the labeling period. These investigators concluded that SP cells may represent TA cells although the heterogeneous expression of many markers such as integrins and the differentiation-specific keratin, K10, are perplexing and suggest that further work is required to clearly define the SP population isolated from skin epidermis.

Ultimately, stem cells are defined by their functionality. Therefore, epidermal SPs must be challenged in vitro and in vivo in a variety of assays under various conditions before their stem cell status can be definitively ascertained. Ideally, a rigorous test of keratinocyte stemness should include in vivo tissue regeneration that demonstrates the appropriate spatial and temporal genetic program to make a therapeutically meaningful contribution to the target tissue. It would also be informative to examine SPs for retention of tritiated thymidine at various short- and medium-term time points to see if they retain label for a moderate period as expected of more rapidly cycling interfollicular stem cells, and determine whether they resist pulse labeling due to deep quiescence. The lack of substantial functional data for or against epidermal SPs as a robust stem cell population suggests that perhaps these cells are not easily recruited into the available in vitro assays and could equally be attributed to deep quiescence or commitment to differentiation. Alternatively, it remains possible that this minor population of skin residents is merely a confounding issue for KSC biology. An arguable scenario is that epidermal SPs are a specialized subset of cells in skin whose role is to efflux toxins.

CONCLUSION

It is an exciting time in the study of KSC biology and we are several steps closer to answering some fundamental questions about epidermal tissue renewal. Experimental approaches have as usual raised even more questions than answers, throwing us into uncertainty about how we define an epidermal stem cell once it is removed from its niche in vivo. Assays thought to measure epidermal stem cell activity merely scratch the surface and much work is needed to find out how stem cells are maintained as such in vivo for the lifespan of an organism, while daughter cells are rapidly expelled to terminally differentiate and die. Perhaps the most relevant issue that needs to be addressed is what is in the immediate environment of a stem cell that makes up its niche. The molecular cell surface composition of stem cells and their neighbors should prove useful, although going from enumerating these to sifting out functional components will be a challenge. An area that needs to be investigated is that of understanding the complex cellular and molecular makeup of the dermis, and its specific interaction with distinct classes of basal keratinocytes. To date, the evidence points to the dermis acting as a supportive microenvironment for epidermal stem cells and their progeny, and it is very likely that a close functional analogy can be drawn between these two compartments of the skin and the stromal:hemopoietic interactions essential to the regulation of blood stem cells. Finally, the early indications are that vast proliferative potential resides within the entire basal layer of the epidermis, throwing into doubt the assumption that only stem cells are capable of tissue regeneration. A major unanswered question is whether stem cells are, indeed, the preferred target for carcinogenic agents. The means now exist to discard all assumptions and embark on a quest for greater understanding of stem cell function and regulation.

REFERENCES

1. Eckert RL. Structure, function, and differentiation of the keratinocyte. Physiol Rev 1989; 69(4):1316–1346.
2. Chuong CM, Nickoloff BJ, Elias PM, Goldsmith LA, Macher E, Maderson PA, Sundberg JP, Tagami H, Plonka PM, Thestrup-Pederson K, et al. What is the 'true' function of skin? Exp Dermatol 2002; 11(2):159–187.
3. Fuchs E, Raghavan S. Getting under the skin of epidermal morphogenesis. Nat Rev Genet 2002; 3(3):199–209.
4. Mackenzie IC. Ordered structure of the stratum corneum of mammalian skin. Nature 1969; 222(196):881–882.
5. Fuchs E. Epidermal differentiation: the bare essentials. J Cell Biol 1990; 111(6 Pt 2): 2807–2814.
6. Kalinin AE, Kajava AV, Steinert PM. Epithelial barrier function: assembly and structural features of the cornified cell envelope. Bioessays 2002; 24(9):789–800.
7. Potten CS. Epidermal transit times. Br J Dermatol 1975; 93(6):649–658.
8. Hill MW, Berg JH, Mackenzie IC. Quantitative evaluation of regional differences between epithelia in the adult mouse. Arch Oral Biol 1981; 26(12):1063–106 7.
9. Potten CS, Wichmann HE, Dobek K, Birch J, Codd TM, Horrocks L, Pedrick M, Tickle SP. Cell kinetic studies in the epidermis of mouse. III. The percent labeled mitosis (PLM) technique. Cell Tissue Kinet 1985; 18(1):59–70.

10. Potten CS, Saffhill R, Maibach HI. Measurement of the transit time for cells through the epidermis and stratum corneum of the mouse and guinea pig. Cell Tissue Kinet 1987; 20(5):461–472.

11. Halprin KM. Epidermal "turnover time"—a re-examination. Br J Dermatol 1972; 86(1): 14–19.

12. Bergstresser PR, Taylor JR. Epidermal "turnover time"—a new examination. Br J Dermatol 1977; 96(5):503–509.

13. Fuchs E. Keratins and the skin. Annu Rev Cell Dev Biol 1995; 11:123–153.

14. Potten CS, Booth C. Keratinocyte stem cells: a commentary. J Invest Dermatol 2002; 119(4):888–899.

15. Potten CS. Epidermal cell production rates. J Invest Dermatol 1975; 65(6):488–500.

16. Potten CS. Identification of clonogenic cells in the epidermis and the structural arrangement of the epidermal proliferative unit (EPU). In: Cairnie AB, Lala PK, Osmond DG, eds. Stem Cells of Renewing Populations. New York: Academic Press, 1976:389.

17. Niemann C, Watt FM. Designer skin: lineage commitment in postnatal epidermis. Trends Cell Biol 2002; 12(4):185–192.

18. Stenback F, Peto R, Shubik P. Initiation and promotion at different ages and doses in 2200 mice. I. Methods, and the apparent persistence of initiated cells. Br J Cancer 1981; 44(1):1–14.

19. Montagna W, Chase HB. Histology and cytochemistry of human skin. X. X-irradiation of the scalp. Am J Anat 1956; 99(3):415–445.

20. Eisen AZ, Holyoke JB, Lobitz WC Jr. Responses of the superficial portion of the human pilosebaceous apparatus to controlled injury. J Invest Dermatol 1956; 25(3):145–156.

21. Oliver RF. Ectopic regeneration of whiskers in the hooded rat from implanted lengths of vibrissa follicle wall. J Embryol Exp Morphol 1967; 17(1):27–34.

22. Withers HR. The dose–survival relationship for irradiation of epithelial cells of mouse skin. Br J Radiol 1967; 40(471):187–194.

23. Spangrude GJ. When is a stem cell really a stem cell? Bone Marrow Transplant 2003; 32(suppl 1):S7–S11.

24. Lajtha LG. Stem cell concepts. Differentiation 1979; 14(1–2):23–34.

25. Lajtha LG. Stem cell concepts. In: Potten CS, ed. Stem Cells: Their Identification and Characterization. London: Churchill Livingston, 1983:1–11.

26. Weissman IL. Stem cells: units of development, units of regeneration, and units in evolution. Cell 2000; 100(1):157–168.

27. Potten CS. Stem cells in epidermis from the back of the mouse. In: Potten CS, ed. Stem Cells: Their Identification and Characterization. London: Churchill Livingston, 1983: 200–232.

28. Curry JL, Trentin JJ. Hemopoietic spleen colony studies. I. Growth and differentiation. Dev Biol 1967; 15(5):395–413.

29. Schofield R. The relationship between the spleen colony-forming cell and the haemopoietic stem cell. Blood Cells 1978; 4(1–2):7–25.

30. Potten CS, Schofield R, Lajtha LG. A comparison of cell replacement in bone marrow, testis and three regions of surface epithelium. Biochim Biophys Acta 1979; 560(2):281–299.

31. Potten CS, Loeffler M. Stem cells: attributes, cycles, spirals, pitfalls and uncertainties. Lessons for and from the crypt. Development 1990; 110(4):1001–1020.

32. Lavker RM, Sun TT. Epidermal stem cells: properties, markers, and location. Proc Natl Acad Sci USA 2000; 97(25):13473–13475.

33. Kaur P, Li A, Redvers RP, Bertoncello I. Keratinocyte stem cell assays: an evolving science. J Invest Dermatol Symp Proc 2004; 9(3):238–247.

34. Weinstein GD, Frost P. Abnormal cell proliferation in psoriasis. J Invest Dermatol 1968; 50(3):254–259.

35. Pinkus H. Examination of the epidermis by the strip method. II. Biometric data on regeneration of the human epidermis. J Invest Dermatol 1952; 19(6):431–447.

36. Schultze B, Oehlert W. Autoradiographic investigations of incorporation of H3-thymidine into cells of the rat and mouse. Science 1960; 131:737–738.

37. Fukuyama K, Bernstein IA. Autoradiographic studies of the incorporation of thymidine-H3 into deoxyribonucleic acid in the skin of young rats. J Invest Dermatol 1961; 36:321–326.

38. Weinstein GD, Van Scott EJ. Autoradiographic analysis of turnover times of normal and psoriatic epidermis. J Invest Dermatol 1965; 45(4):257–262.

39. Leblond CP, Messier B, Kopriwa B. Thymidine-H3 as a tool for the investigation of the renewal of cell populations. Lab Invest 1959; 8(1):296–306; discussion 8.

40. Leblond CP, Greulich RC, Pereira JPM. Relationship of cell formation and cell migration in the renewal of stratified squamous epithelia. Adv Biol Skin 1964; 5(1):39–67.

41. Marques-Pereira JP, Leblond CP. Mitosis and differentiation in the stratified squamous epithelium of the rat esophagus. Am J Anat 1965; 117:73–87.

42. Iversen OH, Bjerknes R, Devik F. Kinetics of cell renewal, cell migration and cell loss in the hairless mouse dorsal epidermis. Cell Tissue Kinet 1968; 1:351–367.

43. Hibbs RG, Clark WH Jr. Electron microscope studies of the human epidermis: the cell boundaries and topography of the stratum malpighii. J Biophys Biochem Cytol 1959; 6(1):71–76.

44. Hamilton E, Potten CS. Influence of hair plucking on the turnover time of the epidermal basal layer. Cell Tissue Kinet 1972; 5(6):505–517.

45. Potten CS, Hendry JH. Clonogenic cells and stem cells in epidermis. Int J Radiat Biol Relat Stud Phys Chem Med 1973; 24(5):537–540.

46. Potten CS. The epidermal proliferative unit: the possible role of the central basal cell. Cell Tissue Kinet 1974; 7(1):77–88.

47. Christophers E. Cellular architecture of the stratum corneum. J Invest Dermatol 1971; 56(3):165–169.

48. Menton DN, Eisen AZ. Structure and organization of mammalian stratum corneum. J Ultrastruct Res 1971; 35(3):247–264.

49. Mackenzie IC. Relationship between mitosis and the ordered structure of the stratum corneum in mouse epidermis. Nature 1970; 226(246):653–655.

50. Allen TD, Potten CS. Fine-structural identification and organization of the epidermal proliferative unit. J Cell Sci 1974; 15(2):291–319.

51. Christophers E, Wolff HH, Laurence EB. The formation of epidermal cell columns. J Invest Dermatol 1974; 62(6):555–559.

52. Kam E, Melville L, Pitts JD. Patterns of junctional communication in skin. J Invest Dermatol 1986; 87(6):748–753.

53. Potten CS, Allen TD. Control of epidermal proliferative units (EPUs): an hypothesis based on the arrangement of neighbouring differentiated cells. Differentiation 1975; 3(1–3): 161–165.

54. Potten CS, Wichmann HE, Loeffler M, Dobek K, Major D. Evidence for discrete cell kinetic subpopulations in mouse epidermis based on mathematical analysis. Cell Tissue Kinet 1982; 15(3):305–329.

55. Lavker RM, Sun TT. Heterogeneity in epidermal basal keratinocytes: morphological and functional correlations. Science 1982; 215(4537):1239–1241.

56. Lavker RM, Sun TT. Epidermal stem cells. J Invest Dermatol 1983; 81(suppl 1): 121s–127s.

57. Cairns J. Mutation selection and the natural history of cancer. Nature 1975; 255(5505): 197–200.

58. Potten CS, Owen G, Booth D. Intestinal stem cells protect their genome by selective segregation of template DNA strands. J Cell Sci 2002; 115(11):2381–2388.

59. Cotsarelis G, Sun TT, Lavker RM. Label-retaining cells reside in the bulge area of pilosebaceous unit: implications for follicular stem cells, hair cycle, and skin carcinogenesis. Cell 1990; 61(7):1329–1337.

60. Kobayashi K, Rochat A, Barrandon Y. Segregation of keratinocyte colony-forming cells in the bulge of the rat vibrissa. Proc Natl Acad Sci USA 1993; 90(15):7391–7395.

61. Yang JS, Lavker RM, Sun TT. Upper human hair follicle contains a subpopulation of keratinocytes with superior in vitro proliferative potential. J Invest Dermatol 1993; 101(5):652–659.

62. Morris RJ, Potten CS. Highly persistent label-retaining cells in the hair follicles of mice and their fate following induction of anagen. J Invest Dermatol 1999; 112(4):470–475.

63. Taylor G, Lehrer MS, Jensen PJ, Sun TT, Lavker RM. Involvement of follicular stem cells in forming not only the follicle but also the epidermis. Cell 2000; 102(4):451–461.

64. Oshima H, Rochat A, Kedzia C, Kobayashi K, Barrandon Y. Morphogenesis and renewal of hair follicles from adult multipotent stem cells. Cell 2001; 104(2):233–245.

65. Morris RJ, Liu Y, Marles L, Yang Z, Trempus C, Li S, Lin JS, Sawicki JA, Cotsarelis G. Capturing and profiling adult hair follicle stem cells. Nat Biotechnol 2004; 22(4):411–417.

66. Miller SJ, Lavker RM, Sun TT. Keratinocyte stem cells of cornea, skin and hair follicle: common and distinguishing features. Semin Dev Biol 1993; 4:217–240.

67. Lavker RM, Sun TT, Oshima H, Barrandon Y, Akiyama M, Ferraris C, Chevalier G, Favier B, Jahoda CA, Dhouailly D, Panteleyev AA, Christiano AM. Hair follicle stem cells. J Investig Dermatol Symp Proc 2003; 8(1):28–38.

68. Reynolds AJ, Jahoda CA. Hair follicle stem cells? A distinct germinative epidermal cell population is activated in vitro by the presence of hair dermal papilla cells. J Cell Sci 1991; 99(Pt 2):373–385.

69. Reynolds AJ, Jahoda CA. Cultured dermal papilla cells induce follicle formation and hair growth by transdifferentiation of an adult epidermis. Development 1992; 115(2): 587–593.

70. Miller SJ, Burke EM, Rader MD, Coulombe PA, Lavker RM. Re-epithelialization of porcine skin by the sweat apparatus. J Invest Dermatol 1998; 110(1):13–19.

71. Ghazizadeh S, Taichman LB. Multiple classes of stem cells in cutaneous epithelium: a lineage analysis of adult mouse skin. EMBO J 2001; 20(6):1215–1222.

72. Panteleyev AA, Jahoda CA, Christiano AM. Hair follicle predetermination. J Cell Sci 2001; 114(Pt 19):3419–3431.

73. Ito M, Kizawa K, Toyoda M, Morohashi M. Label-retaining cells in the bulge region are directed to cell death after plucking, followed by healing from the surviving hair germ. J Invest Dermatol 2002; 119(6):1310–1316.

74. DasGupta R, Rhee H, Fuchs E. A developmental conundrum: a stabilized form of beta-catenin lacking the transcriptional activation domain triggers features of hair cell fate in epidermal cells and epidermal cell fate in hair follicle cells. J Cell Biol 2002; 158(2): 331–344.

75. Braun KM, Niemann C, Jensen UB, Sundberg JP, Silva-Vargas V, Watt FM. Manipulation of stem cell proliferation and lineage commitment: visualisation of label-retaining cells in wholemounts of mouse epidermis. Development 2003; 130(21):5241–5255.

76. Butcher EO. The specificity of the hair papilla in the rat. Anat Rec 1965; 151:231–237.

77. Oliver RF. Whisker growth after removal of the dermal papilla and lengths of follicle in the hooded rat. J Embryol Exp Morphol 1966; 15(3):331–347.

78. Oliver RF. Histological studies of whisker regeneration in the hooded rat. J Embryol Exp Morphol 1966; 16(2):231–244.

79. Oliver RF. The experimental induction of whisker growth in the hooded rat by implantation of dermal papillae. J Embryol Exp Morphol 1967; 18(1):43–51.

80. Fukuda O, Ezaki T. Complications in reconstructive surgery for microtia. Jpn Plast Reconstr Surg 1975; 18:109–114.

81. Inaba M, Anthony J, McKinstry C. Histologic study of the regeneration of axillary hair after removal with subcutaneous tissue shaver. J Invest Dermatol 1979; 72(5):224–231.

82. Ibrahim L, Wright EA. A quantitative study of hair growth using mouse and rat vibrissal follicles. I. Dermal papilla volume determines hair volume. J Embryol Exp Morphol 1982; 72:209–224.

83. Oliver RF, Jahoda CAB. The dermal papilla and maintenance of hair growth. In: Rogers GE, Reis PJ, Ward KA, Marshall RC, eds. The Biology of Wool and Hair. London: Chapman & Hall, 1989:51–67.

84. Bishop GH. Regeneration after experimental removal of skin in man. Am J Anat 1945; 76:153–181.

85. Albert RE, Burns FJ, Heimbach RD. The effect of penetration depth of electron radiation on skin tumor formation in the rat. Radiat Res 1967; 30(3):515–524.

86. Albert RE, Burns FJ, Heimbach RD. The association between chronic radiation damage of the hair follicles and tumor formation in the rat. Radiat Res 1967; 30(3):590–599.

87. Withers HR. Recovery and repopulation in vivo by mouse skin epithelial cells during fractionated irradiation. Radiat Res 1967; 32(2):227–239.

88. Oduye OO. Effects of various induced local environmental conditions and histopathological studies in experimental *Dermatophilus congolensis* infection on the bovine skin. Res Vet Sci 1975; 19(3):245–252.

89. Argyris T. Kinetics of epidermal production during epidermal regeneration following abrasion in mice. Am J Pathol 1976; 83(2):329–340.

90. Al-Barwari SE, Potten CS. Regeneration and dose–response characteristics of irradiated mouse dorsal epidermal cells. Int J Radiat Biol Relat Stud Phys Chem Med 1976; 30(3):201–216.

91. Potten CS. Cell replacement in epidermis (keratopoiesis) via discrete units of proliferation. Int Rev Cytol 1981; 69:271–318.

92. Morris R, Argyris TS. Epidermal cell cycle and transit times during hyperplastic growth induced by abrasion or treatment with 12-*O*-tetradecanoylphorbol-13-acetate. Cancer Res 1983; 43(10):4935–4942.

93. Lenoir MC, Bernard BA, Pautrat G, Darmon M, Shroot B. Outer root sheath cells of human hair follicle are able to regenerate a fully differentiated epidermis in vitro. Dev Biol 1988; 130(2):610–620.

94. Rochat A, Kobayashi K, Barrandon Y. Location of stem cells of human hair follicles by clonal analysis. Cell 1994; 76(6):1063–1073.

95. Lane EB, Wilson CA, Hughes BR, Leigh IM. Stem cells in hair follicles: cytoskeletal studies. Ann NY Acad Sci 1991; 642:197–213.

96. Lavker RM, Miller S, Wilson C, Cotsarelis G, Wei ZG, Yang JS, Sun TT. Hair follicle stem cells: their location, role in hair cycle, and involvement in skin tumor formation. J Invest Dermatol 1993; 101(suppl 1):16S–26S.

97. Morris RJ, Potten CS. Slowly cycling (label-retaining) epidermal cells behave like clonogenic stem cells in vitro. Cell Prolif 1994; 27(5):279–289.

98. Sun TT, Cotsarelis G, Lavker RM. Hair follicular stem cells: the bulge-activation hypothesis. J Invest Dermatol 1991; 96(5):77S–78S.

99. Wilson C, Cotsarelis G, Wei ZG, Fryer E, Margolis-Fryer J, Ostead M, Tokarek R, Sun TT, Lavker RM. Cells within the bulge region of mouse hair follicle transiently proliferate during early anagen: heterogeneity and functional differences of various hair cycles. Differentiation 1994; 55(2):127–136.

100. Wilson CL, Sun TT, Lavker RM. Cells in the bulge of the mouse telogen follicle give rise to the lower anagen follicle. Skin Pharmacol 1994; 7(1–2):8–11.

101. Miller SJ, Sun TT, Lavker RM. Hair follicles, stem cells, and skin cancer. J Invest Dermatol 1993; 100(3):288S–294S.

102. Morris RJ, Coulter K, Tryson K, Steinberg SR. Evidence that cutaneous carcinogen-initiated epithelial cells from mice are quiescent rather than actively cycling. Cancer Res 1997; 57(16):3436–3443.

103. Chase HB. Growth of the hair. Physiol Rev 1954; 34(1):113–126.

104. Bullough WS. Mitotic control in adult mammalian tissues. Biol Rev Camb Philos Soc 1975; 50(1):99–127.

105. Ahmad W, Panteleyev AA, Christiano AM. The molecular basis of congenital atrichia in humans and mice: mutations in the hairless gene. J Investig Dermatol Symp Proc 1999; 4(3):240–243.

106. Moll I. Proliferative potential of different keratinocytes of plucked human hair follicles. J Invest Dermatol 1995; 105(1):14–21.

107. Moll I. Differential epithelial outgrowth of plucked and microdissected human hair follicles in explant culture. Arch Dermatol Res 1996; 288(10):604–610.

108. Fuchs E, Byrne C. The epidermis: rising to the surface. Curr Opin Genet Dev 1994; 4(5): 725–736.

109. Akiyama M, Dale BA, Sun TT, Holbrook KA. Characterization of hair follicle bulge in human fetal skin: the human fetal bulge is a pool of undifferentiated keratinocytes. J Invest Dermatol 1995; 105(6):844–850.

110. Pinkus H, Mehregan AH. A Guide to Dermatohistopathology. New York: Appleton-Century-Crofts, 1981:458–459.

111. Epstein WL, Maibach HI. Cell proliferation and movement in human hair bulbs. In: Montagna W, Dobson RL, eds. Advances in the Biology of Skin. Oxford: Pergamon Press, 1969:88–97.

112. Chapman RE. Cell migration in wool follicles of sheep. J Cell Sci 1971; 9(3):791–803.

113. Van Scott EJ, Ekel TM, Auerbach R. Determinants of rate and kinetics of cell division in scalp hair. J Invest Dermatol 1963; 41:269–273.

114. Malkinson FD, Keane JT. Hair matrix cell kinetics: a selective review. Int J Dermatol 1978; 17(7):536–551.

115. Kligman AM. The human hair cycle. J Invest Dermatol 1959; 33:307–316.

116. Kligman AM. Neogenesis of human hair follicles. Ann NY Acad Sci 1959; 83:507–511.

117. Moffat GH. The kinetics of cell populations in the growing hair follicle of the mouse as revealed by autoradiography using H-Tdr. In: Baccaraeda Boy A, Moretti G, Fray JR, eds. Biopathology of Pattern Alopecia. New York: Karger, 1968:90–106.

118. Niemann C, Owens DM, Hulsken J, Birchmeier W, Watt FM. Expression of DeltaNLef1 in mouse epidermis results in differentiation of hair follicles into squamous epidermal cysts and formation of skin tumours. Development 2002; 129(1):95–109.

119. De Weert J. Embryogenesis of the hair follicle and hair cycle. In: Van Neste D, Lachapelle JM, Antoine JL, eds. Trends in Human Hair Growth and Alopecia Research. Dordrecht: Kluwer Academic Publishers, 1989:3–10.

120. Jahoda CA, Horne KA, Mauger A, Bard S, Sengel P. Cellular and extracellular involvement in the regeneration of the rat lower vibrissa follicle. Development 1992; 114(4):887–897.

121. Reynolds AJ, Jahoda CA. Hair fibre progenitor cells: developmental status and interactive potential. Semin Dev Biol 1993; 4:241–250.

122. Commo S, Gaillard O, Bernard BA. The human hair follicle contains two distinct K19 positive compartments in the outer root sheath: a unifying hypothesis for stem cell reservoir? Differentiation 2000; 66(4–5):157–164.

123. Cotsarelis G. The hair follicle: dying for attention. Am J Pathol 1997; 151(6):1505–1509.

124. Morris RJ, Fischer SM, Slaga TJ. Evidence that the centrally and peripherally located cells in the murine epidermal proliferative unit are two distinct cell populations. J Invest Dermatol 1985; 84(4):277–281.

125. Mackenzie IC. Retroviral transduction of murine epidermal stem cells demonstrates clonal units of epidermal structure. J Invest Dermatol 1997; 109(3):377–383.

126. Gambardella L, Barrandon Y. The multifaceted adult epidermal stem cell. Curr Opin Cell Biol 2003; 15(6):771–777.

127. Kolodka TM, Garlick JA, Taichman LB. Evidence for keratinocyte stem cells in vitro: long-term engraftment and persistence of transgene expression from retrovirus-transduced keratinocytes. Proc Natl Acad Sci USA 1998; 95(8):4356–4361.

128. Topilko P, Schneider-Maunoury S, Levi G, Trembleau A, Gourdji D, Driancourt MA, Rao CV, Charnay P. Multiple pituitary and ovarian defects in Krox-24 (NGFI-A, Egr-1)-targeted mice. Mol Endocrinol 1998; 12(1):107–122.

129. Wan H, Stone MG, Simpson C, Reynolds LE, Marshall JF, Hart IR, Hodivala-Dilke KM, Eady RA. Desmosomal proteins, including desmoglein 3, serve as novel negative markers for epidermal stem cell-containing population of keratinocytes. J Cell Sci 2003; 116(Pt 20): 4239–4248.

130. Jones PH, Harper S, Watt FM. Stem cell patterning and fate in human epidermis. Cell 1995; 80(1):83–93.

131. Jones PH. Epithelial stem cells. Bioessays 1997; 19(8):683–690.
132. Watt FM. Epidermal stem cells: markers, patterning and the control of stem cell fate. Philos Trans R Soc Lond B Biol Sci 1998; 353(1370):831–837.
133. Jensen UB, Lowell S, Watt FM. The spatial relationship between stem cells and their progeny in the basal layer of human epidermis: a new view based on whole-mount labelling and lineage analysis. Development 1999; 126(11):2409–2418.
134. Watt FM. Stem cell fate and patterning in mammalian epidermis. Curr Opin Genet Dev 2001; 11(4):410–417.
135. Watt FM. The stem cell compartment in human interfollicular epidermis. J Dermatol Sci 2002; 28(3):173–180.
136. Legg J, Jensen UB, Broad S, Leigh I, Watt FM. Role of melanoma chondroitin sulphate proteoglycan in patterning stem cells in human interfollicular epidermis. Development 2003; 130(24):6049–6063.
137. Morris RJ, Tryson KA, Wu KQ. Evidence that the epidermal targets of carcinogen action are found in the interfollicular epidermis of infundibulum as well as in the hair follicles. Cancer Res 2000; 60(2):226–229.
138. Ferraris C, Bernard BA, Dhouailly D. Adult epidermal keratinocytes are endowed with pilosebaceous forming abilities. Int J Dev Biol 1997; 41(3):491–498.
139. Ferraris C, Chevalier G, Favier B, Jahoda CA, Dhouailly D. Adult corneal epithelium basal cells possess the capacity to activate epidermal, pilosebaceous and sweat gland genetic programs in response to embryonic dermal stimuli. Development 2000; 127(24): 5487–5495.
140. Gat U, DasGupta R, Degenstein L, Fuchs E. De Novo hair follicle morphogenesis and hair tumors in mice expressing a truncated beta-catenin in skin. Cell 1998; 95(5):605–614.
141. Huelsken J, Vogel R, Erdmann B, Cotsarelis G, Birchmeier W. β-Catenin controls hair follicle morphogenesis and stem cell differentiation in the skin. Cell 2001; 105(4):533–545.
142. Andl T, Reddy ST, Gaddapara T, Millar SE. WNT signals are required for the initiation of hair follicle development. Dev Cell 2002; 2(5):643–653.
143. van Genderen C, Okamura RM, Farinas I, Quo RG, Parslow TG, Bruhn L, Grosschedl R. Development of several organs that require inductive epithelial–mesenchymal interactions is impaired in LEF-1-deficient mice. Genes Dev 1994; 8(22):2691–2703.
144. Morris RJ, Fischer SM, Klein-Szanto AJ, Slaga TJ. Subpopulations of primary adult murine epidermal basal cells sedimented on density gradients. Cell Tissue Kinet 1990; 23(6): 587–602.
145. Barrandon Y, Green H. Cell size as a determinant of the clone-forming ability of human keratinocytes. Proc Natl Acad Sci USA 1985; 82(16):5390–5394.
146. Staiano-Coico L, Higgins PJ, Darzynkiewicz Z, Kimmel M, Gottlieb AB, Pagan-Charry I, Madden MR, Finkelstein JL, Hefton JM. Human keratinocyte culture: identification and staging of epidermal cell subpopulations. J Clin Invest 1986; 77(2):396–404.
147. Pavlovitch JH, Rizk-Rabin M, Gervaise M, Metezeau P, Grunwald D. Cell subpopulations within proliferative and differentiating compartments of epidermis. Am J Physiol 1989; 256(5 Pt 1):C977–C986.
148. Poot M, Rizk-Rabin M, Hoehn H, Pavlovitch JH. Cell size and RNA content correlate with cell differentiation and proliferative capacity of rat keratinocytes. J Cell Physiol 1990; 143(2):279–286.
149. Pavlovitch JH, Rizk-Rabin M, Jaffray P, Hoehn H, Poot M. Characteristics of homogeneously small keratinocytes from newborn rat skin: possible epidermal stem cells. Am J Physiol 1991; 261(6 Pt 1):C964–C972.
150. Tani H, Morris RJ, Kaur P. Enrichment for murine keratinocyte stem cells based on cell surface phenotype. Proc Natl Acad Sci USA 2000; 97(20):10960–10965.
151. Morhenn VB, Wood GS, Engleman EG, Oseroff AR. Selective enrichment of human epidermal cell subpopulations using monoclonal antibodies. J Invest Dermatol 1983; 81(suppl 1): 127s–131s.
152. Tozeren A, Kleinman HK, Wu S, Mercurio AM, Byers SW. Integrin alpha 6 beta 4 mediates dynamic interactions with laminin. J Cell Sci 1994; 107 (Pt 11):3153–3163.

153. Moles J-P, Watt FM. The epidermal stem cell compartment: variation in expression levels of E-cadherin and catenins within the basal layer of human epidermis. J Histochem Cytochem 1997; 45(6):867–874.

154. Jones PH, Watt FM. Separation of human epidermal stem cells from transit amplifying cells on the basis of differences in integrin function and expression. Cell 1993; 73(4):713–724.

155. Li A, Simmons PJ, Kaur P. Identification and isolation of candidate human keratinocyte stem cells based on cell surface phenotype. Proc Natl Acad Sci USA 1998; 95(7):3902–3907.

156. Bickenbach JR, Chism E. Selection and extended growth of murine epidermal stem cells in culture. Exp Cell Res 1998; 244(1):184–195.

157. Fortunel NO, Hatzfeld JA, Rosemary PA, Ferraris C, Monier MN, Haydont V, Longuet J, Brethon B, Lim B, Castiel I, Schmidt R, Hatzfeld A. Long-term expansion of human functional epidermal precursor cells: promotion of extensive amplification by low TGF-beta1 concentrations. J Cell Sci 2003; 116(Pt 19):4043–4052.

158. Pouliot N, Saunders NA, Kaur P. Laminin 10/11: an alternative adhesive ligand for epidermal keratinocytes with a functional role in promoting proliferation and migration. Exp Dermatol 2002; 11(5):387–397.

159. Visser JW, Bauman JG, Mulder AH, Eliason JF, de Leeuw AM. Isolation of murine pluripotent hemopoietic stem cells. J Exp Med 1984; 159(6):1576–1590.

160. Civin CI, Strauss LC, Brovall C, Fackler MJ, Schwartz JF, Shaper JH. Antigenic analysis of hematopoiesis. III. A hematopoietic progenitor cell surface antigen defined by a monoclonal antibody raised against KG-1a cells. J Immunol 1984; 133(1):157–165.

161. Spangrude GJ, Heimfeld S, Weissman IL. Purification and characterization of mouse hematopoietic stem cells. Science 1988; 241(4861):58–62.

162. Kaur P, Li A. Adhesive properties of human basal epidermal cells: an analysis of keratinocyte stem cells, transit amplifying cells, and postmitotic differentiating cells. J Invest Dermatol 2000; 114(3):413–420.

163. Tumbar T, Guasch G, Greco V, Blanpain C, Lowry WE, Rendl M, Fuchs E. Defining the epithelial stem cell niche in skin. Science 2004; 303(5656):359–363.

164. Trempus CS, Morris RJ, Bortner CD, Cotsarelis G, Faircloth RS, Reece JM, Tennant RW. Enrichment for living murine keratinocytes from the hair follicle bulge with the cell surface marker CD34. J Invest Dermatol 2003; 120(4):501–511.

165. Lansdorp PM, Dragowska W. Long-term erythropoiesis from constant numbers of CD34+ cells in serum-free cultures initiated with highly purified progenitor cells from human bone marrow. J Exp Med 1992; 175(6):1501–1509.

166. Fuchs E, Tumbar T, Guasch G. Socializing with the neighbors: stem cells and their niche. Cell 2004; 116(6):769–778.

167. Rheinwald JG, Green H. Serial cultivation of strains of human epidermal keratinocytes: the formation of keratinizing colonies from single cells. Cell 1975; 6(3):331–343.

168. Barrandon Y, Green H. Three clonal types of keratinocyte with different capacities for multiplication. Proc Natl Acad Sci USA 1987; 84(8):2302–2306.

169. Schneider TE, Barland C, Alex AM, Mancianti ML, Lu Y, Cleaver JE, Lawrence HJ, Ghadially R. Measuring stem cell frequency in epidermis: a quantitative in vivo functional assay for long-term repopulating cells. Proc Natl Acad Sci USA 2003; 100(20): 11412–11417.

170. Gallico GG III, O'Connor NE, Compton CC, Kehinde O, Green H. Permanent coverage of large burn wounds with autologous cultured human epithelium. N Engl J Med 1984; 311(7):448–451.

171. Compton CC, Gill JM, Bradford DA, Regauer S, Gallico GG, O'Connor NE. Skin regenerated from cultured epithelial autografts on full-thickness burn wounds from 6 days to 5 years after grafting: a light, electron microscopic and immunohistochemical study. Lab Invest 1989; 60(5):600–612.

172. Li A, Pouliot N, Redvers R, Kaur P. Extensive tissue-regenerative capacity of neonatal human keratinocyte stem cells and their progeny. J Clin Invest 2004; 113(3):390–400.

173. Baines P, Visser JW. Analysis and separation of murine bone marrow stem cells by H33342 fluorescence-activated cell sorting. Exp Hematol 1983; 11(8):701–708.

174. Goodell MA, Brose K, Paradis G, Conner AS, Mulligan RC. Isolation and functional properties of murine hematopoietic stem cells that are replicating in vivo. J Exp Med 1996; 183(4):1797–1806.
175. Goodell MA, Rosenzweig M, Kim H, Marks DF, DeMaria M, Paradis G, Grupp SA, Sieff CA, Mulligan RC, Johnson RP. Dye efflux studies suggest that hematopoietic stem cells expressing low or undetectable levels of CD34 antigen exist in multiple species. Nat Med 1997; 3(12):1337–1345.
176. Parmar K, Sauk-Schubert C, Burdick D, Handley M, Mauch P. Sca+CD34− murine side population cells are highly enriched for primitive stem cells. Exp Hematol 2003; 31(3):244–250.
177. Hulspas R, Quesenberry PJ. Characterization of neurosphere cell phenotypes by flow cytometry. Cytometry 2000; 40(3):245–250.
178. Bhattacharya S, Jackson JD, Das AV, Thoreson WB, Kuszynski C, James J, Joshi S, Ahmad I. Direct identification and enrichment of retinal stem cells/progenitors by Hoechst dye efflux assay. Invest Ophthalmol Vis Sci 2003; 44(6):2764–2773.
179. Murayama A, Matsuzaki Y, Kawaguchi A, Shimazaki T, Okano H. Flow cytometric analysis of neural stem cells in the developing and adult mouse brain. J Neurosci Res 2002; 69(6):837–847.
180. Kim M, Morshead CM. Distinct populations of forebrain neural stem and progenitor cells can be isolated using side-population analysis. J Neurosci 2003; 23(33):10703–10709.
181. Asakura A, Rudnicki MA. Side population cells from diverse adult tissues are capable of in vitro hematopoietic differentiation. Exp Hematol 2002; 30(11):1339–1345.
182. Hierlihy AM, Seale P, Lobe CG, Rudnicki MA, Megeney LA. The postnatal heart contains a myocardial stem cell population. FEBS Lett 2002; 530(1–3):239–243.
183. Oh H, Chi X, Bradfute SB, Mishina Y, Pocius J, Michael LH, Behringer RR, Schwartz RJ, Entman ML, Schneider MD. Cardiac muscle plasticity in adult and embryo by heart-derived progenitor cells. Ann NY Acad Sci 2004; 1015:182–189.
184. Shimano K, Satake M, Okaya A, Kitanaka J, Kitanaka N, Takemura M, Sakagami M, Terada N, Tsujimura T. Hepatic oval cells have the side population phenotype defined by expression of ATP-binding cassette transporter ABCG2/BCRP1. Am J Pathol 2003; 163(1):3–9.
185. Uchida N, Fujisaki T, Eaves AC, Eaves CJ. Transplantable hematopoietic stem cells in human fetal liver have a CD34(+) side population (SP)phenotype. J Clin Invest 2001; 108(7):1071–1077.
186. Wulf GG, Luo KL, Jackson KA, Brenner MK, Goodell MA. Cells of the hepatic side population contribute to liver regeneration and can be replenished with bone marrow stem cells. Haematologica 2003; 88(4):368–378.
187. Summer R, Kotton DN, Sun X, Ma B, Fitzsimmons K, Fine A. Side population cells and Bcrp1 expression in lung. Am J Physiol Lung Cell Mol Physiol 2003; 285(1):L97–L104.
188. Giangreco A, Shen H, Reynolds SD, Stripp BR. Molecular phenotype of airway side population cells. Am J Physiol Lung Cell Mol Physiol 2004; 286(4):L624–L630.
189. Summer R, Kotton DN, Sun X, Fitzsimmons K, Fine A. The origin and phenotype of lung side population cells. Am J Physiol Lung Cell Mol Physiol 2004; 286(4):L624–L630.
190. Welm BE, Tepera SB, Venezia T, Graubert TA, Rosen JM, Goodell MA. Sca-1(pos) cells in the mouse mammary gland represent an enriched progenitor cell population. Dev Biol 2002; 245(1):42–56.
191. Alvi AJ, Clayton H, Joshi C, Enver T, Ashworth A, Vivanco MM, Dale TC, Smalley MJ. Functional and molecular characterisation of mammary side population cells. Breast Cancer Res 2003; 5(1):R1–R8.
192. Clarke RB, Anderson E, Howell A, Potten CS. Regulation of human breast epithelial stem cells. Cell Prolif 2003; 36(suppl 1):45–58.
193. Welm B, Behbod F, Goodell MA, Rosen JM. Isolation and characterization of functional mammary gland stem cells. Cell Prolif 2003; 36(suppl 1):17–32.
194. Clayton H, Titley I, Vivanco M. Growth and differentiation of progenitor/stem cells derived from the human mammary gland. Exp Cell Res 2004; 297(2):444–460.

195. Liu BY, McDermott SP, Khwaja SS, Alexander CM. The transforming activity of Wnt effectors correlates with their ability to induce the accumulation of mammary progenitor cells. Proc Natl Acad Sci USA 2004; 101(12):4158–4163.

196. Gussoni E, Soneoka Y, Strickland CD, Buzney EA, Khan MK, Flint AF, Kunkel LM, Mulligan RC. Dystrophin expression in the mdx mouse restored by stem cell transplantation. Nature 1999; 401(6751):390–394.

197. Jackson KA, Mi T, Goodell MA. Hematopoietic potential of stem cells isolated from murine skeletal muscle. Proc Natl Acad Sci USA 1999; 96(25):14482–14486.

198. McKinney-Freeman SL, Jackson KA, Camargo FD, Ferrari G, Mavilio F, Goodell MA. Muscle-derived hematopoietic stem cells are hematopoietic in origin. Proc Natl Acad Sci USA 2002; 99(3):1341–1346.

199. Asakura A, Seale P, Girgis-Gabardo A, Rudnicki MA. Myogenic specification of side population cells in skeletal muscle. J Cell Biol 2002; 159(1):123–134.

200. Asakura A. Stem cells in adult skeletal muscle. Trends Cardiovasc Med 2003; 13(3): 123–128.

201. Majka SM, Jackson KA, Kienstra KA, Majesky MW, Goodell MA, Hirschi KK. Distinct progenitor populations in skeletal muscle are bone marrow derived and exhibit different cell fates during vascular regeneration. J Clin Invest 2003; 111(1):71–79.

202. McKinney-Freeman SL, Majka SM, Jackson KA, Norwood K, Hirschi KK, Goodell MA. Altered phenotype and reduced function of muscle-derived hematopoietic stem cells. Exp Hematol 2003; 31(9):806–814.

203. Tamaki T, Akatsuka A, Okada Y, Matsuzaki Y, Okano H, Kimura M. Growth and differentiation potential of main- and side-population cells derived from murine skeletal muscle. Exp Cell Res 2003; 291(1):83–90.

204. Alison MR. Tissue-based stem cells: ABC transporter proteins take centre stage. J Pathol 2003; 200(5):547–550.

205. Montanaro F, Liadaki K, Volinski J, Flint A, Kunkel LM. Skeletal muscle engraftment potential of adult mouse skin side population cells. Proc Natl Acad Sci USA 2003; 100(16):9336–9341.

206. Terunuma A, Jackson KL, Kapoor V, Telford WG, Vogel JC. Side population keratinocytes resembling bone marrow side population stem cells are distinct from label-retaining keratinocyte stem cells. J Invest Dermatol 2003; 121(5):1095–1103.

207. Triel C, Vestergaard ME, Bolund L, Jensen TG, Jensen UB. Side population cells in human and mouse epidermis lack stem cell characteristics. Exp Cell Res 2004; 295(1): 79–90.

208. Bickenbach JR. Identification and behavior of label-retaining cells in oral mucosa and skin. J Dent Res 1981; 60(spec C):1611–1620.

209. Mackenzie IC, Bickenbach JR. Label-retaining keratinocytes and Langerhans cells in mouse epithelia. Cell Tissue Res 1985; 242(3):551–556.

210. Bickenbach JR, McCutecheon J, Mackenzie IC. Rate of loss of tritiated thymidine label in basal cells in mouse epithelial tissues. Cell Tissue Kinet 1986; 19(3):325–333.

211. Chen FD, Hendry JH. The radiosensitivity of microcolony- and macrocolony-forming cells in mouse tail epidermis. Br J Radiol 1986; 59(700):389–395.

212. Mackenzie I. The ordered structure of the mammalian epidermis. In: Maibach HI, Rovee DT, eds. Epidermal Wound Healing. Chicago: Year Book Med. Publ., 1972:5–25.

213. Potten CS. Sensitivity of follicular melanoblasts in newborn mouse skin to tritiated thymidine: evidence for a long-term retention of label. Experientia 1982; 38(12): 1464–1468.

214. Cotsarelis G, Cheng SZ, Dong G, Sun TT, Lavker RM. Existence of slow-cycling limbal epithelial basal cells that can be preferentially stimulated to proliferate: implications on epithelial stem cells. Cell 1989; 57(2):201–209.

215. Wei ZG, Cotsarelis G, Sun TT, Lavker RM. Label-retaining cells are preferentially located in fornical epithelium: implications on conjunctival epithelial homeostasis. Invest Ophthalmol Vis Sci 1995; 36(1):236–246.

216. Stasiak PC, Purkis PE, Leigh IM, Lane EB. Keratin 19: predicted amino acid sequence and broad tissue distribution suggest it evolved from keratinocyte keratins. J Invest Dermatol 1989; 92(5):707–716.

217. Michel M, Torok N, Godbout M, Lussier M, Gaudreau P, Royal A, Germain L. Keratin 19 as a biochemical marker of skin stem cells in vivo and in vitro: keratin 19 expressing cells are differentially localized in function of anatomic sites, and their number varies with donor age and culture stage. J Cell Sci 1996; 109(5):1017–1028.

218. Ghali L, Wong S-T, Tidman N, Quinn A, Philpott MP, Leigh IM. Epidermal and hair follicle progenitor cells express melanoma-associated chondroitin sulfate proteoglycan core protein. J Invest Dermatol 2004; 122(2):433–442.

219. Lyle S, Christofidou-Solomidou M, Liu Y, Elder DE, Albelda S, Cotsarelis G. The C8/144B monoclonal antibody recognizes cytokeratin 15 and defines the location of human hair follicle stem cells. J Cell Sci 1998; 111(Pt 21):3179–3188.

220. Zhu AJ, Haase I, Watt FM. Signaling via beta1 integrins and mitogen-activated protein kinase determines human epidermal stem cell fate in vitro. Proc Natl Acad Sci USA 1999; 96(12):6728–6733.

221. Brakebusch C, Grose R, Quondamatteo F, Ramirez A, Jorcano JL, Pirro A, Svensson M, Herken R, Sasaki T, Timpl R, Werner S, Fassler R. Skin and hair follicle integrity is crucially dependent on beta 1 integrin expression on keratinocytes. EMBO J 2000; 19(15): 3990–4003.

222. Waikel RL, Kawachi Y, Waikel PA, Wang XJ, Roop DR. Deregulated expression of c-Myc depletes epidermal stem cells. Nat Genet 2001; 28(2):165–168.

223. Okumura T, Shimada Y, Imamura M, Yasumoto S. Neurotrophin receptor p75(NTR) characterizes human esophageal keratinocyte stem cells in vitro. Oncogene 2003; 22(26): 4017–4026.

224. Polakowska RR, Piacentini M, Bartlett R, Goldsmith LA, Haake AR. Apoptosis in human skin development: morphogenesis, periderm, and stem cells. Dev Dyn 1994; 199(3):176–188.

225. Akiyama M, Smith LT, Holbrook KA. Growth factor and growth factor receptor localization in the hair follicle bulge and associated tissue in human fetus. J Invest Dermatol 1996; 106(3):391–396.

226. Combates NJ, Chuong CM, Stenn KS, Prouty SM. Expression of two Ig family adhesion molecules in the murine hair cycle: DCC in the bulge epithelia and NCAM in the follicular papilla. J Invest Dermatol 1997; 109(5):672–678.

227. Tseng H, Green H. Association of basonuclin with ability of keratinocytes to multiply and with absence of terminal differentiation. J Cell Biol 1994; 126(2):495–506.

228. Weiner L, Green H. Basonuclin as a cell marker in the formation and cycling of the murine hair follicle. Differentiation 1998; 63(5):263–272.

229. Lehrer MS, Sun TT, Lavker RM. Strategies of epithelial repair: modulation of stem cell and transit amplifying cell proliferation. J Cell Sci 1998; 111(Pt 19):2867–2875.

230. Cotsarelis G, Kaur P, Dhouailly D, Hengge U, Bickenbach J. Epithelial stem cells in the skin: definition, markers, localization and functions. Exp Dermatol 1999; 8(1):80–88.

231. Jih DM, Lyle S, Elenitsas R, Elder DE, Cotsarelis G. Cytokeratin 15 expression in trichoepitheliomas and a subset of basal cell carcinomas suggests they originate from hair follicle stem cells. J Cutan Pathol 1999; 26(3):113–118.

232. Liu Y, Lyle S, Yang Z, Cotsarelis G. Keratin 15 promoter targets putative epithelial stem cells in the hair follicle bulge. J Invest Dermatol 2003; 121(5):963–968.

233. Krajewska M, Krajewski S, Zapata JM, Van Arsdale T, Gascoyne RD, Berern K, McFadden D, Shabaik A, Hugh J, Reynolds A, Clevenger CV, Reed JC. TRAF-4 expression in epithelial progenitor cells: analysis in normal adult, fetal, and tumor tissues. Am J Pathol 1998; 152(6):1549–1561.

234. Terstappen LW, Huang S, Safford M, Lansdorp PM, Loken MR. Sequential generations of hematopoietic colonies derived from single nonlineage-committed CD34+CD38− progenitor cells. Blood 1991; 77(6):1218–1227.

235. Botchkarev VA, Metz M, Botchkareva NV, Welker P, Lommatzsch M, Renz H, Paus R. Brain-derived neurotrophic factor, neurotrophin-3, and neurotrophin-4 act as "epitheliotro-phins" in murine skin. Lab Invest 1999; 79(5):557–572.

236. Ess KC, Witte DP, Bascomb CP, Aronow BJ. Diverse developing mouse lineages exhibit high-level c-Myb expression in immature cells and loss of expression upon differentiation. Oncogene 1999; 18(4):1103–1111.

237. Mills AA, Zheng B, Wang XJ, Vogel H, Roop DR, Bradley A. p63 is a p53 homologue required for limb and epidermal morphogenesis. Nature 1999; 398(6729):708–713.

238. Yang A, Schweitzer R, Sun D, Kaghad M, Walker N, Bronson RT, Tabin C, Sharpe A, Caput D, Crum C, McKeon F. p63 is essential for regenerative proliferation in limb, cranio-facial and epithelial development. Nature 1999; 398(6729):714–718.

239. Pellegrini G, Dellambra E, Golisano O, Martinelli E, Fantozzi I, Bondanza S, Ponzin D, McKeon F, De Luca M. p63 identifies keratinocyte stem cells. Proc Natl Acad Sci USA 2001; 98(6):3156–3161.

240. DasGupta R, Fuchs E. Multiple roles for activated LEF/TCF transcription complexes during hair follicle development and differentiation. Development 1999; 126(20): 4557–4568.

241. Merrill BJ, Gat U, DasGupta R, Fuchs E. Tcf3 and Lef1 regulate lineage differentiation of multipotent stem cells in skin. Genes Dev 2001; 15(13):1688–1705.

242. Ivanova NB, Dimos JT, Schaniel C, Hackney JA, Moore KA, Lemischka IR. A stem cell mol-ecular signature. Science 2002; 298(5593):601–604.

243. Sander G, Bawden CS, Hynd PI, Nesci A, Rogers G, Powell BC. Expression of the homeobox gene, Barx2, in wool follicle development. J Invest Dermatol 2000; 115(4):753–756.

244. Lowell S, Jones P, Le Roux I, Dunne J, Watt FM. Stimulation of human epidermal differen-tiation by delta-notch signalling at the boundaries of stem cell clusters. Curr Biol 2000; 10(9):491–500.

245. Bull JJ, Muller-Rover S, Patel SV, Chronnell CM, McKay IA, Philpott MP. Contrasting local-ization of c-Myc with other Myc superfamily transcription factors in the human hair follicle and during the hair growth cycle. J Invest Dermatol 2001; 116(4):617–622.

246. Dunnwald M, Tomanek-Chalkley A, Alexandrunas D, Fishbaugh J, Bickenbach JR. Isolating a pure population of epidermal stem cells for use in tissue engineering. Exp Dermatol 2001; 10(1):45–54.

247. Ito M, Kizawa K. Expression of calcium-binding S100 proteins A4 and A6 in regions of the epithelial sac associated with the onset of hair follicle regeneration. J Invest Dermatol 2001; 116(6):956–963.

248. Kullander K, Klein R. Mechanisms and functions of Eph and ephrin signalling. Nat Rev Mol Cell Biol 2002; 3(7):475–486.

249. Yu Y, Flint A, Dvorin EL, Bischoff J. AC133-2, a novel isoform of human AC133 stem cell antigen. J Biol Chem 2002; 277(23):20711–20716.

250. Matic M, Evans WH, Brink PR, Simon M. Epidermal stem cells do not communicate through gap junctions. J Invest Dermatol 2002; 118(1):110–116.

251. Matic M, Simon M. Label-retaining cells (presumptive stem cells) of mice vibrissae do not express gap junction protein connexin 43. J Investig Dermatol Symp Proc 2003; 8(1):91–95.

252. Couchman JR, McCarthy KJ, Woods A. Proteoglycans and glycoproteins in hair follicle development and cycling. Ann NY Acad Sci 1991; 642:243–251; discussion 51–52.

253. Li L, Mignone J, Yang M, Matic M, Penman S, Enikolopov G, Hoffman RM. Nestin expression in hair follicle sheath progenitor cells. PNAS 2003; 100(17):9958–9961.

254. Albert MR, Foster RA, Vogel JC. Murine epidermal label-retaining cells isolated by flow cytometry do not express the stem cell markers CD34, Sca-1, or Flk-1. J Invest Dermatol 2001; 117(4):943–948.

255. Lu QL, Poulsom R, Wong L, Hanby AM. Bcl-2 expression in adult and embryonic non-haematopoietic tissues. J Pathol 1993; 169(4):431–437.

256. Rodriguez-Villanueva J, Colome MI, Brisbay S, McDonnell TJ. The expression and localiz-ation of bcl-2 protein in normal skin and in non-melanoma skin cancers. Pathol Res Pract 1995; 191(5):391–398.

257. Iuchi S, Green H. Basonuclin, a zinc finger protein of keratinocytes and reproductive germ cells, binds to the rRNA gene promoter. Proc Natl Acad Sci USA 1999; 96(17):9628–9632.

258. Iuchi S, Easley K, Matsuzaki K, Weiner L, O'Connor N, Green H. Alternative subcellular locations of keratinocyte basonuclin. Exp Dermatol 2000; 9(3):178–184.

259. Whitbread LA, Powell BC. Expression of the intermediate filament keratin gene, K15, in the basal cell layers of epithelia and the hair follicle. Exp Cell Res 1998; 244(2):448–459.

260. Porter RM, Lunny DP, Ogden PH, Morley SM, McLean WH, Evans A, Harrison DL, Rugg EL, Lane EB. K15 expression implies lateral differentiation within stratified epithelial basal cells. Lab Invest 2000; 80(11):1701–1710.

261. Waseem A, Dogan B, Tidman N, Alam Y, Purkis P, Jackson S, Lalli A, Machesney M, Leigh IM. Keratin 15 expression in stratified epithelia: downregulation in activated keratinocytes. J Invest Dermatol 1999; 112(3):362–369.

262. Withers HR. The effect of oxygen and anaesthesia on radiosensitivity in vivo of epithelial cells of mouse skin. Br J Radiol 1967; 40(473):335–343.

263. Powis G, Mustacich D, Coon A. The role of the redox protein thioredoxin in cell growth and cancer. Free Radic Biol Med 2000; 29(3–4):312–322.

8

A Perspective on In Vitro Clonogenic Keratinocytes: A Window into the Regulation of the Progenitor Cell Compartment of the Cutaneous Epithelium

Rebecca J. Morris
Department of Dermatology, Columbia University Medical Center,
New York, New York, U.S.A.

INTRODUCTION

Although there have been some recent advances toward the prospective identification of keratinocyte stem cells (KSCs), particularly those of the hair follicle (1–4), stem cells in the cutaneous epithelium have usually been identified by their functions, such as retention of [^{3}H]thymidine label, in vitro colony formation, and the ability to reconstitute a graft. We will discuss here the functional properties of in vitro clonogenic keratinocytes from mice and will show how they provide a window into the regulation of the stem-cell compartment.

CUTANEOUS EPITHELIUM AS A CONTINUALLY RENEWING TISSUE CONTAINING IN VITRO CLONOGENIC KERATINOCYTES

Stem Cells and Transit Amplifying Cells

The cutaneous epithelium is a continuously renewing tissue consisting of a large population of transit amplifying (TA) keratinocytes having limited proliferative capacity and a much smaller population of KSCs with high proliferative and clonogenic potential (5–7). Under the steady-state conditions of normal homeostasis, stem cells divide to produce TA cells and to renew the stem-cell population; whereas the TA cells divide a limited number of times and are displaced to the differentiating suprabasal layers, where they are lost by terminal differentiation (5–8). The cutaneous epithelium may also contain conditional stem cells that would normally undergo terminal differentiation, but that could be recruited as stem cells in situations where the stem cells are damaged (for review, see Ref. 9).

The Stem-Cell Niche

Stem cells usually reside in a niche where they are protected from damage or injury (8,10,11). In the cutaneous epithelium, stem cells are thought to reside in the center of the epidermal proliferative units (EPUs), in the interfollicular epidermis (12), and in the bulge region of the hair follicle (13; for review, see Ref. 9). The existence of stem cells in the EPUs is surmised from cell kinetic data including retention of [3H]thymidine label in autoradiographs (14–16) and mathematical modeling studies (17). Using a method that harvests keratinocytes principally from the interfollicular epidermis of mouse ears, the Bickenbach laboratory (18) has demonstrated that label-retaining cells (LRCs) adhere rapidly to dishes coated with type IV collagen, may be clonogenic in vitro, and may reconstitute an epithelium in an in vitro grafting procedure.

In contrast to the relative paucity of data on interfollicular epidermal stem cells, several observations implicate the bulge as a source of potent multipotential progenitors. First, the bulge is a site for [^{3}H]thymidine LRCs (10) some of which are remarkably persistent (19). Second, bulge keratinocytes enriched by expression of several selectable determinants such as CD34+ (2), K15 promoter expression driving enhanced green fluorescent protein (EGFP) (3), and high levels of histone B1 (1,4) form large colonies in vitro, and in the case of K15 and histone B1, are able to reconstitute a graft.

FUNCTIONAL EVIDENCE OF STEM CELLS IN EPIDERMIS AND HAIR FOLLICLES

In Vitro Colony Formation by Freshly Harvested Keratinocytes

We have focused on colony formation in vitro by freshly harvested epidermal cells because this is a well-recognized, quantifiable indicator of both the number of cells with high growth potential relative to other proliferative cells and also the relative growth potential of single cells (20,21). Although we have refined the culture conditions and media over the years, our assay for clonogenic keratinocytes has in principle remained much the same. Briefly, epidermal cells including those from the hair follicles are harvested by a mild, low-temperature (32°C) trypsinization procedure that we optimized for reproducible yields of highly culturable single cells. The cells are seeded at a clonal density of 1×10^3 trypan-blue-excluding cells per 60 mm dish together with 1×10^6 irradiated 3T3 feeder cells, and are cultured at 32°C for intervals of two and four weeks. The dishes are then fixed with neutral buffered formalin and stained with rhodamine B. Colonies greater than 0.5 mm in diameter are scored and their sizes measured.

In Vitro Clonogenic Keratinocytes Co-sediment on Density Gradients with Other Aspects of Progenitor Activity

We formulated continuous density gradients of Percoll designed to separate basal cells of different buoyant density (22). We collected five fractions from the gradients and characterized them with regard to the number of cells present, their viability, and their basal origin. We determined that suprabasal keratinocytes remained primarily at the top of the gradients, whereas basal cells sedimented throughout. We observed that basal keratinocytes with progenitor activity sedimented with increasing density. Hence, basal keratinocytes within the density range of 1.097 to 1.143 g/mL were enriched for slowly cycling [^{3}H]thymidine label-retaining and [^{3}H]benzo[a]pyrene retaining cells, for keratinocytes that could proliferate in vitro in the continuous presence of 0.1 mg/mL of the tumor promoter, 12-O-tetradecanoylphorbol-13-acetate (TPA), for keratinocytes that were resistant to calcium-induced terminal differentiation, and for clonogenic keratinocytes. This co-sedimentation of activities associated with high in vitro proliferative potential and relative immaturity suggested that basal keratinocytes including clonogenic cells were enriched for progenitor cells including stem cells.

Slowly Cycling (Label-Retaining) Keratinocytes Behave Like Clonogenic Stem Cells In Vitro

We provided further evidence that clonogenic keratinocytes were potent progenitor cells by generating LRCs as well as pulse-labeled cells in mice, then harvesting the keratino-cytes, culturing them at low density on feeder layers for various intervals, and then per-forming light microscopic autoradiography on the culture dishes (23). When we quantified the distribution of labeled nuclei, we found that on day 2 following seeding, keratinocytes from both the label-retaining as well as the pulse-labeled mice were present as single cells. However, after five days, the LRCs were found as pairs and clusters having a grain count consistent with their division. In contrast, pulse-labeled cells remained as single cells that enlarged considerably but did not divide. These results suggested that LRCs in vivo are clonogenic in vitro, whereas pulse-labeled cells are rarely clonogenic. Hence, label-retaining keratinocytes are not only persistent in the epidermis and hair follicles, but also have relatively greater proliferative potential than pulse-labeled cells and may be stem cells.

Two Factors That Do Not Appear to Change the Number of In Vitro Clonogenic Keratinocytes

Normal Aging of Adult Mice

We prepared keratinocytes from the cutaneous epithelium of normal, untreated CD-1 female mice 9 to 69 weeks of age (24). Single-cell suspensions of freshly harvested ker-atinocytes were seeded at a clonal density onto Swiss 3T3 feeders cells, cultivated for two weeks in SPRD-105 medium, fixed, and stained. As shown in Figure 1, the number of primary epidermal colonies in this culture system remained essentially unchanged during adult life with an average cloning efficiency of 0.45%. As an internal technical

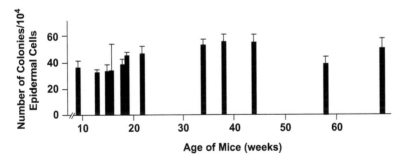

Figure 1 The number of primary clonogenic keratinocytes from normal, untreated adult CD-1 female mice, 9 to 69 weeks of age. Freshly harvested keratinocytes, including those from the hair follicles, were seeded at clonal density onto irradiated 3T3 feeder cells. Values represent the mean of 5 to 10 dishes plus the standard deviation. These data demonstrate that the number of in vitro clonogenic keratinocytes remains essentially constant for an extended period of adulthood. In 10 of 11 experiments performed together with matched acetone-treated controls, differences were not statistically significant ($P > 0.1074$). *Source*: From Ref. 24.

control, some of the determinations were made simultaneously, with epidermal cells harvested from age-matched control mice treated with 0.2 mL of acetone one month earlier. As demonstrated in Figure 1, any differences in cloning efficiency from 10 of 11 such experiments were not statistically significant.

Skin Tumor Initiation

To determine whether a single initiating application of 200 nmoL of the carcinogen 7,12-dimethylbenz[a]anthracene (DMBA) could bring about a change in the number of primary colonies, CD-1 female mice were exposed at eight weeks of age either to 0.2 ml of acetone or to 200 nmol of DMBA (24). At intervals between 7 and 61 weeks thereafter, the number of colonies remained within the control values for the duration of the experiment (Fig. 2). In 9 of 13 experiments, any small difference in the average number of keratinocyte colonies from acetone- or DMBA-exposed mice was not statistically significant. We noted that some of the colonies from the DMBA-treated mice tended to be larger and more densely stained than those from the acetone-controlled mice. This stable number of primary colonies for more than a year following treatment with DMBA argues against a morphologically undetectable expansion of initiated cells, but raises the question of when and where the "latent neoplastic lesion" occurs.

Several In Vivo Factors That Influence the Number of In Vitro Clonogenic KSCs

In Vivo Application of a Single Dose of TPA Induces a Transient Increase in the Number of In Vitro Clonogenic Keratinocytes

TPA is a powerful tumor promoter of carcinogen-exposed mouse skin (25). TPA is thought to work by providing an environment for the clonal expansion of carcinogen-initiated

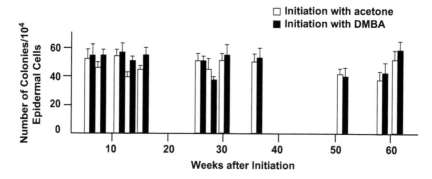

Figure 2 The number of primary in vitro clonogenic keratinocytes from CD-1 female mice exposed at eight weeks of age to a topical application of either 0.2 mL acetone or to 200 nmol of DMBA, and harvested at 7 to 61 weeks thereafter. Single-cell suspensions of epidermal keratinocytes were harvested from groups of mice, were seeded at clonal density onto irradiated 3T3 feeder cells, cultivated for two weeks in SPRD-105 medium, fixed, stained, and counted. The bars represent the mean of 4 to 10 dishes plus the standard deviation. These data demonstrate that initiation of mice with DMBA did not detectably affect the number of clonogens for 61 weeks. Qualitative differences in colony growth were observed such that many colonies from the DMBA-treated mice were larger and more densely staining than those from acetone-treated mice. In 9 of 13 separate experiments, there was no statistically significant difference ($P > 0.05$) between colony numbers from acetone- or DMBA-treated mice. *Abbreviation*: DMBA, 7,12-dimethylbenz[a]anthracene. *Source*: From Ref. 24.

cells (25). We tested the effects of a single application of TPA to CD-1 female mice on the number of in vitro clonogenic keratinocytes (Morris, unpublished observations). Our approach was to treat the mice with either TPA or with acetone when 54 days of age and then to harvest keratinocytes from the cutaneous epithelium every day following TPA treatment for 10 days, and to determine the number of keratinocyte colonies in vitro. As demonstrated in Figure 3, the significant increase in the number of colonies was not in the early

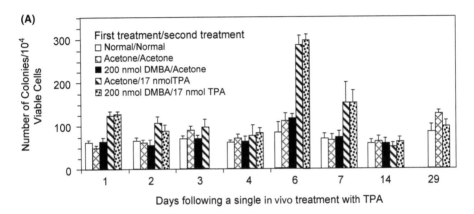

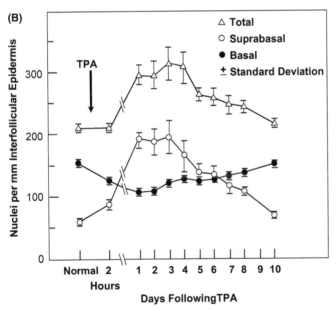

Figure 3 The number of primary in vitro clonogenic keratinocytes from mice following a single topical application of TPA: (**A**) Keratinocytes from CD-1 female mice were first treated with either 0.2 mL of acetone or 200 nmol of DMBA. For the second treatment, mice were exposed to 17 nmol of TPA or to acetone as a control. The bars represent the mean of 4 to 10 dishes plus the standard deviation. Note the marked increase in the number of colonies at six days following treatment with TPA. (**B**) Timecourse of epidermal hyperplastic growth in CD-1 female mice in vivo following a single treatment of TPA. Interfollicular epidermal cells were counted in hematoxylin-stained paraffin sections. Note the rapid increase in the total number of interfollicular cells at one day following treatment and that the hyperplastic response decreases after six days. Points represent the mean of at least six mice plus or minus the standard deviation. *Abbreviations*: TPA, 12-*O*-tetradecanoylphorbol-13-acetate; DMBA, 7,12-dimethylbenz[a]anthracene.

intervals following TPA treatment of the mice, but instead was at six days. These results are surprising and interesting in light of the TPA-induced hyperplastic response in vivo (Fig. 3B). These results are also interesting because they demonstrate that clonogenic activity is normally tightly regulated.

The epidermis responds to most types of skin damage by hyperplastic growth. Hyperplastic growth is characterized by a rapid increase in epidermal thickness and cell number followed by slower return to normal thickness and cell number. Comparison of Figures 3A and B demonstrates that the increase in epidermal colonies occurs not during the production phase of the hyperplastic response, but during its regression. This suggests that the in vitro clonogenic population may not respond directly to the damaging effects of TPA, but may either have a much delayed reaction or perhaps an indirect reaction such as a response to a cytokine made by other rapidly proliferating keratinocytes or by infiltrating inflammatory cells.

To test whether the increase in in vitro clonogenic keratinocytes represents a true increase in the number of progenitor cells in vivo, we pretreated mice topically with TPA either two days (when clonogenic activity was not significantly increased) or six days (when in vitro clonogenic activity was significantly increased) before an initiating application of N-methyl-N′-Nitro-N-Nitrosoguanidine (MNNG) (Morris, unpublished observations). One week following MNNG treatment, we treated all the mice with twice weekly tumor promotion with TPA for 15 weeks. As shown in Figure 4, mice pretreated with TPA six days before tumor initiation, when in vitro clonogenic activity was high, developed more papillomas than the control group pretreated two days prior to tumor

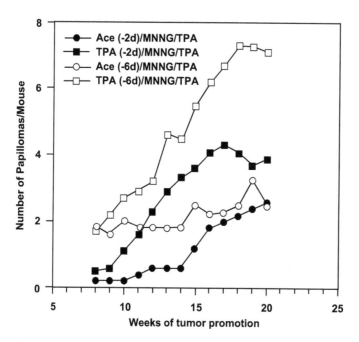

Figure 4   Effects of pretreating mice with TPA either two or six days prior to treatment with MNNG and subsequent promotion with TPA. Note the two-fold increase in the number of papillomas when mice are pretreated with TPA six days prior to tumor initiation, a time when the number of keratinocyte colonies is maximal (Fig. 3A). *Abbreviations*: MNNG, N-methyl-N′-Nitro-N-Nitrosoguanidine; TPA, 12-*O*-tetradecanoylphorbol-13-acetate.

initiation. These results suggest that in vitro clonogenic activity reflects a true change in the number of progenitors in the cutaneous epithelium. It also follows that when the clonogenic keratinocytes are removed from their in vivo environment, they express a growth potential not expressed in vivo.

In Vivo Application of Multiple Treatments of TPA

We determined the number of primary clonogenic keratinocytes from mice exposed to either acetone or to DMBA and promoted (in vivo) 1, 4, or 12 times with either TPA or with acetone as a control (24). Four weeks after the final in vivo treatment, we counted keratinocyte colonies and found them to be significantly increased in number in the cultures from TPA-treated mice over those from acetone-treated mice (Fig. 5). The increase was also significantly greater in the DMBA-initiated groups than in the acetone-initiated groups. Many colonies derived from TPA-treated epidermis tended to be pale staining and characterized by fuzzy edges or irregular margins. It is significant that a single application of TPA to mice induces little obvious persistent change in the number of primary clonogenic keratinocytes from mice treated with either acetone or DMBA. However, the number of primary in vitro clonogenic keratinocytes from control as well as DMBA-treated mice remained elevated following multiple applications of TPA. This is not surprising in light of the considerable evidence that tumor promoters are substances or treatments capable of inducing a chronic-regenerative epidermal hyperplastic growth upon repeated application.

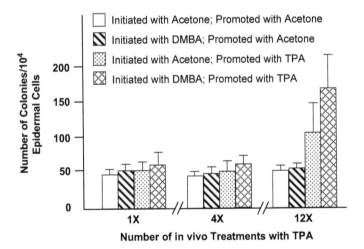

Figure 5 The number of primary in vitro clonogenic keratinocytes from groups of CD-1 mice treated first with either acetone or DMBA and second with either acetone or TPA at 1, 4, or 12 times. Four weeks after the last treatment, freshly harvested keratinocytes were seeded at clonal density onto irradiated 3T3 feeder cells, cultivated for two weeks in SPRD-105 medium, fixed, and stained. The bars represent the average number of keratinocyte colonies in 18 to 42 dishes from three to five separate experiments plus the standard error of the mean. These data demonstrate that promotion with TPA significantly ($P < 0.05$) increased the number of clonogenic keratinocytes from mice exposed to acetone as well as DMBA, but that the increase was greater when the mice were treated with DMBA. *Abbreviations*: TPA, 12-*O*-tetradecanoylphorbol-13-acetate; DMBA, 7,12-dimethylbenz[a]anthracene. *Source*: From Ref. 24.

We conclude from the foregoing experiments that the number of primary in vitro clonogenic keratinocytes is transiently and tightly regulated during epidermal hyperplasia, but is deregulated in skin carcinogenesis. These observations suggest the importance of identifying the genes regulating the number of clonogenic progenitor cells in the cutaneous epithelium. As described subsequently, we have taken a genetic approach toward the identification of genes regulating the number of keratinocyte progenitors.

Although it is possible that KSCs might be regulated by the same genes as TA cells, our observation that the number of clonogenic keratinocytes increases, not during the production phase of a hyperplastic response as we expected, but instead during the regression phase when cell proliferation subsides suggests different regulatory processes. There are three possible reasons for this. The first reason is purely technical due to differences in ease of trypsinization, cellular damage, or differences in adhesiveness. Comparison of 24-hour attached cells suggested that this is probably not the case. Second, the clonogenic keratinocytes might have a delayed response to the damaging stimulus and take longer to be released from the G0 phase of the cell cycle. Third, the clonogenic keratinocytes might not respond to the damage at all, but respond instead to growth factors and cytokines produced either by the hyperproliferative keratinocytes themselves or by some aspect of the inflammatory response. This would implicate specific stem-cell-regulatory genes.

Although we have noted increased adhesiveness during the production phase of the hyperplastic response, this would not account for an increased clonal growth during the regression phase. Moreover, cell viabilities as reflected by exclusion of trypan blue dye are high throughout the hyperplastic response.

The Number of In Vitro Clonogenic Keratinocytes Is a Function of Mouse Strain Differences

The number and colony size of clonogenic keratinocytes are influenced by mouse strain (26). Because experimental results described earlier appeared to implicate a deregulation of clonogenic keratinocytes during cutaneous carcinogenesis, we investigated whether there might be mouse-strain-dependent differences in the number of clonogenic keratinocytes. We initially hypothesized that mouse strains such as CD-1, FVB, or to a lesser extent DBA/2 and BALB/c sensitive to skin carcinogenesis would have more colonies and those resistant strains (C57BL/6) would have fewer colonies. As described subsequently, this was clearly not the case. To avoid potential bias in colony number associated with day-to-day variation, we performed a balanced incomplete block design to obtain 12 replicates from each strain. This design was "balanced" because every strain was compared with every other strain an equal number of times, "incomplete" because each block included less than the total number of strains on a given day, and "block" because each experiment had the same number of strains in a random order. When we performed this analysis, we found three subsets of mice giving significantly different numbers of colonies: C57BL/6 $\gg$ C3H = DBA/2 = SENCAR = BALB/c > FVB = CD-1, all under culture conditions optimized for the growth of keratinocytes from CD-1 mice. These results are shown in Figure 6. These strain-dependent differences in colony number were not related in any obvious way to the number of cells per millimeter of interfollicular epidermis, number of hair follicles per square centimeter of skin, or the number of cells in mitosis or DNA synthesis. However, studies of other cell kinetic parameters such as epidermal transit time or the number of LRCs need to be determined. Preliminary experiments suggest that the number of LRCs may differ between C57BL/6 and BALB/c mice; however, further work is needed to confirm these observations.

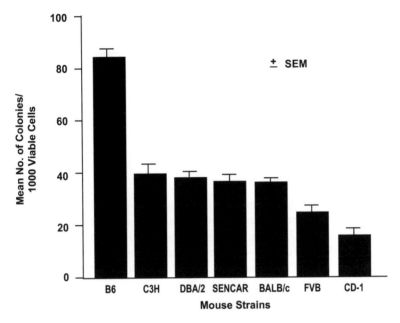

Figure 6 Mean number of keratinocyte colonies per 1000 viable cells in 57 individual C57BL/6, 24 C3H, 27 DBA/2, 30 SENCAR, 54 BALB/c, 24 FVB, and 24 CD-1 female mice (plus standard error of the mean). Freshly harvested keratinocytes were seeded onto irradiated 3T3 feeder cells and cultured for two weeks in supplemented Williams Medium E, prior to fixing, staining, and counting the colonies. Note that the colony counts fall into three groups: C57BL/6 > C3H = DBA/2 = SENCAR = BALB/c > FVB = CD-1. *Source*: From Ref. 26.

IN MICE, KERATINOCYTE COLONY NUMBER IS GENETICALLY DEFINED AND QUANTITATIVELY COMPLEX

Our observed deregulation of clonogenic keratinocytes in skin carcinogenesis suggested that identification of genes controlling KSC number might provide new insights into skin tumor development as well as other conditions where stem-cell regulation might be implicated. We chose to take a genetic approach toward gene identification because this approach has led to the identification of many important disease genes and other regulatory genes. Alternative approaches such as reverse genetics of screening mice with naturally occurring or induced skin mutations at the time appeared to be high risk or more expensive. Nevertheless, we are currently using gene expression studies to augment our genetic analysis.

The genetic approach to gene identification involves the identification of a phenotype, in our case, keratinocyte colony number, demonstrating that the phenotype is genetically defined and quantitative, and then using linkage analysis to map the phenotype to increasingly smaller chromosomal segments. Gene identification is accomplished by a candidate gene approach where interesting genes are resequenced and a sequence variant or mutation is noted, or by direct or in silico positional cloning and finding a sequence variant or a mutation.

In the cell kinetic and carcinogenesis experiments described earlier, we had always used CD-1 mice because they are fairly sensitive to skin carcinogenesis and because large numbers of them are readily available. Hence, our in vitro assay for clonogenic keratinocytes was optimized for CD-1 female mice.

Table 1 Genetics of Keratinocyte Colony Number

Mouse strain	Characterization	Number of mice	Number of colonies[a]
BALB/c	Parent strain	54	36.4 ± 12.2
C57BL/6	Parent strain	57	84.3 ± 24.2
CB6F1	F1 (hybrid)	30	53 ± 22.1
BALB/c × CB6F1	Backcross	44	42.9 ± 16.3
C57BL/6 × CB6F1	Backcross	45	72.2 ± 27.2
CB6F1	Intercross	104	65.5 ± 26

[a]Values represent the mean keratinocyte colony number ± S.D.

We chose C57BL/6 and BALB/c mice for further analysis because they differed significantly ($P < 0.01$) in the number of keratinocyte colonies and because they were highly inbred and genetically distinct (27). Table 1 shows the results of the various genetic crosses between C57BL/6 and BALB/c mice on keratinocyte colony number. These results demonstrate that the mean number of keratinocyte colonies in the F1 hybrid (CB6F1) between the two parental strains was intermediate between the two parents. This result indicates that keratinocyte colony number is a multigenic trait (27). When we investigated the two backcrosses (C57BL/6 × CB6F1 and BALB/c × CB6F1), we found segregation of colony number to the high and low parent such that the difference between the two backcrosses was significant ($P < 0.001$). The intercross mice (CB6F2) had a mean colony number that fell between the two backcrosses. These results reflected segregation of the trait of keratinocyte colony number. Further genetic analysis indicated that the number of keratinocyte colonies probably are not associated with a single-locus autosomal model and suggested that the trait is regulated by two or more loci having additive but not equal effects. These results suggested that we could use linkage analysis as a tool for identification of stem-cell-regulatory loci. When we performed linkage analysis according to Kruglyak and Lander (28,29), we found several loci with single-point significance but not genome-wide significance. As we had analyzed a sufficient number of animals, this finding suggested that our phenotype needed to be refined.

KERATINOCYTE COLONY SIZE IS ALSO GENETICALLY DEFINED

The observation that there were obvious size differences in the keratinocyte colonies in BALB/c and C57BL/6 mice did not escape our notice (27). We found two phenotypes: one characterized by a high number of small colonies in BALB/c mice and one characterized by a high number of large colonies in C57BL/6 mice. When we analyzed colony size in these mice and their genetic crosses, we found that colony size was also genetically inherited. Taking into account this refined phenotype, our linkage analysis disclosed a locus on chromosome 9 (*Ksc1*) with genome-wide significance and linked to the number of small colonies, and a locus on chromosome 4 (*Ksc2*) with single-point significance associated with the number of large colonies. Two additional suggestive loci were found on chromosomes 6 and 7. These results indicated the strong likelihood that one or more genes within the locus on chromosome 9 regulates the trait of a high number of small colonies and that a gene or genes within the locus on chromosome 4 may regulate the number of large colonies. Surprisingly, the locus on chromosome 9 and the loci on chromosomes 6 and 7 map close to loci mapped in other laboratories as skin tumor susceptibility loci (30–32). This observation bears close watching, as the laboratories

Table 2   In Vitro Clonogenic Keratinocytes as Stem Cells and as Target Cells in Carcinogenesis

High proliferative potential in vitro
Include label retaining cells
Among the smallest and most dense of basal cells
Remain constant in number for most of adult lifespan (mouse)
Remain constant in number following in vivo carcinogen exposure
Increase transiently during the regression phase of in vivo epidermal hyperplasia
Increase in number during skin tumor promotion
Number and size are genetically defined quantitative complex traits

involved proceed toward gene identification. Additional studies directed toward gene identification are currently ongoing in our laboratory.

Our current model for how colony size and number relate to susceptibility or resistance to skin carcinogenesis is that susceptible mouse strains have a population of conditional stem cells as represented by a high number of small colonies. Although this population would normally undergo terminal differentiation, it can be recruited into papilloma development during tumor promotion. Identification and cloning of the genes in Ksc1 and Ksc2 may lead to the identification of genes regulating the number of KSCs. Moreover, as suggested by our data, the mechanisms regulating the intrinsic number of stem cells undoubtedly underlie the responses of the cells to extrinsic manipulation. The ability to manipulate these genes in vivo raises exciting possibilities for stem-cell-focused treatments for skin diseases including cancer. Finally, one of the interesting problems for the future is whether the stem-cell-regulatory genes are themselves targets for carcinogens and tumor promoters.

SUMMARY

We have discussed here the properties of in vitro clonogenic keratinocytes that make them candidates for a stem-cell population (summarized in Table 2). These features provide a window on the regulation of the stem-cell compartment of the cutaneous epithelium of mice. Finally, we have shown that the size and number of keratinocyte colonies are genetically defined quantitative complex traits amenable to linkage analysis and, in the future, identification of stem-cell-regulatory genes.

REFERENCES

1. Tumbar T, Guasch G, Greco V, Blanpain C, Lowry WE, Rendl M, Fuchs E. Defining the epithelial stem cell niche in skin. Science 2004; 303:359–363.
2. Trempus C, Morris RJ, Bortner CD, Cotsarelis G, Faircloth RS, Reece JM, Tennant RW. Enrichment for living keratinocytes from the hair follicle bulge with the cell surface marker CD34. J Invest Dermatol 2003; 120:501–511.
3. Morris RJ, Liu Y, Marles L, Yang Z, Trempus C, Li S, Lin JS, Sawicki JA, Cotsarelis G. Capturing and profiling adult hair follicle stem cells. Nat Biotechnol 2004; 22:411–417.
4. Blanpain C, Lowry WE, Geogheagan A, Polak L, Fuchs E. Self-renewal, multipotency, and the existence of two cells populations within an epithelial niche. Cell 2004; 118:635–648.
5. Potten CS. Stem Cells: Their Identification and Characterisation. Edinburgh: Churchill Livingstone, 1983; 200–232.
6. Watt FM. Epidermal stem cells: markers, patterning and control of stem cell fate. Phil Trans R Soc Lond 1988; 353:831–837.

7. Lavker RM, Sun T-T. Epidermal stem cells: properties, markers, and location. Proc Natl Acad Sci USA 2000; 97:13473–13475.

8. Morris RJ, Potten, CS. Slowly cycling label-retaining epidermal cells behave like clonogenic stem cells in vitro. Cell Prolif 1994; 27:279–289.

9. Potten CS, Booth C. Keratinocye stem cells: a commentary. J Invest Dermatol 2002; 4:888–899.

10. Cotsarelis G, Kaur P, Dhouailly D, Hengge U, Bickenbach J. Epithelial stem cells in the skin: definition, markers, localization, and functions. Exp Dermatol 1999; 8:80–88.

11. Spradling A, Drummond-Barbosa D, Kai T. Stem cells find their niche. Nature 2001; 414:98–104.

12. Allen TD, Potten CS. Fine structural identification and organization of the epidermal proliferative unit. J Cell Sci 1974; 15:291–319.

13. Cotsarelis G, Sun T-T, Lavker RM. Label-retaining cells reside in the bulge area of pilosebaceous units: implications for follicular stem cells, hair cycle, and skin carcinogenesis. Cell 1990; 6:1329–1337.

14. Bickenbach J. Identification and behavior of label-retaining cells in oral mucosa and skin. J Dent Res 1981; 60:1611–1620.

15. Morris RJ, Fischer SM, Slaga TJ. Evidence that the centrally and periperally located cells in the murine epidermal proliferative unit are two distinct cell populations. J Invest Dermatol 1985; 34:277–281.

16. Potten CS. Cell cycles in cell hierarchies. Int J Radiat Biol 1986; 49:257–278.

17. Potten CS, Loeffler M. Stem cells. epidermal proliferation. I. Changes with time in the proportion of isolated, paired and clustered labelled cells in sheets of murine epidermis. Virchows Arch B Cell Pathol 1987; 53:279–285.

18. Bickenbach JR, Chism E. Selection and extended growth of murine epidermal stem cells in culture. Exp Cell Res 1998; 244:184–195.

19. Morris RJ, Potten CS. Highly persistent label-retaining cells in the hair follicles of mice and their fate following induction of anagen. J Invest Dermatol 1999; 112:470–474.

20. Steel GG. Growth and survival of tumor stem cells. In: Growth Kinetics of Tumors. Oxford: Oxford University Press, 1977:218–219, 228.

21. Buick RN. Perspectives on clonogenic tumor cells, stem cells, and oncogenes. Cancer Res 1984; 44:4909–4918.

22. Morris RJ, Fischer SM, Klein-Santo AJP, Slaga TJ. Subpopulations of primary adult murine epidermal basal cells sedimented on density gradients. Cell Tissue Kinet 1991; 23:587–602.

23. Morris RJ, Potten CS. Slowly cycling (label-retaining) epidermal cells behave like clonogenic stem cells in vitro. Cell Prolif 1994; 27:279–289.

24. Morris RJ, Tacker KC, Fischer SM, Slaga TJ. Quantitation of primary in vitro clonogenic keratinocytes from normal murine epidermis, following initiation, and during promotion of epidermal tumors. Cancer Res 1988; 48:6285–6290.

25. DiGiovanni J. Multistage skin carcinogenesis in mice. In: Waalks MP, Ward JM, eds. Carcinogenesis. New York: Raven Press, 1994:265–299.

26. Popova NV, Tryson K, Wu K, Morris R. Evidence that keratinocyte colony number is genetically controlled. Exp Dermatol 2002; 11:503–508.

27. Popova NV, Teti KA, Wu KQ, Morris RJ. Identification of two keratinocyte stem cell regulatory loci implicated in carcinogenesis. Carcinogenesis 2003; 24:417–425.

28. Kruglyak L, Lander E. A nonparametric approach for mapping quantitative trait loci. Genetics 1995; 139:1421–1428.

29. Lander E, Kruglyak L. Genetic dissection of complex traits: guidelines for interpreting and reporting linkage results. Nat Genet 1995; 11:241–247.

30. Nagase H, Bryson S, Cordell K, Kemp C, Fee F, Balmain A. Distinct genetic loci control development of benign and malignant skin tumors in mice. Nat Genet 1995; 10:424–429.

31. Angel J, Beltran L, Minda K, Rupp T, DiGiovanni J. Association of murine chromosome 9 locus (Ps11) with susceptibility to mouse skin tumor promotion by 12-O-tetradecanoylphorbol-13-acetate. Mol Carcinog 1997; 20:162–167.

32. Mock B, Lowry D, Rehman I, Padlan C, Yuspa S, Hennings H. Multigenic control of skin tumor susceptibility in SENCAR/pt mice. Carcinogenesis 1998; 19:1109–1115.

9

Hepatic Stem Cells and the Liver's Maturational Lineages: Implications for Liver Biology, Gene Expression, and Cell Therapies

Eva Schmelzer, Randall E. McClelland, and Aloa Melhem
Department of Cell and Molecular Physiology, UNC School of Medicine, Chapel Hill, North Carolina, U.S.A.

Lili Zhang
Department of Infectious Diseases, Nanjing Medical University, Nanjing, China

Hsin-lei Yao
Department of Biomedical Engineering, UNC School of Medicine, Chapel Hill, North Carolina, U.S.A.

Eliane Wauthier
Department of Cell and Molecular Physiology, UNC School of Medicine, Chapel Hill, North Carolina, U.S.A.

William S. Turner
Department of Biomedical Engineering, UNC School of Medicine, Chapel Hill, North Carolina, U.S.A.

Mark E. Furth
Institute for Regenerative Medicine, Wake Forest Medical Center, Winston Salem, North Carolina, U.S.A.

David Gerber
Department of Surgery, UNC School of Medicine, Chapel Hill, North Carolina, U.S.A.

Sanjeev Gupta
Departments of Medicine and Pathology, Albert Einstein College of Medicine, Bronx, New York, U.S.A.

Lola M. Reid
Department of Cell and Molecular Physiology, Department of Biomedical Engineering and Program in Molecular Biology and Biotechnology, UNC School of Medicine, Chapel Hill, North Carolina, U.S.A.

THE LIVER AS A MATURATIONAL LINEAGE SYSTEM

Numerous excellent articles and reviews have been published within the last several years on developmental biology of the liver (1,2), on hepatic precursors found in bone marrow

161

(3–6), and on oval cells and oval cell lines (7–10). In this review, we have focused on studies on normal hepatic stem-cell and liver lineage biology not covered by these prior reviews. The readers should refer to the prior reviews for summaries of the literature ignored here. Table 1 provides definitions of terms used throughout the review.

The liver is being recognized increasingly as a maturational lineage system, including the presence of a stem-cell compartment, similar to those in the bone marrow, skin, and gut (11–20). The liver's lineage is organized physically within the acinus, the structural and functional unit of the liver (Fig. 1) (21). In a two-dimensional cross-section, the acinus is organized conceptually like a wheel around two distinct vascular beds: six sets of portal triads, each with a portal venule, hepatic arteriole, and a bile duct form the periphery, and the central vein forms the hub (Fig. 1). The parenchyma, effectively the "spokes" of the wheel, consists of single-parenchymal cell plates lined on either side by fenestrated sinusoidal endothelium. By convention, the liver is demarcated into three zones: zone 1 is periportal, zone 2 is mid-acinar, and zone 3 is pericentral (Figs. 1 and 2). Blood enters the liver from the portal venules and hepatic arterioles at the portal triads, flows through sinusoids that line the plates of parenchyma, and exits from the central vein, known also as the terminal hepatic venule. Hepatocytes display marked morphological, biochemical, and functional heterogeneity based on their zonal location (22–28). Their size increases from zone 1 to zone 3, and one can observe distinctive zonal variations in morphological features of the cells such as mitochondria, endoplasmic reticulum, and glycogen granules (24).

An indicator of the maturational lineages is ploidy (Tables 2 and 3; Figs. 2–4) (29–37). Hepatocytes show dramatic differences in DNA content from zone 1 to zone 3 with periportal cells being diploid and with a gradual shift to polyploid cells in the mid-acinar zone (rats and mice) to the pericentral zone (all mammalians) (38). Subpopulations of the polyploid cells in the pericentral zone show evidence of apoptosis, and the classic markers for apoptosis are pericentrally located (39–42). The extent of hepatic polyploidy varies with mammalian species. In young adult rats, four to five weeks of age, 90% of the parenchyma are polyploid (tetraploid and octaploid), whereas in young adult humans (20 to 30 years of age), at least 50% to 70% of the parenchyma are diploid. The extent of polyploidy also changes with age. All parenchymal cells in fetal and neonatal livers of all mammals are diploid, but they transit to the adult profile by three to four weeks of age in rats and mice and by late teenage years in humans. The fraction of the liver cells that are polyploid continues to increase with age. By six months of age, the livers of rodents are less than 2% to 3% diploid; by 50 to 60 years of age, human livers are less than half diploid (Note: rigorous estimates of the extent of polyploidy in humans are not available, as polyploid cells are intolerant of ischemia and are selectively eliminated within an hour of death in warm ischemia and within a few hours of death in cold ischemia.). It is assumed that the steady loss of diploid subpopulations with age is related to the well-known reduction in regenerative capacity of the liver with age (31–33,43).

Another representative function demonstrating lineage dependence is the cell division potential of parenchymal cells in vitro and in vivo. The diploid periportal cells demonstrate the maximum growth, whereas the pericentral parenchymal cells demonstrate the least (44). Only the diploid parenchymal cells are capable of undergoing complete cell division (45); these comprise the subpopulations of stem cells and unipotent progenitors (all less than 15 μm in diameter) and the diploid adult hepatocytes (the "small hepatocytes"), with an average diameter of 18 to 22 μm (46–48). Moreover, there remains a difference in cell division potential between the diploid subpopulations. For example, a single small hepatocyte will yield 120 daughter cells in a 20-day time period, whereas

Table 1 Glossary of Terms

Canals of Hering	Rod-like structures around the portal triads of the liver acinus are found to be the reservoir of stem cells in pediatric and adult livers; assumed to be derived from the ductal plates
Clonogenic expansion	Cells that can expand from a single cell and that can be repeatedly passaged at single-cell seeding densities; only the pluripotent progenitors (and possibly the unipotent, committed progenitors) are able to undergo clonogenic expansion
Colony formation	Cells that can form a colony of cells when seeded at low densities; diploid subpopulations, both progenitors and adult diploid cells, are able to form colonies of cells, but the adult diploid cells are limited in the numbers of divisions and are not able to undergo passaging
Committed progenitors	Unipotent progenitors capable of maturing into only one adult fate
Determined stem cells	Pluripotent cells that can develop into some, but not all, adult cell types
Ductal plate (also called limiting plate)	A plate of cells surrounding the portal triads in the liver acinus and separating the connective tissue associated with the portal triads from the parenchyma; found in the fetal and neonatal liver tissues
ES cell	Totipotent cells derived from pre- or post-implantation embryos and that can be maintained in their undifferentiated (unspecialized) state ex vivo under specific conditions
Oval cells	Small cells ($\sim$10 μm diameter) with oval-shaped nuclei and related to the stem cells and committed progenitors in the liver; they are located near the portal trials and expand in the livers of animals exposed to oncogenic insults; the insults result in stem cells or committed progenitor cells that are partially or completely transformed; characterization of them has been derived almost entirely from animals exposed to such treatments (the term is often used as a synonym for the liver's stem cells and progenitors; however, although they derive from the cells of the stem-cell compartment, they are distinguishable phenotypically and in their growth-regulatory requirements from their normal counterparts)
Pluripotent cells	Cells capable of producing more than one mature cell type
Progenitors or precursors	Broad terms encompassing both stem cells and committed progenitors
Stem cells	Totipotent or pluripotent cells that are capable of clonogenic expansion and self-replication (i.e., capable of producing daughter cells identical to the parent)
Totipotent stem cells	Cells capable of producing all cell types from all embryonic germ layers (ectoderm, mesoderm, and endoderm)

Abbreviation: ES, embryonic stem.

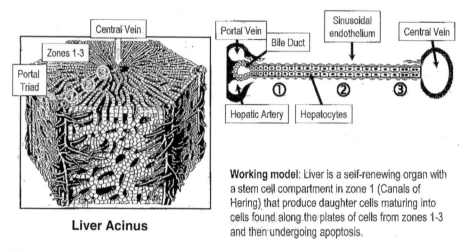

Liver Acinus

Working model: Liver is a self-renewing organ with a stem cell compartment in zone 1 (Canals of Hering) that produce daughter cells maturing into cells found along the plates of cells from zones 1-3 and then undergoing apoptosis.

Figure 1   The liver acinus. The nomenclature of the liver zones is indicated. *Source*: From Refs. 16, 21.

a single hepatoblast will yield 4000 to 5000 daughter cells in the same time period and under the same conditions. Matured, polyploid cells can undergo DNA synthesis but have limited, if any, cytokinesis under even the most optimal expansion conditions in culture due to down-regulation of factors regulating cytokinesis (45). The findings in division potential are summarized in Table 4.

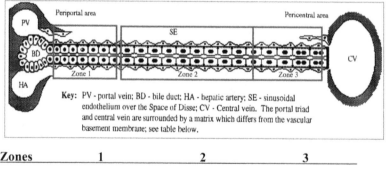

Key: PV - portal vein; BD - bile duct; HA - hepatic artery; SE - sinusoidal endothelium over the Space of Disse; CV - Central vein. The portal triad and central vein are surrounded by a matrix which differs from the vascular basement membrane; see table below.

Zones		1	2	3
Ploidy	rats	2N	4N	4N & 8N
	mice	2N & 4N	4N & 8N	up to 32N
	humans	2N	2N	2N & 4N
Growth		maximum	limited	negligible
ECM		Type IV & III collagen, laminin, HS-PG*	>>>>>>>>>>>>>>>>>> *gradient*	Type I & III collagens, Fibronectin, HP-PG*
Genes		early	intermediate	late

Size (μ)   **2N** <20   ; **4N**=~25-35; **8N and above**=>35

*HS-PG = heparan sulfate proteoglycan; HP-PG = heparin proteoglycan

Figure 2   Lineage-dependent properties of the liver. The table accompanying the figure indicates the known properties that are distributed across zones in the liver acinus. Further descriptions of these properties are provided in several reviews. *Source*: From Refs. 16, 22–24, 44, 82, 328, 329.

Table 2 Maturational Lineages Varying in Kinetics

Rapidly regenerating tissues (rapid kinetics)[a]	Quiescent tissues (slow kinetics)[b]
% Polyploid cells low (e.g., 5–10%)	% Polyploid cells intermediate (e.g., 30%) to high levels (e.g., 95%)
Representative tissues	Representative tissues
Hemopoietic cells	Lung, liver, pancreas, and other internal organs
Epidermis	Blood vessels
Intestinal epithelia	Skeletal muscle
Hair	Nerve cells, including the brain
	Heart muscle

Note: Hypothesis: kinetics of lineage inversely correlated with extent of polyploidy.
[a]Turnover in days to weeks.
[b]Turnover in months to years.

Studies of the proliferation potential of diploid and polyploid cells have been conducted in rodents. Transplantation of cells isolated from the normal liver followed by their fractionation into diploid or polyploid cells with either centrifugal elutriation or fluorescence-activated cell sorting showed that both cell fractions could proliferate in intact animals, although reconstitution of the livers occurred only with the diploid subpopulations (48,49). However, a study of rat hepatocytes isolated from the liver of animals several days after partial hepatectomy (PH), which induces hepatic polyploidy, showed that the proliferation capacity of polyploid cells was extensively attenuated compared with cells from the normal rat liver (50).

Tissue-specific gene expression has long been known to occur in distinct patterns associated with the three zones of the liver acinus. Several excellent reviews, particularly those of Gebhardt et al. (22,26,27,44,51–54) and Gumucio et al. (23,24,55), have summarized these investigations. We will mention only a few of the reports as representative of these studies.

Representative zone 1 (periportal) genes. The periportal parenchymal cells are diploid and typically approximately 18 to 22 μm in diameter and express genes associated with gluconeogenesis, such as the glucose transporter 2 (GLUT 2) (56) and phosphoenolpyruvate carboxykinase (PEPCK) (57,58), and specific fetal forms of P450s, such as

Table 3 Age Effects on Ploidy Profiles of Parenchymal Cells

Rodents	Humans[a]
Fetal and neonatal: entirely diploid	Fetuses, neonates, and children up to teenage years; entirely diploid
Young adults: 4–5 weeks of age; 10% diploid; 80% tetraploid; 10% octaploid	Young adults: 20–40 years of age; 50–70% diploid; 30–50% tetraploid
Six months and older: <5% diploid; >95% polyploid; polyploid cells are a mix of mononucleated and binucleated cells	>50 years of age: steady increase of polyploid cells with age; polyploid cells are mostly (entirely?) binucleated (tetraploid)

[a]Estimates of ploidy profiles in human parenchyma vary due to the effects of ischemia: polyploid cells are lost selectively with ischemia, especially warm ischemia.

DNA content/nucleus + number of
nuclei/cell defines ploidy

Ploidy analysis requires robust
assays such as flow cytometry.
Below: ploidy analysis of rat
hepatocytes treated with DNA
dye, Hoechst 33342, and
analyzed

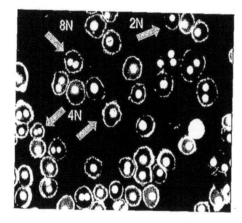

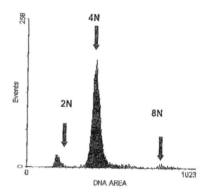

Figure 3 Ploidy of liver cells is a definitive indicator of the maturational lineage. Polyploid cells can be mononucleated or binucleated. Shown are rodent liver cells stained with a DNA dye, Hoechst 33342, and then photographed. Note the sizes of the nuclei, an indicator of the DNA content. One must utilize robust assays, such as flow cytometry, to determine the ploidy profile in liver cells. In rodent livers of young animals, that profile is, as shown, approximately 10% diploid, 80% tetraploid, and 10% octaploid *Source*: From Ref. 37.

CYP3A7 (59,60). The GLUT 2 mRNA has been determined to be 1.9-fold higher in periportal than perivenous hepatocytes, corresponding with higher protein levels in the periportal hepatocytes (61). Similarly, PEPCK mRNA expression in the adult rat, mouse, and hamster is predominantly restricted to the periportal region as shown by in situ hybridization (62) and Northern blot analyses (twofold higher in periportal than in perivenous hepatocytes). Starvation can increase the expression in the periportal region and can cause slightly increased expression in the intermediate zone in the mouse. Also, fructose-1,6-biphosphatase mRNA is detected solely in the periportal region of the rat liver and is unaffected by feeding conditions (63).

Gap junctions are formed by one of the large family of connexin genes with lineage-dependent isoforms (64–66). For example, connexin 26 has been found expressed by periportal hepatocytes, and its expression declines by mid-acinus to be replaced by connexin 32 (67). However, functional relevance of the different connexin isoforms is not known.

Representative zone 2 (mid-acinar) genes. Distinctive mid-acinar gene expression occurs primarily in mammals, such as in mice and rats, in which ploidy changes occur. Several genes, including transferrin (68) and tyrosine aminotransferase, are expressed optimally in the polyploid cells (69). It is unknown if variables other than ploidy elicit maximal expression of genes in this region of the acinar plates of parenchymal cells.

Representative zone 3 (pericentral) genes. Many genes are expressed uniquely in zone 3, the pericentral zone of the liver acinus, which include major urinary protein (MUP), alpha-1-antitrypsin (AAT), glutamine synthetase (GS) (27,70,71), a heparin proteoglycan (72), and specific isoforms of P450s such as CYP3A4 (60,73). Enzymatic activities of phase I and II enzymes (ECOD and GST activities) are restricted to mature

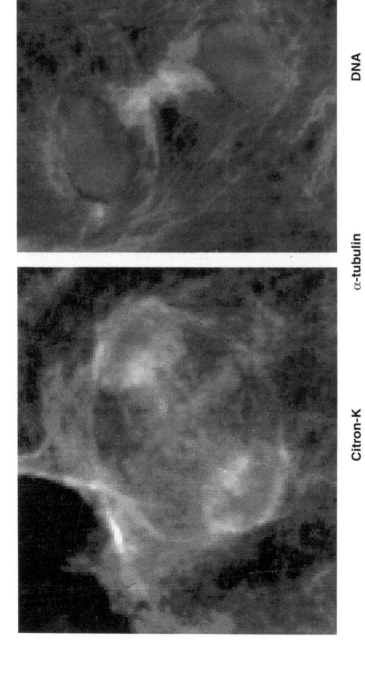

Citron-K   α-**tubulin**   **DNA**

Figure 4 Citron kinase is one of the enzymes required for cytokinesis that is down-regulated in polyploidy cells. The figure shows an hepatocyte undergoing division and indicating the localization of citron kinase during and at the end of cytokinesis. *Source:* From Ref. 37. (*See color insert.*)

Table 4 Ex Vivo Growth Potential for the Known Lineage Stages

Hepatic stem cells	Division rates of ~1/day under optimal conditions; can be subcultured repeatedly; one cell can generate >40,000 daughter cells in ~3 wk (144,148)
Hepatoblasts and committed progenitors	Division rates of ~1 every few days; ~12 divisions in three weeks; one cell can generate 4000–5000 daughter cells in three weeks (132)
Diploid adult cells ("small hepatocytes")	One cell yields ~130 daughter cells in three weeks (~5–7 divisions total); limited ability to be subcultured (132)
Polyploid adult cells	Attach, survive; DNA synthesis but limited or no cytokinesis (45)

hepatocytes in the perivenous region in vivo (74). Proliferating progenitor cells that appear in regenerating liver after PH in the uPA/RAG-2 mouse lack expression of cytochrome P450 enzymes (75). In vivo quantitative analyses of expression of CYP450 mRNA isoforms revealed similar expression levels of CYP3A4 and CYP3A7 in the fetal liver, but 10 times higher expression of CYP3A4, and 10 times lower expression of CYP3A7 in the adult liver (59).

Both in vivo and in vitro studies have implicated microenvironment, ploidy, and/or other lineage-dependent properties of the parenchymal cells in defining the zonal gene expression. Most importantly, they indicate a unidirectional, maturational process going from zone 1 to zone 3 (for review, see 16,19,20,76). Surgical rerouting of the blood flow through the liver from portal vein to central vein can alter the expression of some genes (e.g., gluconeogenesis) implicating gradients in signals in blood (77,78). Expression of GS, normally restricted to a single-layer hepatocytes around the central vein (74), has been found regulated by specific paracrine interactions between the terminal hepatocyte and the endothelial cells of the central vein and can be artificially induced by seeding parenchyma onto feeder layers of endothelia (28,79–81). Culture studies testing the influence of extracellular matrix components known to be in zone 1 (e.g., type IV collagen, laminin, and heparan sulfate proteoglycans) versus zone 3 (e.g., type I collagen, fibronectin, and heparin proteoglycans) have indicated that cells from zone 1 can be differentiated to ones with a phenotype similar to those from zone 3, and the cells from zone 3 can become muted in their zone 3 tissue-specific gene expression when plated onto matrix components found in zone 1; however, the zone 3 cells cannot be reprogramed into cells with a ploidy and phenotype identical to those from zone 1 (summarized in several reviews: 82,83). Studies of transplanted cells exhibiting zone-1-predominant (glucose-6-phosphatase, glucose content) are able to acquire all the functions typical of zone 3 functions, whereas zone 3 cells are able to acquire some, but not all the functions of zone 1 cells (84). Together, these findings support the interpretation that some genes (e.g., those regulating gluconeogenesis and GS) are regulated by microenvironment; others are affected by the ploidy state of the cells (e.g., transferring and tyrosine aminotransferase) and yet others [e.g., α-fetoprotein (AFP) and MUP] by some other aspect of the cells that is maturationally dependent.

GENERAL COMMENTS ON STEM CELLS

Stem cells and unipotent progenitors have unique biological properties with respect to their capacity for self-renewal and the ability to regenerate tissue and organ systems (76,85–89). There are two major families of stem cells being evaluated for clinical and commercial programs: *embryonic stem* (ES) *cells*, totipotent stem cells derived from early embryos, are capable of giving rise to all adult cell types and are able to undergo indefinite self-renewal (90,91). Although there is considerable interest in developing toti-potent stem cells as a universal solution for cell therapy ("one cell fits all purposes"), the ability to use such stem cells clinically is constrained by their propensity to form tumors when injected into ectopic sites, that is, sites other than in utero (92). The hope for their future use is in identifying conditions to lineage restrict them into progenitors that have lost the tumorigenic potential but retained the capacity to mature in normal tissues.

Determined stem cells, stem cells that give rise to some but not all adult cell types, are capable of self-renewal and do not demonstrate tumorigenic potential when trans-planted (93,94). In general, they are small (less than 15 μm) with low side scatter in flow cytometric analyses, with high nucleus to cytoplasmic ratios, with expression of telo-merase resulting in stability of the telomere lengths, with loosely packed chromatin, and with the presence of export pumps, such as MDR1, that reduce the presence of dyes. The pumps result in cells that flow cytometrically sort as a "side pocket" (SP cells) cell popu-lation relative to other cells within the tissue (3,95–97).

HEPATIC STEM CELLS

General Comments

The formation of the liver is initiated by an endodermal stem-cell population in the embryo-nic foregut (1,98,99) and with processes leading to the subsequent formation of mature hepatocytes, cholangiocytes, and other hepatic cell types. Shiojiri and co-workers (8,100–102) established that uncommitted hepatoblasts are capable of developing into biliary progenitors, apparently in response to paracrine signals from mesenchymal tissues surround-ing the portal vasculature. Commitment to the biliary lineage has been linked to HNF1 and HNF6b signaling in a highly localized response to cells immediately adjacent to the portal tracts (103,104) and leading to the formation of the ductal plate or limiting plate, shown now to be the reservoir of the hepatic stem cells, and having characteristic intense staining with cytokeratin 19 (CK19) and with neural cell adhesion molecule (NCAM; 102,105). The ductal plate transitions to become the Canals of Hering in adult livers (106). Adjacent to the ductal plates are hepatoblasts, recognizable by their intense expression of AFP and being the dominant parenchymal cell population in fetal and neonatal livers, and shown to be bipotent giving rise to the committed biliary and hepatocytic progenitors. The number of hepatoblasts declines in the livers of hosts of increasing age; they are difficult to find in adult livers except in the presence of ongoing disease such as cirrhosis or hepatitis.

Although the most well studied of the hepatic precursors are those located within the liver, there has been considerable excitement about the pioneering discovery by Petersen et al. (3,4,88,107) of the bone marrow as an alternate source of progenitors that give rise to hepatocytes by a phenomenon called "transdifferentiation." The possibility of transdiffer-entiation was bolstered by the studies of LaGasse et al. (88), who purified hemopoietic stem cells from bone marrow, transplanted them into mice with a genetic condition model-ing tyrosinemia, and showed that the cells were able to form hepatocytes. Transdifferen-tiation has been suggested by studies in multiple tissues. Many of these studies have now been refuted by evidence indicating that the donor cells fused with the host tissue. The issue of plasticity with data still accepted has been narrowed to that between cell types of the same embryological germ layer. Thus, plasticity does indeed appear to occur between mesodermal to mesodermal fates, or ectodermal to ectodermal fates but not across germ layers. Cells from fetal mouse livers can differentiate into hepatocytes, bile duct cells, pancreatic cells, gastric epithelial cells, and intestinal epithelial cells (108–111). Analyses of transdifferentiation have demonstrated that it is due primarily to cell fusion (112–114). Yet, there remain findings still supporting transdifferentiation such as those by Verfaille and coworkers (5) in which a rare stem cell in the bone marrow has been found to be multipotent giving rise to cell types of all the germ layers. Unfortunately, bone marrows contain such small numbers of these multipotent adult progenitor cells that bone marrow transplants result in exceedingly low efficacy (1% or less) with respect to reconstitution of damaged liver (112). Thus, the initial excitement of the remarkable discoveries of "transdifferentiation" has waned due to the low efficacy at which it occurs and the fact that even that observed has been found due primarily to fusion of the donor cells to the parenchymal cells of the liver (113,114). Although the transdifferentiation issue remains an area of ongoing controversy and research, the general consensus is that it is a minor pathway with little hope to be utilized in clinical programs. Therefore, the liver remains as the primary source of progenitor populations capable of significant reconstitution of liver. The known maturational stages of parenchymal cells are summarized in Table 5 and in Figures 5 and 6 and the known markers for the hepatic stem cells and other progenitors are given in Table 6.

Murine and Rodent Progenitors—Oval Cells, Progenitors in the Livers of Injury Models

Most of the initial knowledge of the stem-cell compartment in mice and rats has been derived from the voluminous literature on "oval cells," small cells with oval-shaped nuclei, identified in analyses of livers following a variety of oncogenic insults to the liver (7,9,115–117). The studies, especially from the 1960s to 1990s, have made use of carcinogenic injury models including: (*i*) administering the carcinogen, 2-acetylamino-fluorene (2-AAF), followed by a two-thirds partial hepatectomy (AAF/PH model), (*ii*) a necrogenic dose of the hepatotoxin, carbon tetrachloride, (*iii*) feeding a choline-deficient diet supplemented with etluonine, (*iv*) treating animals with the toxins, 3′-methyl-diaminobenzidine, galactosamine, or furan, (*v*) treating with the DNA-alkylating agent Dipin (1,4-bis[*N,N*′-di(ethylene)-phosphamide]-piperazine) intraperitoneally followed by two-thirds PH (118), (*vi*) treatment with the biliary toxin, 3,5-diethoxycarbonyl-1,4-dihydrocollidine (DDC), and (*vii*) liver injury developed in the albumin-urokinase-type plasminogen-activator transgenic (AL-uPA) mice (49,119–121).

These various oncogenic insults result in the expansion of oval cells, localized primarily to the periportal region. Oval cells express markers of both the hepatocytic (e.g., albumin and AFP) and biliary lineages (e.g., CK19). In addition, a number of investigators have generated monoclonal antibodies to antigens on oval cells, and the antibodies have been instrumental in the characterization of oval cell phenomena and in the identification of normal hepatic progenitors related to oval cells (18,20,122–127). These antibodies to oval cell antigens identify both hepatic and hemopoietic subpopulations and, to date, none of the antigens recognized by the antibodies has been purified and fully characterized, a fact that has limited the usefulness of these antibodies. It is hoped that this limitation will be overcome soon with ongoing research to define these antigens. For example, Ov 6, a monoclonal antibody raised against cells isolated from carcinogen-treated rat livers, is a popular marker for identifying murine oval cells. However, in addition to the fact that the antigen is not known, it reacts with normal bile duct epithelia in rats and humans and with hepatocyte and ductal reactive cells in diseased human tissue (124,125).

In vitro oval cells can be differentiated into cells with some of the characteristics of either biliary or hepatocytic cells, but the oval cells behave more like partially transformed or sometimes completely transformed cells and are able to expand in cultures on culture plastic, with medium supplemented with serum, without signals from embryonic

Table 5   Known Stages in the Liver's Maturational Lineages

Stage 1	Hepatic stem cells	Thought to be multipotent; give rise to hepatoblasts and also possibly other endodermal cell types
		Have cell divisions that can be symmetric (self-replication) or asymmetric (differentiation) depending on the conditions
		Present in the ductal plates of fetal and neonatal livers and in the Canals of Hering in adult livers
		Express albumin, cytokeratins 7, 8, 18, 19, EpCAM, NCAM, and CD133/1
Stage 2	Hepatoblasts	Bipotent; give rise to committed progenitors for hepatocytes and biliary epithelia
		Unknown if they can go through symmetric divisions
		Present throughout the parenchyma in fetal and neonatal livers and as single cells or small aggregates of cells attached to the ends of Canals of Hering in pediatric and adult livers
		Express albumin, cytokeratins 7, 8, 18, 19, EpCAM, ICAM1, CD133/1, AFP, and P450A7
Stage 3	Committed hepatocytic progenitors	Small parenchymal cells, typically $12-15\,\mu m$ in diameter, with low side scatter and expressing EpCAM, ICAM1, cytokeratins 8 and 18, AFP, and albumin
		Evident in significant numbers only in fetal and neonatal livers
	Committed biliary progenitors	Small parenchymal cells, typically $12-15\,\mu m$ in diameter and expressing EpCAM, ICAM1, cytokeratins 7 and 19, fetal forms of aquaporins
		Evident in significant numbers only in fetal and neonatal livers
Stage 4	Diploid hepatocytes	Hepatocytes that are approximately $18-22\,\mu m$ in diameter and expressing ICAM1, cytokeratins 8 and 18, albumin, PEPCK, and connexin 26
		Form plates of cells, blanketed by endothelia extending from the portal triads to the central veins of the liver acinus
	Diploid biliary epithelia	Bile duct epithelia, approximately $18-22\,\mu m$ in diameter, and expressing ICAM1, cytokeratins 7, 8, 18, and 19, aquaporins, and MDR3
		Form ducts running from the portal triads to the bile duct connecting to the gall bladder and to the gut
Stage 5	Tetraploid hepatocytes	Evident pericentrally in human livers and mid-acinar and pericentrally in rodent livers
		They are $25-35\,\mu m$ in diameter and with high side scatter
		They express TAT, transferrin, connexin 32, and late (adult-specific) P450s

(Continued)

Table 5 Known Stages in the Liver's Maturational Lineages (*Continued*)

		They have lost some of the regulatory mechanisms involved with cytokinesis (e.g., citron kinase) and so undergo only hypertrophic growth responses to stimuli for regeneration
		Produce soluble signals (unidentified) that inhibit the growth of stem cells and progenitors (feedback loop signals)
	Tetraploid biliary epithelia	These are unknown but assumed to exist
Higher stages	Octaploid (and higher levels of ploidy) parenchymal cells	To date, these have been found in rats and mice but not human livers; polyploid hepatocytes occur in some mammals (mice and rats) and can have DNA content of 8–32 N
		They express MUP and late P450s
		Produce feedback loop signals

Abbreviations: EpCAM, epithelial cell adhesion molecule; NCAM, neural cell adhesion molecule; AFP, α-fetoprotein; ICAM, intercellular cell adhesion molecule; MUP, major urinary protein.

mesenchymal feeder cells, and with few, if any, of the known mitogens requisite for normal hepatic progenitors. Although oval cells demonstrate characteristics of partially or completely transformed cells, they are still able to form liver tissue when transplanted. Wang et al. (2003) using Nycodenz gradient centrifugation to isolate oval cells from DDC-treated mice were able to use them to rescue recipient mice with lethal hepatic failure resulting

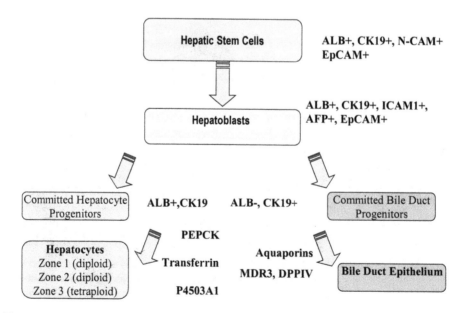

Figure 5 Schematic representation of the known stages of human liver lineage and representative genes expressed by the cells at those stages. *Abbreviations*: Alb, albumin; AFP, alpha-fetoprotein; CK19, cytokeratin 19; EpCAM, epithelial cell adhesion molecule; NCAM, neuronal cell adhesion molecule; ICAM, intercellular cell adhesion molecule; MDR3, multidrug resistance gene 3 (involved in biliary functions).

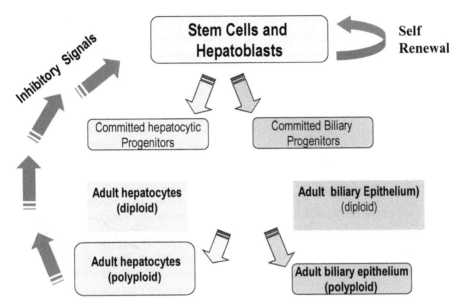

Figure 6 Schematic representation of the presumptive feedback loop in which one or more signals from the mature cells inhibit the proliferation of the cells from the stem-cell compartment.

from homozygous deletion of the gene for fumarylacetoacetate (fah$^{-/-}$) (77). The recipient mice had significant donor-derived hepatocyte repopulation and phenotypic rescue.

In Long Evans Cinnamon (LEC) rats, oval cells expand in the course of liver injury induced by excessive accumulation of copper, a phenomenon that models Wilson's disease. The oval cells are positive for gamma-glutamyl transpeptidase, AFP, and for CK18 and CK19 but are negative for albumin (128–130). The cells were transduced ex vivo with a reporter gene, β-galactosidase, transplanted into LEC/Nagase analbuminemic double-mutant rats, and were found to differentiate into mature parenchymal cells.

In summary, oval cell studies have made evident the presence of a stem-cell compartment in livers, and oval cells share many of the antigens and gene expression with normal hepatic progenitors. Yet, their regulation ex vivo and in vivo and some aspects of their phenotype can be distinct from that of their normal counterparts and can indicate a partially or completely malignantly transformed state due to mutational events caused by the injuries used to induce their expansion.

Murine and Rodent Hepatic Stem Cells from Normal Hosts

More recent investigations have attempted to identify normal hepatic progenitors in animals not subjected to any method of liver injury. The earliest reports are those in which monoclonal antibodies developed to antigens on oval cells (122) were used to flow cytometrically sort hepatic progenitors from embryonic rat livers (126) and subsequently from neonatal and adult rat livers (18,20,131). As these antibodies identify antigens on both hepatic mid-hemopoietic progenitors, it was essential to do multiparametric flow cytometric sorts for cells negative for hemopoietic markers (glycophorin A, OX43, and OX44) and then positive for one of the oval cell antigens. The one used most extensively in these early studies was OC3, an antigen identified by the monoclonal antibody 374.3. As with all other known oval cell antigens, OC3 has yet to be cloned and identified, limiting its utility in characterizing the hepatic progenitors. The OC3$^+$ cells from normal

Table 6 Markers for Hepatic Progenitors

Marker	Species	Comments/references
Hepatic stem-cell markers that are cloned and sequenced		
Albumin	All species	Found in the hepatic stem cells, hepatoblasts, and hepatocytic lineage (1,300)
AFP	All species	AFP has long been a protein considered definitive for endodermal progenitors and within the liver lineages is definitive for hepatoblasts (300); a variant form of AFP is expressed by hemopoietic progenitors (213) and is identical to that in hepatic cells except for exon-1-encoded sequences
Cytokeratins 7/19	All species	Cytokeratins 7/19 are found in the hepatic stem cells, the hepatoblasts, and biliary epithelia but not the hepatocytic parenchyma (106,138,144,209,301,302)
CD133 (prominin)	Humans	A transmembrane protein found on hepatic and hemopoietic stem cells (201,202)
EpCAM	Humans	Present on hepatic stem cells, hepatoblasts, and committed progenitors but not on mature hepatocytes
CD44H (hyaluronan receptor)	Rats and humans	Present on rat hepatoblasts and on human hepatic stem cells (133,134)
MDR1	Rats	Present on hepatoblasts (303,304)
ICAM1	Rats and humans	Present on hepatoblasts, committed progenitors, and mature parenchymal cells; not expressed by hepatic stem cells
NCAM	Humans	Present on hepatic stem cells but not on any lineage stage thereafter (144,305)
DLK-Pref-1	Mice	(138)
Telomerase	All species	Essential for maintenance of telomere length (96,97,306)
Wnt/beta-catenin pathway	All species	A pathway that appears to be generic for stem-cell populations (307,308)
Markers that have variably been found on hepatic progenitors		
CD117 (ckit)	All species	Receptor for stem-cell factor; expressed by progenitors of mesodermal lineages and by some subpopulations of hepatic progenitors; it has been found on hepatic stem cells but not hepatoblasts and with considerable variability (e.g., sorting for it does not yield clonogenic hepatic progenitors); therefore, an alternative interpretation is that it is on endothelial progenitors (angioblasts) tightly associated with the hepatic stem cells, an hypothesis still under investigation (211,219,222,309)
CD146	Human	Antigen expressed on mesenchymal cells; related to NCAM; cells tightly associated with the hepatic progenitors (endothelial progenitors) are positive for this antigen

(Continued)

Table 6 Markers for Hepatic Progenitors (*Continued*)

Marker	Species	Comments/references
KDR	All species	VEGF receptor present on endothelial progenitors; it is possible that the findings of KDR on hepatic progenitors are actually for tightly associated endothelial cells
CD34	Rodents and mice	Expression of CD34 has been reported to be on hepatic progenitors in various murine and rat species, and all data are studied on liver injury model systems; the data have not proven credible, as sorts for CD34+ cells (310) do not yield clonogenic populations capable of liver reconstitution or of forming liver tissue in vitro (125,209,311)
Transcription factors		
Prox1	All species	Homeobox gene defining pancreatic and liver fates (172,173)
Hex	All species	Homeobox gene found in early liver (175,176)
HLX	All species	Gene required for endoderm to migrate into the cardiac mesenchyme (177,311)
HNF1, HNF3, HNF4, HNF6	All species	(313–315)
C/EBP	All species	(186,187,190,316)
DBP	All species	(186)
c-jun proto-oncogene	All species	Defining transcriptional element for liver developments (179,317,318)
Markers defining epithelial cells		
E-cadherin	All species	Cell adhesion molecule on parenchymal cells but not on mesenchymal cell types (139,195,319)
CD8/18	All species	Cytokeratins evident in all forms of epithelia (320)
Oval cell antigens		
Oval cell antigens (general comments)	All species	Identified in the livers of various injury models; present on both hepatic and hemopoietic cells; none of the antigens have been identified making it difficult to know if on inflammatory cells or the hepatic progenitors (7,10,122,123)
A6	Murine	(124,125)
OC2 and OC3	rat	(122,126,127,321,322)
Cloned and sequenced markers not found in/on hepatic progenitors		
CD45	All species	Common leukocyte antigens (132,137,242)
Glycophorin A	All species	Red blood cell antigen (20,126,132)
CD14	All species	(323)

Abbreviations: AFP, α-fetoprotein; EpCAM, epithelial cell adhesion molecule; MDR, multidrug resistance; ICAM, intercellular cell adhesion molecule; NCAM, neural cell adhesion molecule; Dlk/Pref-1[+], Delta-like/ Preadipocyte factor-1.

rodent livers were able to expand ex vivo if cultured on purified embryonic matrix substrata layered onto porous surfaces, in serum-free medium supplemented with purified hormones and growth factors, and with feeders of liver stroma derived from embryonic livers of E14 to E17 hosts. The OC3+ progenitors isolated from the livers were able to mature in vitro (20) or in vivo (131) to mature liver cells.

Recently, a more complete antigenic profile of rat hepatoblasts has been defined rigorously in flow cytometric analyses utilizing monoclonal antibodies to well-characterized antigens and showing that rat hepatoblasts are negative for hemopoietic markers (glycophorin A, CD45, OX43, and OX44), negative for class 1a major histocompatibility complex (MHC) antigens, dull for class 1b MHC antigens, and positive for ICAM1 and CD44H (132–134). Highly purified hepatoblasts isolated by flow cytometric sorts for this antigenic profile were able to form colonies from single cells, with clonal efficiencies up to 50%, when seeded onto SIM (Sandoz inbred Swiss mouse) mouse embryonic fibroblasts, selected for 6-thioguanine and ouabain resistance (STO) feeder cells and in a serum-free medium supplemented only with lipids, insulin, and transferrin/fe. The individual cells gave rise to colonies expressing both hepatocytic and biliary markers (113,132) (Fig. 9).

As these early studies are on hepatic progenitors in normal, untreated rats, others have obtained parallel results with purification of hepatic progenitors from murine livers. Azuma et al. (135) developed an enrichment system to isolate hepatic progenitor cells from adult mouse livers using their cadherin-dependent cell–cell adhesion properties. A procedure of two-hour hypoxic suspension culture with constant shaking eliminated almost entirely the mature hepatocytes that are more sensitive to ischemia and resulted in aggregates of progenitor cells in Ca^{2+}-containing medium. About 5% of these cell aggregates proliferated and formed colonies that expressed AFP, albumin, and E-cadherin, but not CK19. Suzuki et al. (108,136) utilized the fluorescence-activated cell sorter (FACS) and fluorochrome-conjugated antibodies against a set of cell surface markers to isolate the clonogenic hepatic stem cells from Balb/cA ED 13.5 fetal mice. In vitro colony assays showed that cell populations with an antigenic profile of c-Met^{+} CD49f$^{+/low}$ c-Kit^{-} CD45^{-} TER119^{-} formed colonies on laminin-coated plates. Flow cytometrically sorted c-Kit^{low} CD45 TER119 hepatic progenitor cells isolated from ED 11 fetal mouse livers have been shown to form colonies (>50 cells/colony) in which 28% expressed both albumin and CK19 (137). Tanimizu et al. (138) used both FACS and an automatic magnetic cell sorter (AutoMACS) to enrich for cells positive for Delta-like/Preadipocyte factor-1 (Dlk/Pref-1^{+}) and reported formation of large colonies (>100 cells/colony) from the Dlk/Pref-1^{+} population. Unfortunately, this antibody is not yet commercially available and its limited availability has prevented others from reproducing these experiments. Nitou et al. (139) used magnetic bead separation methods to purify mouse E-cadherin^{+} (ECCD-1) hepatoblasts from ED 12.5 fetal mouse livers and subsequently obtained monolayer cell sheets expressing AFP, albumin, and cytokeratins on glass slides at day 5 in cell culture. However, they have not yet reported the clonogenic ability of these MACS-sorted E-cadherin^{+} cells. A novel rat monoclonal antibody, called anti-Liv2, specifically recognizing murine hepatoblasts has been produced by immunizing adult WKY/NCrj female rats with ED 11.5 murine fetal liver lysate (140). Ongoing investigations are assessing whether the antibody to Liv2 protein can be utilized to purify hepatoblasts from fetal and adult mouse livers.

The Stem-Cell Compartment of Human Livers

The number of studies on identification and isolation of human hepatic progenitors has been limited due to the costs and the difficulties in obtaining normal human liver tissue.

Moreover, many of the initial efforts have not been particularly successful due to the use of classical fractionation protocols (Ficoll or Percoll fractionation) that select for only one parameter (e.g., cell density) rather than the more successful multiparametric purification strategies especially those using immunoselection (76). Their success has been limited also by (*i*) the use of culture conditions consisting of tissue culture plastic and serum supplemented medium, conditions that are not conducive to survival and growth of the progenitors and (*ii*) the use of cultures containing both mature cells and progenitors (83). Mature parenchymal cells, particularly those that are polyploid, produce soluble signals present in the conditioned medium that inhibit the growth of hepatic stem cells (Reid and associates, unpublished observations). Thus, there is a feedback loop signal(s) by which "old" cells control the production of "young" cells (Table 7 and Figs. 6 and 12). Expansion of the progenitors ex vivo requires the use of purified progenitors separated from the mature cells and in vivo requires selective loss of pericentral parenchymal cells to create a "cellular vacuum" (reviewed in 141).

Hepatic stem cells in human livers have been hypothesized to be present in the Canals of Hering, small ducts that are present in zone 1 of the liver acinus, forming connections between hepatocytes and bile ducts and demonstrating strong expression for certain cytokeratins (CK), particularly CK7 and 19 (89,142). The pioneering work of Strain and coworkers (26,143) is noteworthy in identifying human hepatic progenitor cells that express CD117 (c-kit). More recently, greater success has been achieved by using multiparametric flow cytometric sorts for cells with antigenic profiles negative for hemopoietic markers and positive for certain epithelial markers, enabling the

Table 7 Feedback Loop: Relevance to Studies on Reconstitution of Livers

Findings	Hypothesis	Predictions
Stem cells or progenitors do not grow ex vivo when co-cultured with mature parenchymal cells or with conditioned medium from the mature cells	Mature parenchymal cells (e.g., polyploid cells) produce soluble signals constituting a feedback loop that regulates stem-cell compartment	The signals do not exist in peritoneum; site is permissive for expansion and maturation of human liver cells
		Other hosts (e.g., sheep and pig) that have higher proportion of diploid cells will be better models for studies of human hepatic progenitors
		Strategies for clinical programs must take feedback loops into account
		Transplant-purified human hepatic stem cells or progenitors (therefore avoiding feedback loop from mature human cells)
		Hosts with high polyploidy will cause liver injury to mature cells (zones 2/3)

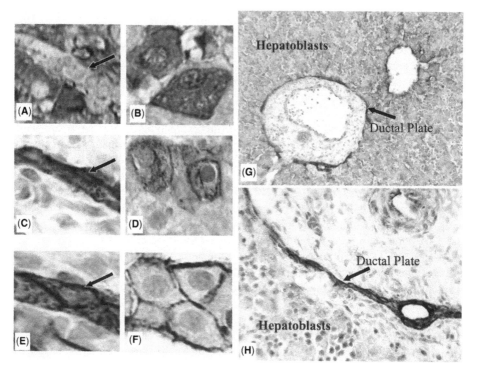

Figure 7   Histological sections of human, fetal livers stained for AFP (**A** and **B**), CK19 (**C**, **D**, and **H**), and for EpCAM (**E**, **F**, and **G**). The *arrows* indicate the ductal plate (also called limiting plate). The figures in (**G**) and (**H**) are low magnification (10×) and the rest are high magnifications (40×) to indicate the hepatic stem cells present in the ductal plate (**A**, **C**, and **E**) and the hepatoblasts present adjacent to the ductal plates and throughout the parenchyma of the fetal livers. *Abbreviation*: AFP, α-fetoprotein. *Source*: From Ref. 147. (*See color insert.*)

identification and isolation of two pluripotent progenitors (hepatic stem cells and hepatoblasts) and two unipotent progenitors (committed biliary mid-hepatocytic progenitors) from human fetal livers (144,145) and from pediatric and adult human livers (141,146) (Figs. 5–7, Table 5). All four populations have proven wholly negative for hemopoietic markers (CD45, CD34, CD38, CD14, and glycophorin A), making them distinct from hepatocyte precursors from the bone marrow (4,5,38), and all share expression of epithelial cell adhesion molecule (EpCAM), cytokeratins 8, 18 and cadherin and CD133/1, also called prominin. The size of the EpCAM+ populations, 7 to 12 μm in diameter, is strikingly different from that of adult liver cells, 18 to 22 μm for the diploid parenchymal cells, and 25 to 35 μm for the polyploid ones. The two pluripotent populations, the hepatic stem cells and hepatoblasts, are distinguishable from each other by differential expression of N-CAM, ICAM1, AFP, P450 7A, whether the EpCAM is expressed cytoplasmically and/or on the plasma membrane and by the intensity of expression of CK19 (Fig. 7 and Table 5). Both populations form human liver tissue when transplanted into immunocompromised hosts (144) (Melhem et al., in preparation). Purified populations of the hepatic stem cells will lineage restrict to hepatoblasts when placed under specific culture conditions (144).

The antigenic profiles defined for the two pluripotent progenitors and the two unipotent ones have been used to define the hepatic stem-cell compartment in vivo (147). The hepatic stem cells are found in the ductal plates (also called limiting plates) of fetal

Serum-free basal medium with <u>low</u>
 calcium (< 0.5 mM) and <u>no</u> copper

Lipids: HDL + free fatty acids bound to
 albumin)

Hormones/growth factors

 --Stem cells: insulin and transferrin/Fe

 --Diploid adult cells ("small
  hepatocytes"): insulin,
  transferrin/Fe, EGF

Feedback Loop: Absence of soluble
 signal(s) from mature cells

Feeders:

 --Stem cells: angioblasts

 --Hepatocytes: endothelia

 --Biliary cells: stroma

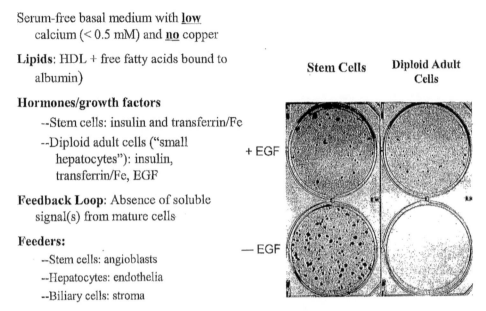

Figure 8   Conditions for ex vivo expansion of diploid subpopulations of parenchymal cells. These conditions consist of serum-free basal medium with low calcium (less than 0.5 mM), no copper, a mixture of lipids (high-density lipoprotein and a mixture of free fatty acids bound to albumin), insulin, and transferrin/fe (and for the diploid adult hepatocytes EGF), signals from embryonic feeders and the absence of the signal(s) from mature parenchymal cells. Details of the preparation of these conditions are given in the methods review. *Abbreviation*: EGF, epidermal growth factor. *Source*: From Ref. 83.

and neonatal livers and in the Canals of Hering in the pediatric and adult livers (Fig. 7). The hepatoblasts are the dominant parenchymal cell population in fetal and neonatal livers and then dwindle in numbers with age of the hosts such that in adults they are found as single cells or small aggregates of cells physically connected, tethered, to the ends of the Canals of Hering. The numbers of the hepatoblasts are dramatically higher in diseased livers, especially cirrhotic livers.

Expansion of the hepatic stem cells and progenitors occurs ex vivo if plated onto appropriate embryonic mesenchymal feeders (144–146) and/or on embryonic matrix substrata (148) and in a serum-free, hormonally defined medium that was developed for rodent hepatoblasts (132,133) (Tables 6 and 10). On the basis of these findings, the known maturational lineage stages, the differential antigenic profiles of the stem cells, the unipotent progenitors, and two of the marine liver cell subpopulations, are summarized in Tables 4 and 5 and Figures 8–11.

Contributions of the Stem-Cell Compartment in Liver Regeneration

Two forms of liver regeneration have long been known, and the stem-cell compartment plays roles, albeit distinct ones, in both (Fig. 12). *Liver regeneration following toxic injuries* (chemicals, viruses, and radiation) involves selective loss of the mature parenchymal cells in zones 2 and 3 and with secondary proliferation of the hepatic progenitors periportally; subsequently, these differentiate into the mature cells typically found in the pericentral zone. This phenomenon is characteristic of the findings from the many investigations

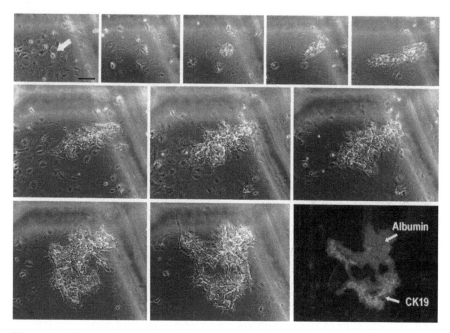

Figure 9 Clonogenic expansion of a rat hepatoblast under the conditions specified in Figure 8. A single cell is able to expand into a colony of cells in 20 days, and the cells express markers for both the hepatocytic lineage (albumin) and for the biliary lineage, CK19. Many of the cells have undergone lineage restriction to become committed progenitors of one of the two lineages (the cells at the periphery of the colony), whereas those at the center are cells co-expressing both markers and are, therefore, hepatoblasts. *Source*: From Ref. 132. (*See color insert.*)

on oval cells. In culture studies, the mature parenchymal cells, particularly those that are polyploid, produce soluble signals present in the conditioned medium and that inhibit the growth of hepatic stem cells (Reid and associates, unpublished observations). Thus, there is a feedback loop, signal(s) by which "old" cells control the production of young cells. The feedback loop explains why purification of diploid subpopulations away from polyploid ones is required to observe clonal growth of diploid cells in culture and why

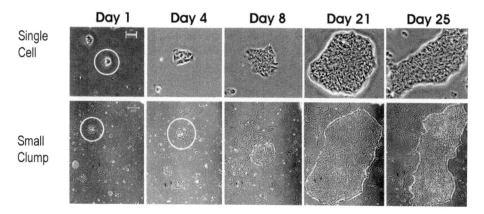

Figure 10 Human hepatic stem cells from human fetal livers and plated under serum-free conditions found requisite for expansion ex vivo. *Source*: From Ref. 144.

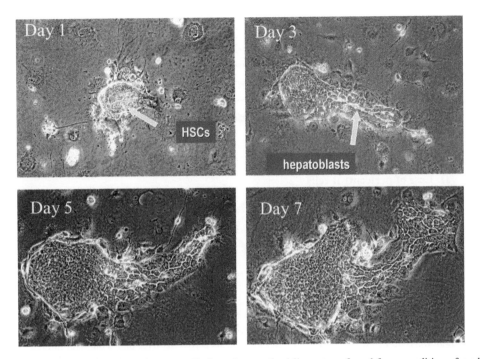

Figure 11   Human hepatic stem cells from human fetal livers transferred from conditions found for self-replication to STO feeder, found to promote differentiation of the cells. Note the hepatic stem-cell colony one day after transfer and then after several days. Within 24 hours, there are cords of cells erupting from the edges of the colonies and with a phenotype of hepatoblasts. *Source*: From Ref. 144.

significant expansion of transplanted liver cells occurs only in hosts in which there is a "cellular vacuum" in the pericentral zone.

   Liver regeneration after PH, surgical removal of a portion of the liver, has long been thought mediated only by mature liver cells (149). However, it has now been shown to involve the stem-cell compartment (150). In the first 24 hours after PH, there is a wave of DNA synthesis across the liver plates, but with limited cytokinesis, resulting in elevated polyploidy and a sharp decline in the diploid subpopulations. The ploidy profile of the parenchymal cells is restored slowly and gradually over several weeks by contributions from the stem cells.

The Stem-Cell Niche

The microenvironment of the stem-cell niche is that found within the ductal plates in fetal and neonatal livers and that within the Canals of Hering in pediatric and adult livers. It is assumed to be comprised the matrix components and soluble factors exchanged as paracrine signals between the hepatic stem cells and their native mesenchymal partners, angioblasts. Little is known of these other than some of the extracellular matrix components. Extracellular matrix chemistry is known to be age- and tissue-specific and to regulate the cell morphology, growth, and cellular gene expression (151–156). The extracellular matrix components are present in the Space of Disse, between the parenchyma and the endothelia, and form a gradient in their composition extending from the portal triads to the central vein (19,76,157). The periportal zone contains specific embryonic matrix

I. Partial Hepatectomy

↓

Acute response is an increase in
polyploidy followed by cellular
hypertrophy; with time, stem cells
restore diploid subpopulations

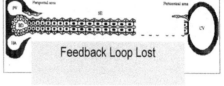

Feedback Loop Intact

**II. Loss of pericentral cells
with chemicals, viruses, or
radiation**

↓

Hyperplasia of periportal cells
(oval cell response) and gradual
restoration of mature cells

Feedback Loop Lost

Figure 12 The two forms of liver regeneration. There are distinct mechanisms involved in liver
regeneration following PH versus that following toxic injury (caused by viruses, chemicals,
radiation). With PH, there is a wave of DNA synthesis across the liver plates but with limited cytokinesis. This results in elevated ploidy for the liver. The normal ploidy profile of the cells is restored
over several weeks by contributions from the stem-cell compartment. With toxic injury, there is
selective loss of the cells pericentrally and, therefore, loss of the feedback loop. The periportal
cells, including the cells of the stem-cell compartment, proliferate extensively and then mature gradually to cells typical of zones 2 and 3. *Abbreviation*: PH, partial hepatectomy. *Source*: From Ref. 83.

components such as hyaluronans, type III and IV collagen, laminin, and fetal forms of proteoglycans. This microenvironment can be mimicked ex vivo by using culture conditions
that comprised the embryonic matrix components, now available commercially, coated
onto porous and flexible surfaces to permit polarization of the cells and critical cell
shape changes (83). This is especially important for hepatic stem cells and progenitors
that empirically show an intolerance for attachment to impervious and rigid surfaces.
The soluble components, comprising nutrients and soluble signals, are mimicked, in
part, by the use of serum-free basal media with no copper, low calcium (below
0.5 mM), trace elements (selenium and zinc), a mixture of lipids (free fatty acids bound
to albumin and high-density lipoprotein), insulin, and transferrin/fe (76,83,132). The
signals, as yet unidentified, are provided by the use of embryonic stromal feeders that
ideally are derived from embryonic livers (18,127) but can be substituted, in part, by
the use of STO feeders (132).

Optimal survival, expansion, and differentiation of the cells depend on use of serum-
free medium conditions, as serum drives the cells toward biochemical and antigenic
responses appropriate for wound formation (fibrosis or cirrhosis, i.e., scar formation)
and, in parallel, loss of tissue-specific functions (76,83,158,159). Serum-free, basal
media supplemented with defined mixtures of purified nutrients, lipids, trace elements,
hormones, growth factors, and matrix components can be tailored to elicit an appropriate
response, either growth or differentiation, of the cells (reviewed in 82). Thus, there are hormonally defined media for expansion and others for differentiation of a given maturational
stage of parenchymal cell. The details for preparation of such media are given in methods
review (83).

There is preliminary evidence to indicate that the mature liver cells produce a soluble signal(s) that inhibits the expansion of stem/progenitor cells and constitutes a feedback loop signal; co-culture of hepatic stem cells with mature cells or with conditioned medium from mature cells results in lack of growth by the hepatic stem cells (Reid and associates, unpublished data). Therefore, hepatic stem/progenitor cells must be kept separate from mature liver cells in order to observe their maximal potential growth in culture. This phenomenon can also explain the well-known need for selective loss of zone 2 and 3 cells, mature host liver cells, creating a "cellular vacuum" for transplanted donor cells to undergo expansion in a recipient (reviewed in 160).

GENE EXPRESSION

The Developing Liver and Genes Defining Hepatic Fates

The formation of the liver begins with the budding of the visceral endoderm into the cardiac mesenchyme (1,2,161) and is associated with various complex changes of gene expression patterns (for reviews, see 141,162). In vivo data have been obtained mostly in studies of in situ mRNA hybridization of liver or whole embryo sections and of comparison of normal versus knockout gene mouse models. In the rat, AFP mRNA can be detected first in the epithelium of the yolk sac (163) and cells of the ventral foregut at embryonic day 10.5 (E10.5). One day later, protein expression can be determined (80,164). In normal mouse development, starting from E9 to E15, homogeneous distribution of albumin and AFP mRNA in the liver can be detected (165). AFP mRNA expression is present with considerable intensity of expression in the liver and the yolk sac, but low levels can be found in the epithelium of the small intestine, the heart, and the renal tubes (163,166,167).

In the rat, albumin mRNA is expressed a day later than AFP mRNA, at E11.5 (164) and its levels remain lower than AFP mRNA until E19. Albumin mRNA levels increase steadily during development, whereas AFP mRNA levels remain stable up to E21. By E19, both levels are equal. AFP expression remains constant up to day 14 after parturition and declines to zero thereafter. Albumin mRNA can be detected at E20 in the kidney (163). In the normal adult rat liver, first evidence for a zonal distribution pattern is detectable at E20; albumin transcripts are expressed at a lower level in the hepatocytes of the centrilobular than in the periportal areas, and the bile duct epithelium does not show any expression of it (80,168).

It is known that early liver development in vivo from the ventral foregut endoderm requires interaction with the cardiac mesoderm (161). Cells of the septum transversum contribute to hepatic induction. In particular, it has been shown that bone morphogenic protein (BMP) signaling from mesenchymal cells of the septum transversum is necessary to induce albumin expression in the endoderm and to exclude pancreatic fate; findings demonstrated using a Bmp4 null mutant mouse model (169). Also, fibroblast growth factor is necessary to induce liver-specific gene expression and proliferation through cardiac induction (170,171). BMP signaling regulates positively the expression of the transcription factor GATA4 in the hepatic endoderm.

One of the earliest genes in the development of the early endoderm that can be associated with pancreatic and hepatic fates is the homeobox gene Prox1, which had been shown to be restricted to early regions (mouse E8.5) of the pancreas and liver in vivo (172). Although it is not necessary for hepatic differentiation, it is a prerequisite for the migration of hepatocytes into the septum transversum (173). The homeobox gene Hex is expressed by early liver cells (174,175) and mutants lacking expression of Hex do not show migration of hepatocyte precursors into the septum transversum (175) and development of the liver bud (176). Hlx, the murine homeobox gene, is required for liver expansion but not for morphogenesis or differentiation (177).

Proto-oncogene expression has been shown as essential for normal liver development. Homozygous mice lacking the c-jun proto-oncogene show impaired liver formation and die at E16.5 at the latest; the mice demonstrate reduced mitotic and increased apoptotic rates in primary murine hepatic cell cultures derived from c-Jun$^{-/-}$ fetuses (178). ES cells, also $-/-$ for c-jun, can contribute to all somatic cells except liver cells, demonstrating the important role of c-jun in liver formation (178). Passegue et al. (179) showed that hepatoblasts derived from Jun-deficient ED 12.5 mouse livers show reduced proliferation and increased apoptosis when grown in vitro and that JunB can rescue Jun-dependent proliferation defect of hepatoblasts. The Jun proteins, Fos proteins, and some members of the activating transcription factor (ATF) and cAMP (cyclic adenosine monophosphate)-responsive element (CRE)-binding protein (CREB) protein families are the essential components of the AP-1 transcription factor. Watanabe et al. (180) reported more severe impairment of hepatoblast proliferation in sek1$^{-/-}$ (stress-activated protein-kinase-1-deficient) mouse livers from ED10.5 compared with c-Jun-deficient and wild-type mouse embryos. They proposed that sek1, a direct activator of the stress-activated protein kinase [SAPK, also called c-Jun N-terminal kinase (JNK)], appears to play a crucial role in hepatoblast proliferation and survival in a manner different from NF-κB or c-Jun. TGF-alpha has been shown to stimulate c-jun expression in rat liver cell cultures (181). Also, mice lacking in NF-κB (nuclear factor) expression die at E16 at the latest and exhibit severe liver degeneration (182,183).

Hepatoma-derived growth factor (HDGF) has been reported to be strongly expressed in the fetal mouse liver at mid-gestation stage, and significantly down-regulated near birth (184). Supplementation of recombinant HDGF significantly enhanced the growth of primary cultures of fetal liver cells from ED14.5 mouse livers, although the effect was small (11% increase in cell number). Adenoviral introduction of HDGF antisense cDNA into the fetal hepatic cells significantly suppressed their proliferation. The inhibitory effect of HDGF antisense virus was reversed by exogenous HDGF. HDGF is a heparin-binding protein purified from the conditioned media of HuH-7 hepatoma cells.

Expression of six families of liver-specific transcription factors has been identified (for review, see 185). Although some of them are expressed in other tissues, especially the epithelium of the gut and kidney (163), HNF1, HNF3, HNF4, HNF6, C/EBP, and DBP are found predominantly in liver. Interestingly, during early development, expression of transcription factors is significantly higher in the entire embryo, and is shut off or enhanced during further development (186). During normal rat embryonic development, the expression of different transcription factors is unequally regulated (187). From E16 and E12 throughout adult expression of HNF3 and HNF2 (C/EBP) (186) respectively, is relatively constant in the liver with a slight increase of C/EBP expression around birth (188). Expression of HNF1, HNF3, HNF4, and C/EBP alpha remains stable up to late gestational period and decreases after birth. In rat embryos, C/EBP mRNA expression cannot be detected before E12, and then in the same cells that express albumin and AFP; at E20, C/EBP mRNA can be detected in epithelial cells of the gut and kidney (163). DBP expression can be detected at E14 in the rat (186), its expression increases with further development, and to the greatest extent in the adult liver. Other aspects of DBP expression observed are at E14 in certain nerves, and at E20 in gut and kidney.

The C/EBP transcription factor has been implicated in lineage restriction of hepatic progenitors into the hepatocytic lineage. Tomizawa et al. (189) showed that the normal formation of hepatic plates was disrupted by the abundant formation of pseudoglandular structures starting from ED16.5 in C/EBP$\alpha^{-/-}$ (CCAAT/Enhancer Binding Protein

α-deficient) mice. These pseudoglandular structures co-expressed AFP and A6, antigens for hepatic progenitors. In the C/EBP alpha knockout mouse, an impaired energy metabolism can be observed, with no storage of glycogen and lipid, and death after eight hours of birth (190). The mRNA for glycogen synthase was up to 70% lower and those of PEPCK and glucose-6-phosphatase was significantly delayed. In the HNF4alpha$^{-/-}$ mice, it has been shown that HNF4 alpha is not necessary for early liver development but is indispensable for expression of liver-specific functions as well as expression of HNF1alpha (191). Murine hepatic progenitors flow cytometrically sorted from the E13.5 mouse liver to be negative for c-kit, CD45, and Terr119, and low positivity for CD49f, showed strongly increased proliferation and stopped hepatic differentiation in culture when C/EBP function was inhibited (192).

In the normal liver, expression of transcription factors shows a zonal distribution (55). HNF3 mRNA appears slightly more abundant periportally, whereas C/EBP, HNF1, and HNF4 transcripts are slightly higher in perivenous hepatocytes. Homozygous mice with deletion of the HNF3 gene showed decreased expression of hepatic genes PEPCK, transferrin, and TAT (193).

In the normal fetal mouse liver, the proto-oncogene c-myc RNA is expressed homogeneously throughout the cell population from d9 to d15 with a maximum at d13 (165). After d15, mRNA expression decreased drastically but with higher expression in "prehepatocytes" than in other cell types.

The Wnt/beta-catenin pathway is known to play a major role in oncogene activation of tumor growth of various organs and it has been shown that mutations in beta-catenin can be detected frequently in oncogene activation in hepatocarcinomas and hepatoblastomas (reviewed in 194). Beta-catenin also mediates cell–cell adhesion by the interaction with E-cadherin (for review, see 195), which is involved in normal bile duct morphogenesis (196). At E10 in the mouse liver, nearly all the non-hematopoietic cells express beta-catenin protein, whereas the expression pattern in the cell itself shifts with development (197). Most of the cells express beta-catenin membraneously, some express it in the cytoplasm or in the nucleus at E10; at E14, nearly all cells express beta-catenin at the membrane, with an overall decrease in expression. Mature hepatocytes show lower protein expression than progenitors. Inhibition of beta-catenin protein expression leads to a decrease in cell proliferation, an increase in apoptosis, and promoted c-kit immunoreactivity of hepatocytes. Nevertheless, cyclin Dl and c-myc could be excluded as potential targets of beta-catenin.

Gene Expression in Cells in Liver Injury Models

Gene expression analyses of hepatic progenitors have been dominated by analyses of oval cells in liver injury models, such as the AAF/PH model systems (9,168), meaning that the relevance of some of the reports of specific gene expression patterns observed to that in normal hepatic stem cells is still unknown. Moreover, now that hepatic progenitors have been isolated from normal, untreated, animals, and it has been realized that expression patterns of genes in oval cells can differ from those found in normal tissues or in normal embryonic development (Section 4.3). The phenomenology of patterns of gene expression in oval cells overlaps with that of hepatic progenitors in normal development, such as elevation in the numbers of hepatoblasts, cells expressing AFP but there are other patterns with striking distinctions such that oval cells are not dependent upon paracrine signals from feeders of endothelial progenitors for growth. Some of the phenomena must be ascribed to either aberrant progenitors or the inflammatory processes associated with the injuries. Therefore, the phenomena must be noted descriptively and, at present, without an ability to interpret it fully or to clarify the mechanisms.

In general, the highest proliferation of oval cells is observed between 7 and 13 days after the liver injuries by AAF and PH. This is also the time period in which expression of HNF1 and HNF3 has shown higher expression with maximal expression of HNF3 after 16 days and an association with a strong albumin mRNA level (198). Interestingly, during the period of the highest cell proliferation, between days 11 and 13, albumin expression is dramatically reduced and is not observed again until day 16. In contrast, increased expression of HNF4 and HNF3 goes up over the period up to nine days after PH and thereafter undergoes a decrease. C/EBPb and DBP displayed elevated steady-state levels throughout the observed time period. Transcripts for all analyzed transcription factors are higher than those in the surrounding hepatocytes, with the exception of HNF4. According to Ott et al., cells positive for the oval cell antigen A6 do not show expression of albumin or AFP, but induction with sodium butyrate leads to induction of albumin and AFP mRNA expression (199), whereas Peterson and coworkers (200) report positive AFP protein expression of 75% of MACS A6+ sorted cells, and Evarts et al. (198) observed a strong positive expression in groups of oval cells nine days after AAF/PH treatment.

Green fluorescent protein (GFP)-expressing, d13.5 embryonic mouse progenitor cells that were transplanted in the uPA/RAG-2 mouse showed increased AFP gene expression at the early stage (d3 to d7) and decreased AFP gene expression after two weeks (75), compared with normal hepatocytes. After four and six weeks, AFP expression was no longer detectable, whereas albumin expression slightly increased.

Genes Associated with Normal Hepatic Stem Cells, Hepatoblasts, and Committed Progenitors

The markers used to identify normal hepatic stem cells, hepatoblasts, and committed progenitors (Table 5, Fig. 5) consist of long known *cytoplasmic markers* of hepatic progenitors (albumin, AFP, and CK19), the *hepatic-specific transcription factors* noted above (e.g., homeobox genes, jun, and cEBP), and the more recently identified *surface markers* consisting of embryonic cell adhesion molecules (EpCAM, NCAM, and E-cadherin), embryonic matrix receptors (hyaluronan receptor CD44H and an integrin isoform, CD49f), and CD133, also called prominin, a transmembrane antigen expressed both at RNA and protein levels by a number of stem-cell types including hepatic progenitors (201,202). Hepatocyte growth factor (HGF) is also a prerequisite for liver development; mice lacking in HGF expression failed to develop and die before birth, having significantly reduced liver development (203). Also, it is increased 10-fold after addition of HGF to fetal liver cells after 30 minutes, and its expression returns to normal levels after 48 hours (204). These findings are among those that helped to identify cMet, the receptor for HGF, as a marker for murine hepatic progenitors (136). Suzuki and coworkers used an antibody to cMet in combination with an antibody to an embryonic integrin, CD49f, to flow cytometrically sort murine hepatic progenitors.

The recent recognition of well-characterized and purified surface antigens on hepatic stem cells, such as cMet and EpCAM, and the availability of monoclonal antibodies to these antigens should launch a new era in investigations of hepatic stem cells in normal and diseased tissues. However, it is important to note that, as for other stem-cell types, there is no one antigen or marker that uniquely defines the subpopulations of hepatic progenitors. Combinations of markers are required to ascertain which lineage stage of cell is being purified or analyzed. For example, CK19 and EpCAM are expressed by hepatic stem cells, hepatoblasts, and by biliary epithelia, but only the hepatic stem cells and hepatoblasts also express albumin, and only the hepatoblasts express AFP.

Regulators of Cell Cycle and Cytokinesis

Gene expression of cyclins, known as important cell-cycle-dependent proteins of mitosis (205) and also often overexpressed in cancer, can be induced by thyroid hormone (T3). In particular, T3 induces expression of cyclin D1, E, E2F, and $_p$107 and enhances phosphorylation of $_p$Rb, the substrate in the pathway leading to transition from G1 to S phase (206). Transfection with cyclin D1 in vivo leads to hepatocyte proliferation, that is reduced after several days by up-regulation of $_p$21 (207). Cyclin D1 and D3 genes are strongly expressed in the E12.5 and E14.5 mouse embryo liver, but not in the neonate and adult liver (208). Aim-1 protein and citron kinase are enzymes critically involved in cytokinesis and both are down-regulated in cells undergoing polyploidization (Fig. 4) (45). Their protein levels are highest in embryonic liver and are lost or below the level of detectability in polyploid cells. Citron kinase has been shown to be a cell-cycle-dependent protein regulating the G2/M transition cytokinesis in parenchymal cells by phosphorylation of myosin II. Citron-K is found as either cytosolic aggregates or nuclear protein during interphase and concentrates at the cleavage furrow and mid-body during anaphase, telophase, and cytokinesis. However, mutant mice, which are null for citron kinase, survive embryonic development and die within a couple of weeks of postnatal life. The hypothesis is that there are regulators other than Citron kinase that regulate cytokinesis in embryonic development, that citron kinase is a postnatal regulator of cytokinesis, and that its absence in the mutant mice results in death due to early apoptosis of the cells (45).

Markers for Which There Is Debate About Their Existence on Hepatic Stem Cells

There are several antigenic markers for which there has been considerable debate about whether they are expressed by hepatic progenitors or rather by inflammatory cells or endothelia in close association with the hepatic progenitors (125,209,210). The most noteworthy of these are CD34, CD45 (common leukocyte antigen), CD90 (Thy-1), and CD117 (ckit). CD34 has been found on murine and rat oval cells in various liver injury models and it has been claimed that in situ hybridization and immunohistochemistry confirm the expression on hepatic progenitors, as well as on hepatic endothelial cells (125,210,211). Cells positive for the oval cell antigen A6 have been shown to be positive for the expression of the hemopoietic stem-cell marker Thy-1 on their cell surface as well as at the transcriptional level (210,212). However, identification of oval cells is by antibodies to oval cell antigens, and all known oval cell antigens are expressed also on various hemopoietic and mesenchymal cell subpopulations such as endothelia. Therefore, it has put into question the prior studies and claims about the expression of these markers on hepatic progenitor cells. In addition, the hepatic progenitors are identified either with oval cell antigens, such as A6, or with expression of AFP. Some subpopulations of hemopoietic progenitors have been found to express a form of AFP identical with that in hepatic cells except for exon-1-encoded sequences (213). Therefore, proof of characterizing hepatic progenitors requires that clonal cell populations derived from immunoselected cells can give rise to mature liver cells in vitro or in vivo. The analyses in which these more demanding criteria have been used indicated that hemopoietic and mesenchymal antigens are not expressed by the hepatic stem cells, hepatoblasts, and committed progenitors from the livers of any species (108,126,132,144,145,214). The stem cells and hepatic progenitors have proven negative for glycophorin A (red blood cell antigen), CD45 (common leukocyte antigen), CD34, CD14, CD38, and all surveyed lymphocytic antigens.

The data with respect to CD117 (c-kit) remain debatable but more convincing. High levels of gene expression for stem-cell factor can be observed in human early-stage liver (day 34) (215). Immunocytochemistry reveals that the expression of the receptor for stem-cell factor, c-kit, in normal human liver is present on presumptive stem cells at the portal triads and in the development of bile ducts (89,216). What remains unclear is whether c-kit is associated with the hepatic progenitors or with tightly associated endothelial progenitors. These two alternative interpretations have not been resolved in most cases leading to some studies in which c-kit is found, especially studies in rodent livers (137,208,210,216,217), and others in which it has not been found, for example, in murine and human hepatic progenitors (108,132,134,144,145,214,218). As many of the studies in which c-kit has been reportedly found on hepatic progenitors have been with liver injury models (218), it remains possible that it is identifying inflammatory cells induced by the liver injury processes. This option is especially plausible given that the antigen is present on hemopoietic, mesenchymal, and endothelial progenitors (215,220). In the Ws/Ws rat lacking c-kit activity caused by deletion in the kinase domain after application of the combined AAF/PH model, the development of oval cells is clearly suppressed. Nevertheless, stem cells that develop in the Ws/Ws rats show similar protein expression and proliferation capacity to normal +/+ phenotypes indicating the important role of c-kit in stem-cell development but not proliferation or maintenance of phenotype, findings that might also be interpreted as diminished paracrine signaling by endothelia (221). In normal and cirrhotic liver levels of c-kit, mRNA is consistent, but elevated in fulminant hepatic failure (222).

RECONSTITUTION OF LIVER BY CELL TRANSPLANTATION
Animal Models

Numerous animal models are now available for studying reconstitution of livers by transplanted cells (Table 9). To simulate the varying states of proliferation that are observed in diseased livers, several studies have been designed to promote proliferation naive mice in order to induce repopulation of the transplanted hepatocytes (223–225). It is clear that the primary rate-limiting component toward better results of hepatocyte transplantation is the low level of expansion of transplanted cells in the host liver (226). This can be explained by at least three factors: (*i*) the very low level of cell turnover present in the normal adult liver, providing no significant driving force for the growth of the transplanted cells, (*ii*) the lack of a selective growth advantage for transplanted cells over resident cells, whereby they could differentially respond to specific stimuli and preferentially expand, and (*iii*) the probable existence of a feedback loop in which a soluble signal(s) from mature parenchyma in the pericentral zone inhibits the proliferation of the parenchymal cells of zone 1, including those of the stem-cell compartment.

Studies using dipeptidyl-peptidase-IV (DPPIV)-deficient rats have been particularly helpful in elucidating the biology of transplanted cells (226). DPPIV is abundantly expressed in bile canaliculi, which provide methods to demonstrate whether transplanted cells are integrated in the liver parenchyma (227,228). However, transplanted cells do not proliferate in normal rat or mouse livers (229–231), with the exceptions being the livers of either very young or old F344 rats, in which transplanted cells exhibit spontaneous proliferative activity (231). This translates into the repopulation of only 0.5% to 1% of the liver following transplantation of 20 million cells in the rat or 2 million cells in the mouse liver,

Table 8   Requirements for Ex Vivo Growth of Parenchymal Cells

Requirements for all lineage stages	Cells of the stem-cell compartment	Adult cells
Nutrient-rich basal media (e.g., RPMI 1640) Lipids: high-density lipoprotein and mixture of free fatty acids (bound to carrier molecules such as albumin)	Transferrin/fe Embryonic/fetal matrix substrata Avoid copper (drives differentiation) Avoid EGF—its lineage restricts to hepatocytic lineage Avoid type I collagen—its lineage restricts to biliary cells	Mature matrix substrata (chemistry of the matrix distinct for growth vs. differentiation) Epidermal growth factor No need for transferrin/fe (because they make it)
Trace elements; zinc, selenium Insulin	Unidentified signals from embryonic stroma or endothelial progenitors (STO cells can substitute partially)	

Abbreviation: EGF, epidermal growth factor. *Source*: From Refs. 76, 82, 324.

and correction of specific diseases might be incomplete with this magnitude of liver reconstitution (232).

Several laboratory animal models have been used to model correction of metabolic diseases. For example, transplantation of normal hepatocytes to Nagase analbuminemic rats with low levels of serum albumin due to an albumin gene defect results in alleviation of the metabolic abnormality (233). Similarly, transplantation of normal liver cells into Gum rats, that model Crigler-Najjar syndrome type-1, results in normalization of functions (234). Hepatocyte transplantation into the Watanabe heritable hyperlipidemic rabbit, which lacks cell surface receptors for low-density lipoproteins and models familial hypercholesterolemia (235) or into the Long-Evans Cinnamon (LEC) rat, an animal model for Wilson's disease (129), shows that transplantation of hepatocytes can alleviate the conditions. However, one must transplant relatively large numbers of mature hepatocytes, as the transplanted cells do not expand significantly.

Induction of transplanted cell proliferation in the liver requires selective ablation of pericentral parenchymal cells. The AL-uPA transgenic mouse model was the first to be used to demonstrate massive liver repopulation (119,120). In this animal, the transgene

Table 9   Strategies: Method of Transplantation

Transplantation into blood stream
    Currently used methods; based on long history of studies with hemopoietic cells
    Less successful for therapies for cells from solid organs such as liver
    Problems: emboli, cells carried to inappropriate sites, difficulties for engraftment, cells not in ideal environment
Transplantation by grafting
    Ideal for cells from solid organs
    Requires implanting aggregated cells or, ideally, cells on scaffolding [e.g., polylactide meshes developed by Langer and co-workers (264,325,326)]
    Optimal results require mix of epithelial and mesenchymal cell partners (e.g., hepatic stem cells + embryonic stroma) or use of the paracrine signals they produce
    Laparascopic procedures can be used, therefore, minor surgery (can even be outpatient procedure)

Table 10 Properties of Cells in Liver Cell Therapies

Mature liver cells	Hepatic progenitors
Must be isolated from livers exposed to limited cold ischemia	Can be isolated from livers exposed to cold and some warm ischemia
Difficult to cryopreserve	Readily cryopreserved
Logistical strategies of getting cells from donor to recipient are difficult	Logistical strategies of getting cells from donor to recipient are relatively easy
Large cell volume for transplant	Small cell volume for transplant
Little growth potential	Full lineage potential with time
Emboli formation	Maximum proliferative potential
Immunogenicity problems	Reduced immunogenicity (little or no expression of MHC antigens)
Rapid restoration of adult functions	Restoration of adult functions takes longer but is more stable
Progenitors from liver	Progenitors from bone marrow
Large numbers of progenitors	Technologies fully established for sourcing, cryopreservation, and use
High degree of efficacy in reconstitution of livers	Efficacy of unfractionated bone marrow is extremely low due to the rarity of the liver precursors in bone marrow
No ex vivo expansion required	
Sourcing straightforward as can be obtained from cadaveric livers	Ex vivo expansion required to overcome limitation in number of precursors (forcing complications in regulation by FDA)
Minimal complications for FDA regulation	Fusion of bone marrow cells with liver cells provides much of the effects

Abbreviation: MHC, major histocompatibility complex.

is expressed in albumin-expressing cells and results in toxicity to those cells. Most of the transgenic mice die at birth except for those in which the transgene is lost in one or more cells and these cells are able to proliferate selectively to regenerate the liver (119). Subsequent studies by Rhim et al. (120,236) showed that transplantation of both syngeneic and xynogeneic (rat and human) adult hepatocytes into the new born transgenic pups could reconstitute the damaged livers. Four to five weeks later, up to 80% of hepatocytes in the recipient livers were found to be of donor origin (120,236), confirming that the constitutive expression of the uPA transgene in resident hepatocytes generates a selective environment which favors the growth of cells with a normal phenotype.

A similar general principle has been used as the basis for the development of the fumaryl-acetoacetate hydrolase (FAH) null mouse model for the human disease hereditary tyrosinemia type I, which is due to a lack of the enzyme FAH involved in the tyrosine catabolic pathway (112). Transplanted liver cells repopulated nearly the entire diseased recipient liver requiring an estimated 12 to 18 rounds of cell division. Using the FAH null mouse model, Overturf et al. (237) found that normal male adult hepatocytes, when transplanted to female FAH null recipients, could repopulate the recipient animal's liver to >90% within six to eight weeks. Rescue of FAH-deficient animals and restoration of liver function required as few as 1000 donor cells.

Other models have been developed to allow reconstitution of livers by donor liver cells, all using mechanisms that selectively eliminate mature, pericentral parenchymal cells. They include: (*i*) Fas ligand-induced apoptosis (225), (*ii*) prodrug activation of herpes simplex virus thymidine kinase (HSV-TK; 49), (*iii*) expression of the cell-cycle regulator Mad1 (238), (*iv*) use of toxic bile salts in mice deficient in the *mdr 2* gene, which impairs biliary phospholipid excretion with hepatobiliary injury (239), and (*v*) a retrorsine model in which retrorsine alkylates DNA and induces extensive hepatic polyploidy and more rapid liver turnover in rats (240). Retrorsine's effects mimic those in animals after two-thirds PH and then treatment with the thyroid hormone, triiodothyronine (T3) (150,241,242).

Multiple animal models with acute liver failure show that hepatocyte transplantation can reduce mortality, but it is not clear that this improvement is due to the cell transplantation or another mechanism such as mitogenic stimulation (129,243–245). The same results of reduction in mortality were achieved by transplanted cells to a genetic model in which acute liver failure was induced by activation of ganciclovir by HSV-TK or Mad1 expression (49,246). Animals with cirrhosis, induced by repeated CC14 administration, develop significant liver fibrosis, portal hypertension, and ascites (247). Studies show that transplanted cells could integrate in the liver parenchyma despite extensive fibrosis in cirrhotic animals, but cell proliferation of the donor cells in the host liver was limited. Yet, there were no differences in mortality among the experimental groups over a 12-month period. In contrast, creation of an additional reservoir of cells by intrasplenic cell transplantation in extremely sick, cirrhotic rats was associated with improvement in liver tests but coagulation abnormality (248); it is unclear why transplanted hepatocytes demonstrated superior functions compared with those of the host livers of these animals unless the microenvironment of the spleen was more supportive of the cells than that in the host liver or the host hepatocytes had become aberrant due to the disease.

Hepatic Stem Cells and Hepatoblasts

Hepatic stem cells and hepatoblasts are the source of cells most likely able to provide the maximum reconstitution of livers after transplantation given their maximum potential for proliferation (Table 10). Transplantation of hepatoblasts from embryonic days 12 to 14 into the livers of young adult rats subjected to PH or following retrorsine treatment and PH results in donor cells that differentiate into hepatocytes and into cholangiocytes that form bile ducts continuous with host bile ducts. However, cells derived from embryonic day 18 livers did not give rise to cholangiocytes because either a distinct parenchymal progenitor subpopulation was isolated or their ability to integrate was altered by that developmental stage (249,250). Hepatoblasts sorted from ED13.5 murine fetal liver and that express the antigenic profile c-kit$^-$ CD45$^-$ TER119$^-$ CD49f$^+$ CD29$^+$ engraft and mature into hepatocytes when transplanted into the livers of congenic hosts exposed to retrorsine and CCl$_4$ (136). Similarly, hepatoblasts that are c-kit$^-$ CD45$^-$ TER119$^-$ c-met$^+$ CD49f$^{+/low}$ also engraft and give rise to both hepatocytes and cholangiocytes when transplanted into the liver (214).

Indeed, these cells differentiate also into acinar and duct cells of the pancreas and into intestinal epithelial cells of crypts and villi of the small intestine (251). Using uPA transgenic mice as a model of liver injury shows a differentiation of mouse fetal liver (ED13.5) cells and proliferation of these cells after engraftment; differentiation correlates with increased expression of albumin and decreased expression of AFP (200).

Transplantation of Precursors from Sources Other than Liver

As noted in an earlier section of this review, bone marrow cells can be a source of precursors for hepatocytes (3,4,252) as can purified hemopoietic stem cells (88) and multipotential progenitors isolated from bone marrow (5). However, the hepatocytes have been shown to arise from cell fusion and not by differentiation of hematopoietic stem cells or other bone marrow cells (113,253). Although experimental studies will go on in exploring transdifferentiation, they are likely to provide a minor contribution toward future efforts in establishing liver cell therapies other than investigations of bone marrow cells or factors, such as those involved in inflammation, that might participate in aspects of liver regeneration.

ES cells, in contrast, could prove a major new source of cells for liver cell therapies if methods for lineage restricting them into endodermal fates prove successful and if they can be modified to eliminate concerns about tumorigenicity. Liver progenitor cells purified from lineage-restricted ES cells from culture using AFP as a marker differentiate into hepatocytes when transplanted into partially hepatectomized lacZ-positive ROSA26 mice (254). GFP(+) cells engrafted and differentiated into lacZ-negative and albumin+ cells. Differentiation into hepatocytes also occurred after transplantation of GFP(+) cells in apolipoprotein-E- (ApoE) or haptoglobin-deficient mice as demonstrated by the presence of ApoE-positive hepatocytes and ApoE mRNA in the liver of ApoE-deficient mice or by haptoglobin in the serum and haptoglobin mRNA in the liver of haptoglobin-deficient mice. This study describes the first isolation of ES-cell-derived liver progenitor cells that are viable mediators of liver-specific functions in vivo (254). Cellular uptake of indocyanine green (ICG) was used to evaluate liver function in the cultures of lineage-restricted murine ES cells; ICG-stained cells appeared around 14 days after the formation of embryoid bodies and formed distinct three-dimensional structures. They were immunoreactive to albumin and expressed mRNAs such as albumin, AFP, transthyretin, HNF3 beta, AAT, tryptophan-2,3-dioxygenase, urea cycle enzyme, and gluconeogenic enzyme. After transplantation of ICG+ cells into the portal veins of female mice, they incorporate into hepatic plates and produce albumin. Bipotential mouse embryonic liver (BMEL) cell lines present a mixed morphology, derived from E14 embryos, containing both epithelial and palmate-like cells, and an uncoupled phenotype, expressing hepatocyte transcription factors (HNF1alpha, HNF4alpha, and GATA4) but not liver-specific functions (apolipoproteins and albumin). The BMEL stem-cell lines participate in liver regeneration in albumin-urokinase plasminogen activator/severe combined immunodeficiency disease (Alb-uPA/SCID) transgenic mice (255). After transplantation into the spleen, they engraft into the liver and then proliferate and differentiate into both hepatocytes and bile ducts, forming small to large clusters detected throughout the three to eight weeks analyzed after transplantation. They participate in the repair of damaged tissue without evidence of cell fusion (255).

Method of Inoculation of Parenchymal Cells

The host liver represents an ideal "home" for the transplanted hepatocytes in terms of the unique hepatic organization and interactions with non-parenchymal liver cells (256). All the experiments summarized above utilized methods of transplantation involving infusing cells to the liver through intraportal or intrasplenic routes. For transplanted hepatocytes to engraft, the most important criterion for these cells was to translocate from the portal pedicle into the liver microenvironment, as described by several groups (227,229,257). This process begins after hepatocyte infusion and takes approximately 20 hours until the cells finally join adjacent host hepatocyte, the transplanted cells become stacked at the portal vein radicals, which results in some of the cells being deposited at the hepatic sinusoid. Although the majority of the hepatocytes are cleared from these areas, a portion of the cells start to translocate into the space of Disse by disrupting the sinusoidal endothelium (227).

The spleen is also a viable target tissue for transplantation of hepatocytes because it offers the ability to form differentiated cord structures and to reform nearly normal hepatic architectures (258–260). The major limitation of the transplanted hepatocyte procedure via the portal vein or intrasplenic route is that the number of viable cells that can be engrafted without causing complications is in the range of 2% to 5% of the host hepatocytes (261,262). The major complication has been found to be portal vein thrombosis, which results in liver failure and severe portal hypertension, hemorrhage, and migration of cells to the lungs leading to pulmonary embolism (263). Portal hypertension is associated with a high probability of intrapulmonary deposition of hepatocytes, as shown in a previous study in the rat (264).

Efforts are needed to evaluate grafting methods in which donor cells are transplanted as a graft while embedded in forms of extracellular matrix (76,265) or onto biodegradable scaffolds (266–270) (Table 10). Grafting could increase the numbers of cells that can be transplanted, could avoid the problems of portal hypertension, and of the spread of cells to sites other than liver, and could be done with a microenvironment within the graft designed to optimize initial expansion of the cells.

LIVER CELL THERAPIES—CLINICAL PROGRAMS

Introduction

Liver failure is a serious health problem. Each year, there are an estimated 300,000 hospitalizations and 30,000 deaths in the United States due to liver diseases, and approximately 18,000 patients are on the liver transplant waiting list, an increase of more than 100% over the last four years. Currently, the only cure available for many of these liver diseases is a liver transplant. However, the vast majority of patients with liver diseases cannot rely on organ transplantation as a solution in the coming years.

Efforts by numerous investigators are ongoing to develop liver cell therapies (Tables 9–12) as alternatives to organ transplantation for dysfunctional livers (141). The two major forms of liver cell therapies are injections, implantations or transplantation of cells (141), and extracorporeal bioartificial livers used as liver assistance devices (271–274). A major anticipated advantage of cell therapy, in light of the well-known regenerative capacity of the liver, is that cells obtained from a single donated liver might be used to treat many patients. Furthermore, the surgical procedures for cell therapy are less drastic, potentially safer, and more economical than whole-organ transplantation (275). However, unless the dose of liver cells sufficient to treat an individual patient turns out to be surprisingly small, the current level of organ donation will remain inadequate to support widespread clinical investigation or future implementation of liver cell therapy. The only real hope of solving the "sourcing" problem is to use stem cells with their renowned capacity for expansion and differentiation (Table 9).

Liver Cell Therapies Using Liver Assistance Devices (Bioartificial Organs)

Bioartificial livers are being developed as extracorporeal liver assistance devices to support patients in liver failure (76,276–279). They are likely to be used as adjuncts to transplantation of liver cells to enable a patient to have liver functions even while transplanted donor cells are reconstituting normal liver tissue. Although there have been

Table 11   Sourcing of Human Liver Tissue

Fetal livers (16–20 wk gestation)	High percentage of stem cells and progenitors; ease in isolation; economical	Ability to procure and use them depends on political and cultural attitudes
Liver resections	Pediatric and adult livers	Difficult to obtain; highly variable quality of tissue; small amounts
Organ donors ("brain-dead but beating heart donors")	Pediatric and adult livers; ~1–2% of deaths; ~5000/yr in United States; cold ischemia only	Highly variable quality of tissue; considerable competition for organs rejected from transplant programs
Cadaveric livers (asystolic donors)	Neonatal, pediatric, and adult livers; all neonatal deaths and 98–99% of pediatric and adult deaths; all are available for research and cell therapy programs; warm and cold ischemia; stem cells survive for 6–8 hr; neonatal livers survive as an organ for 6–8 hr, as so rich in stem cells	The organs cannot be used for transplantation, mature liver cells are lost within ~1 hr of death

clinical trials with hepatic cell lines (Hepatics, San Diego, CA) and porcine liver cells (277), the only ones that have achieved success have been those with human liver cells inoculated into bioreactors with efficient supply of oxygen and nutrients (280,281). The ones with cell lines failed clinical trial due to poor functioning of the cells and those with porcine liver cells partially succeeded but have been constrained by potential severe immunological reactions with long-term use and concerns about pathogens that might derive from porcine cells with unknown effects on humans (280,282,283). The clinical trials with mature human liver cells (284) have offered the best results to date in ability to support patients in liver failure; the patient and organ survival rate has been 100% with an observation period of three years (285). However, the limiting issue for liver support still depends on the availability of fresh, normal human liver cells. The present sources are from discarded organs intended for transplantation. Thus, although liver assistance devices are an attractive technology with therapeutic potential, the limited availability of normal human liver cells has prevented the technology from being utilized in clinical settings.

Therefore, one of the great hopes is that hepatic stem cells will be able to make possible the expansion of the bioartificial organ technologies into widespread clinical programs.

Liver Cell Therapies Using Injection or Implantation of Cells

The idea of liver cell transplantation for the treatment of liver disease was first touted in 1977, when it was noted that liver cells could be isolated and transplanted into animal models to ameliorate liver insufficiency (286,287). Recent research has demonstrated the ability of donor liver cells to repopulate the diseased liver in animal models of metabolic liver disease (120) and fulminant liver failure (288), whereas from the

Table 12   Representative Strategies for Cell Therapies

Group 1. Adults with cirrhosis and who do not qualify for organ transplantation
   Large, underserved patient population
      6–18 months life expectancy
      Co-morbidities keep patients off transplant list
   Assess
      Safety, engraftment, proliferation (scans, donor HLA)
      Functions (MEG-X, ammonia challenge, etc.)
      Clinical complications of end stage of liver disease; quality of life
   Concerns
      Scar tissue in the liver will block engraftment and maturation of donor cells
      Immunosuppression may lead to expansion of tumor cells that are pre-existing in
      recipient's liver
   Ideal future therapies: autologous therapies with hepatoblasts prepared from one of the liver lobes
      and seeded onto polylactide meshes that are then grafted onto the residual liver; should
      minimize (or eliminate) the need for immunosuppression

Group 2. Children with inborn errors of metabolism
   Small, underserved patient population
      Children typically die before they can be transplanted
      Many difficult to manage clinically
      Livers are normal except for the effects of the defective gene
   Strategies unique to this patient population
      Liver's feedback loop will be intact so must transplant large numbers of stem cells
      Monitor function(s) missing due to genetic condition
   Concerns
      Immunological issues: will the children reject the cells after the cells mature?
   Future for these patients: should be ideal patients for stem-cell therapies; may be able to modulate
      immunology to be able to avoid immunosuppression

Group 3. Children and adults with acute liver failure
   Acute crisis and requires rapid response
      Low dosage of cells should work, as feedback loop will be inactivated
      May require cell therapy with diploid adult cells to give rapid response of adult-specific
         functions
      May require adjunct therapy with bioartificial liver to give cells time to become established and
         mature
   Assess
      Safety, engraftment, proliferation (scans, donor HLA)
      Functions (MBG-X, ammonia challenge)
      Clinical complications, quality of life
   Concerns
      Will the cells engraft and mature sufficiently fast to overcome liver failure?
      Will the cells be rejected once the cells mature?
   Future ideal therapies for these patients: most likely stem cells plus temporary support with
      bioartificial liver; alternatively, large graft of diploid adult cells on polylactide meshes

Group 4. Children and adults with viral infections
   Lineage-dependent viruses
      Hepatitis C is representative; it is hypothesized to replicate in stem cells and early progenitors
         and then matures along with host cells producing mature virions only in mature cells (327)
      Cannot use stem-cell therapy, as stem cells will become infected

---

(Continued)

Table 12 Representative Strategies for Cell Therapies (*Continued*)

Viruses that are not lineage-dependent
  Grafts with cells modified by gene therapies to protect cells from virus
Concerns
  It must use a lineage stage with limited growth potential, must use large numbers of cells or large graft, or must do the treatment repeatedly
  Will the cells be rejected once the cells mature?
  The effect of immunosuppression may allow the virus to flourish
Future ideal treatments for these patients: grafts with lineage stage(s) still capable of hyperplastic growth and yet not infectable by virus or grafts with stem cells modified by gene therapies to block the virus

Abbreviation: HLA, human leukocyte antigens.

other side researchers have shown the ability of non-liver-derived cells such as bone marrow cells to differentiate into functional hepatocytes under the condition of liver injury (3,4,88).

Liver cell therapies in humans during the past decade have made use of suspensions of mature liver cells and have resulted in benefits to patients with fulminant liver failure (260,289–290) and are able to bridge some patients until whole-organ transplant is possible (292). Cell therapies for metabolic liver disorders in humans, such as familial hypercholesterolemia (261), Crigler-Najjar (293), and ornithine transcarbamylase (OTC) deficiency (294), have shown proof of principle in that the donor liver cells have the potential to survive and function long term in patients with good safety profiles clinically. However, the ability to expand these early studies into a widespread clinical program is minimal for many reasons (Table 11): (*i*) the mature liver cells must be obtained from the rare livers that are rejected from organ transplant programs, (*ii*) they cannot be cryopreserved with any degree of success meaning that there are limits on testing for diseases and limits on how far one can transport them, (*iii*) the cells do not proliferate after transplantation resulting in the need to transplant large numbers of cells, (*iv*) they rapidly form balls of cells, spheroids, that can cause potentially lethal emboli, and (*v*) the cells are highly immunogenic requiring significant immunosuppression of the recipients. These difficulties will be alleviated or solved by use of stem cells, especially probably in combination with grafting methods (Table 10), because the progenitor cells can be cryopreserved, have dramatic expansion potential, and have low or negligible immunogenic antigens (although these will appear with differentiation of the cells) that can possibly be managed with minimal need for immunosuppressive drugs. The problems with portal hypertension and with emboli formation are solvable by utilizing grafting rather than inoculation into blood vessels of the liver.

Current clinical trials of liver cell transplantation are underway for the treatment of fulminant liver failure. Therapy for fulminant liver failure is effective if patient survival is significantly improved. Success can be achieved in several ways: (*i*) bridging patients to organ transplant, (*ii*) bridging them to recovery of liver function of the native liver with concurrent disappearance of the donor liver cells, (*iii*) by engraftment and long-term function of the liver cell transplant. This third possibility cannot be achieved yet even with multiple infusions of mature cells, increasing portal hypertension, and pulmonary dysfunction cause the maximum number of cells that can be transplanted to be only 2% to 5% of the recipient original liver mass. In one clinical trial, an attempt to treat OTC-deficient patients who received 10^9 hepatocytes via portal vein injection showed an increase of approximately 70% in portal pressure (295). Another study (291) reported a transient increase in oxygen requirements due to the cell migration into the lungs and ventilation/perfusion mismatch after transplantation. Another limitation is that the transplanted cells do not grow and do not survive long term in the host. There are similar complications after intrasplenic transplantation, because 80% of the cells migrate out of the spleen into the portal circulation (259,264,296).

Several recent clinical trials of liver cell transplantation (LCT) for the treatment of acute liver failure have been reported in the literature. Strom et al. (292) describes the use of LCT as a bridge to whole-organ transplantation in five patients with grade IV hepatic encephalopathy and multisystem organ failure. Those who received an arterial splenic perfusion of a mixture of liver cells (freshly isolated and cryopreserved liver cells) maintained normal cerebral perfusion and cardiac stability, with withdrawal of medical support 2 to 10 days before whole-organ transplantation. Blood ammonia levels decreased significantly, and three of the five patients successfully bridged to whole-organ transplant were alive and well at 20 months follow-up compared to four control patients who died within three days. Other trials that have been reported (289,297,298) had some degrees of success after transplantation of fresh and frozen human hepatocytes into the portal vein of patients with liver failure. Some of the critically ill patients recovered spontaneously, whereas other patients demonstrated some improvement in ammonia, prothrombin time, encephalopathy, cerebral perfusion pressure, and cardiovascular stability, but there has been no evidence that the demonstrated engraftment of transplanted liver cells in these patients was responsible for the clinical improvements. Trials of LCT in metabolic liver diseases have been reported using autologous hepatocytes transfected in vitro with a human low-density lipoprotein-expressing recombinant retrovirus in a patient with familial hypercholesterolemia (261) or allogeneic clinical trial of LCT (299). In the first trial as described by Grossman et al. (261) cells could be harvested and safely infused into the recipient's liver, with significantly decreased serum cholesterol levels for a prolonged period (18 months). Fox et al. (299) described the first allogeneic liver cell transplantation in a 11-year-old girl with Crigler-Najjar metabolic disorder and showed that the patient's total serum bilirubin decreased from 26.1 to 14 mg/dL, and bilirubin conjugates measured in bile increased from a trace to 33%. Bilirubin uridyl glucuronyl transferase activity measured in a liver biopsy sample increased from 0.4% to 5.5% of normal activity. Furthermore, phototherapy treatment could be reduced from 12 to 6 hours per day—an outcome that would significantly improve this patient's quality of life. Long-term evidence of liver cell transplant engraftment and function in this patient was demonstrated for more than 18 months; this study demonstrated the proof of principle that donor liver cells have the potential to survive and function long term in patients but with a limitation of art being able to repopulate the host liver. Thus, further liver cells would need to be infused that could increase the possibility to develop portal vein thrombosis or portal hypertention and pulmonary dysfunction.

The ideal outcome of liver cell therapy is not just the engraftment but the coordinated and orderly expansion of donor cells so that a new liver can be created in the architecture of the old with the smallest number of donor cells. Alternatively, engraftment without proliferation could be insufficient support for a metabolic disorder for long-term outcome, and the need for multiple cell perfusion increases the risk for sepsis, hemodynamic instability, and developing portal vein thrombosis and parenchymal ischemia, which could be minimized by using a small number of cells and a slow perfusion speed, but the effect of a small number of cells could show minor improvement in the patient without repopulating the host liver.

Cell Sources

Cell sourcing remains among the most critical difficulties in the development of cell therapies, whether for bioartificial organs or for cell transplantation (Table 9). To date, the
studies have made use of mature liver cells derived from organs rejected for transplantation. However, the quality of these cells, the inability to cryopreserve them with sufficient preservation of functions, and the limitation on proliferation of the cells after
transplanting have caused investigators to focus on alternative sources such as ES cells
that are differentiated into cells of the liver lineage or even porcine hepatocytes (Tables
9–11). The studies to date on transplantation with progenitors and the findings that progenitors survive even warm ischemia provide hope that cadaveric organs, such as those
from neonates, may alleviate the sourcing problems.

Strategies for Patients

Future considerations for liver cell therapies must incorporate the realization that the strategies will be different for different diseases. Some representative examples of these are
indicated in Table 12. For example, the use of purified hepatic stem cells and hepatoblasts
should be ideal for those with inborn errors of metabolism but the numbers of the cells to
be injected or grafted must be large given that the feedback loop will be intact in these
patients. In contrast, patients with lineage-dependent viruses, such as hepatitis C,
cannot be treated with stem cells because the virus can enter and undergo some stages
of replication in the stem cells; transplanted stem cells will be obvious targets for the
endogenous virus. The two options for these patients are either to identify the lineage
stages in which the virus cannot enter the cells and if it is a stage at which the cells can
still undergo significant replication, use cells of that stage to transplant the patients. Alternatively, the stem cells can be modified by gene therapy mechanisms to protect them from
infection after transplantation. The patients with acute liver failure can be treated with
stem cells and hepatoblasts but will surely require adjunct support from bioartificial
livers while the transplanted cells are expanding and maturing.

Perhaps the most difficult category of patients will be that with end-stage cirrhosis.
The microenvironment of the cirrhotic livers will limit engraftment and proliferation of
transplanted cells. Moreover, these patients are known to have transformed cells that
could flourish into tumors with immunosuppression. The patients with cirrhosis, especially
those with the end-stage disease, are likely to be the patients who in the future will be
transplanted. An alternative that, theoretically could work, is to do a form of autologous
cell therapy: isolate the large numbers of hepatoblasts, known to be in cirrhotic livers,
from a portion of the person's liver; graft the cells onto biodegradable scaffolds; and
graft the scaffolds to the remaining liver. Although this strategy is quite exotic and
would require support from a bioartificial liver while the graft is maturing, it would
have the advantage of not requiring immunosuppression.

CONCLUSION

The liver is a tissue comprised maturational lineages of cells with lineage-dependent size,
morphology, growth potential, gene expression and functions. These phenomena have
ramifications for liver biology, liver regeneration, and various liver diseases and strategies
for cell and gene therapies. The coming years offer great hope of exploiting the stem-cell
and lineage biology phenomena in experimental studies and in the treatment of patients
with liver diseases.

ACKNOWLEDGMENTS

Funding for the research has been provided by grants to L.M.R. that include a sponsored research grant from Vesta Therapeutics (Research Triangle Park, NC), NIH grants (DK52851, AA014243, and IP30-DK065933), and by a Department of Energy Grant (DE-FGO2-02ER-63477); grants to S.G. of NIH grants (DK46952, P30-DK41296, P01-DK052956, and M01-RR1224), and grants to D.G. consisting of a Roche Organ Transplant Research Foundation Award, an American Liver Foundation Scholars award, and an NIH K08 award (DK059302).

REFERENCES

1. Zaret KS. Regulatory phases of early liver development: paradigms of organogenesis. Nat Rev Genet 2002; 3:499–512.
2. Zaret K. Early liver differentiation: genetic potentiation and multilevel growth control. Curr Opin Genet Dev 1998; 8:526–531.
3. Petersen BE, Bowen WC, Patrene KD, Mars WM, Sullivan AK, Murase N, Boggs SS, Greenberger JS, Goff JP. Bone marrow as a potential source of hepatic oval cells. Science 1999; 284:1168–1170.
4. Theise ND, Nimmakayalu M, Gardner R, Illei PB, Morgan G, Teperman L, Henegariu O, Krause DS. Liver from bone marrow in humans. Hepatology 2000; 32:11–16.
5. Jian Y, Jahagirdar BN, Reinhardt RL, Schwartz RE, Keene GD, Ortiz-Gonzalez XR, Largaespada DA, Verfaille CM. Pluripotency of mesenchymal stem cells derived from adult marrow. Nature 2002; 418:41–49.
6. Grompe M. Adult Liver Stem Cells. In: Lanza R, Blau H, Melton DA, Moore MA, Thomas E, Verfaille C, Weissman IL, West M, eds. Handbook of stem Cells, New York: Elsevier Academic Press, 2004.
7. Grisham JW, Thorgeirsson SS. In: Potter CS, ed. Stem Cells. London: Academic Press, 1997:233–282.
8. Fausto N, Lemire JM, Shiojiri N. Cell lineages in hepatic development and the identification of progenitor cells in normal and injured liver. Proc Soc Exp Biol Med 1993; 204:237–241.
9. Fausto N. Liver regeneration. J Hepatol 2000; 32:19–31.
10. Thorgeirsson S, Factor V, Grisham J. In: Lanza R, Blau H, Melton DA, Moore DD, Thomas E, Verfaille CM, Weissman IL, West M, eds. Handbook of Stem Cells. Vol. 2. New York: Elsevier, 2004:497–512.
11. Sell S. Liver stem cells. Mod Pathol 1994; 7:105–112.
12. Zajicek G, Schwartz-Arad D, Bartfeld E. The streaming liver. V: Time and age-dependent changes of hepatocyte DNA content, following partial hepatectomy. Liver 1989; 9:164–171.
13. Zajicek G, Arber N, Schwartz-Arad D. Streaming liver. VIII: Cell production rates following partial hepatectomy. Liver 1991; 11:347–351.
14. Jungermann K. Dynamics of zonal hepatocyte heterogeneity. Perinatal development and adaptive alterations during regeneration after partial hepatectomy, starvation and diabetes. Acta Histochem Suppl 1986; 32:89–98.
15. Reid LM. Stem cell biology, hormone/matrix synergies and liver differentiation. Curr Opin Cell Biol 1990; 2:121–130.
16. Sigal SH, Brill S, Fiorino AS, Reid LM. The liver as a stem cell and lineage system. Am J Physiol 1992; 263:G139–G148.
17. Sigal S, Brill S, Fiorino A, Reid LM. In: Zern MA, Reid LM, eds. Extracellular Matrix: Chemistry, Biology, and Pathobiology with Emphasis on the Liver. New York: Marcel Dekker, 1993:507–538.
18. Brill S, Holst P, Sigal S, Zvibel I, Fiorino A, Ochs A, Somasundaran U, Reid LM. Hepatic progenitor populations in embryonic, neonatal, and adult liver. Proc Soc Exp Biol Med 1993; 204:261–269.

19. Reid LM, Fiorino AS, Sigal SH, Brill S, Holst PA. Extracellular matrix gradients in the space of Disse: relevance to liver biology [editorial]. Hepatology 1992; 15:1198–1203.

20. Sigal SH, Gupta S, Gebhard DF Jr, Holst P, Neufeld D, Reid LM. Evidence for a terminal differentiation process in the rat liver. Differentiation 1995; 59:35–42.

21. Weiss L. Histology, Cell and Tissue Biology. 5th ed. New York: Elsevier Biomedical, 1983.

22. Gebhardt R, Bellemann P, Mecke D. Metabolic and enzymatic characteristics of adult rat liver parenchymal cells in non-proliferating primary monolayer cultures. Exp Cell Res 1978; 112:431–441.

23. Gumucio JJ, ed. Hepatocyte Heterogeneity and Liver Function. Vol. 19. Madrid: Springer International, 1989.

24. Traber PG, Chianale J, Gumucio JJ. Physiologic significance and regulation of hepatocellular heterogeneity. Gastroenterology 1988; 95:1130–1143.

25. Gebhardt R, Marti U. Heterogeneous distribution of the epidermal growth factor receptor in rat liver parenchyma. Prog Histochem Cytochem 1992; 26:164–168.

26. Gebhardt R, Jonitza D. Different proliferative responses of periportal and perivenous hepatocytes to EGF. Biochem Biophys Res Commun 1991; 181:1201–1207.

27. Gebhardt R, Mecke D. Heterogeneous distribution of glutamine synthetase among rat liver parenchymal cells in situ and in primary culture. EMBO J 1983; 2:567–570.

28. Gaasbeek Janzen JW, Gebhardt R, ten Voorde GH, Lamers WH, Charles R, Moorman AF. Heterogeneous distribution of glutamine synthetase during rat liver development. J Histochem Cytochem 1987; 35:49–54.

29. Anatskaya OV, Vinogradov AE, Kudryavtsev BN. Hepatocyte polyploidy and metabolism/life-history traits: hypotheses testing. J Theoret Biol 1994; 168:191–199.

30. Anti M, Marra G, Rapaccini GL, Rumi C, Bussa S, Fadda G, Vecchio FM, Valenti A, Percesepe A, Pompili M, et al. DNA ploidy pattern in human chronic liver diseases and hepatic nodular lesions. Flow cytometric analysis on echo-guided needle liver biopsy. Cancer 1994; 73:281–288.

31. Brodsky WY, Uryvaeva IV. Cell polyploidy: its relation to tissue growth and function. Int Rev Cytol 1977; 50:275–332.

32. Feldmann G. Liver ploidy. J Hepatol 1992; 16:7–10.

33. Gupta S. Hepatic Polyploidy and Liver Growth Control. Semin Cancer Biol 2000; 10:161–171.

34. Severin E, Willers R, Bettecken T. Flow cytometric analysis of mouse hepatocyte ploidy. II. The development of polyploidy pattern in four mice strains with different life spans. Cell Tissue Res 1984; 238:649–652.

35. Severin E, Meier EM, Willers R. Flow cytometric analysis of mouse hepatocyte ploidy. I. Preparative and mathematical protocol. Cell Tissue Res 1984; 238:643–647.

36. Rubin EM, DeRose PB, Cohen C. Comparative image cytometric DNA ploidy of liver cell dysplasia and hepatocellular carcinoma. Mod Pathol 1494; 7:677–680.

37. Liu H, DiCunto F, Imarisio S, Reid LM. Citron kinase is a cell cycle-dependent, nuclear protein required for G2/M transition of hepatocytes. J Biol Chem 2003; 278:2541–2548.

38. Michalopoulos GK, DeFrances MC. Liver regeneration. Science 1997; 276:60–66.

39. Chiarugi V, Magnelli L, Cinelli M, Basi G. Apoptosis and the cell cycle. Cell Mol Biol Res 1994; 40:603–612.

40. Tomei LD, Cope FO, eds. Apoptosis: The Molecular Basis of Cell Death. New York: Cold Spring Harbor Laboratory Press, 1991.

41. Meikrantz W, Schlegel R. Apoptosis and the cell cycle. J Cell Biochem 1995; 58:160–174.

42. Orrenius S. Apoptosis: molecular mechanisms and implications for human disease. J Intern Med 1995; 237:529–536.

43. Gomez-Lechon MJ, Barbera E, Gil R, Baguena J. Evolutive changes of ploidy and polynucleation in adult rat hepatocytes in culture. Cell Mol Biol 1981; 27:695–701.

44. Gebhardt R. Different proliferative activity in vitro of periportal and perivenous hepatocytes. Scand J Gastroenterol Suppl 1988; 151:8–18.

45. Liu H, Di Cunto F, Imarisio S, Reid LM. Citron kinase is a cell cycle-dependent, nuclear protein required for G2/M transition of hepatocytes. J Biol Chem 2003; 278:2541–2548.

46. Tateno C, Yoshizato K. Growth potential and differentiation capacity of adult rat hepatocytes in vitro. Wound Repair Regen 1999; 7:36–44.

47. Tateno C, Takai-Kajihara K, Yamasaki C, Sato H, Yoshizato K. Heterogeneity of growth potential of adult rat hepatocytes in vitro. Hepatology 2000; 31:65–74.

48. Overturf K, al-Dhalimy M, Finegold M, Grompe M. The repopulation potential of hepatocyte populations differing in size and prior mitotic expansion. Am J Pathol 1999; 155:2135–2143.

49. Braun KM, Degen JL, Sandgren EP. Hepatocyte transplantation in a model of toxin-induced liver disease: variable therapeutic effect during replacement of damaged parenchyma by donor adult liver cells. Nat Med 2000; 6:320–326.

50. Gorla GR, Malhi H, Gupta S. Polyploidy Associated with Oxidative Injury Attenuates Proliferative Potential of Cells. J Cell Sci 2001; 114:2943–2951.

51. Gebhardt R, Lindros K, Lamers WH, Moorman AF. Ear J Cell Biol 1991; 56:464–467.

52. Marti U, Gebhardt R. Acinar heterogeneity of the epidermal growth factor receptor in the liver of male rats. Eur J Cell Biol 1991; 55:158–164.

53. Gebhardt R, Alber J, Wegner H, Mecke D. Different drug metabolizing capacities in cultured periportal and pericentral hepatocytes. Biochem Pharmacol 1994; 48:761–766.

54. Volk A, Michalopoulos G, Weidner M, Gebhardt R. Different proliferative responses of periportal and pericentral rat hepatocytes to hepatocyte growth factor. Biochem Biophys Res Commun 1995; 207:578–584.

55. Lindros KO, Oinonen T, Issakainen J, Nagy P, Thorgeirsson SS. Zonal distribution of transcripts of four hepatic transcription factors in the mature rat liver. Cell Biol Toxicol 1997; 13:257–262.

56. Ogawa A, Kurita K, Ikezawa Y, Igarashi M, Kuzumaki T, Daimon M, Kato T, Yamatani K, Sasaki H. Functional localization of glucose transporter 2 in rat liver. J Histochem Cytochem 1996; 44:1231–1236.

57. Bartels H, Herbort H, Jungermann K. Predominant periportal expression of the phosphoenolpyruvate carboxykinase and tyrosine aminotransferase genes in rat liver. Dynamics during the daily feeding rhythm and starvation-refeeding cycle demonstrated by in situ hybridization. Histochemistry 1990; 94:637–644.

58. Nauck M, Wolfle D, Katz N, Jungermann K. Modulation of the glucagon-dependent induction of phosphoenolpyruvate carboxykinase and tyrosine aminotransferase by arterial and venous oxygen concentrations in hepatocyte cultures. Eur J Biochem 1981; 119:657–661.

59. Nishimura M, Yaguti H, Yoshitsugu H, Naito S, Satoh T. Tissue distribution of mRNA expression of human cytochrome P450 isoforms assessed by high-sensitivity real-time reverse transcription PCR. Yakugaku Zasshi 2003; 123: 369–375.

60. LeCluyse EL, Madan A, Hamilton G, Carroll K, DeHaan R, Parkinson A. Expression and regulation of cytochrome P450 enzymes in primary cultures of human hepatocytes. J. Biochem. Molec. Toxicol 2000; 14(4):177–188.

61. Juang JH, Bonner-Weir S, Ogawa Y, Vacanti JP, Weir GC. Outcome of subcutaneous islet transplantation improved by polymer device. Transplantation 1996; 61: 1557–1561.

62. Bartels H, Freimann S, Jungermann K. Predominant periportal expression of the phosphoenolpyruvate carboxykinase gene in liver of fed and fasted mice, hamsters and rats studied by in situ hybridization. Histochemistry 1993; 99:303–309.

63. Eilers F, Modaressi S, Jungermann K. Predominant periportal expression of the fructose 1,6-bisphosphatase gene in rat liver: dynamics during the daily feeding rhythm and starvation-refeeding cycle. Histochem Cell Biol 1995; 103:293–300.

64. Kojima T, Yamamoto M, Tobioka H, Mizuguchi T, Mitaka T, Mochizuki Y. Changes in cellular distribution of connexins 32 and 26 during formation of gap junctions in primary cultures of rat hepatocytes. Exp Cell Res 1996; 223:314–326.

65. Stumpel F, Ott T, Willecke K, Jungermann K. Connexin 32 gap junctions enhance stimulation of glucose output by glucagon and noradrenaline in mouse liver. Hepatology 1998; 28:1616–1620.

66. Seseke FG, Gardemann A, Jungermann K. Signal propagation via gap junctions, a key step in the regulation of liver metabolism by the sympathetic hepatic nerves. FEBS Lett 1992; 301:265–270.

67. Rosenberg E, Faris RA, Spray DC, Monfils B, Abreu S, Danishefsky I, Reid LM. Correlation of expression of connexin mRNA isoforms with degree of cellular differentiation. Cell Adhes Commun 1996; 4:223–235.

68. Foucrier J, Pechinot D, Rigaut JP, Feldmann G. Transferrin secretion and hepatocyte ploidy: analysis at the single cell level using a semi-automatic image analysis method. Biol Cell 1988; 62:125–131.

69. Van Noorden CJ, Vogels IM, Fronik G, Houtkooper JM, James J. Ploidy class-dependent metabolic changes in rat hepatocytes after partial hepatectomy. Exp Cell Res 1985; 161:551–557.

70. Gebhardt R, Ebert A, Bauer G. Heterogeneous expression of glutamine synthetase mRNA in rat liver parenchyma revealed by in situ hybridization and Northern blot analysis of RNA from periportal and perivenous hepatocytes. FEBS Lett 1988; 241:89–93.

71. Shiojiri N, Wada JI, Tanaka T, Noguchi M, Ito M, Gebhardt R. Heterogeneous hepatocellular expression of glutamine synthetase in developing mouse liver and in testicular transplants of fetal liver. Lab Invest 1995; 72:740–747.

72. Roskams T, Moshage H, De Vos R, Guido D, Yap P, Desmet V. Heparan sulfate proteoglycan expression in normal human liver. Hepatology 1995; 21: 950–958.

73. Arlotto MP, Parkinson A. Identification of cytochrome P450a (P450IIA1) as the principal testosterone 7 alpha-hydroxylase in rat liver microsomes and its regulation by thyroid hormones. Arch Biochem Biophys 1989; 270(2):441–457.

74. Christa L, Simon MT, Flinois JP, Gebhardt R, Brechot C, Lasserre C. Overexpression of glutamine synthetase in human primary liver cancer. Gastroenterology 1994; 106:1312–1320.

75. Gordon GJ, Coleman WB, Grisham JW. Temporal analysis of hepatocyte differentiation by small hepatocyte-like progenitor cells during liver regeneration in retrosine-exposed rats. Am J Pathol 2000; 157:771–786.

76. Xu A, Luntz T, Macdonald J, Kubota H, Hsu E, London R, Reid LM. In: Lanza R, Langer R, Vacanti J, eds. Principles of Tissue Engineering. 2nd ed. San Diego: Academic Press, 2000.

77. Jungermann K, Heilbronn R, Katz N, Sasse D. The glucose/glucose-6-phosphate cycle in the periportal and perivenous zone of rat liver. Eur J Biochem 1982; 123:429–436.

78. Jungermann K, Katz N. Functional hepatocellular heterogeneity. Hepatology 1982; 2:385–395.

79. Kuo FC, Darnell JE Jr. Evidence that interaction of hepatocytes with the collecting (hepatic) veins triggers position-specific transcription of the glutamine synthetase and ornithine aminotransferase genes in the mouse liver. Mol Cell Biol 1991; 11:6050–6058.

80. Fahrner J, Labruyere WT, Gaunitz C, Moorman AF, Gebhardt R, Lamers WH. Identification and functional characterization of regulatory elements of the glutamine synthetase gene from rat liver. Eur J Biochem 1993; 213:1067–1073.

81. Schols L, Mecke D, Gebhardt R. Reestablishment of the heterogeneous distribution of hepatic glutamine synthetase during regeneration after CCl4-intoxication. Histochemistry 1990; 94:49–54.

82. Brill S, Holst PA, Zvibel L, Fiorino A, Sigal SH, Somasundaran U, Reid LM. In: Arias IM, Boyer JL, Fausto N, Jakoby WB, Schachter D, Shafritz DA, eds. Liver Biology and Pathobiology. 3rd ed. New York: Raven Press, 1994:869–897.

83. Macdonald JM, Xu ASL, Hiroshi K, LeCluyse E, Hamillton G, Liu H, Rong YW, Moss N, Lodestro C, Luntz T, Wolfe SP, Reid L. In: Lanza WL, Langer R, Vacanti J, eds. Methods of Tissue Engineering. San Diego: Academic Press, 2002:151–201.

84. Gupta S, Rajvanshi P, Sokhi RP, Vaidya S, Irani AN, Gorla GR. Position-specific gene expression in the liver lobule is directed by the microenvironment and not by the previous cell differentiation state. J Biol Chem 1999; 274:2157–2165.

85. Potten C, Wilson J. In: Lanza R, Blau H, Melton DA, Moore MA, Thomas E, Verfaille CM, Weissman IL, West M, eds. Handbook on Stem Cells. Vol. 2. New York: Elsevier, 2004: 1–12.

86. Morrison SJ, Shah NM, Anderson DJ. Regulatory mechanisms in stem cell biology. Cell 1997; 88:287–298.

87. Reya T, Morrison SJ, Clarke MF, Weissman IL. Stem cells cancer and cancer stem cells. Nature 2001; 414:105–111.

88. LeGasse E, Connors H, al-Dhalimy M, Reitsma M, Dohse M, Osborne L, Wang X, Finegold M, Weissman IL, Grompe M. Purified hemopoietic stem cells can differentiate into hepatocytes in vivo. Nat Med 2000; 11:1229–1234.

89. Theise N. The Canals of Hering and Hepatic Stem cells in Humans. Hepatology 1999; 30:1425–1433.

90. Niwa H, Miyazaki J, Smith AG. Quantitative expression of Oct-3/4 defines differentiation, dedifferentiation or self-renewal of ES cells. Nature Genetics 2000; 24:372–376.

91. Burdon T, Smith A, Savataier P. Signalling, cell cycle and pluripotency in embryonic stem cells. Trends Cell Biol 2002; 12:432–438.

92. Martin GR. Isolation of a pluripotent cell line from early mouse embryos cultured in medium conditioned by teratocarcinoma stem cells. Proc Natl Acad Sci USA 1981; 78:7634–7638.

93. Fuchs E, Raghavan S. Getting under the skin of epidermal morphogenesis. Nat Rev Genet 2002; 3:199–209.

94. Antonchuk J, Sauvageau G, Humphries RK. HOXB4-induced expansion of adult hematopoietic stem cells ex vivo. Cell 2002; 109:39–45.

95. Harley CB, Kim NW, Prowse KR, Weinrich SL, Hirsch KS, West MD, Bacchetti S, Hirte HW, Counter CM, Greider CW, et al. Telomerase, cell immortality, and cancer. Cold Spring Harbor Symp Quant Biol 1994; 59:307–315.

96. Harley CB, Villeponteau B. Telomeres and telomerase in aging and cancer. Curr Opin Genet Dev 1995; 5:249–255.

97. Bodnar AG, Ouellette M, Frolkis M, Holt SE, Chiu CP, Morin GB, Harley CB, Shay JW, Lichtsteiner S, Wright WE. Extension of life-span by introduction of telomerase into normal human cells. Science 1998; 279:349–352.

98. Mobest D, Goan SR, Junghahn I, Winkler J, Fichtner I, Hermann V, Becker M, de Lima-Hahn E, Henschler R. Differential kinetics of primitive hematopoietic cells assayed in vitro and in vivo during serum-free suspension culture of CD34(+) blood progenitor cells. Stem Cells 1999; 17:152–161.

99. Matsumoto K, Yoshitomi H, Rossant J, Zaret K. Liver organogenesis promoted by endothelial cells prior to vascular function. Sciencexpress 2001.

100. Matsumoto K, Yoshitomi H, Rossant J, Zaret K. Liver organogenesis promoted by endothelial cells prior to vascular function. Science 2001; 294(5542):559–563.

101. Shiojiri N. The origin of intrahepatic bile duct cells in the mouse. J Embryol Exp Morphol 1984; 79:25–39.

102. Ruebner BH, Blankenberg TA, Burrows DA, SooHoo W, Lund JK. Development and transformation of the ductal plate in the developing human liver. Pediatr Pathol 1990; 10:55–68.

103. Coffinier C, Gresh L, Fiette L, Tronche F, Schutz G, Babinet C, Pontoglio M, Yaniv M, Barra J. Bile system morphogenesis defects and liver dysfunction upon targeted deletion of HNF1beta. Development 2002; 129:1829–1838.

104. Clotman F, Lannoy VJ, Reber M, Cereghini S, Cassiman D, Jacquemin P, Roskams T, Rousseau GG, Lemaigre FP. The onecut transcription factor HNF6 is required for normal development of the biliary tract. Development 2002; 129:1819–1828.

105. Fabris L, Strazzabosco M, Crosby HA, Ballardini G, Hubscher SG, Kelly DA, Neuberger JM, Strain AJ, Joplin R. Characterization and isolation of ductular cells coexpressing neural cell adhesion molecule and Bcl-2 from primary cholangiopathies and ductal plate malformations. Am J Pathol 2000; 156:1599–1612.

106. Theise ND, Saxena R, Portmann BC, Thung SN, Yee H, Chiriboga L, Kumar A, Crawford JM. The canals of Hering and hepatic stem cells in humans. Hepatology 1999; 30:1425–1433.

107. Saxena T, Theise ND Canals of Hering: recent insights and current knowledge. Seminars in Liver Disease 2004; 24(1):43–48.

108. Suzuki A, Zheng Y-W, Kaneko S, Monodera M, Fukao K, Nakauchi H, Taniguchi H. Clonal identification and characterization of self-renewing pluripotent stem cells in the developing liver. J Cell Biol 2002; 156:173–185.

109. Shen C, Horb ME, Slack JMW, Tosh D. Transdifferentiation of pancreas to liver. Mech Dev 2003; 120:107–116.

110. Berishvili E, Liponava E, Kochlavashvili N, Kalandarishvili K, Benashvili L, Gupta S, Kakabadze Z. Heterotopic Auxiliary Liver in an Isolated and Vascularized Segment of the Small Intestine in Rats. Transplantation 2003; 75:1827–1832.

111. Grompe M. Pancreatic-hepatic switches in vivo. Mech Dev 2003; 120:99–106.

112. Overturf K, al-Dhalimy M, Ou CN, Finegold M, Grompe M. Serial transplantation reveals the stem-cell-like regenerative potential of adult mouse hepatocytes. Am J Pathol 1997; 151: 1273–1280.

113. Wang X, Willenbring H, Akkari Y, Torimaru Y, Foster M, Al-Dhalimy M, Lagasse E, Finegold M, Olson S, Grompe M. Cell fusion is the principal source of bone-marrow-derived hepatocytes. Nature 2003; 422:897–901.

114. Lucas J, Terada N. Spontaenous cell fusion. In: Lanza R, Blau H, Melton DA, Moore MA, Thomas E, Verfaillie C, Weissman I, West M, eds. Handbook of Stem Cells. Vol. 2. New York: Elsevier, 2004:153–158.

115. Grisham JW. Hepatocyte lineages: of clones, streams, patches, and nodules in the liver. Hepatology 1997; 25:250–252.

116. Fausto N. Growth factors in liver development, regeneration and carcinogenesis. Prog Growth Factor Res 1991; 3:219–234.

117. Thorgeirsson SS. Hepatic stem cells in liver regeneration. FASEB J 1996; 10:1249–1256.

118. Factor V, Radeva S, Thorgeirsson S. Origin and fate of oval cells in Dipin-induced hepatocarcinogenesis in the mouse. Am J Pathol 1994; 145:409–422.

119. Sandgren EP, Palmiter RD, Heckel JL, Daugherty CC, Brinster RL, Degen JL. Complete hepatic regeneration after somatic deletion of an albumin-plasminogen activator transgene. Cell 1991; 66:245–256.

120. Rhim JA, Sandgren EP, Degen JL, Palmiter RD, Brinster RL. Replacement of diseased mouse liver by hepatic cell transplantation. Science 1994; 263:1149–1152.

121. Sandgren EP, Palmiter RD, Heckel JL, Brinster RL, Degen JL. DNA rearrangement causes hepatocarcinogenesis in albumin-plasminogen activator transgenic mice. Proc Natl Acad Sci USA 1992; 89:11523–11527.

122. Hixson DC, Faris RA, Thompson NL. An antigenic portrait of the liver during carcinogenesis. Pathobiology 1990; 58:65–77.

123. Dunsford H, Sell S. Production of monoclonal antibodies to preneoplastic liver cell populations induced by chemical carcinogens in rats and to transplantable Morris hepatomas. Cancer Res 1989; 49:65–77.

124. Engelhardt NV, Factor VM, Medvinsky AL, Baranov VN, Lazareva MN, Poltoranina VS. Common antigen of oval and biliary epithelial cells (A6) is a differentiation marker of epithelial and erythroid cell lineages in early development of the mouse. Differentiation 1993; 55:19–26.

125. Petersen BE, Grossbard B, Hatch H, Pi L, Deng J, Scott EW. Mouse A6-positive hepatic oval cells also express several hematopoietic stem cell markers. Hepatology 2003; 37:632–640.

126. Sigal SH, Brill S, Reid LM, Zvibel I, Gupta S, Hixson D, Faris R, Holst PA. Characterization and enrichment of fetal rat hepatoblasts by immunoadsorption ("panning") and fluorescence-activated cell sorting. Hepatology 1994; 19:999–1006.

127. Reid LM, Agelli M, Ochs A. Method of expanding hepatic precursor cells. U.S. Patent number 5,576,207, 1994.

128. Malhi H, Irani AL, Volenberg I, Schilsky ML, Gupta S. Early Cell Transplantation in LEC Rats Modeling Wilson's Disease Eliminates Hepatic Copper with Reversal of Liver Disease. Gastroenterology 2002; 122: 438–447.

129. Yoshida Y, Tokusashi Y, Lee G, Ogawa K. Intrahepatic transplantation of normal hepatocytes prevents Wilson's disease in Long-Evans cinnamon rats. Gastroenterology 1996; 111:1654–1660.

130. Irani AL, Malhi H, Slehria S, Giridhar GR, Volenberg I, Schilsky ML, Gupta S. Correction of Liver Disease Following Transplantation of Normal Rat Hepatocytes into long-Evans Cinnamon Rats Modeling Wilson's Disease. Mol Ther 2001; 3:302–309.

131. Sigal SH, Rajvanshi P, Reid LM, Gupta S. Demonstration of differentiation in hepatocyte progenitor cells using dipeptidyl peptidase IV deficient mutant rats. Cell Mol Biol Res 1995; 41:39–47.

132. Kubota H, Reid LM. Clonogenic hepatoblasts, common precursors for hepatocytic and biliary lineages, are lacking classical major histocompatiblity complex class I antigen. Proc Natl Acad Sci USA 2000; 97:12132–12137.

133. Kubota H, Reid LM. Processes for clonal growth of hepatic progenitor cells. U.S. Patent Application number 113918.500, 1999.

134. Kubota H, Reid LM. Methods of isolating bipotent hepatic progenitor cells. U.S. Patent Application number 113918.400, 1999.

135. Azuma H, Hirose T, Fujii H, Oe S, Yasuchika K, Fujikawa T, Yamaoka Y. Enrichment of hepatic progenitor cells from adult mouse liver. Hepatology 2003; 37:1385–1394.

136. Suzuki A, Zheng Y, Kondo R, Kusakabe M, Takada Y, Fukao K, Nakauchi H, Taniguchi H. Flow cytometric separation and enrichment of hepatic progenitor cells in the developing mouse liver. Hepatology 2000; 32:1230–1239.

137. Minguet S, Cortegano I, Gonzalo P, Martinez-Marin J, Andres B, Salas C, Melero D, Gaspar M, Marcos M. A population of c-kit((low), CD45-, TER119- hepatic cell progenitors of 11 day postcoitus mouse embryo liver reconstitutes cell-depleted liver organoids. J Clin Invest 2003; 112:1152–1163.

138. Tanimizu N, Nishikawa M, Saito H, Tsujimura T, Miyajima A. Isolation of hepatoblasts based on the expression of Dlk/Pref-1. J Cell Sci 2003; 116: 1775–1786.

139. Nitou M, Sugiyama Y, Ishikawa K, Shiojiri N. Purification of fetal mouse hepatoblasts by magnetic beads coated with monoclonal anti-E-cadherin antibodies and their in vitro culture. Exp Cell Res 2002; 279:330–343.

140. Watanabe T, Nakagawa K, Ohata S, Kitagawa D, Nishitai G, Seo J, Tanemura S, Shimizu N, Kishimoto H, Eada T, et al. SEK1/MKK4-mediated SAPK/JNK signlaing partipates in embyonic hepatoblast proliferation via pathway different from NF-KB-induced anti-apoptosis. Dev Biol 2002; 250:332–347.

141. Cho J-J, Malhi H, Wang R, Joseph B, Ludlow JW, Susick R, Gupta S. Enzymatically Labeled Chromosomal Probes for in situ Identification of Human Cells in Xenogeneic Transplant Models. Nat Med 2002; 8:1033–1036.

142. Saxen, Theise, N. Seminars in Liver Disease 2004; 24(1):43–48.

143. Strain A. J. (1999) Ex vivo liver cell morphogenesis: one step nearer to the bioartificial liver? Hepatology 29(1), 288-290

144. Schmelzer E, Zhang L, Melhem A, Moss N, Wauthier E, McClelland RE, Bruce A, Yao H, Turner WS, Cheng N, Furth ME, Reid LM. Human ductal plates contain hepatic stem cells that are precursors to hepatoblasts. 2006; submitted.

145. Reid LM, Moss N, Kubota H. Human hepatic progenitors. U.S. Patent Application # 20050148072, 1999.

146. Bruce A, Zhang L, Ludlow J, Schmelzer E, Melhem A, Kulik M, Reid LM, Furth ME. Hepatic stem cells from cadaveric postnatal human liver. 2005; submitted.

147. Zhang L, Theise ND, Woosley J, Chua M, Reid L. The Stem Cell Compartment of Human Livers. 2006; In preparation.

148. McClelland R, Schmelzer E, Wauthier E, Reid L. Matrix substrata required for self-replication of human hepatic stem cells. 2006; submitted.

149. Michalopoulos GK, Eckl PM, Cruise JL, Novicki DL, Jirtle RL. Mechanisms of rodent liver carcinogenesis. [Review] [27 refs]. Toxicol Ind Health 1987; 3:119–128.

150. Sigal SH, Rajvanshi P, Gorla GR, Sokhi RP, Saxena R, Gebhard DR Jr, Reid LM, Gupta S. Partial hepatectomy-induced polyploidy attenuates hepatocyte replication and activates cell aging events. Am J Physiol 1999; 276:G1260–G1272.

151. Ingber DE, Folkman J. Mechanochemical switching between growth and differentiation during fibroblast growth factor-stimulated angiogenesis in vitro: role of extracellular matrix. J Cell Biol 1989; 109:317–330.

152. Reid LM. Defining hormone and matrix requirements for differntiated epithelia. In: Pollard JW, Walker JM, eds. Basic Cell Culture Protocols. Vol. 75. Chapter 21. Totowa, NJ: Humana Press, 1990:237–262.

153. Brill S, Zvibel I, Reid LM. Maturation-dependent changes in the regulation of liver-specific gene expression in embryonal versus adult primary liver cultures [published erratum appears in Differentiation. Differentiation 1995; 59:95–102.

154. Singhvi R, Stephanopoulos G, Wang DIC. Effects of substratum morphology on cell physiology. Biotech Bioeng 1943; 43:764–771.

155. Powers MJ, Rodrigueze RE, Griffith LG. Cell-substratum adhesion strength as determinant of hepatocyte aggregate morphology. Biotech Bioeng 1997; 53:415–426.

156. Mooney D, Hansen L, Vacanti J, Langer R, Farmer S, Ingber D. Switching from differen-tiation to growth in hepatocytes: control by extracellular matrix. J Cell Physiol 1992; 151:497–505.

157. Zern MA, Reid LM, eds. Extracellular Matrix Chemistry and Biology. New York: Academic Press, 1993.

158. Jefferson DM, Clayton DF, Darnell JE Jr, Reid LM. Post-transcriptional modulation of gene expression in cultured rat hepatocytes. Mol Cell Biol 1984; 4:1929–1934.

159. Enat R, Jefferson DM, Ruiz-Opazo N, Gatmaitan Z, Leinwand LA, Reid LM. Hepatocyte pro-liferation in vitro: its dependence on the use of serum- free hormonally defined medium and substrata of extracellular matrix. Proc Natl Acad Sci USA 1984; 81:1411–1415.

160. Susick R, Moss N, Kubota H, Leduyse E, Hamitton G, Luntz T, Ludlow J, Fair J, Gerber D, Bergstrand K, White J, Bruce A, Drury O, Gupta S, eid M. Hepatic progenitors and strategies for liver cell therapies. Ann NY Acad Sci 2001; 944:398–419.

161. Houssaint E. Differentiation of the mouse hepatic primordium. I. An analysis of tissue inter-actions in hepatocyte differentiation. Cell Differ 1980; 9:269–279.

162. Bossard P, Zaret KS. GATA transcription factors as potentiators of gut endoderm differen-tiation. Development 1998; 125:4909–4917.

163. Geller SA, Nichols WS, Kim S, Tolmachoff T, Lee S, Dycaico MJ, Felts K, Sorge JA. Hepa-tocarcinogenesis is the sequel to hepatitis in Z:2 alpha 1-antitrypsin transgenic mice: histo-pathological and DNA ploidy studies. Hepatology 1994; 19:389–397.

164. Liao WS, Conn AR, Taylor JM. Changes in rat alpha 1-fetoprotein and albumin mRNA levels during fetal and neonatal development. J Biol Chem 1980; 255:10036–10039.

165. Schmid P, Schulz WA. Coexpression of the c-myc protooncogene with alpha-fetoprotein and albumin in fetal mouse liver. Differentiation 1990; 45:96–102.

166. Sellem CH, Frain M, Erdos T, Sala-Trepat JM. Differential expression of albumin and alpha-fetoprotein genes in fetal tissues of mouse and rat. Dev Biol 1984; 102:51–60.

167. Sell S, Longley MA, Boulter J. alpha-Fetoprotein and albumin gene expression in brain and other tissues of fetal and adult rats. Brain Res 1985; 354:49–53.

168. Evarts RP, Nagy P, Marsden E, Thorgeirsson SS. A precursor-product relationship exists between oval cells and hepatocytes in rat liver. Carcinogenesis 1987; 8:1737–1740.

169. Rossi JM, Dunn NR, Hogan BL, Zaret KS. Distinct mesodermal signals, including BMPs from the septum transversum mesenchyme, are required in combination for hepatogenesis from the endoderm. Genes Dev 2001; 15:1998–2009.

170. Fennekohl A, Schieferdecker HL, Jungermann K, Puschel GP. Differential expression of prostanoid receptors in hepatocytes, Kupffer cells, sinusoidal endothelial cells and stellate cells of rat liver. J Hepatol 1999; 30:38–47.

171. Deutsch G, Jung J, Zheng M, Lora J, Zaret KS. A bipotential precursor population for pancreas and liver within the embryonic endoderm. Development 2001; 128:871–881.

172. Burke Z, Oliver G. Prox1 is an early specific marker for the developing liver and pancreas in the mammalian foregut endoderm. Mech Dev 2002; 118:147–155.

173. Sosa-Pineda B, Wigle JT, Oliver G. Hepatocyte migration during liver development requires Prox1. Nat Genet 2000; 25:254–255.

174. Hromas R, Radich J, Collins S. PCR cloning of an orphan homeobox gene (PRH) preferentially expressed in myeloid and liver cells. Biochem Biophys Res Commun 1993; 195:976–983.

175. Martinez Barbera JP, Clements M, Thomas P, Rodriguez T, Meloy D, Kioussis D, Beddington RS. The homeobox gene Hex is required in definitive endodermal tissues for normal forebrain, liver and thyroid formation. Dev Suppl 2000; 127:2433–2445.

176. Keng VW, Yagi H, Ikawa M, Nagano T, Myint Z, Yamada K, Tanaka T, Sato A, Muramatsu I, Okabe M, Sato M, Noguchi T. Homeobox gene Hex is essential for onset of mouse embryonic liver development and differentiation of the monocyte lineage. Biochem Biophys Res Commun 2000; 276:1155–1161.

177. Hentsch B, Lyons I, Li R, Hartley L, Lints TJ, Adams JM, Harvey RP. Hlx homeo box gene is essential for an inductive tissue interaction that drives expansion of embryonic liver and gut. Genes Dev 1996; 10:70–79.

178. Eferl E, Sibilia M, Hilberg F, Fuchsbichler A, Kufferath I, Guertl B, Zenz R, Wagner E, Azatloukal K. Functions of c-jun in liver development. J Cell Biol 1999; 145:1049–1061.

179. Passegue E, Jochum W, Behrens A, Ricci R, Wagner E. Jun B can substitute for jun in mouse development and cell proliferation. Nat Genet 2002; 30:158–166.

180. Watanabe T, Nakagawa K, Ohata S, Kitagawa D, Nishitai G, Seo J, Tanemura S, Shimizu N, Kishimoto H, Wada T, et al. SEK1/MKK4-mediated SAPK/JNK signaling participates in embryonic hepatoblast proliferation via a pathway different from NF-kappaB-induced anti-apoptosis. Dev Biol 2002; 250:332–347.

181. Brenner DA, Koch KS, Leffert HL. Transforming growth factor-alpha stimulates proto-oncogene c-jun expression and a mitogenic program in primary cultures of adult rat hepatocytes. DNA 1989; 8:279–285.

182. Beg AA, Sha WC, Bronson RT, Ghosh S, Baltimore D. Embryonic lethality and liver degeneration in mice lacking the RelA component of NF-kappa B. Nature 1995; 376:167–170.

183. Doi TS, Marino MW, Takahashi T, Yoshida T, Sakakura T, Old LJ, Obata Y. Absence of tumor necrosis factor rescues RelA-deficient mice from embryonic lethality. Proc Natl Acad Sci USA 1999; 96:2994–2999.

184. Enomoto H, Yoshida K, Kishima Y, Kinoshita T, Yamamoto M, Everett A, Miyajima A, Nakamura H. Hepatoma-derived growth factor is highly expressed in developing liver and promotes fetal hepatocyte proliferation. Hepatology 2002; 36:1519–1527.

185. Schrem H, Klempnauer J, Borlak J. Liver-enriched transcription factors in liver function and development. Part I: the hepatocyte nuclear factor network and liver-specific gene expression. Pharmacol Rev 2002; 54:129–158.

186. Van den Hoff MJ, Vermeulen JL, De Boer PA, Lamers WH, Moorman AF. Developmental changes in the expression of the liver-enriched transcription factors LF-B1, C/EBP, DBP and LAP/LIP in relation to the expression of albumin, alpha-fetoprotein, carbamoylphosphate synthase and lactase mRNA. Histochem J 1994; 26:20–31.

187. Nagy P, Bisgaard HC, Thorgeirsson SS. Expression of hepatic transcription factors during liver development and oval cell differentiation. J Cell Biol 1994; 126:223–233.

188. Birkenmeier EH, Gwynn B, Howard S, Jerry J, Gordon JI, Landschulz WH, McKnight SL. Tissue-specific expression, developmental regulation, and genetic mapping of the gene encoding CCAAT/enhancer binding protein. Genes Dev 1989; 3:1146–1156.

189. Tomizawa M, Garfield S, Factor VM, Xanthopoulos K. Hepatocytes deficient in CCAAT/Enhancer binding protein alpha(C/EBPalpha)exhibit both hepatocyte and biliary epithelial cell character. Biochem Biophys Res Commun 1998; 249:1–5.

190. Wang ND, Finegold MJ, Bradley A, Ou CN, Abdelsayed SV, Wilde MD, Taylor LR, Wilson DR, Darlington GJ. Impaired energy homeostasis in C/EBP alpha knockout mice. Science 1995; 269:1108–1112.

191. Li J, Ning G, Duncan SA. Mammalian hepatocyte differentiation requires the transcription factor HNF-4alpha. Genes Dev 2000; 14:464–474.

192. Suzuki A, Iwama A, Miyashita H, Nakauchi H, Taniguchi H. Role for growth factors and extracellular matrix in controlling differentiation of prospectively isolated hepatic stem cells. Development 2003; 130: 2513–2524.

193. Kaestner KH, Hiemisch H, Schutz G. Targeted disruption of the gene encoding hepatocyte nuclear factor 3gamma results in reduced transcription of hepatocyte-specific genes. Mol Cell Biol 1998; 18:4245–4251.

194. Buendia MA. Genetic alterations in hepatoblastoma and hepatocellular carcinoma: common and distinctive aspects. Med Pediatr Oncol 2002; 39:530–535.

195. Aberle H, Schwartz H, Kemler R. Cadherin-catenin complex: protein interactions and their implications for cadherin function. J Cell Biochem 1996; 61:514–523.

196. Tatsuka M, Katayama H, Ota T, Tanaka T, Odashima S, Suzuki F, Terada Y. Multinuclearity and increased ploidy caused by overexpression of the aurora- and Ipl1-like midbody-associated protein mitotic kinase in human cancer cells. Cancer Res 1998; 58:4811–4816.

197. Monga SP, Monga HK, Tan X, Mule K, Pediaditakis P, Michalopoulos GK. Beta-catenin antisense studies in embryonic liver cultures: role in proliferation, apoptosis, and lineage specification. Gastroenterology 2003; 124:202–216.

198. Evarts RP, Nagy P, Nakatsukasa H, Marsden E, Thorgeirsson SS. In vivo differentiation of rat liver oval cells into hepatocytes. Cancer Res 1989; 49: 1541–1547.

199. Bisgaard HC, Muller S, Nagy P, Rasmussen LJ, Thorgeirsson SS. Modulation of the gene network connected to interferon-gamma in liver regeneration from oval cells. Am J Pathol 1999; 155:1075–1085.

200. Cantz T, Zuckerman DM, Burda MR, Dandri M, Goricke B, Thalhammer S, Heckl WM, Manns MP, Petersen J, Ott M. Quantitative gene expression analysis reveals transition of fetal liver progenitor cells to mature hepatocytes after transplantation in uPA/RAG-2 mice. Am J Pathol 2003; 162:37–45.

201. Weigmann A, Corbeil D, Hellwig A, Huttner WB. Prominin, a novel microvilli-specific polytopic membrane protein of the apical surface of epithelial cells, is targeted to plasmalemmal protrusions of non-epithelial cells. Proc Natl Acad Sci USA 1997; 94: 12425–12430.

202. Corbeil D, Roper K, Hannah MJ, Hellwig A, Huttner WB. Selective location of the polytopic membrane protein prominin in microvilli of epithelial cells-a combination of apical sorting and retention in plasma membrane protrusion. J Cell Sci 1999; 112:1023–1033.

203. Schmidt C, Bladt F, Goedecke S, Brinkmann V, Zschiesche W, Sharpe M, Gherardi E, Birchmeier C. Scatter factor/hepatocyte growth factor is essential for liver development. Nature 1995; 373:699–702.

204. de Juan C, Sanchez A, Nakamura T, Fabregat I, Benito M. Hepatocyte growth factor up-regulates met expression in rat fetal hepatocytes in primary culture. Biochem Biophys Res Commun 1994; 204:1364–1370.

205. Hunt T, Luca FC, Ruderman JV. The requirements for protein synthesis and degradation, and the control of destruction of cyclins A and B in the meiotic and mitotic cell cycles of the clam embryo. J Cell Biol 1992; 116:707–724.

206. Pibiri M, Ledda-Columbano GM, Cossu C, Simbula G, Menegazzi M, Shinozuka H, Columbano A. Cyclin D1 is an early target in hepatocyte proliferation induced by thyroid hormone (T3). FASEB J 2001; 15:1006–1013.

207. Nelsen CJ, Rickheim DG, Timchenko NA, Stanley MW, Albrecht JH. Transient expression of cyclin D1 is sufficient to promote hepatocyte replication and liver growth in vivo. Cancer Res 2001; 61:8564–8568.

208. Matsui T, Kinoshita T, Hirano T, Yokota T, Miyajima A. STAT3 down-regulates the expression of cyclin D during liver development. J Biol Chem 2002; 277: 36167–36173.

209. Crosby HA, Kelly DA, Strain AJ. Human hepatic stem-like cells isolated using c-kit or CD34 can differentiate into biliary epithelium. Gastroenterology 2001; 120:534–544.
210. Petersen BE, Goff JP, Greenberger JS, Michalopoulos GK. Hepatic oval cells express the hematopoietic stem cell marker Thy-1 in the rat. Hepatology 1998; 27:433–445.
211. Omori M, Evarts RP, Omori N, Hu Z, Marsden ER, Thorgeirsson SS. Expression of alpha-fetoprotein and stem cell factor/c-kit system in bile duct ligated young rats. Hepatology 1997; 25:1115–1122.
212. Fiegel HC, Kluth J, Lioznov MV, Holzhuter S, Fehse B, Zander AR, Kluth D. Hepatic lineages isolated from developing rat liver show different ways of maturation. Biochem Biophys Res Commun 2003; 305:46–53.
213. Kubota H, Storms RW, Reid LM. Variant forms of alpha-fetoprotein transcripts expressed in human hemopoietic progenitors. J Biol Chem 2002; 277:27629–27635.
214. Taniguchi H, Suzuki A, Zheng R, Kondon Y, Takada Y, Fukunaga K, Seino K, Yuzawa K, Otsuka M, Fukao K, Nakauchi H. Usefullness of flow-cytometric cell sorting for enrichment of hepatic stem and progenitor cells in the liver. Transplant Proc 2000; 32:249–251.
215. Teyssier-Le Discorde M, Prost S, Nandrot E, Kirszenbaum M. Spatial and temporal mapping of c-kit and its ligand, stem cell factor expression during human embryonic haemopoiesis. Br J Haematol 1999; 107: 247–253.
216. Fujio K, Hu Z, Evarts RP, Marsden ER, Niu CH, Thorgeirsson SS. Coexpression of stem cell factor and c-kit in embryonic and adult liver. Exp Cell Res 1996; 224:243–250.
217. Baumann U, Crosby HA, Ramani P, Kelly DA, Strain AJ. Expression of the stem cell factor receptor c-kit in normal and diseased pediatric liver: identification of a human hepatic progenitor cell? Hepatology 1999; 30:112–117.
218. Suzuki A, Zheng Y-W, Fukao K, Nakauchi K, Tniguchi H. Hepatic stem/progenitor cells with high proliferative potential in liver organ formation. Transplant Proc 2001; 33: 585–586.
219. Matsusaka S, Tsujimura T, Toyosaka A, Nakasho H, Sugihara A, Okamoto E, Uematsu K, Terada N. Role of c-kit receptor tyrosine kinase in development of oval cells in the rat 2-acetylaminofluorene/partial hepatectomy model. Hepatology 1999; 29:670–676.
220. Lian ZX, Feng B, Sugiura K, Inaba M, Yu CZ, Jin TN, Fan TX, Cui YZ, Yasumizu R, Toki J, Adachi Y, Hisha H, Ikehara S. c-kit(< low) pluripotent hemopoietic stem cells form CFU-S on day 16.Stem Cells 1999; 17:39–44.
221. Matsusaka S, Tsujimura T, Toyosaka A, Nakasho K, Sugihara A, Okamoto E, Uematsu K, Terada N. Role of c-kit receptor tyrosine kinase in development of oval cells in the rat 2-acetylaminofluorene/partial hepatectomy model. Hepatology 1999; 29:670–676.
222. Baumann U, Crosby HA, Ramani P, Kelly DA, Strain A. Expression of the stem cell factor receptor c-kit in normal and diseased pediatric liver: identification of a human hepatic progenitor cell? Hepatology 1999; 30:112–117.
223. Mitchell C, Mignon A, Guidott J, Besnard S, Fabre M, Duverger N, Parlier D, Tedgui A, Kahn A, Gilgenkrantz H. Therapeutic liver repopulation in a mouse model of hypercholesterolemia. Hum Mol Genet 2000; 9:1587–1602.
224. Vrancken Peeters MJ, Patijn G, Lieber A, Perkins J, Kay M. Expansion of donor hepatocytes after recombinant adenovirus-induced liver regeneration in mice. Hepatology 1997; 25:884–888.
225. Mignon A, Guidotti J, Mitchell C, Fabre M, Wernet A, DeLaCoste A, Soubrane O, Gilgenkrantz H, Kahn A. Selective repopulation of normal mouse liver by Fas/CD95-resistant hepatocytes. Nat Med 1998; 4:1185–1188.
226. Thompson N, Hixson D, Callanan H, Panzica M, Flanagan D, Faris R, Hong W, Hartel-Schenk S, Doyle D. A Fischer rat substrain deficient in dipeptidyl peptidase IV activity makes normal steady-state RNA levels and an altered protein. Use as a liver-cell transplantation model. Biochem J 1991; 273:497–502.
227. Gupta S, Rajvanshi P, Lee C. Integration of transplanted hepatocytes in host liver plates demonstrated with dipeptidyl peptidase IV deficient rats. Proc Natl Acad Sci USA 1995; 92:5860–5864.

228. Rajvanshi P, Kerr A, Bhargava K, Burk R, Gupta S. Studies of liver repopulation using the dipeptidyl peptidase IV deficient rat and other rodent recipients: cell size and structure relationships regulate capacity for increased transplanted hepatocyte mass in the liver lobule. Hepatology 1996; 23:482–496.

229. Ponder KP, Gupta S, Leland F, Darlington G, Finegold M, DeMayo J, Ledley FD, Chowdhury JR, Woo SL. Mouse hepatocytes migrate to liver parenchyma and function indefinitely after intrasplenic transplantation. Proc Natl Acad Sci USA 1991; 88:1217–1221.

230. Gupta S, Aragona E, Vemuru R, Bhargava K, Burk R, Roy Chowdhury J. Permanent engraftment and function of hepatocytes delivered to the liver: implications for gene therapy and liver repopulation. Hepatology 1991; 14:144–149.

231. Sokhi R, Rajvanshi P, Gupta S. Transplanted reporter cells help in defining onset of hepatocyte proliferatin during the life of F344 rats. Am J Physiol Gastrointest Liver Physiol 2000; 279:631–640.

232. Rajvanshi P, Kerr A, Bhargava K, Burk R, Gupta S. Efficacy and safety of repeated hepatocyte transplantation for significant liver repopulation in rodents. Gastroenterology 1996; 111:1092–1102.

233. Oren R, Dabeva M, Petkov P, Hurston E, Laconi E, Shafritz D. Restoration of serum albumin levels in nagase analbuminemic rats by hepatocyte transplantation. Hepatology 1999; 29:75–81.

234. Demetriou A, Levenson S, Novikoff P, Novikoff A, Roy Chowdhury N, Whiting J, Reisner A, Roy Chowdhury J. Survival, organization and function of microcarrier-attahced hepatocytes transplanted in rats. Proc Natl Acad Sci USA 1986; 83:7475–7479.

235. Gunsalus JR, Brady D, Coulter SM, Gray B, Edge A. Reduction of serum cholesterol in Watanabe rabbits by xenogeneic hepatocellular transplantation. Nat Med 1997; 3:48–49.

236. Rhim JA, Sandgren EP, Palmiter RD, Brinster RL. Complete reconstitution of mouse liver with xenogeneic hepatocytes. Proc Natl Acad Sci USA 1995; 92:4942–4946.

237. Overturf K, al-Dhalimy M, Tanguay R, Brantly M, Ou CN, Finegold M, Grompe M.) Hepatocytes corrected by gene therapy are selected in vivo in a murine model of hereditary tyrosinaemia type I [see comments] [published erratum appears in Nat Genet 1996 Apr;12(4):458]. Nat Genet 1996; 12:266–273.

238. Gagandeep S, Sokhi R, Slehria S, Gorla GR, Furgiuele J, DePinho RA, Gupta S. Hepatocyte transplantation improves survival in mice with liver toxicity induced by hepatic overexpression of Mad1 transcription factor. Mol Ther 2000; 1:358–365.

239. De Vree J, Ottenhoff R, Bosma P, Smith A, Aten J, Oude Elferink R. Correction of liver disease by hepatocyte transplantation in a mouse model of progressive failial intrahepatic cholestasis. Gastroenterology 2000; 119:1720–1730.

240. Laconi E, Oren R, Mukhopadhyay DK, Hurston E, Laconi S, Pani P, Dabeva MD, Shafritz DA. Long-term, near total liver replacement by transplantation of isolated hepatocytes in rats treated with retrorsine. Am J Pathol 1998; 153:319–329.

241. Torres S, Diaz BP, Cabrera JJ, Diaz-Chico JC, Diaz-Chico BN, Lopez-Guerra A. Thyroid hormone regulation of rat hepatocyte proliferation and polyploidization. Am J Physiol 1999; 276:G155–G163.

242. Oren R, Dabeva MD, Karnezis AN, Petkov PM, Rosencrantz R, Sandhu JP, Moss SF, Wang S, Hurston E, Laconi E, Holt PR, Thung SN, Zhu L, Shafritz DA.Role of thyroid hormone in stimulating liver repopulation in the rat by transplanted hepatocytes. Hepatology 1999; 30:903–913.

243. Chamuleau R. In: Mito M, Sawa M, eds. Hepatocyte Transplantation. Basel, Switzerland: Karger Landes Systems, 1999:159–167.

244. Makowka L, Rotstein L, Falk R, Falk J, Langer B, Nossal N, Blendis L, Phillips J. Reversal of toxic and anoxic induced hepatic failure by syngeneic, allogeneic and xenogeneic hepatocyte transplantation, allogeneic and xenogeneic hepatocyte transplantation. Surgery 1980; 88:2243–2253.

245. Miyazaki M, Makowka L, Falk R, Falk W, Ventura D. Reversal of lethal, chemotherapeutically induced acute hepatic necrosis in rats by regenerating liver cytosol. Surgery 1983; 94:142–150.

246. Gagandeep S, Sokhi RP, Slehria S, Gorla GR, Furgiuele J, DePinho RA, Gupta S. Hepatocyte Transplantation Improves Survival in Mice with Liver Toxicity Induced by Hepatic Overexpression of Mad1 Transcription Factor. Mol Ther 2000; 1:358–365.

247. Gagandeep S, Rajvanshi P, Sokhi RP, Slehria S, Palestro CJ, Bhargava KK, Gupta S. Transplanted hepatocytes engraft, survive, and proliferate in the liver of rats with carbon tetrachloride-induced cirrhosis. J Pathol 2000; 191:78–85.

248. Baumgartner D, LaPlante-O'Neill P, Sutherland D, Najarian J. Effects of intrasplenic injection of hepatocytes, hepatocyte fragments and hepatocyte culture supernatans on d-galactosamine-induced liver failure in rats. Eur Surg Res 1983; 15: 129–135.

249. Dabeva MD, Petkov PM, Sandhu J, Oren R, Laconi E, Hurston E, Shafritz DA. Proliferation and differentiation of fetal liver epithelial progenitor cells after transplantation in adult rat liver. Am J Pathol 2000; 156:2017–2031.

250. Sandhu J, Petcov P, Dabeva M, Shafritz D. Stem cell prop0erties and repopulation of the rat liver by fetal liver epithelial progenitor cells. Am J Pathol 2001; 159:1323–1334.

251. Suzuki A, Taniguchi H, Zheng YW, Takada Y, Fukunaga K, Seino K, Yazawa K, Otsuka M, Fukao K, Nakauchi H. Clonal colony formation of hepatic stem/progenitor cells enhanced by embryonic fibroblast conditioning medium. Transplant Proc 2000; 32:2328–2330.

252. Theise N, Badve S, Saxena R, Henegariu O, Sell S, Crawford J, Krause D. Derivation of hepatocytes from bone marrow cells in mice after radiation-induced myeloablation. Hepatology 2000; 31:235–240.

253. Vassilopoulos G, Wang P, Russell D. Transplanted bone marrow regenerates liver by cell fusion. Nature 2003; 422:901–904.

254. Yin Y, Lim Y, Salto-Tellez M, Ng S, Lin C, Lim S. ESC-derived cells engraft and differentiate into hepatocytes in vivo. Stem Cells 2002; 20:338–346.

255. Strick-Marchand H, Morosan S, Charneau P, Kremsdorf D, Weiss M. Bipotential mouse embryonic liver stem cell lines contribute to liver regeneration and differentiate as bile ducts and hepatocytes. Proc Natl Acad Sci USA 2004; 101:8360–8365.

256. Gupta S, Chowdhary J. Hepatocyte transplantation: back to the future. Hepatology 1992; 15:156–162.

257. Gupta S, Rajvanshi P, Sokhi R, Slehria S, Yam A, Kerr A, Novikoff P. Entry and integration of transplanted hepatocytes in liver plates occur by disruption of hepatic sinusoidal endothelium. Hepatology 1999; 29:505–519.

258. Mito M, Ebata H, Kusano M, Onishi T, Saito T, Sakamoto S. Morphology and function of isolated hepatocytes transplanted into rat spleen Transplantation 1979; 28: 499–505.

259. Kusano M, Mito M. Observations on the fine structure of long-survived isolated hepatocytes inoculated into rat spleen. Gastroenterology 1982; 82:616–628.

260. Strom S, Fisher RA, Thompson MT, Sanyal AJ, Cole PE, Ham JM, Posner MT. Hepatocyte transplantation as a bridge to orthotopic liver transplantation in terminal liver failure. Transplantation 1997; 63:559–569.

261. Grossman M, Raper S, Kozarsky K, Stein E, Engelhardt J, Muller D, et al. Successful ex vivo gene therapy directed to liver in a patient with familial hypercholesterolemia. Nat Genet 1994; 6:335–341.

262. Benedetti E, Kirby J, Asolati M, Blanchard J, Ward M, Williams R, Hewett T, Fontaine M, Pollak R. Intrasplenic hepatocyte allotransplantation in dalmation dogs with and without cyclosporine immunosuppression. Transplantation 1997; 63:1206–1209.

263. Gupta S, Chowdhury J. Hepatocyte transplantation. In: Arias I, Boyer J, Fausto N, Jakoby W, Schachter D, Shafritz D, eds. The Liver: Biology and Pathology. 3rd ed. New York: Raven Press, 1994:1533–1536.

264. Gupta S, Yermeni P, Vemuru R, Lee C, Yellin E, Bhargava, K. Studies on the safetyof intrasplenic hepatocyte transplantation: relevance to ex vivo gene therapy and liver repopulation in acute hepatic failure. Hum Gene Ther 1993; 4: 249–257.

265. Brill S, Zvibel I, Halpern Z, Oren R. The role of fetal and adult hepatocyte extracellular matrix in the regulation of tissue-specific gene expression in fetal and adult hepatocytes. Eur J Cell Biol 2002; 81:43–50.

266. Xu AS, Reid LM. Soft, porous poly (D,L-lactide-co-glycotide) microcarriers designed for ex vivo studies and for transplantation of adherent cell types including progenitors. Ann N Y Acad Sci 2002; 944:144–159.

267. Mikos AG, Sarakinos G, Leite SM, Vacanti JP, Langer R. Laminated three-dimensional bio-degradable foams for use in tissue engineering. Biomaterials 1993; 14:323–330.

268. Gilbert JC, Takada T, Stein JE, Langer R, Vacanti JP. Cell transplantation of genetically altered cells on biodegradable polymer scaffolds in syngeneic rats. Transplantation 1993; 56:423–427.

269. Langer R, Vacanti JP. Artificial organs. Sci Am 1995; 273:130–133.

270. Mooney DJ, Baldwin DF, Suh NP, Vacanti JP, Langer R. Novel approach to fabricate porous sponges of poly(D,L-lactic-co-glycolic acid) without the use of organic solvents. Biomaterials 1996; 17:1417–1422.

271. Tharakan J, Chau P. A Radial Flow Hollow Fiber Bioreactor for the Large-Scale Culture of Mammalian Cells. Biotechnol Bioeng 1986; XVIII:329–342.

272. Gerlach JC. Use of hepatocyte cultures for liver support bioreactors. Adv Exp Med Biol 1994; 368:165–171.

273. Catapano G, Bartolo LD. Importance of the Kinetic Characterization of Liver Cell Metabolic Reactions to the Design of Hybrid Liver Support Devices. J Artif Organs 1996; 19:670–676.

274. Nyberg S, Shatford R, Hu W, Payne W, Cerra F. Hepatocyte Culture Systems for Artificial Liver Support: Implications for Critical Care Medicine (Bioartificial Liver Support). Crit Care Med 1992; 20:1157–1168.

275. Chowdhury JR, Chowdhury NR, Strom SC, Kaufman SS, Horslen SM, Fox IJ. Human hep-atocyte transplantation: gene therapy and more? Pediatrics 1998; 102:647–648.

276. Jauregui HO. In: Lanza WL, Langer R, Vacanti J, eds. Principles of Tissue Engineering. 2nd ed. San Diego: Academic Press, 2000:541–552.

277. Mullon C, Solomon BA. HepatAssist liver support system. In: Lanza R, Langer R, Vacanti J, eds. Principles of Tissue Engineering. San Diego, CA: Academic Press, 2000.

278. Macdonald JM, Griffin J, Kubota H, Griffith L, Fair J, Reid L. Bioartificial Livers. Boston: Birkhauser, 1999.

279. MacDonald JM, Wolfe SP, Roy-Chowdhury I, Kubota H, Reid LM Effect of flow configur-ation and membrane characteristics on membrane fouling in a novel multicoaxial hollow-fiber bioartificial liver. Ann N Y Acad Sci 2001; 944:334–343.

280. Sauer I, Zeilinger K, Pless G, Kardassis D, Theruvath T, Pascher A, Mueller A, Steinmueller T, Neuhaus P, Gerlach J. Extracorporeal liver support based on primary human liver cells and albumin dialysis–treatment of a patient with primary graft nonfunction. J Hepatol 2003; 39:649–653.

281. Gerlach J, TJorres A, Trost O, Hole O, Vienken J, Courtney J, Gahl G, Neuhaus P. Side effects of hybrid liver support therapy: TNF-alphaliberation in pigs, associated with extracorporeal bioreactors. Int J Artif Organs 1993; 16:604–608.

282. Mullon CM, Solomon BA. Hepatic Assist Liver Support System. In: Lanza WL, Langer R, Vacanti J, eds. Principles of Tissue Engineering, 2nd ed, San Diego: Academic Press, 2000; 553–558.

283. Gerlach J, Brombacher J, Courtney J, Neuhaus P. Nonenzymatic versus enzymatic hepatocyte isolation from pig livers for larger scale investigations of liver cell perfusion systems. Int J Artif Organs 1993; 16:677–681.

284. Gerlach JC, Lemmens P, Schon M, Janke J, Rossaint R, Busse B, Puhl G, Neuhaus P. Exper-imental evaluation of a hybrid liver support system. Transplant Proc 1997; 29:852.

285. Gerlach J. Development of a Hybrid Liver Support System: A Review. Int J Artif Organs 1996; 19:645–654.

286. Groth G, Arborgh B, Bjorken C, Sundberg B, Lundgren G. Correction of hyperbilirubinemia in the glucuronyltransferase-deficient rat by intraportal hepatocyte transplantation. Transplant Proc 1977; 9:313–316.

287. Sutherland D, Numata M, Matas A, Simons R, Najarian J. Hepatocellular transplantation in acute liver failure. Surgery 1977; 82:124–132.

288. Ribeiro J, Nordlinger B, Ballet F, Cynober L, Lucas C, Baudrimont M, et al. Intrasplenic hepatocellular transplantation corrects hepatic encephalopathy in portalcaval shunted rats. Hepatology 1992; 15:12–18.

289. Soriano H, Wood R, Kang D, Ozaki C, Finegold M, Bischoff F, et al. Hepatocellular transplantation (HCT) in children with fulminant liver failure. Hepatology 1997; 30:239A.

290. Bilir B, Kumpe D, Krysl J, Guenette D, Ostrowska A, Eberson GT, Sresthra R, Lin TC, Cole W, Lear J, Durham JD. Hepatocyte transplantation in patients with liver cirrhosis. Digestive Diseases (published conference abstracts) AASLD Meetings, Abstract number LOO56, May 16–22, 1998.

291. Bilir BM, Guenette D, Karrer F, Kumpe DA, Krysl J, Stephens J, McGavran L, Ostrowska A, Durham J. Hepatocyte transplantation in acute liver failure. Liver Transplant 2000; 6:41–43.

292. Strom SC, Chowdhury JR, Fox IJ. Hepatocyte transplantation for the treatment of human disease (Review). Semin Liver Dis 1999; 19:39–48.

293. Fox IJ, Chowdhury JR, Kaufman S, Goertzen TC, Chowdhury NR, Warkentin P, Dorko K, Saulter BV, Strom S. Brief Report: Treatment of the Craigler-Najar Syndrome Type I with hepatocyte transplantation. N Engl J Med 1998; 338:1422–1426.

294. Horslen S, McCowan T, Goertzen T, Warkentin P, Strom S, IJ F. Isolated hepatocyte transplantation: a new approach to the treatment of urea cycle disorders. J Inherit Metab Dis 1999; 1(suppl):123.

295. Strom S, Fisher R, Rubinstein W, Barranger J, Towbin R, Charron M, Mieles L, Pisarov L, Dorko K, Thompson M, Reyes J. Transplantation of human hepatocytes. Transplant Proc 1997; 29:2103–2106.

296. Bohnen NL, Charron M, Reyes J, Rubinstein W, Strom SC, Swanson D, Towbin R. Use of indium-111-labeled hepatocytes to determine the biodistribution of transplanted hepatocytes through portal vein infusion. Clin Nucl Med 2000; 25:447–450.

297. Bilir B, Guenette D, Ostrowska A, Durham J, Kumpe D, Krystal J. Percutaneous hepatocyte transplantation in liver failure. Hepatology 1997; 26:252A.

298. Bilir B, Durham J, Krystal J. Transjugular intraportal transplantation of cryopreserved humanhepoatyctes in a patient with acute liver failure. Hepatology 1996; 24:728A.

299. Fox IJ, Chowdhury JR, Kaufman SS, Goertzen TC, Chowdhury NR, Warkentin P, Dorko K, Sauter BV, Strom S. Treatment of the Crigler-Najjar Syndrome Type I with hepatocyte transplantation. N Engl J Med 1998; 338:1422–1426.

300. Zaret K. Developmental competence of the gut endoderm: genetic potentiation by GATA and HNF3/fork head proteins. Dev Biol (Orlando) 1999; 209:1–10.

301. Shiojiri N. Transient expression of bile-duct-specific cytokeratin in fetal mouse hepatocytes. Cell Tissue Res 1994; 278:117–123.

302. Van Eyken P, Sciot R, Desmet V. Intrahepatic bile duct development in the rat: a cytokeratin-immunohistochemical study. Lab Invest 1988; 59:52–59.

303. Ros JE, Libbrecht L, Geuken M, Jansen PL, Roskams TA. High expression of MDR1, MRP1, and MRP3 in the hepatic progenitor cell compartment and hepatocytes in severe human liver disease. J Pathol 2003; 200:553–560.

304. Joseph B, Bhargava K, Malhi H, Schilsky ML, Jain D, Palestro C, Gupta S. Sestamibi is a Substrate for MDR1 and MDR2 P-glycoprotein Genes. Eur J Nucl Med Mol Imaging 2003; 30:1024–1031.

305. Van Den Heuvel M, Sloof M, Visser L, Muller M, De Jong K, Poppema S, Gouw A. Expression of anti-OV6 antibody and anti-N-CAM antibody along the biliary line of normal and diseased human livers. Hepatology 2001; 33:1387–1393.

306. Morrison SJ, Prowse KR, Ho P, Weissman IL. Telomerase activity in hematopoietic cells is associated with self- renewal potential. Immunity 1996; 5:207–216.

307. Austin TW, Solar GP, Ziegler FC, Liem L, Matthews W. A role for the Wnt gene family in hematopoiesis: expansion of multilineage progenitor cells. Blood 1997; 89:3624–3635.

308. Plescia CP, Rogler CE, Rogler LE. Genomic expression analysis implicates Wnt signaling pathway and extracellular matrix alterations in hepatic specification and differentiation of murine hepatic stem cells. Differentiation 2001; 68:254–269.

309. Fujio K, Evarts RP, Hu Z, Marsden ER, Thorgeirsson SS. Expression of stem cell factor and its receptor, c-kit, during liver regeneration from putative stem cells in adult rat. Lab Invest 1994; 70:511–516.

310. Kobari L, Giarratana MC, Pflumio F, Izac B, Coulombel L, Douay L. CD133$^+$ cell selection is an alternative to CD34$^+$ cell selection for ex vivo expansion of hematopoietic stem cells. J Hematother Stem Cell Res 2001; 10:273–281.

311. Omori N, Omori M, Evarts RP, Teramoto T, Miller MJ, Hoang TN, Thorgeirsson SS. Partial cloning of rat CD34 cDNA and expression during stem cell-dependent liver regeneration in the adult rat. Hepatology 1997; 26:720–727.

312. Lints TJ, Hartley L, Parsons LM, Harvey RP. Mesoderm-specific expression of the divergent homeobox gene Hlx during murine embryogenesis. Dev Dyn 1996; 205:457–470.

313. Rollini P, Fournier RE. The HNF-4/HNF-1alpha transactivation cascade regulates gene activity and chromatin structure of the human serine protease inhibitor gene cluster at 14q32.1. Proc Natl Acad Sci USA 1999; 96:10308–10313.

314. Monaghan AP, Kaestner KH, Grau E, Schutz G. Postimplantation expression patterns indicate a role for the mouse forkhead/HNF-3 alpha, beta and gamma genes in determination of the definitive endoderm, chordamesoderm and neuroectoderm. Development 1993; 119:567–578.

315. Runge D, Runge DM, Drenning SD, Bowen WC Jr, Grandis JR, Michalopoulos GK. Growth and differentiation of rat hepatocytes: changes in transcription factors HNF-3, HNF-4, STAT-3, and STAT-5. Biochem Biophys Res Commun 1998; 250:762–768.

316. Timchenko NA, Wilde M, Darlington GJ. C/EBPalpha regulates formation of S-phase-specific E2F-p107 complexes in livers of newborn mice. Mol Cell Biol 1999; 19:2936–2945.

317. Eferl R, Sibilia M, Hilberg F, Fuchsbichler A, Kufferath I, Guertl B, Zenz R, Wagner EF, Zatloukal K. Functions of c-Jun in liver and heart development. J Cell Biol 1999; 145:1049–1061.

318. Hilberg F, Aguzzi A, Howells N, Wagner EF. c-jun is essential for normal mouse development and hepatogenesis. Nature 1993; 365:179–181.

319. Terada T, Ashida K, Kitamura Y, Matsunaga Y, Takashima K, Kato M, Ohta T. Expression of epithelial-cadherin, alpha-catenin and beta-catenin during human intrahepatic bile duct development: a possible role in bile duct morphogenesis. J Hepatol 1995; 28:263–269.

320. Germain L, Goyette R, Marceau N. Differential cytokeratin and alpha-fetoprotein expression in morphologically distinct epithelial cells emerging at the early stage of rat hepatocarcinogenesis. Cancer Res 1985; 45:673–681.

321. Reid LM, Agelli M. Compositions comprising hepatocyte precursors. U.S. Patent number 5,789,246, 1996.

322. Eghbali B, Kessler JA, Reid LM, Roy C, Spray DC. Involvement of gap junctions in tumorigenesis: transfection of tumor cells with connexin 32 cDNA retards growth in vivo. Proc Natl Acad Sci USA 1991; 88:10701–10705.

323. Yin M, Bradford BU, Wheeler MD, Uesugi T, Froh M, Goyert SM, Thurman RG. Reduced early alcohol-induced liver injury in cd14-deficient mice. J Immunol 2001; 166:4737–4742.

324. MacDonald J, Xu A, Kubota H, LeCluyse E, Hamilton G, Liu H, Rong Y, Moss N, Lodestro C, Luntz T, Wolfe S, Reid L. Liver Cell Culture and Lineage Biology. In: Atala A, Lanza R, eds. Methods of Tissue Engineering. San Diego: Academic Press, 2001:151–201.

325. Langer R, Vacanti JP. Tissue engineering. Science 1993; 260:920–926.

326. Mikos AG, Thorsen AJ, Czerwonka LA, Bao Y, Langer R. Preparation and characterization of poly (L-lactic acid) foam. Polymer 1994; 35:1068–1077.

327. Kwong A, Byrn R, Reid L. Composition and Methods useful for HCV Infection. U.S. Patent Application number PCT/US02/09685, 2001.

328. Gebhardt R. Heterogeneous intrahepatic distribution of glutamine synthetase. Acta Histochem Suppl 1990; 40:23–28.

329. Gebhardt R. Metabolic zonation of the liver: regulation and implications for liver function. Pharmacol Ther 1992; 53:275–354.

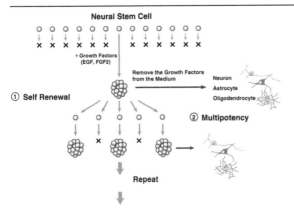

Figure 4.1 See text page 56.

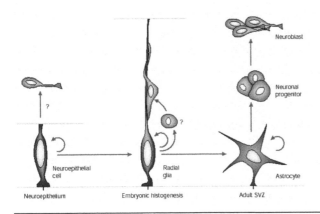

Figure 4.2 See text page 56.

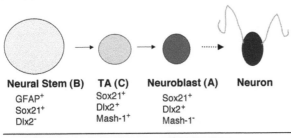

Figure 4.3 See text page 58.

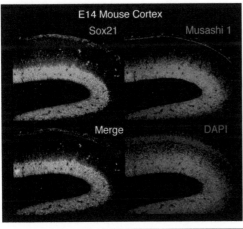

Figure 4.4 See text page 59.

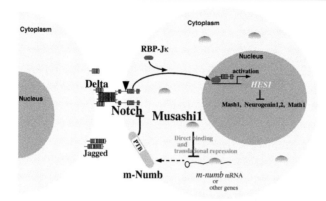

Figure 4.6 See text page 65.

(A)

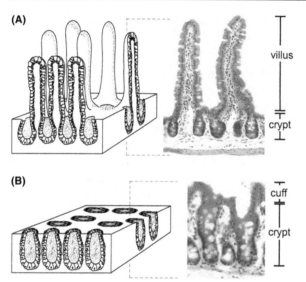

villus

crypt

(B)

cuff

crypt

Figure 6.1 See text page 90.

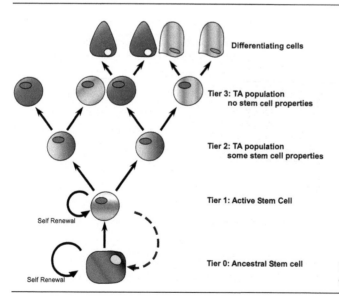

Differentiating cells

Tier 3: TA population
no stem cell properties

Tier 2: TA population
some stem cell properties

Tier 1: Active Stem Cell

Self Renewal

Tier 0: Ancestral Stem cell

Self Renewal

Figure 6.2 See text page 91.

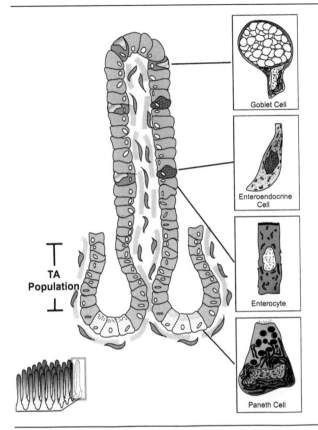

Goblet Cell

Enteroendocrine Cell

Enterocyte

Paneth Cell

TA Population

Figure 6.3 See text page 92.

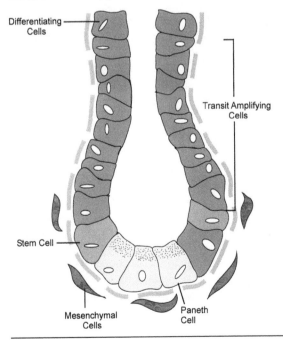

Differentiating Cells

Transit Amplifying Cells

Stem Cell

Mesenchymal Cells

Paneth Cell

Figure 6.5 See text page 95.

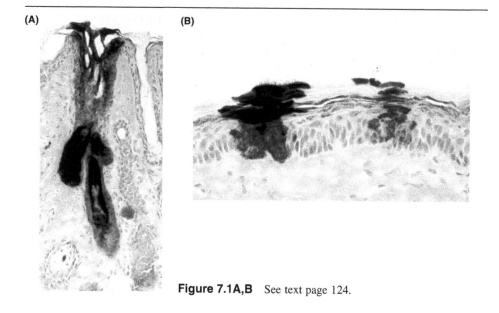

Figure 7.1A,B See text page 124.

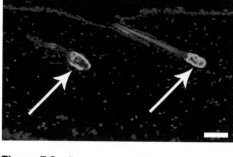

Figure 7.3 See text page 130.

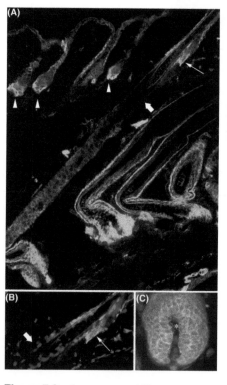

Figure 7.2 See text page 129.

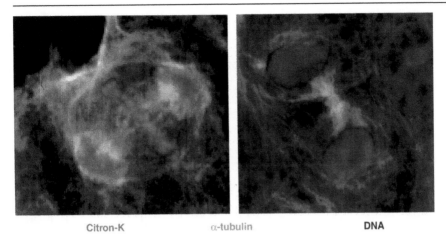

Citron-K α-tubulin DNA

Figure 9.4 See text page 167.

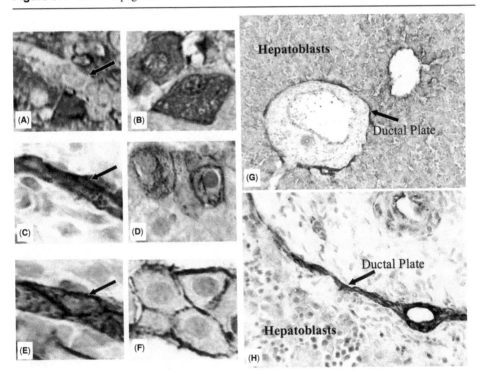

(A) (B) (C) (D) (E) (F)

Hepatoblasts

Ductal Plate

(G)

Ductal Plate

Hepatoblasts

(H)

Figure 9.7 See text page 178.

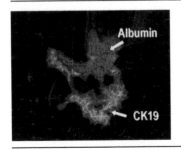

Albumin

CK19

Figure 9.9 See text page 180.

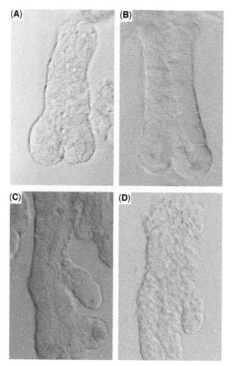

Figure 11.3 See text page 240.

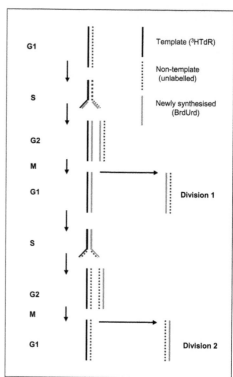

G1

S

G2

M

G1

S

G2

M

G1

Template (³HTdR)

Non-template
(unlabelled)

Newly synthesised
(BrdUrd)

Division 1

Division 2

Figure 11.4 See text page 241.

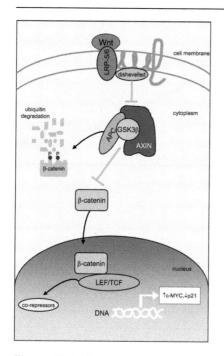

Wnt

LRP-5/6

dishevelled

cell membrane

ubiquitin
degradation

cytoplasm

APC GSK3β

AXIN

P P

β-catenin

β-catenin

β-catenin

nucleus

LEF/TCF

co-repressors

DNA

↑c-MYC,↓p21

Figure 11.5 See text page 243.

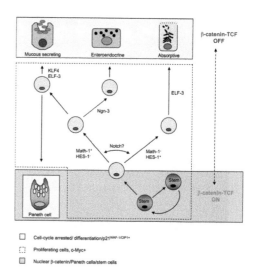

β-catenin-TCF
OFF

Mucous secreting

Enteroendocrine

Absorptive

↑ KLF4
ELF-3

ELF-3

Ngn-3

Notch?

Math-1⁺
HES-1⁻

Math-1⁻
HES-1⁺

Stem

Paneth cell

Stem

β-catenin-TCF
ON

☐ Cell-cycle arrested/ differentiation/p21ᵂᴬᶠ⁻ᵁᶜᴵᴾ¹⁺

☐ Proliferating cells, c-Myc+

☐ Nuclear β-catenin/Paneth cells/stem cells

Figure 11.7 See text page 246.

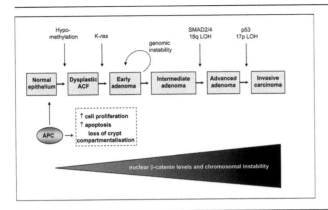

Figure 11.8 See text page 252.

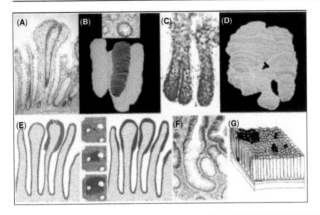

Figure 11.9 See text page 255.

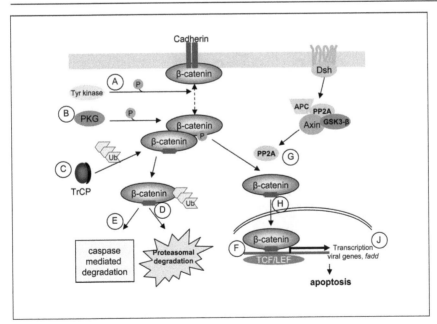

Figure 11.10 See text page 257.

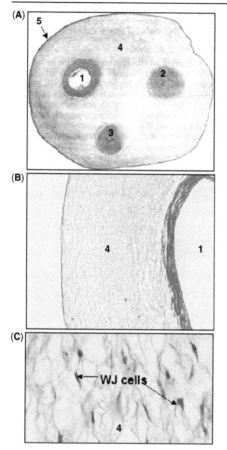

(A)

(B)

(C)

WJ cells

Figure 12.3 See text page 278.

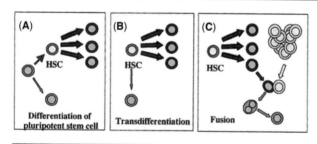

(A)

HSC

Differentiation of
pluripotent stem cell

(B)

HSC

Transdifferentiation

(C)

HSC

Fusion

Figure 12.4 See text page 279.

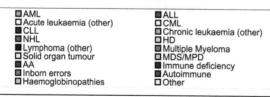

☐ AML
☐ Acute leukaemia (other)
☐ CLL
☐ NHL
☐ Lymphoma (other)
☐ Solid organ tumour
☐ AA
☐ Inborn errors
☐ Haemoglobinopathies

☐ ALL
☐ CML
☐ Chronic leukaemia (other)
☐ HD
☐ Multiple Myeloma
☐ MDS/MPD
☐ Immune deficiency
☐ Autoimmune
☐ Other

Figure 14.1 See text page 340.

10

Multistage Carcinogenesis: From Intestinal Stem Cell to Colon Cancer in the Population

E. Georg Luebeck
Public Health Sciences Division, Fred Hutchinson Cancer Research Center, Seattle, Washington, U.S.A.

INTRODUCTION

The organization of a high-turnover epithelial tissue into self-renewing crypt units that are maintained by a small number of resident but pluripotent stem cells provides a protective environment for cancer induction. In the intestine, two important principles appear to be at work: the first one, suggested by Cairns (1,2), relates to the preferential segregation of DNA strands in the stem cells preventing (or minimizing) the accumulation of DNA replication errors. Experimental evidence for this mechanism has recently been proffered by Potten et al. (3). The genomic integrity of stem cells may be further protected by an overriding apoptotic response to DNA damage, rather than being permissive to misrepair of DNA damage. The second principle relates to the transport of stem-cell progeny by the crypt "conveyor belt", which is driven by proliferating transient cells above the stem cell compartment (4). The implications of these two principles, their incorporation into quantitative models of crypt dynamics and carcinogenesis, and how cancer circumvents these protective mechanisms, are important issues that have only come into focus more recently in experimental studies and their mathematical analysis. Here we describe a simple multistage carcinogenesis model for the development of colon cancer, however a model that is consistent with the role of stem cells in maintaining tissue (and crypt) architecture (Chap. 6).

It is generally believed that mutations that arise in the transient amplifying cell (TAC) compartment, or in differentiated cells, are lost when the cells carrying the mutations undergo apoptosis or are sloughed off into the lumen within a few days. Thus, only a few (resident) stem cells at the base of the crypt are assumed susceptible to accumulating mutations over extended periods of time. In situations of normal (intact) crypt architecture, a mutant stem cell is physically isolated in its niche at the bottom of the crypt, and this tissue architecture provides a physical barrier to clonal expansion of the mutant. However, if mutations disrupt orderly maturation and differentiation in the crypt or if stem cells can migrate and populate adjacent crypts during clonal expansion, it is conceivable that a mutant colony is formed that continues to proliferate, escaping the cryptal barrier. Alternatively, abnormal (mutant) crypts may possibly undergo bifurcation as a result of an increase in stem-cell number or an increase in the size of the TAC compartment.

Intriguing clues on the origins of disruption of crypt structure and cell migration and differentiation were recently reported in two papers by Clevers and co-workers (5,6). In the colon, the transition of an epithelial stem cell into a fully transformed cancer cell is thought to require mutations in multiple proto-oncogenes and tumor suppressors. Although the exact number of mutational targets is not known, there is consensus that tumor-initiating mutations occur in the Wnt pathway, two members of which are recognized as primary targets: the adenomatous polyposis coli (APC) gene and the β-catenin gene. The interaction between these two components in particular as well as their interactions with other components of the Wnt pathway (GSK3β, axin/conductin) regulate a cell proliferation/differentiation switch as described by van de Wetering et al. (6). Briefly, specific defects in Wnt signaling lead to the accumulation of nuclear β-catenin and the β-catenin/TCF4 transcription factor complex even when the signal is turned off. Defective cells, therefore, are stuck in a proliferative state. In contrast, normal proliferating crypt cells respond to the change in Wnt signaling and undergo differentiation as they leave the proliferative zone near the top of the crypt. The outcome of this disruption appears to be the accumulation of proliferating cells at the luminal end of the crypt, crypt elongation, aberrant invaginations into the intercrypt space (7), and possibly the formation of aberrant crypt foci (ACF) (8–12).

This model of tumor initiation in human colon has an interesting dynamic consequence. If the APC-inactivating mutation occurs in a (quasi)immortal stem cell, then there will be a constant production of mutant progeny (amplified in the proliferative zone) that accumulate at the boundary between the proliferative zone and the differentiation zone. This "flux" of mutant progeny may be substantial. At any given time, a stem cell maintains several hundred transient cells in the proliferative compartment of a crypt (4). Even if only 10% of these cells divide every day, we would expect several thousand cells to emerge from the mutant progenitor annually. However, it is not clear what fraction of these cells will "stick" around before they undergo apoptosis or succumb to mechanical pressure and are sloughed off into the lumen.

Among the earliest premalignant lesions observed in colorectal cancer are the so-called ACF of both hyperplastic and dysplastic histology (13). The dysplastic ACF appears to play an important role in cancer development and are also referred to as an *adenomatous crypt* or a *microadenoma* (14). These early lesions frequently show loss of heterozygosity (LOH) on chromosome 5q, the locus of the APC gene (15,16) and are believed to be precursors to the adenomatous polyps, the characteristic lesion in people afflicted with familial adenomatous polyposis (FAP).

The number of necessary genomic changes required for malignant transformation of an adenoma is not known with certainty, although it is thought to be at least two (invoking Knudson's "two-hit" hypothesis). In fact, the transition from an adenoma to high-grade dysplasia (HGD) appears frequently accompanied by LOH on 17p, implicating the *TP53* tumor suppressor gene, generally considered a *guardian* of the genome. Once HGD is activated, "genetic chaos" may ensue setting the stage for malignant transformation (17). In contrast, hyperplastic polyps in colon have long been considered to have no, or only low, neoplastic potential. Recent studies, however, also appear to contradict this view (18,19).

Finally, colorectal tumors can be geno- and allelo-typed and assigned to either one of the two categories: the so-called LOH-positive cancers (believed to comprise 80% to

90% of all colorectal cancers) and the so-called microsatellite instability (MIN) prone cancers [e.g., see Ref. 20]. These two categories appear to be mutually exclusive, but are known to share common pathways as discussed by Laurent-Puig et al. (21). MIN-positive cancers are usually associated with defects in the DNA-mismatch repair system, involving mutations (or epigenetic silencing) in a number of genes, among them human homologues of the *MSH2*, *MLH1*, *MSH6*, and *PMS2* genes. These defects are now known to give rise to a *mutator phenotype* as postulated early on by Loeb (22,23). The inheritance of a requisite step of this form of cancer (in colon) is referred to as *heritable non-polyposis colon cancer* (HNPCC) or Lynch syndrome [for a review, see Ref. 24]. In contrast, LOH-positive cancers, which show abundant allelic losses and gains at numerous loci (14,17,25), frequently present biallelic inactivation of the APC tumor suppressor gene.

A BIOLOGICALLY BASED MULTISTAGE MODEL FOR COLON CANCER

In a 1954 landmark paper, Armitage and Doll (26) observed that the age-specific incidence of many solid tumors appeared to increase, at least roughly so, with the power of age. The power (slope on a log–log plot of incidence versus age) could be related mathematically, at some level of approximation, to the number of rate-limiting steps in the sequence of transformations from a normal cell to the formation of a malignant tumor under the assumption that the target cells in which the rate-limiting changes occur do not proliferate [e.g., see Ref. 27]. However, considering the large body of evidence on the importance of premalignant precursor lesions during carcinogenesis (e.g., enzyme-altered liver foci in rodents, mouse skin papillomas, adenomas in mouse intestine, and human colon), this assumption is not tenable. Clonal expansion of intermediate (premalignant) cells may greatly amplify the number of cells at risk for malignant transformations. Therefore, the number of required rate-limiting transitions may well be much lower than predicted by models that do not take cell proliferation into account. However, it has also been hypothesized that the rate at which mutations occur in the genome increases with advanced neoplastic progression, either as a result of the development of a mutator phenotype or as a result of ensuing genomic instability (28,29). In either case, clonal evolution is thought to favor specific cancer pathways that include a number of critical genomic targets in addition to neutral and opportunistic mutations (30).

The emerging picture of carcinogenesis is one of clonal evolution. It considers cancer as the outcome of a sequence of (epi)genetic events that lead to heterogeneous cell populations. Clonal selection, cellular competition, and complex interactions of the cells with environmental factors are viewed to determine the carcinogenic process. This process has also been compared with Darwinian selection (31,32).

The Two-Stage Clonal Expansion Model

Salient features of these concepts have been condensed into an effective model of carcinogenesis, the two-stage clonal expansion (TSCE) model, also known as the Moolgavkar–Venzon–Knudson (MVK) model (33–35). It combines two important aspects of carcinogenesis: First, the idea of recessive oncogenesis, as formulated by Knudson (36) for retinoblastoma, a rare embryonal cancer, explaining the role of tumor suppressor genes that (when inactivated) lead to loss of cellular growth control. The second aspect of this model pertains to the process of clonal expansion after growth control is abrogated. The model represents this process as a (stochastic) birth–death process, and therefore allowing for the possibility of clonal extinction and for random fluctuations in the size of the clones. Incidentally, the latter aspect parallels the "jackpot" phenomenon predicted by the fluctuation analysis of Luria and Delbrück (37), which provides a dramatic illustration of the rapid spread of mutations in a clonally expanding population of cells.

Recently, the TSCE model has been extended to reflect more details of the carcinogenic process and to account for the specific role of stem cells in maintaining crypt renewal in the colon. The basic model framework is shown in Figure 1. Colonic stem cells may undergo a series of pre-initiation steps, accumulating allelic losses and/or mutations in genes participating in critical pathways such as the Wnt pathway. Our model allows resident (immortal) stem cells to amplify mutant progeny that may accumulate to form a nascent lesion of proliferating cells in the crypt (Fig. 2). In addition to the constant accumulation (in the model with rate μ_{k-2}) generated by the mutant progenitor cells, the lesion may also undergo clonal expansion via increased symmetric cell division (with rate α) or a decrease in terminal differentiation or apoptosis (with rate β).

Clonal expansion, as emphasized before, may dramatically increase the risk of malignant transformation. In the current formulation of the model (Fig. 1), this expansion process comprises the entire clonal evolution up to the point of malignant transformation of an initiated (premalignant) stem cell. However, more than a single distinct proliferating compartment can be added to our models. This has been considered by Moolgavkar and Luebeck (39) and by Herrero-Jimenez et al. (40) with the result that a single proliferative compartment at the penultimate stage is sufficient to explain the age-specific incidence of colorectal cancer in the population.

Models of the type shown in Figure 1 have recently been fitted to colorectal cancer incidence data from the surveillance epidemiology and end results (SEER) registry (35). The data appear to be most consistent with a four-stage model that posits two rare events followed by an event that can be interpreted as asymmetric stem-cell divisions that are not mutational, but describe a positional lineage effect in the crypt, that is, the accumulation of $APC^{-/-}$ stem-cell progeny in the differentiation zone of the crypt. This interpretation simply states that a stem cell that has suffered mutations on both copies of the APC gene continues to function as a stem cell and populates the proliferative zone with mutant progeny. Upon entering the differentiation zone, $APC^{-/-}$ cells fail to down-regulate β-catenin/T cell factor (TCF)-mediated transcription resulting in continued cell proliferation, although the model suggests that this failure has only a subtle effect on disturbing the balance between symmetric cell division and apoptosis or terminal

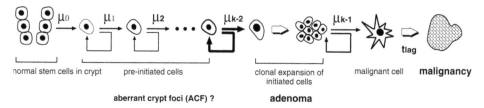

Figure 1 Extension of the TSCE model, which describes the stepwise progression of a normal stem cell to an initiated cell via pre-initiation stages in which mutated stem cells may accumulate, but have not yet acquired, the capacity to proliferate clonally. Note that pre-initiated cells are considered immortal in this model. However, once initiated, stem cells may undergo clonal expansion, which is modeled by stochastic birth and death processes. Initiated cells may also divide asymmetrically with rate μ_{k-1} giving rise to a malignant cell. Cancer progression until detection may be modeled by a fixed or randomly distributed lag time t_{lag} [e.g., see Ref. 38]. *Abbreviation*: TSCE, two-stage clonal expansion.

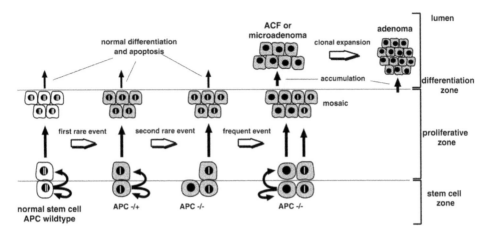

Figure 2 The formation of an adenoma in the Luebeck–Moolgavkar model, schematically. The basic steps involved (from *left* to *right*) in a section of the colonic crypt: normal stem-cell division maintaining the crypt with normal cell differentiation and apoptosis. Step 1: rare mutation in a stem cell inactivates one allele of the APC gene. Unless the normal (APC-wild-type) stem is inactivated or dies, the crypt may become mosaic, that is, may consist of a mixture of APC-wild-type and $APC^{+/-}$ cells. Step 2: second rare event leads to biallelic inactivation of the APC gene in a stem cell. Step 3: frequent asymmetric divisions of the defective stem cell and transient amplification populate proliferative zone with $APC^{-/-}$ progeny. Unresponsive to the change in Wnt signaling, the mutant progeny remains in a proliferative state as it enters the differentiation zone. The constant stream of mutant cells out of the proliferative zone leads to rapid accumulation and subsequent clonal expansion. *Abbreviations*: APC, adenomatous polyposis coli; ACF, aberrant crypt foci.

differentiation. The value of the net growth parameter $\alpha - \beta$ is about 0.15 per year, which leads to a tumor doubling time of about 4.5 years which states that adenoma grow very slowly. In contrast, the cell division rate in adenomas has been estimated to be much larger, about 10 symmetric divisions per year (40,41).

It is intriguing to consider the consequence of our assumption that the stem cell is immortal and possibly can divide many times before a lethal event occurs that leads to its extinction (say, as a result of cytotoxic exposure or exposure to ionizing radiation). Under this assumption, an $APC^{-/-}$ stem cell continues to be lodged in the stem-cell compartment, supplying the nascent polyp with mutant progeny. As long as the mutant stem cell remains in place, the polyp cannot become extinct but continues to be fed mutant progeny at a high rate, possibly several thousand cells per year (4). However, other explanations may be possible. For instance, the high-frequency event may represent an epigenetic phenomenon involved in carcinogenesis, or the consequence of genomic instability in the stem cell (42,43). For completeness, the parameter estimates of this model are provided in Table 1.

Temporal Trends

It is well known that cancer incidence is subject to geographic variation (44) and temporal trends. The joint determination of age, cohort, and calendar-year effects under this model reveals that for colorectal cancers in the SEER database, the calendar-year effects were much stronger than the estimated cohort effects [see Ref. 35 and supporting evidence posted at the PNAS Web site]. No obvious trends of colorectal cancer with birth cohort could be seen with this model. In contrast, for all population segments studied, the

Table 1 Maximum Likelihood Estimates of the Four-Stage Model Parameters from Analyses of Colorectal Cancer Incidence in the SEER registry (1973–1996)

	APC mutation rate (per year)	Initiation index	Malignant transformation rate × α (per year)^2	Adenoma growth rate (per year)
White males	1.4×10^{-8}	9.0	5.2×10^{-7}	0.15
Black males	1.2×10^{-6}	4.3	1.8×10^{-6}	0.15
White females	1.3×10^{-6}	0.7	1.2×10^{-8}	0.13
Black females	1.1×10^{-6}	2.9	5.2×10^{-6}	0.13

Note: With one exception (black females), the four-stage model gave the best fits.
Source: From Ref. 35.

incidence of colorectal cancer rises significantly with calendar-year until 1985, and then decreases modestly (Figs. 3 and 4). This increase in incidence by calendar-year may possibly be related to improved population screening for colon cancer (related to fecal occult blood tests, and/or wider use of colonoscopies and sigmoidoscopies) in the United States, whereas the drop seen after 1985 could be due to the gradual wearing-off of a "harvesting" effect (45), or alternatively to a reduction of cancers from increased opportunistic polypectomies following screening (A. Renehan, personal communication).

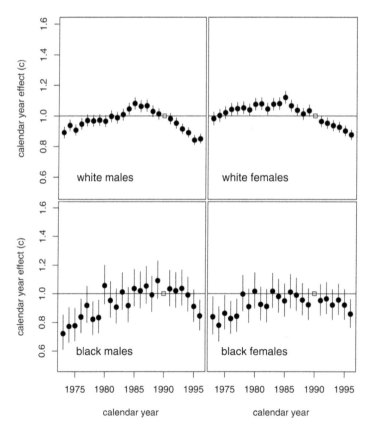

Figure 3 Adjustment of the incidence of colorectal cancer in the SEER registry (1973–1996) for calendar-year effect. *Error bars* reflect 95% confidence intervals of estimated coefficients. *Squares* indicate normalization points where the coefficients are anchored to 1.

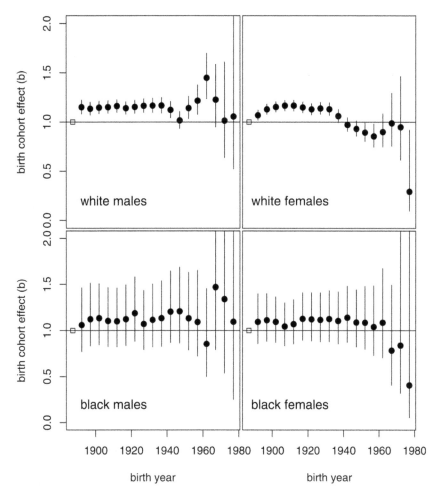

Figure 4 Adjustment of the incidence of colorectal cancer in the SEER registry (1973–1996) for birth-cohort effect. *Error bars* reflect 95% confidence intervals of estimated coefficients. *Squares* indicate normalization points where the coefficients are anchored to 1.

A particular advantage of the biologically motivated model described here (in contrast to purely statistical descriptions of carcinogenesis) is that one can compute the distribution of numbers and sizes of clones in intermediate compartments. This feature is useful when the model is used to derive predictions for colon cancer screening and interventions that involve the detection and surgical removal of adenomatous polyps. It is also useful when one desires to analyze quantitatively, rather than descriptively, polyp data from endoscopic screening studies.

ADENOMATOUS POLYPS—OBSERVATIONS AND MODEL PREDICTIONS

The particular colon cancer model described here was exclusively fitted to cancer incidence data without the inclusion of quantitative information on precursor lesions such as adenomatous polyps (35). It is interesting to compare the model prediction for the number of polyps with the observed number of polyps in asymptomatic subjects of

Table 2 Predicted Prevalence of Adenomas by Males and Females per Age Group

Average no. of polyps	Age (range)			
	35 (30–39)	45 (40–49)	55 (50–59)	65 (60–69)
Observed no. (videoendoscopic)	0.07	0.18	0.32	0.37
Predicted (males)	0.13	0.22	0.33	0.46
Predicted (females)	0.10	0.17	0.25	0.35

Source: From Refs. 46 and 48.

different age. Table 2 shows the numbers observed in a Japanese study using videoendoscopy summarized recently by Iwama (46). These observations are compared with the model-generated numbers for males and females in SEER. The agreement is surprising given the expected ethnic differences in cancer risks between these two populations. Note that 60% to 70% of the observed adenoma were hyperplastic (therefore of low neoplastic potential) and hence were left out in this comparison between endoscopic observation and model prediction.

Individuals afflicted with FAP present clinically colons with hundreds and sometimes thousands of dysplastic polyps at an early age with a high degree of variability (46). The same multistage model, modified to accommodate the fact that individuals with FAP require one less step in the process of initiating an adenoma, predicts about 3200 polyps (at age 20) to about 6300 polyps (at age 40) in an FAP colon.

Our model does not yet provide an explanation for the observed strong variability, but yields expectations that are in the right range. However, studies in the APC^{Min} mouse show that the severity of the *multiple intestinal neoplastia* (Min) phenotype is sensitive to the location of the truncating Apc mutation. For example, mice that carry the Apc$^{\Delta716}$ mutation (when present in a C57BL/6J background) develop a severe Min phenotype with hundreds of adenomas in their intestinal tract, whereas Apc1638 N mice will develop only a few during the first six months of life (47).

Several important questions regarding the intermediate endpoints can be addressed with the multistage model. For example, for colon cancer we may want to explore the consequences of the model concerning polyp prevalence, their size distribution, their risk of malignant transformation, and the sojourn time distribution before a polyp turns malignant and becomes the first malignancy in the tissue.

To demonstrate the utility of the multistage model, we provide four illustrations. Using the four-stage model for colon cancer (with the parameters given in Table 1, for white males), we predict the size distribution of adenomas at different times given that they all arose at time 0 from an APC$^{-/-}$ stem cell (Fig. 5). This calculation shows that the distribution can be very long-tailed after three decades of growth. Figure 6 gives the simulated adenoma prevalences (i.e., percent of individuals with one or more adenomas) and shows the expected size distribution in individuals with at least one adenoma at age 60, 70, and 80 years. Note, however, that in clinical studies involving sigmoidoscopies and/or colonoscopies, typically only polyps larger than a few millimeters in diameter are detectable. According to an estimate by Pinsky (49), a polyp of size 1 cm in diameter may consist of hundreds of thousands of cells. This raises the question as to the fraction of cells in a polyp that are actively dividing and are not yet committed to differentiation or apoptosis. Although adenoma size usually refers to its physical diameter, in the context of the mathematical model described here, this term needs to be translated into the size of the pool of actively dividing (undifferentiated) cells in a lesion.

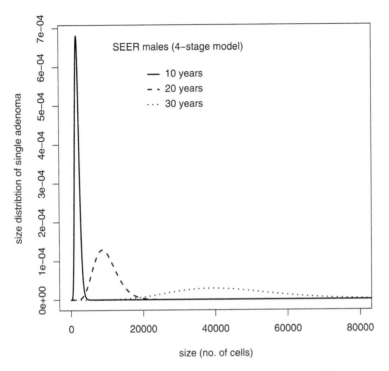

Figure 5 The predicted size distribution of adenomas at different times after birth of the adenomas.

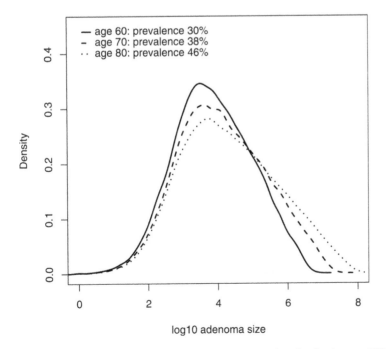

Figure 6 The simulated adenoma prevalences and size distributions at different ages.

Figures 7 and 8 show the predicted probability of detectable colon cancer arising from a single adenoma born at time 0 and the simulated distribution of sojourn times of adenomas that gave rise to detectable colon cancers up to age 80, respectively. What is striking is that the model predicts that most colon cancers arise in adenomas that have been around for five to six decades. If true, interventions that aim to retard or reverse the growth of adenoma [e.g., by using non-steroidal anti-inflammatory drugs (NSAIDs)] or seek to identify and remove polyps altogether (via polypectomy) may be very effective prevention strategies. However, removal of polyps just larger than some detection threshold may leave smaller polyps in place, as well as *pre-initiated* cells that have not yet turned into polyps. For example, such pre-initiated cells may be stem cells that have acquired a mutated allele at the APC locus. Again, the model presented here can be used to compute the risks associated with latent precursors, in addition to lesions that are already detectable.

The predictions and the hypotheses that follow from our model could clearly be strengthened substantially if we were to include data on adenomatous polyps in our analyses, for example, polyp number and sizes in individuals of known age and gender, as well as data on malignancies (absence, presence, number, sizes, etc.) in individuals who have developed this cancer. Multistage models that are consistent with both incidence and intermediate lesion data may then be considered validated models for the prediction of risk (modifications) in response to cancer screening (by conditioning on outcome), secondary preventions such as surgical removal of benign lesions, and chemopreventions using NSAIDs, for example.

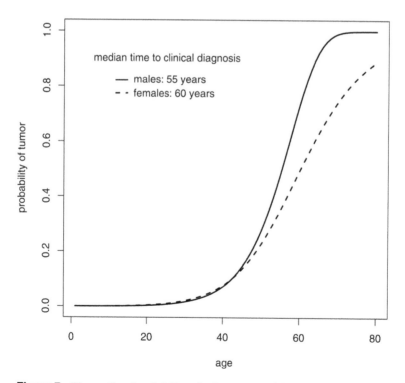

Figure 7 The predicted probability of colon cancer arising from a single adenoma as a function of time since the adenoma first appeared.

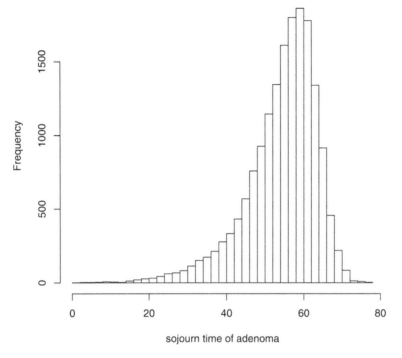

Figure 8 The simulated distribution of *sojourn* times of adenomas giving colon cancers.

SUMMARY

The model of colorectal cancer presented here is an example of how colonic tissue organ-ization, crypt structure and maintenance, and stem-cell kinetics contribute to our under-standing of tumor development in this organ (Fig. 9). There are many important (but not yet fully characterized) details that we wish to incorporate into this model, for example, the spatial development of ACF, or the development of polyps of different mor-phology (villous vs. tubulovillous), or the dynamics of clonal evolution in an adenoma that

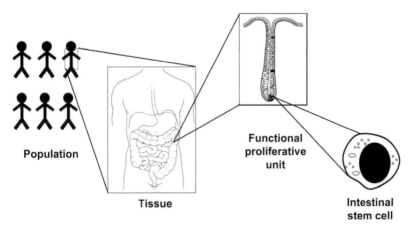

Figure 9 The model of colorectal cancer from intestinal stem-cell kinetics, crypt structure, and function to tissue organization and a common cancer within a population.

leads to malignant transformation. As these details become clearer and their impact on carcinogenesis better understood, our cancer models will become more sophisticated and hopefully more specific and predictive.

The ability to compute and predict the distribution of numbers and sizes of clones in intermediate stages of tumor development (as demonstrated here) is a major advantage over purely statistical descriptions of carcinogenesis, which do not impose biological constraints on the predictions of one type of lesion given the observation of another (subsequent) type of lesion. Colorectal cancer and its prevention via screening for precursors and specific interventions targeted to block the development of such precursors also present the unique opportunity to use these models to plan optimal screening schedules and to optimize prevention strategies that directly improve public health.

REFERENCES

1. Cairns J. Mutation selection and the natural history of cancer. Nature 1975; 255:197–200.
2. Cairns J. Somatic stem cells and the kinetics of mutagenesis and carcinogenesis. Proc Natl Acad Sci USA 2002; 99:10567–10570.
3. Potten CS, Owen G, Booth D. Intestinal stem cells protect their genome by selective segregation of template DNA strands. J Cell Sci 2002; 115:2381–2388.
4. Potten CS, Loeffler M. Stem cells: attributes, cycles, spirals, pitfalls and uncertainties. Lessons for and from the crypt. Development 1990; 110:1001–1020.
5. Batlle E, Henderson JT, Beghtel H, van den Born MM, Sancho E, Huls G, Meeldijk J, Robertson J, van de Wetering M, Pawson T, Clevers H. Beta-catenin and TCF mediate cell positioning in the intestinal epithelium by controlling the expression of EphB/ephrinB. Cell 2002; 111(2):251–263.
6. van de Wetering M, Sancho E, Verweij C, de Lau W, Oving I, Hurlstone A, van der Horn K, Batlle E, Coudreuse D, Haramis AP, et al. The beta-catenin/TCF-4 complex imposes a crypt progenitor phenotype on colorectal cancer cells. Cell 2002; 111(2):241–250.
7. Oshima H, Oshima M, Kobayashi M, Tsutsumi M, Taketo MM. Morphological and molecular processes of polyp formation in Apc(delta716) knockout mice. Cancer Res 1997; 57:1644–1649.
8. Rosenberg DW, Liu Y. Introduction of aberrant crypts in murine colon with varying sensitivity to colon carcinogenesis. Cancer Lett 1995; 92:209–214.
9. Morin OJ, Sparks AB, Korinek V, et al. Activation of beta-catenin-Tcf signaling in colon cancer by mutations in beta-catenin or APC. Science 1997; 275:1787–1790.
10. Korinek V, Barker N, Moerer P, et al. Depletion of epithelial stem-cell compartments in the small intestine of mice lacking Tcf-4. Nat Genet 1998; 19:379–383.
11. Wong MH, Rubinfeld B, Gordon JI. Effects of forced expression of an NH2-terminal truncated beta-catenin on mouse intestinal epithelial homeostasis. J Cell Biol 1998; 141:7665–7677.
12. Satoh S, Daigo Y, Furukaway Y, et al. ASIN1 mutations in hepatocellular carcinomas and growth suppression in cancer cells by virus-mediated transfer of AXIN1. Nat Genet 2000; 24:245–250.
13. Bird RP. Observation and quantification of aberrant crypts in the murine colon treated with a colon carcinogen: preliminary findings. Cancer Lett 1987; 37(2):147–151.
14. Jen J, Powell SM, Papadopoulos N, Smith KJ, Hamilton SR, Vogelstein B, Kinzler KW. Molecular determinants of dysplasia in colorectal lesions. Cancer Res 1994; 54:5523–5526.
15. Takayama T, Katsuki S, Takahashi Y, Ohi M, Nojiri S, Sakamaki S, Kato J, Kogawa K, Miyake H, Niitsu Y. Aberrant crypt foci of the colon as precursors of adenoma and cancer. N Engl J Med 1998; 339:1277–1284.
16. Otori K, Konishi M, Sugiyama K, Hasebe T, Shimoda T, Kikuchi-Yanoshita R, et al. Infrequent somatic mutation of the adenomatous polyposis coli gene in aberrant crypt foci of human colon tissue. Cancer 1998; 83:896–900.

17. Boland CR, Sato J, Appelman HD, Bresalier RS, Feinberg AP. Microallelotyping defines the sequence and tempo of allelic losses at tumour suppressor gene loci during colorectal cancer progression. Nat Med 1995; 1:902–909.

18. Goldstein NS, Bhanot P, Odish E, Hunter S. Hyperplastic-like colon polyps that preceded microsatellite-unstable adenocarcinomas. Am J Clin Pathol 2003; 119(6):778–796.

19. Jass JR. Hyperplastic-like polyps as precursors of microsatellite-unstable colorectal cancer. Am J Clin Pathol 2003; 119(6):773–775.

20. Ward R, Meagher A, Tomlinson I, O'Connor T, Norrie M, Wu R, Hawkins N. Microsatellite instability and the clinicopathological features of sporadic colorectal cancer. Gut 2001; 48:821–829.

21. Laurent-Puig P, Blons H, Cugnenc P-H. Sequence of molecular genetic events in colorectal tumorigenesis. Eur J Cancer Prev 1999; 8:S39–S47.

22. Loeb LA. Cancer cells exhibit a mutator phenotype. Adv Cancer Res 1998; 72:25–56.

23. Loeb LA. A mutator phenotype in cancer. Cancer Res 2001; 61(8):3230–3239.

24. Peltomaki P, de la Chapelle A. Mutations predisposing to hereditary nonpolyposis colorectal cancer. Adv Cancer Res 1997; 71:93–119.

25. Shih IM, Zhou W, Goodman SN, Lengauer C, Kinzler KW, Vogelstein B. Evidence that genetic instability occurs at an early stage of colorectal tumorigenesis. Cancer Res 2001; 61:818–822.

26. Armitage P, Doll R. The age distribution of cancer and a multistage theory of carcinogenesis. Br J Cancer 1954; 8:1–12.

27. Moolgavkar SH. The multistage theory of carcinogenesis and the age distribution of cancer in man. J Natl Cancer Inst 1978; 61:49–52.

28. Hanahan D, Weinberg RA. The hallmarks of cancer. Cell 2000; 100:57–70.

29. Loeb KR, Loeb LA. Significance of multiple mutations in cancer. Carcinogenesis 2000; 21:379–385.

30. Maley CC, Galipeau PC, Li XA, Sanchez CA, Paulson TG, Reid BJ. Selectively advantageous mutations and hitchhikers in neoplasms: p16 lesions are selected in Barrett's esophagus. Cancer Res 2004; 64:3414–3427.

31. Nowell PC. The clonal evolution of tumor cell populations. Science 1976; 194:23–28.

32. Cahill DP, Kinzler KW, Vogelstein B, Lengauer C. Genetic instability and Darwinian selection in tumours. Trends Cell Biol 1999; 9:M57–M60.

33. Moolgavkar SH, Venzon DJ. Two-event models for carcinogenesis: incidence curves for childhood and adult tumors. Math Biosci 1979; 47:55–77.

34. Moolgavkar SH, Luebeck EG. Two-event model for carcinogenesis: biological, mathematical and statistical considerations. Risk Anal 1990; 10:323–341.

35. Luebeck EG, Moolgavkar SH. Multistage carcinogenesis and the incidence of colorectal cancer. Proc Natl Acad Sci 2002; 99:15095–15100.

36. Knudson AG. Mutation and cancer: statistical study of retinoblastoma. Proc Natl Acad Sci 1971; 68:820–823.

37. Luria SE, Delbrück M. Mutations of bacteria from virus sensitivity to virus resistance. Genetics 1943; 28:491–511.

38. Hazelton WD, Luebeck EG, Heidenreich WF, Moolgavkar SH. Analysis of a cohort of Chinese tin miners with arsenic, radon, cigarette and pipe smoke exposures using the biologically based two-stage clonal expansion model. Radiat Res 2001; 156:78–94.

39. Moolgavkar SH, Luebeck EG. Multistage carcinogenesis: population-based model for colon cancer. J Natl Cancer Inst 1992; 84:610–618.

40. Herrero-Jimenez P, Thilly G, Southam PJ, Tomita-Mitchell A, Morgenthaler S, Furth EE, Thilly WG. Mutation, cell kinetics, and subpopulations at risk for colon cancer in the United States. Mutat Res 1998; 400:553–578.

41. Herrero-Jimenez P, Tomita-Mitchell A, Furth EE, Morgenthaler S, Thilly WG. Population risk and physiological rate parameters for colon cancer: the union of an explicit model for carcinogenesis with the public health records of the United States. Mutat Res 2000; 447:73–116.

42. Nowak MA, Komarova NL, Sengupta A, Jallepalli PV, Shih IM, Vogelstein B, Lengauer C. The role of chromosomal instability in tumor initiation. Proc Natl Acad Sci 2002; 99(25):16226–16231.
43. Michor F, Iwasa Y, Nowak MA. Dynamics of cancer progression. Nat Rev Cancer 2004; 4:197–205.
44. Potter JD. Colorectal cancer: molecules and populations. J Natl Cancer Inst 1999; 91:916–932.
45. Mandel JS, Church TR, Bond JH, Ederer F, Geisser MS, Mongin SJ, Snover DC, Schuman LM. The effect of fecal occult-blood screening on the incidence of colorectal cancer. N Engl J Med 2000; 343:1603–1607.
46. Iwama T. Somatic mutation rate of the APC gene. Jpn J Clin Oncol 2001; 31(5):185–187.
47. Smits R, Kartheuser A, Jagmohan-Changur S, Leblanc V, Breukel C, de Vries A, van Kranen H, van Krieken JH, Williamson S, Edelmann W, Kucherlapati R, Khan PM, Fodde R. Loss of Apc and the entire chromosome 18 but absence of mutations at the Ras and Tp53 genes in intestinal tumors from Apc1638 N, a mouse model for Apc-driven carcinogenesis. Carcinogenesis 1997; 18(2):321–327.
48. Mitooka H, Fujimori T, Naeda S, Nagasako K. Minute flat depressed neoplastic lesions of the colon detected by contrast chromoscopy using an indigo carmine capsule. Gastrointest Endosc 1995; 41:453–459.
49. Pinsky PF. A multi-stage model of adenoma development. J Theor Biol 2000; 207:129–143.

11

Intestinal Stem Cells and the Development of Colorectal Neoplasia

Stuart A. C. McDonald
*Digestive Diseases Centre, University Hospitals Leicester, Leicester, U.K.;
Histopathology Unit, Cancer Research U.K., Lincoln's Inn Fields and Department of
Histopathology, Bart's and the London School of Medicine and Dentistry, London, U.K.*

Trevor Graham
*Molecular and Population Genetics Laboratory, Cancer Research U.K., London
Research Institute, London, U.K.*

Christopher S. Potten
*Epistem Ltd. and School of Biological Sciences, University of Manchester,
Manchester, U.K.*

Nicholas A. Wright
*Histopathology Unit, Cancer Research U.K., Lincoln's Inn Fields and Department of
Histopathology, Bart's and the London School of Medicine and Dentistry, London, U.K.*

Ian P. M. Tomlinson
*Molecular and Population Genetics Laboratory, Cancer Research U.K., London Research
Institute, London, U.K.*

Andrew G. Renehan
Department of Surgery, Christie Hospital NHS Trust, Manchester, U.K.

INTRODUCTION

The mammalian intestinal epithelium is a rapidly renewing tissue in which tissue homeostasis is regulated by a balance between cell proliferation, differentiation, and apoptosis.
Over the last three decades, investigators have described the structure and cell kinetics
of the functional unit—the intestinal crypt (known as the *crypt of Lieberkühn* in the
small intestine)—and evidence has accumulated to support the concept that there are principally four differentiated intestinal cell types (enterocytes, mucosecreting or goblet cells,
enteroendocrine cells, and Paneth cells in the small intestine), derived from a common
pluripotent progenitor cell, the *intestinal stem cell*, located at or just above the bottom
of the intestinal crypt. The first half of this chapter will review the evidence behind
these prevailing concepts. Until recently, chapters on intestinal stem cells concluded

with speculation on the molecular regulation of intestinal stem cells (1,2). Over the last five years, however, there has been an explosion in our understanding of the key molecular systems regulating cell proliferation, differentiation and migration, and, in particular, the Wingless (Wnt) pathway, which includes adenomatous polyposis coli (APC) and β-catenin (3). These molecules are not only pivotal to normal crypt homeostasis, but are also frequently mutated as early events of intestinal neoplastic transformation. Colorectal cancer, the third commonest malignancy worldwide (4), is thought to arise from a mutated intestinal stem cell and thus, understanding the genetic mechanisms of this stem-cell system strikes at the very origin of these tumors. The second half of this chapter will thus cover the molecular regulation of intestinal stem cells and the molecular and cellular changes observed in early tumorigenesis.

Much of our knowledge of intestinal stem-cell function is based upon experiments carried out in the mouse, and throughout the chapter we will describe how intestinal stem cells are characterized in this model with human correlates where these are known. Finally, this chapter is built from the frameworks of many previous reviews from the authors (5–16) and other investigators (17–20).

BASIC STRUCTURE AND FUNCTION OF THE INTESTINAL CRYPT

Basic Cell Kinetics and Topography

The intestine is lined by a simple columnar epithelium, which is continually replaced, as cells are shed into the gut lumen (21). The baseline characteristics for the intestinal crypts of the large and small bowels in mice and humans are listed in Table 1. Each new cell will undergo four to six rounds of cell division before it migrates out of the crypt to the mucosal surface—a process that takes five to seven days. Murine small intestinal crypts constitute an average of 250 cells in a test-tube-like structure. When viewed in longitudinal cross-section, there are approximately 22 cells in height, with 16 cells forming an average circumference at the widest point. The vertical dimension, however, is overestimated in cross-section, due to the three-dimensional configuration of the cells, but using *crypt cell positional analysis* (Fig. 1A and B), this value is actually nearer 16 once

Table 1 Baseline Characteristics of the Small and Large Intestines

	Small intestine		Large intestine	
	Mouse	Human	Mouse	Human
Cells/column	25	34	42	82
Cells/circumference	16	22	18	46
Cells/crypt	250	450	300–450	2250
Cell cycle (hr)	12	~33	~35	~34
Stem-cell cycle (hr)	~24	≥36	≥36	≥36
Stem cells/crypt	4–16	NK	1–8	NK
Transit cell generations	4–6	>4–6	5–9	>5–9
Crypts per villus	6–10	~6		
Crypts per intestine	$1–3 \times 10^6$	NK		

Abbreviation: NK, not known.
Source: From Ref. 21.

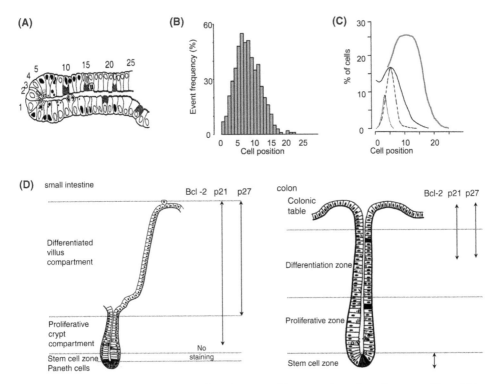

Figure 1 Crypt cell positional analysis and topographic expression of regulatory peptides within crypts of the small and large intestinal crypts. (**A**) Representation of a longitudinal section of an intestinal crypt and illustrates how the position of events up the crypt axis can be determined (position 1 being at the crypt base). When a number of crypt cross-sections are counted, an event frequency at each cell position can be plotted (**B**). (**C**) The apoptosis frequency plot (radiation-induced: *solid line*) can be compared with the theoretical distribution of actual stem cells (*dotted line*), clonogenic, or potential stem cells (*dashed line*) based on the mathematical modeling. The proliferating cells in normal homeostasis are shown as a thick *solid gray line*. (**D**) Representation of the expression of various apoptotic-related and growth arrest peptides within crypts of the small intestine (*left*) and colon (*right*).

these factors are taken into account (22,23). Approximately 30 fully differentiated Paneth cells occupy the very lowest crypt cell positions. The next 150 or so cells are actively proliferating as determined by incorporation of tritiated thymidine (^{3}H-Tdr) or bromodeoxyuridine (BrdU), with 75 of these in the S phase of the cell cycle at any one time. Analysis of the percentage of mitotic cells labeled demonstrates an average cell-cycle time of 12 to 13 hours for these rapidly proliferating cells—referred to as the *proliferation zone* [for review, see Ref. 9]. A small proportion of cells situated at the base of this band has a somewhat slower cell-cycle time of at least 24 hours (though this may be considerably longer) and it is proposed that these may be stem cells (24). The remaining cells occupying positions toward the luminal pole of the crypt are relatively more differentiated—*the differentiation zone*—and will usually undergo only one further cell division before emerging onto the villus surface. Conventional models suggested that there is a gradual transition between proliferating and differentiation zones, but when crypt size is taken into account, this is not the case (25).

In humans, the cell-cycle times of stem cells are less well defined, but they are generally thought to be at least four to eight times longer (26,27).

Apoptotic Activity in the Intestinal Crypt

Using the murine model, and fixing the intestinal tissue rapidly in Carnoy's medium, apoptotic bodies and fragments can be readily identified and reliably distinguished from mitotic and normal cells. Consequently, spontaneous and induced apoptosis can be quantified, as for proliferating cells, using crypt cell positional analysis (23). Over a decade of studies at the Potten laboratory have convincingly demonstrated that the patterns of apoptotic activity differ between the small and large intestines [for review, see Refs. 28–32]. In the small intestine, spontaneous apoptotic cells are readily observed but restricted to the stem-cell region (positions 4 and 5), whereas in colonic crypts, spontaneous apoptosis is very infrequent (Fig. 1C). Critically, few apoptotic cells are observed at the base of the colonic crypts where the stem cells are thought to be located (defined later). This so-called naturally occurring or *spontaneous apoptosis*, which is p53-independent (33,34), has been interpreted as part of the stem-cell homeostasis mechanism. When the process is repressed by Bcl2 (an anti-apoptotic factor), the colonic stem-cell numbers, and hence carcinogen target cells, may gradually drift upwards with time (8). Additionally, in comparison with the small intestine, the DNA-damage-induced apoptosis response in large intestine is blunted and distributed throughout the crypt. These observations of the differential amounts and position of apoptosis in the crypts led to the hypothesis that damaged small intestinal stem cells were deleted by an *altruistic* apoptotic process, thereby protecting this site from genetic and carcinogenic damages, whereas in the colon, damaged cells survive with the consequence of increased susceptibility to neoplastic transformation (31,35,36).

In support of this hypothesis, Bcl2 protein is minimally expressed in the small intestine of both mouse and human, but more strongly expressed at the base of colonic crypts in both species, indicating that this may be involved in overriding the apoptotic (both spontaneous and induced) homeostatic mechanisms in these cells (31,35–39). In addition, in *Bcl2* knockout mice, the incidence of spontaneous and induced apoptosis is dramatically increased in the stem-cell region of the colon, but unchanged in the stem-cell region of the small intestine (37) (Fig. 1D). Expression of cell-cycle regulators [for review, see Ref. 40] is also relevant. On the one hand, expression of proteins associated with cell growth arrest such as $p21^{WAF1/CIP1}$ and $p27^{KIP}$ is restricted to the nonproliferating compartment of the crypt (40–43), whereas on the other hand, expression of cell-cycle promoters such as Cdk2 and cyclin D1 is down-regulated in crypt areas corresponding to terminal differentiation (44).

Problems in Defining Intestinal Stem Cells

Morphological criteria do not exist to identify stem cells in gut mucosa and until very recently, there have been no molecular markers for intestinal stem cells. Instead, intestinal stem cells are defined by their characteristics, namely relatively undifferentiated cell types capable of (*i*) proliferation and self-maintenance, (*ii*) producing a variety of cell lineages, and (*iii*) tissue regeneration following injury (10). It should be remembered that in attempting to measure stem cells, one may find oneself in a circular argument, that is, in order to answer the question whether a cell is a stem cell, one has to alter its circumstances, and in doing so inevitably lose the original cell properties, a situation with marked analogy to *Heisenberg's uncertainty principle* in quantum physics (12).

Stem-Cell Location and Number

Studies measuring cell velocity, as determined by changes in the position of ^{3}H-Tdr-labeled cells with time, show that, under steady-state conditions, the cellular migration pathways of small intestinal crypts arise from positions 4 to 6 (that is, above the Paneth cells), whereas in the colon, they originate from the very base of the crypt (45,46). When large doses of irradiation or cytotoxic drugs (e.g., hydroxyurea or etoposide) are used to induce significant cell death within intestinal crypts, proliferative regenerative responses also arise from the aforementioned positions (24,47). Furthermore, when these basally situated cells are exposed to a lethal dose of radiation derived from the filtered weak beams of beta particles from [147]promethium, the whole crypt is sterilized by radiation doses that spare middle and upper crypt regions, further supporting the hypothesis that regenerative clonogenic cells are located exclusively at the lower pole of the crypt (48).

The number of stem cells located within small intestinal and colonic crypts is not known precisely. However, estimates based upon cell proliferation studies and mathematical modeling of stem-cell division and subsequent crypt fission suggest that in the small intestine, a crypt could be maintained under steady-state conditions by between four and six ultimate stem cells, with six generations of dividing transit cells (48,49). The situation is somewhat different in colonic epithelium, where modeling of ^{3}H-Tdr labeling and mitoses suggests that there is only one stem cell with eight generations of transit cells. However, it should be noted that a larger number of stem cells could also be supported by these data. Therefore, despite the greater size of colonic crypts, it would appear that their stem-cell quota might actually be lower than that of small intestinal crypts, presumably because the turnover of the former is less rapid.

The Intestinal Stem-Cell Niche

Stem cells within many tissues are thought to reside within a niche formed by a group of surrounding cells and their extracellular matrices, which provide an optimal environment for the stem cells to function. The identification of a niche within any tissue involves the knowledge of the location of the stem cells. According to Spradling et al. (50), to prove that a niche is present, the stem cells must be removed and subsequently replaced while the niche persists. Although this has been accomplished in *Drosophila* (51), such manipulations have not yet been possible in mammals. Despite this, the intestinal stem-cell niche is proposed to be as follows. The intestinal crypts are surrounded by a fenestrated sheath of intestinal *subepithelial myofibroblasts* (ISEMFs). These cells exist as a syncytium that extends throughout the lamina propria and merges with the pericytes of the blood vessels. The ISEMFs are closely applied to the intestinal epithelium and play a vital role in epithelial–mesenchymal interactions. ISEMFs secrete hepatocyte growth factor, transforming growth factor β type 2 (TGF-β2) (52), and keratinocyte growth factor (53), but the receptors for these growth factors are located on the epithelial cells. Thus, the ISEMFs are essential for the regulation of epithelial cell differentiation through these growth factors, and possibly others including factors in the Wnt pathway (discussed later in this chapter).

Novel Intestinal Stem-Cell Markers

Until recently, evidence of predicted stem-cell position and numbers relied on indirect experimental approaches, as described earlier. Recently, Potten et al. (54) have observed using immunohistochemistry the expression of *Musashi1* (Msi1)—a gene that encodes an RNA-binding protein associated with asymmetric divisions in neural progenitor cells (55)—in neonatal, adult, and regenerating crypts with a staining pattern consistent with the predicted number and distribution of early lineage cells, including the functional stem cells, in these situations. Early dysplastic crypts and adenomas are also strongly Musashi1-positive. In situ hybridization studies showed similar expression patterns for the Musashi mRNA and real-time quantitative polymerase chain reaction showed dramatically more Msi1 mRNA expression in multiple intestinal neoplasia (Min) mouse adenomas compared with adjacent normal tissue (56).

Notably, Musashi1 and the transcriptional repressor Hes1 were co-expressed in the crypt base columnar cells located between the Paneth cells, findings that suggest that not only the cells just above the Paneth cells, but also the crypt base columnar cells between the Paneth cells have stem-cell characteristics (57). Nishimura et al. (58) have shown similar patterns of expression of Musashi1 in human colon crypts. It should be noted, however, that Hes1 can be expressed in some cells outside the stem-cell zone and that Msi1 persists in all adenomas.

STEM-CELL HIERARCHY

The ability of stem cells to regenerate damaged tissue following injury has been used to study their functional characteristics. The *microcolony clonogenic stem-cell assay* measures the number of intestinal stem cells surviving exposure to radiation or cytotoxic therapy (59). The number of regenerating crypts is measured in cross-sections of murine intestine following a range of enterotoxic treatment dosages. Crypt regeneration occurs where one or more functional stem cells survive the toxic insult. Repopulation of the crypt will begin over the course of three days enabling surviving crypts to be counted at day 4. By this time, crypts without surviving stem cells have largely disappeared or are reproductively sterile. Dose–response curves (survival curves) can then be generated. These data suggest that the number of clonogenic cells present within a crypt is dependent upon the level of damage induced within the crypt. As damage increases, so more cells appear to be recruited into the clonogenic compartment. At low doses of radiation, there are approximately six clonogenic cells per crypt, a figure that corresponds closely to the *ultimate stem-cell* number, under steady-state conditions predicted by the mathematical model discussed earlier (10). At higher doses, this number increases to 36, in both the small intestine and colon (60,61). Similar experiments complement these data following cytotoxic exposure (62,63).

The number of stem cells per crypt is governed by net production versus cell deletion. To maintain the stem-cell population, each stem cell gives rise to one stem-cell daughter plus one daughter cell that will undergo further rounds of division prior to commitment to differentiate—termed *asymmetric division*. If both daughters are stem cells, under normal steady-state conditions, the excess stem cell is thought to be deleted by apoptosis (the niche environment presumably providing a limiting quantity of stem-cell survival factors) and stable stem-cell population is maintained (8). If both stem-cell daughters become committed to a differentiated fate (i.e., they are biologically equivalent)—*symmetric division*—then the stem cell from which they arose will cease to exist. It is probable that stem cells have the ability to switch between these various options in response to environmental conditions, thereby regulating their own number and consequently that of the crypt as a whole.

On the basis of the above observations, a *hierarchical* stem-cell organization has been proposed for mouse small and large intestines (10). A similar system is probably applicable in the human intestine. Three distinct categories of stem cells have been suggested: (*i*) in the steady state, the murine small intestinal crypt contains four to six *actual stem cells* (lineage ancestor cells) located approximately four cells up from the base of the crypt, (*ii*) an area of *clonogenic cells* (regenerative stem cells) that normally divide into transit cells and ultimately differentiate, but retain the ability to act as stem cells if needed, and (*iii*) a further tier of clonogenic cells (approximately 20 cells) that are particularly "hardy" and are the final resource when the first two tiers have died. It should be noted that it is extremely unlikely that the destruction of all stem cells occurs in nature resulting in recruitment from tier 2 and 3 stem cells—this is likely to be a consequence of the experimental situation. Thus, there is a gradual loss of "stemness" along the crypt and although a crypt may be using four to six stem cells normally, it has the ability to call upon about 36 cells to ensure crypt survival (Fig. 2) (60,61,63). There are, in addition to the actual and potential stem cells, about 120 other proliferative cells with no stem-cell attributes, that is, the *dividing transit cells*.

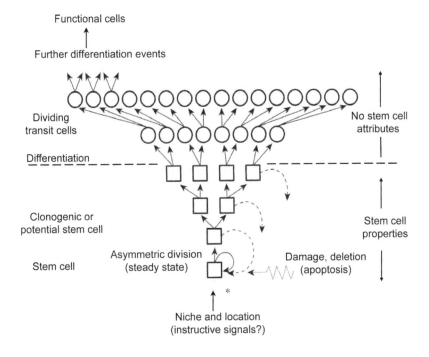

Intestinal stem cell hierarchy

Figure 2   The stem-cell hierarchy of the intestinal crypt. This is a three-tier system. There are four to six actual stem cells per crypt, but many more cells (potential stem cells) are capable of stem-cell function. When a stem cell undergoes a commitment to differentiation, it often first enters a transient state of rapid proliferation. Upon exhaustion of its proliferative potential, the transiently amplifying cell withdraws from the cell cycle and executes terminal differentiation. *Note*: *Approximately 5% may be symmetrical divisions.

Spatial Considerations for Intestinal Stem-Cell Populations

Combined studies from Potten's laboratory and Loeffler's bio-mathematical team have built up a concept of spatial distribution of stem cells within the intestinal crypt (9,64). Specifically, in the mouse small intestine, there are four to six stem cells per crypt, with a preferential location in the annulus of 16 cells at cell position 4. However, in reality, cell position 4 is the average position immediately above the highest Paneth cell, and this, in turn, may vary between cell positions 2 and 7. So, within this undulating annulus, the stem cells are unlikely to be touching, but somehow know to initiate apoptosis when numbers increase, and trigger symmetric division when numbers decrease. This could be achieved by a "shell" of stemness effect (15) and add support to the existence of a stem-cell niche (65).

CRYPT CLONALITY

Somatic mutations at certain loci allow us to study clonal succession of stem cells within intestinal crypts. Mutations in the *Dlb1* gene on chromosome 11 are one good example of this; C57BL/6J/SWR chimeric mice show heterozygous expression of a binding site on intestinal epithelial cells for the *Dolichos biflorus* agglutinin (DBA) lectin. This binding site can be abolished when the *Dlb1* locus becomes mutated either spontaneously or by the chemical mutagen ethyl nitrosourea (ENU). After ENU treatment, crypts emerge that are initially partially and then entirely negative for DBA staining (66). An obvious explanation for this is that the initial mutation is occurring in one of the crypt stem cells which then expands, presumably in a stochastic fashion, until all stem cells within a crypt are mutated and do not bind DBA. A "knock-in" strategy at the *Dlb1* locus can also be used to further explain these findings. If SWR mice do not express a DBA-binding site on their intestinal epithelial cells but can be induced to bind DBA by ENU treatment, wholly Dba+ or Dba− crypts would result if this occurred in the stem cells. From the use of this model, Bjerknes and Cheng (67) proposed that "committed epithelial progenitor" cells exist in mouse intestinal crypts by visualizing the morphology, location, and longevity of mutant clones in crypts and villi of the mouse small intestine. These transitory committed progenitor cells—the *columnar cell progenitors* (C_o) and the *mucus cell progenitors* (M_o)—evolve from pluripotential stem cells and then differentiate further into adult intestinal epithelial cell types.

There remains the possibility that cells from different parental strains of chimeric animals segregate independently during development to produce *monophenotypic*, but not necessarily *monoclonal*, crypts. When mice heterozygous for the X-linked alleles $Pgk1^a$ and $Pgk1^b$ were examined for clonality, no mixed crypts were observed, each being either Pgk1a+ or Pgk1b+ (68). Similar results were found in experiments using mice heterozygous for the glucose-6-phosphate dehydrogenase (*G6pd*) gene which have a crypt-restricted pattern of G6pd expression (69). These experiments show that murine crypt epithelial cells are ultimately derived during development from a single progenitor cell. Park et al. (70) have also shown that ENU-induced mutations in the G6pd gene result in crypts being initially partially then later wholly negative for G6pd staining.

After ENU treatment in both the *G6pd* and the *Dlb1* models, the time taken for a partial deficient crypt to become a wholly deficient crypt is similar. Two weeks after ENU treatment, crypts became wholly negative, reaching a plateau in four weeks in the small intestine and 12 weeks in the large intestine. The difference between the large and small bowels is interesting and cannot be fully explained by cell-cycle differences in each region of the gut. An explanation can be found in the *stem-cell niche hypothesis* that states that multiple stem cells occupy a crypt with random cell loss after division. This was originally formulated as the stem-cell zone hypothesis by Bjerknes and Cheng (71–75). The numbers of stem cells in the small intestine may be greater than that in the large intestine, explaining the speed at which small intestinal crypts can become G6pd-deficient compared with large intestinal crypts within the same mouse. Crypt fission, the process by which a crypt splits to form two daughter crypts (discussed later), also occurs at different rates between the small and large intestines and this may be a further reason why there are time differences.

Crypt clonality in the human has been harder to show. Initial experiments, transferring a human single-cell-derived colorectal carcinoma cell line into nude mice, produced identical tumors to the original tumor it was derived from and contained all the major epithelial cell types. This, of course, is not in any form a normal system, but does highlight that all these cells are multi-potential and can produce all major epithelial cell types. The majority of crypt clonality studies have been performed using patients with traceable mutations, whether they are genetic or somatic. Nine percent of the human Caucasian population has a homozygous ($OAT^{-/-}$) mutation in the O-acetyl transferase gene (O-acetylated mucin is normally expressed by goblet cells). Goblet cells from these patients are positive when stained for mild periodic acid-Schiff (mPAS) stain (76). Forty-two percent of the Caucasian population is heterozygous for the OAT mutation ($OAT^{-/+}$) and mPAS staining of crypts is negative. Loss of the remaining active OAT gene converts the genotype to $OAT^{-/-}$, resulting in the occasional, apparently randomly located positive mPAS-stained crypts with uniform staining of the goblet cells from the base to the luminal surface (77). Similar to the mouse models, when crypts are stained with mPAS from patients who have undergone radiation therapy, over time there is partial then whole crypt staining where the goblet cells are positive (78).

A rare case of an XO/XY patient with familial adenomatous polyposis (FAP) was able to give valuable insight into the monoclonal nature of human colonic crypts (79). Non-isotopic in situ hybridization (NISH) using Y-chromosome-specific probes showed the patient's normal intestinal crypts to be composed almost entirely of either Y-chromosome-positive or Y-chromosome-negative cells with about 20% of crypts being XO. Immunostaining for neuroendocrine-specific markers along with Y-chromosome NISH showed that crypt neuroendocrine cells shared the genotype of other crypt cells. The villous epithelium of the small intestine was, however, a mixture of Y− and Y+ epithelial cells, which follows from the theory that each villus is derived from the stem cells of more than one crypt. The vast majority of the crypts examined in this patient were monoclonal, with only 4 of 12,614 crypts showing a mixed phenotype, but none of these at patch boundaries. Further work by the same group has shown in Sardinian women heterozygous for X-linked mutation of the *G6pd* gene that crypts are either G6PD-positive or -negative and that monoclonal patches can contain up to ~450 crypts (80).

Somatic mutations have also been used to assess clonality and stem-cell hierarchy in the human colon. Mutations within the mitochondrial-encoded enzyme cytochrome coxidase (COX) occur naturally at random and increase in number with age (81,82). Mitochondrial DNA (mtDNA) mutations are thought to occur due to the lack of protective histones, poor DNA repair mechanisms, and the presence of free-radical-generating enzymes (83). They are lifelong, but in order for a mutated genotype to result in a mutated phenotype (such as COX deficiency), most, or all, of the copies of mtDNA within any cell must carry the mutation. Taylor et al. (84) have used the detection of mtDNA mutations by histochemical means to suggest that these mutations occur initially in the colonic crypt stem cell and are passed onto all the subsequent progeny, eventually leading to whole crypt COX deficiency.

All these data have shown that within the normal mammalian intestinal crypt, a single stem cell is able to dominate the entire crypt by the so-called *monoclonal conversion* and that crypts are monoclonal in nature.

INTESTINAL STEM-CELL REPERTOIRE

Closely related to the property of crypt clonality is the question of pluripotentiality, that is, does one intestinal stem cell give rise to all the different cell lineage types in each crypt? At a functional and structural level, there are four main cell lineages in the intestinal epithelium: columnar, mucosecreting, enteroendocrine, and Paneth cells. There are other less abundant lineages, such as caveolated and M cells, but these are not covered in this chapter. The columnar cells are the most populated in the intestine, and most of the time they are termed enterocytes in the small intestine and colonocytes in the large intestine. Comprehensive characterization of these four cell lineages can be found elsewhere [for review, see Ref. 11].

Previous debates about the origin of cell lineages in the intestine have focused around the endocrine cells. Pearse and Takor (85) maintained that these cells were derived from the neural crest, presumably by migration of neuroendocrine stem cells, in the same way that the ultimo-branchial body is colonized by migrating neuroectoderm, eventually to produce the C cell lineage in the thyroid (86). As these cells were amine precursor uptake decarboxylase positive, they were referred to as APUD cells. This concept has now essentially been abandoned in favor of the Unitarian hypothesis, which postulates that a single stem cell gives rise to all cell lineages in the epithelium. A modification of this concept was proposed by Holzer (87), suggesting that pluripotentiality can only be the property of a group of cells rather than a single cell and led to the proposal of "committed progenitor cells," but ultimately all cell lineages take their origin from a single cell.

Evidence for Pluripotentiality

The Unitarian hypothesis has been supported by many investigators (73,88–91) for the following reasons. First, radiation experiments indicate that a single surviving cell can form a regenerative crypt containing all four cell lineages (59–61). The surviving clonogenic cell was therefore probably pluripotent (88). Second, the injection of single cells from rat colonic adenocarcinoma subcutaneously into mice can give rise to tumors containing all cell lineages (92). Third, the human HRA19 cell line has also been shown to produce a variety of cell types from a single cell in vitro (91,93).

However, convincing evidence in support of the Unitarian hypothesis only came about relatively recently with the publication of the work by Bjerknes and Cheng (67). Using the DBA chimeric mouse model described earlier, they demonstrated that mutated progenitor cells give rise to a clone of similarly unstained progeny. The presence of multiple cell lineages within a mutated clone indicates pluripotency of the progenitor, whereas a clone composed of a single cell type is likely to be derived from a unipotent progenitor. Interestingly, during the early weeks of these experiments, crypts appear comprising some mutated and some nonmutated cells. In these experiments, there is induction of a rapid, but transient, increase in the frequency of crypts showing a partial or segmented mutated phenotype. Later on, there is an increase in the frequency of crypts showing a completely or wholly mutated phenotype, an increase which levels off at the same time as partially or segmented crypts disappear. Interestingly, and most importantly, the small intestine and colon show major differences in the timing of these events: the plateau is reached at between five and seven weeks in the colon, but not until some 12 weeks in the small intestine. The reasons for these differences in timing are threefold: (*i*) different durations of the stem-cell-cycle time, (*ii*) the presence of a stem-cell "niche" with differences in the number of stem cells between the two tissues, or (*iii*) the possibility that *crypt fission* plays an important part in the genesis of the wholly mutated phenotype (94).

In order to appreciate this more closely and the implication for colon tumorigenesis (see later), it is necessary to describe the process of crypt fission in more detail. In addition, as we will see later, molecular regulation also plays a role in determining the cell fate specification.

Crypts Grow by Fission

The importance of crypt fission as a mechanism determining crypt number in the small and large intestines has been appreciated for almost two decades (95). Initial studies indicated its pivotal position in two processes: (*i*) the massive increase in crypt numbers that occurs in the postnatal period (96) and (*ii*) during recovery of the intestine from irradiation (97) and cytotoxic chemotherapy (98). The morphology of this process can be followed in histological sections, but is perhaps best seen in bulk-stained microdissected material (Fig. 3). In many instances, crypt fission begins as an indentation in the base of the crypt and advances via a vertical split in the crypt (bifurcation), which continues until two new crypts are produced. In other instances, the process begins asymmetrically with respect to the crypt axis, a process called "budding," which can be seen in apparently normal colonic mucosa, but is more common in precancerous states such as FAP and in rat colonic mucosa after systemic treatment with carcinogens such as 1,2-dimethylhydrazine. In some situations, notably after irradiation, multiple buds can be seen coming off the same crypt (97).

The dynamics of crypt fission have been described in a series of seminal papers by Bjerknes and coworkers (99–101), leading to the concept of the crypt cycle. Crypts, born by fission, gradually increase in size, and after about 108 days in the mouse, the crypt undergoes fission, a process that takes about 12 hours. Measurements of crypt volume or size show that crypts at the upper end of the crypt size distribution initiated fission, suggesting that they have acquired a sufficient number of stem cell increases to a value above which crypt fission is initiated. In the human colon, the fraction of crypts in fission (crypt fission index) is small, of the order of 0.003%, and calculations indicate a crypt cycle time of 17 years (49,64). Despite the good evidence for crypt fission, some data suggest that it may be of limited importance in the normal bowel. Kim and Shibata (102) have shown that CpG methylation patterns of the *MyoD* gene in the normal colon appear to be as dissimilar in neighboring crypts as they do in crypts separated by some distance. This could be explained by the random accumulation of CpG methylation mutations over time masking any relationship two neighboring crypts may have.

THE CONCEPT OF STEMNESS AND "IMMORTAL" DNA STRANDS

A recurrent theme in stem-cell biology is whether stem cells are long-lived progenitors with the intrinsic capability to self-perpetuate their pluripotency or whether "stemness"

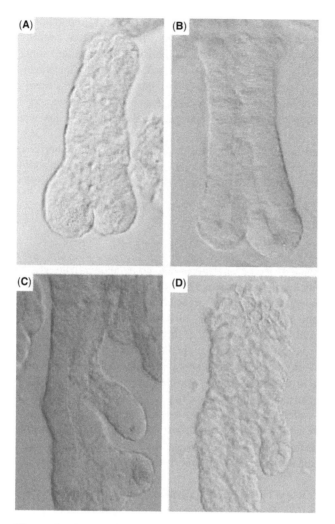

Figure 3 Crypt fission in adult mouse intestines. (**A** and **B**) Examples of "spontaneous" crypt fission in normal adult mouse small intestine. (**C**) Multiple fission after irradiation with 12 Gy. A new crypt can be seen developing from the mid-crypt region as well as from the crypt base. Each neocrypt contains Paneth's cells. (**D**) Similar bifurcation from a higher crypt cell position can be seen in the colon after irradiation (10 Gy). *Source*: From Ref. 6. (*See color insert.*)

is not an intrinsic property, but rather a non-autonomous feature. The latter model suggests that in every adult tissue with renewal capabilities there must be a niche that determines the stem potential of cells within. Due to the general lack of specific markers (until recently), epithelial stem cells have been traditionally identified by their ability to retain radiolabeled thymidine for long periods of time (103). More than 25 years ago, Cairns (103,104) proposed that stem cells selectively retain old (i.e., labeled) replication error-free DNA strands while donating newly synthesized strands to their descendents that will be lost from the tissue after a short time. Although this has long been controversial, recent works by two independent groups have confirmed Cairns's model. Potten et al. (105) demonstrated asymmetric segregation of chromatids in stem cells of small intestine using specific labels for new and old chromatids. This study labeled DNA template strands of proliferative cells by injecting ^{3}H-thymidine into mouse during development or in adult animals that have been irradiated. After several cell generations, the label was retained by only a few stem cells. Newly synthesized DNA strands, labeled with BrdU, segregated to the immediate stem-cell descendents and, therefore, were not retained (Fig. 4). Moreover, retention of old chromatids by stem cells has also been demonstrated using in vitro models (106) and in in vivo breast stem-cell systems (107). The existence of "immortal" DNA strands has implications for understanding the lifespan and mutagenesis dynamics of stem cells, but it also implies that certain stem-cell properties are maintained or inherited autonomously throughout adulthood (108).

It is estimated that the small intestinal stem cells of mice undergo up to 1000 divisions in their lifetime. It has been noted that these cells divide more slowly with a cycle time of approximately twice that seen in their daughters within the transit cell

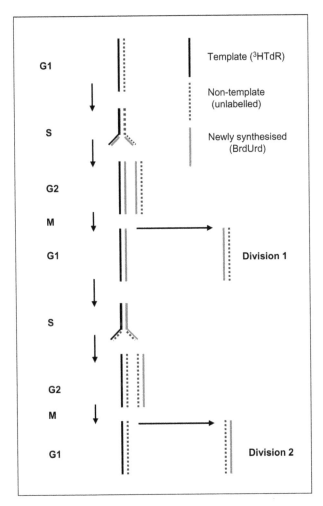

Figure 4 The segregation of template and newly synthesized DNA strands in one chromosome. The Cairns' hypothesis (1975) proposed that all the chromosomes would behave in this way. The template strands are selectively retained by the stem-cell daughter of a cell division, whereas the newly synthesized strands are segregated to the daughter cell destined to enter the dividing transit compartment and be shed from the tissue after a few days, thus removing any replication-induced errors. Label introduced into the newly synthesized strands takes two divisions to be removed from the stem cells. Label in the template strand would persist in the stem-cell-line. *Source*: Adapted from Ref. 105. (*See color insert.*)

compartment. It is possible that this occurs in order to minimize the risk of mutation within the stem-cell compartment and to allow maximum time for detection and correction of replicative errors or the implementation of altruistic apoptosis. Heddle et al. (109) proposed that the continued existence of stem cells throughout an organism's lifetime is not necessary for the purpose of populating tissue compartments. It is calculated that their progeny could create sufficient cells to adequately perform this function. Instead, this group hypothesizes that stem cells by their comparatively low rate of division serve to reduce the rate of spontaneous somatic mutation and therefore the risk of developing cancer. Actively proliferating cell types are more prone to mutating events, but their short lifespans prohibit the development of cancer.

Despite the above observations, intestinal stem-cell populations apparently go through bottlenecks in which a single stem cell survives and repopulates each crypt. This was first demonstrated during the transition from juvenile to adult intestinal epithelium by studying aggregation chimeras (110), and it has also been postulated recently for the adult crypt (111). These authors applied population dynamics algorithms to methylation tag patterns of single crypts to reach the conclusion that intestinal stem cells are not determined to divide asymmetrically, but rather that each stem cell can choose to generate zero, one, or two stem cells every time they divide. This stochastic pattern of division, together with data derived from mutagenesis studies, suggests the existence of a niche in the crypts responsible for determining "stemness," a concept in concert with the model that molecular control local to the crypt region where stem cells reside is of key importance, as outlined next.

MOLECULAR REGULATION OF NORMAL INTESTINAL CRYPT HOMEOSTASIS

So far in this chapter, we have established that stem cells reside at or near the base of the intestinal crypt, from where cells proliferate, migrate, and differentiate toward the lumen of the intestine. Over the past five years, a huge volume of data has emerged elucidating the molecular mechanisms which regulate these tightly controlled processes.

Developmental studies identified a number of genes that affect cell fate specification and proliferation during intestinal development [for review, see Ref. 112], including *Hes1* (113), *Math1* (114), *Rac1* (115), and *Tcf4* (116). Simultaneously, studies in human colon cancers demonstrated that mutations in the *APC* tumor suppressor gene, together with aberrations in other members of the Wnt/β-catenin signaling pathway, are among the commonest and earliest in colorectal carcinogenesis [for review, see Refs. 117–122]. As dysregulation of normal cellular function is a hallmark of early tumorigenesis (123), these observations were the impetus to focus on these molecular factors as key regulators of normal cellular homeostasis within the intestinal crypt (124,125).

This section will describe the Wnt/β-catenin signaling pathway, the bone morphogenetic protein (BMP)/SMAD4 pathway, and other molecular systems as potential regulators of intestinal homeostasis.

Wnt/β-Catenin Signaling Pathway

The so-called canonical Wnt/β-catenin signaling transduction pathway is essentially a network of separate but interacting pathways, characterized by binding of Wnt ligands to membrane receptors of the frizzled family, with subsequent inhibition of the complex that targets β-catenin for degradation and downstream activation of the transcription of Wnt target genes [for review, see Ref. 119] (Fig. 5). This multi-protein destruction complex involves axin and APC as scaffolds which in turn bind both β-catenin and GSK3β, to facilitate phosphorylation of the former by the latter. In turn, phosphorylated β-catenin is ubiquitinated and degraded in proteasomes, whereas unphosphorylated β-catenin accumulates and associates with nuclear transcription factors such as lymphoid enhancer-binding factor (LEF) (126,127) and T-cell factor 4 (TCF4) (128), leading to the eventual transcription and expression of target genes, including *c-MYC* [for review, see Ref. 129], cyclin D1 p21$^{\text{WAF1/CIP1}}$ (reduced expression) (130), matrilysin,

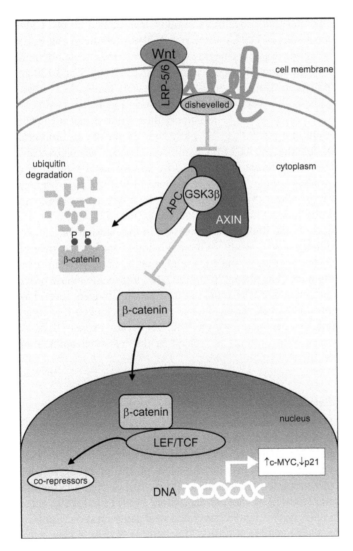

Figure 5 Wnt/β-catenin signaling pathway. Wnt proteins initiate intracellular signaling by binding to a complex containing frizzled and LRP5/6 receptors at the cell surface. This mobilizes the signaling protein dishevelled, which in turn inhibits the activity of the kinase GSK3β function. Wnt signaling reduces phosphorylation and degradation of repressors from LEF/TCF family transcription factors and gene activation. Tumor-associated mutations in APC that disrupts its ability to bind β-catenin and axin stabilize β-catenin and result in dysregulation of the Wnt/β-catenin signaling pathway. *Source*: Adapted from Ref. 205. (*See color insert.*)

CD44, urokinase-type plasminogen activator and, recently described, SOX9 (131). There are several regulatory mechanisms for the down-regulation of the Wnt/β-catenin signal, reflecting the pivotal nature of the pathway and the detrimental consequences of inappropriate activation [for review, see Ref. 3].

Wnt/β-Catenin Pathway and Control of Cell Proliferation

The small intestine of mice deficient for *Tcf4* is populated only by cell-cycle-arrested, differentiated cells (116) and progenitor cells in the epithelium of fetal small intestine accumulate nuclear β-catenin (132). Recent data suggest that TCF/β-catenin also plays a role in the maintenance of intestinal progenitors in adult mammalian crypts. Clevers' laboratory (133) reported the results of DNA micro-array analysis in human colon adenocarcinoma cell lines with inducible dominant-negative TCF4 mutations, which by inhibiting TCF/β-catenin complex formation, induced G1 growth arrest. They found 120 genes with at least a twofold drop in expression, of which five genes were known TCF targets in colorectal cancer. Among these five genes, only the transcription factor c-MYC was capable of overriding the G1 growth arrest. The authors then showed that c-MYC was able to repress the growth inhibitor p21$^{WAF1/CIP1}$, a key coordinator of cell proliferation and differentiation. It is possible that the TCF/β-catenin complex acts as a *master switch* that controls proliferation versus differentiation in healthy and malignant intestinal epithelial cells. Two further studies support these observations (134,135). Consistent with the earlier described labeling studies (25), there appears to be a distinct transition along the intestinal crypt from proliferation to differentiation status (Fig. 6).

Using a different approach, the transgenic expression in the intestine of Dickkopf1 (Dkk1), a secreted Wnt inhibitor (136), results in the reduction of the proliferative compartment and the loss of crypts in adult animals (3), a phenotype largely reminiscent of that present in *Tcf4* knockout mice. The phenotype of the *Dkk1* transgenic mouse provides a strong indication for the presence of a Wnt source in the intestine. Indeed, several Wnts are expressed along the intestinal track during mouse (137) and chicken development (138). However, it remains unclear as to which Wnt genes are expressed in adult crypts or where the source of Wnt proteins is located in the intestinal epithelium or subepithelial tissue.

Wnt/β-Catenin Pathway and Control of Cell Migration

Cell renewal in the small intestine is intimately coupled to bidirectional migration of the precursors of the various differentiated lineages. Although the mucosecreting cells, absorptive enterocytes, and enteroendocrine cells migrate upwards toward the lumen, Paneth cells migrate downwards and locate to the base of the crypt. It is unclear whether cells move passively along the epithelium or are pushed or pulled by attractive or repulsive forces. A second study from Clevers' laboratory (139) tested the hypothesis that TCF/β-catenin up-regulates two receptors associated with cell migration (EphB2 and EphB3) and down-regulates their ligand (Ephrin B). This family of tyrosine kinase receptors is known to control cytoskeletal re-modeling during cell migration (140) and may also be involved in colorectal tumorigenesis, as EphB4 is overexpressed in adenomas (141) (Fig. 6). This group of investigators first showed that β-catenin and TCF inversely control the expression of EphB2/EphB3 receptors and, then, using mice lacking these receptors, showed that their absence results in intermingling of proliferating and differentiating cells. In the intestinal crypt, therefore, the TCF/β-catenin complex appears to couple proliferation and differentiation to the sorting of cell populations through the EphB/ephrin system.

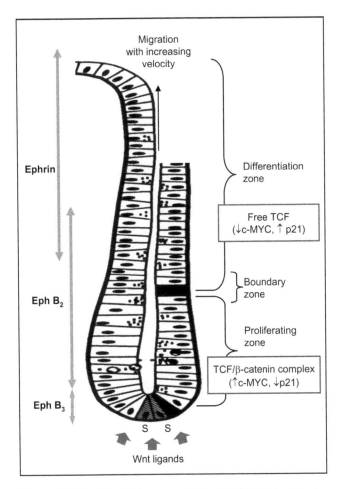

Figure 6 The intestinal crypt and the Wnt/β-catenin signaling pathway. Illustrated are protein expression patterns of TCF/β-catenin target genes EphB/Ephrin and c-MYC/p21. Also shown are the proliferative regions, including stem-cell locations. *Abbreviation*: S, stem cells. *Source*: Adapted from Ref. 124.

Wnt/β-Catenin Pathway and Control of Cell Fate Specification

Recent data suggest that the Wnt signaling pathway and related systems are also responsible, at least in part, for cell fate specification. In the simplest model, the mature cell types in the intestinal epithelium are determined via at least four consecutive binary decisions. Two of the genes acting in this decision chain have been identified, the basic helix–loop–helix (b-HLH) transcription factors Math1 and neurogenin-3 (Ngn-3) (Fig. 7). Math1 expression is required for commitment toward the secretory lineage (Paneth, goblet, and enteroendocrine cells)—the epithelium of *Math1* mutant mice is populated only by enterocytes (114). Neurogenin null (*Ngn-3$^{-/-}$*) mice lack enteroendocrine precursors, yet the other three cell types develop normally (142), indicating that this transcription factor drives the already committed secretory precursor toward the enteroendocrine fate. Interestingly, inhibition of Wnt signaling by Dkk1 results in the down-regulation of Math1 and the consequent loss of all secretory lineages. Transmembrane proteins Kremen1 and Kremen2 are high-affinity Dkk1 receptors that functionally cooperate with Dkk1 to block Wnt/β-catenin signaling (143).

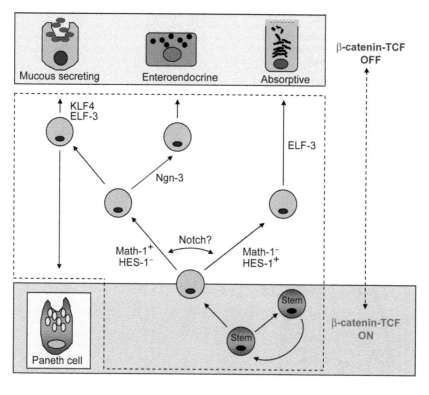

Figure 7   Cell fate specification in intestinal differentiation. The simplest model proposes that four different lineages arise from a common source of pluripotent stem cells through binary decisions. A few "pro-choice" genes involved in these decisions have been characterized. Math1 is required for the commitment toward the secretary lineage (Paneth, mucosecreting, and enteroendocrine cells), and HES1 probably antagonizes Math-1. $Ngn-3^{-/-}$ mice lack enteroendocrine precursors. Other transcription factors appear to play a role in the terminal differentiation of these cells, such as KLF4 an ELF-3. Nuclear β-catenin is detected in the bottom-most positions of the intestinal crypt, that is, in the position occupied by the putative stem cells and Paneth cells. Although in colonic crypts, the domain of β-catenin nuclear staining is extended through part of the transient amplifying compartment, it is not clear whether this is the case in the small intestine. However, the expression of several target genes, including those for c-MYC or EphB2, suggests that Wnt signaling might occur. A perfect complementation exists between the expression of c-MYC in proliferative precursors and $p21^{WAF1/CIP1}$ in the cell cycle arrested, differentiated cells. Paneth cells are cell cycle arrested and differentiated cells, despite showing the highest levels of nuclear β-catenin accumulation. Although in colon crypts the domain of β-catenin nuclear staining is extended through part of the transient amplifying compartment, it is not clear whether this is the case in the small intestine. *Source*: Adapted from Ref. 123. (*See color insert.*)

How does the differential expression of "selector" genes occur in identical precursors? Indirect evidence suggests that, as in many developmental systems, Notch signaling plays an important role. Animals deficient for hairy and enhancer of split1 (*Hes1*), a transcriptional repressor downstream of Notch signaling, show increased numbers of mucosecreting and enteroendocrine cells at the expense of absorptive cells (113). Hes1 represses Math1 expression in the intestine (113) and in other systems (144), suggesting that the absorptive versus secretory fate decision is likely to be established through Notch–HES1 coupling [for review, see Refs. 16,20].

Besides the involvement of b-HLH factors in lineage determination, other transcriptional factors are implicated in terminal differentiation in the intestinal epithelium. Kruppel-like factor 4 (KLF4) mutant mice are impaired in the maturation of goblet cells in the colon (145), whereas E74-like factor (ELF-3)-deficient animals show abnormalities in the maturation of both enterocytes and goblet cells (146). The latter work suggests that, once determined, some lineages might use the same signaling cascades during their terminal differentiation. Interestingly, both KLF4 and ELF-3 are up-regulated on expression of dominant-negative TCFs in colorectal cancer cells (133), indicating that they might participate in the regulation of the proliferation/differentiation transition by the TCF/β-catenin complex.

Finally, terminal differentiation in the intestine appears to be intimately coupled to cell-cycle arrest. It remains unclear how both processes are linked. One of the primary consequences of β-catenin–TCF blockage in colorectal cancer cells is up-regulation of $p21^{WAF1/CIP1}$, which, in turn, mediates cell-cycle arrest (133). Additionally, cell-cycle inhibitors of the $p21^{WAF1/CIP1}/p27^{KIP}$ family have been closely linked to terminal differentiation in other systems [for review, see Ref. 147]. Indeed, enforced expression of these inhibitors triggers intestinal differentiation in vitro (148,149). Although current data point to an important role for these cell-cycle inhibitors in intestinal homeostasis, the lack of phenotype in the intestine of mice with targeted $p21^{WAF1/CIP}$ or $p27^{KIP1}$ deletions is somewhat puzzling (41,150).

Wnt/β-Catenin Pathway and Control of Apoptosis

Other outcomes of Wnt signaling in the intestine are imaginable. A study from the Gordon laboratory (151) demonstrated a fusion protein between the DNA-binding domain of LEF1, a transcription factor from the LEF–TCF family, and the trans-activation domain of β-catenin-induced apoptosis in intestinal stem cells of transgenic animals, suggesting that β-catenin in complex with different LEF/TCF family members may control alternative genetic programs in the gut through stimulation of the expression of *survivin*—a bi-functional regulator of cell death and proliferation expressed during embryonic development but undetectable in healthy adult tissues and re-expressed in many cancers, including colorectal cancer (152,153)—the TCF/β-catenin imposes a stem-cell-like phenotype in colonic crypt epithelium with resistance to apoptosis and thus may contribute to the pathogenesis of colorectal cancer (154).

BMP/SMAD4 Pathway

BMP is a member of the TGF family of proteins and is an important regulating pathway that also has a key role during intestinal development. BMP proteins bind to their type II receptor, recruit type I receptors (BMPR1A or BMPR1B), and the signal is then transduced to the nucleus via SMAD transcription factors. The BMP signal is antagonized by Noggin (Nog), an extracellular protein that binds BMP and prevents its activity (155). During mouse embryonic development, there is high BMP4 expression in the intravillous mesenchyme but not in the precursor cells of the crypts of Lieberkühn. This pattern of expression is confirmed in normal colons of adult mice and humans, with highest expression in differentiating and mature colonocytes (156). Transgenic mice have been created with *Xenopus* cDNA expressing the BMP inhibitor Noggin, under the control of the mouse *villin* gene promoter, which becomes fully active during late gestation. The inhibition of the BMP pathway results in a characteristic pattern of epithelial development characterized by ectopic crypt development perpendicular to the crypt–villus axis. In addition, there is excessive branching and budding of the epithelium along with dilated cysts, an inflammatory infiltrate, and a high rate of dysplastic change (156). The changes closely mimic features seen in human juvenile polyposis (JP), an autosomal-dominant hereditary polyposis syndrome, with increased risk of gastrointestinal malignancy. JP syndrome is heterogeneous, although 25% to 40% of cases have mutations in the gene for BMPR1A and 15% to 20% have germline mutations in the gene for SMAD4 (157). Recent work on conditionally inactivated BMP1RA mice suggests that BMP signaling may have a role in preventing stem-cell renewal by inhibition of β-catenin, thus providing a counterbalance to the Wnt signaling cascade. Mutated BMP1RA mice develop characteristic polyps containing increased numbers of colonic crypts and had a fivefold increase in the number of stem cells in comparison to wild-type mice (158). The inhibitory effect on β-catenin appears to be mediated via the tumor suppressor PTEN (a dual protein and lipid phosphatase). This acts via phosphatidyl-inositol-3 kinase (PI3K) to inhibit the serine–theonine kinase *Akt*, which normally promotes cell-cycle progression, inhibits apoptosis (159), and enhances β-catenin activity (158). Thus, it seems that BMP activity is required for control of duplication of intestinal stem cells and this effect is via suppression of Wnt signaling (160).

Other Molecular Regulators

In both the small intestine and colon, wild-type p53 protein is expressed two to four hours after radiation exposure and in the small bowel its expression, in terms of time and cell position, is coincident with that observed for apoptosis (33). It is not, however, expressed in many of the apoptotic cells but can be found in other cells at the stem cell position. The p53-related gene, p21$^{\text{WAF1/CIP}}$, is also expressed at this time and broadly over the stem cell positions and slightly higher positions within the crypt, suggesting a role for p21$^{\text{WAF1/CIP}}$ in the cellular repair mechanisms of the clonogenic stem cells. In p53 knockout mice, radiation-induced apoptosis (mainly an ultimate stem-cell response) is completely absent, indicating a role for this protein in the detection of DNA damage in the ultimate stem cells (33,34).

The anti-apoptotic gene *Bcl2* is expressed at the base of murine and human colonic crypts while expression is not seen in the small intestine, supporting the view that Bcl2 increases the apoptotic threshold of colonic stem cells (37). Irradiation of *Bcl2* null mice significantly increases apoptotic cell death within the colon, compared with wild-type controls. In human adenomas, Bcl2 expression is increased, whereas low levels are generally found in carcinomas (38), which may indicate that altered expression of the *Bcl2* gene initially confers a survival advantage, but later is superseded by more potent factors, for instance, the strong expression of the survival gene *Bcl-w* in colonic adenocarcinomas (161).

The TGF-β signaling pathway inhibits intestinal epithelial proliferation, particularly in colonic mucosa. In vivo, the role of TGF-β may be to modulate cell-cycle exit and the subsequent differentiation of enterocytes in the upper crypt or villus (162). Alternatively, increased expression has also been reported in the proliferative zone of the crypt and it is hypothesized that this factor mediates the output of cells from this area (163). Loss of responsiveness to TGF-β is commonly seen during the development of colon cancer; indeed, under these circumstances, TGF-β may become a tumor promoter by stimulating angiogenesis, causing immunosuppression, and encouraging the growth of extracellular matrix (164).

Mammalian homeobox genes *Cdx1* and *Cdx2* have specific expression distributions in the developing and mature colon and small intestine. During embryogenesis, *Cdx1* is found in proliferating cells in the crypts and maintains this expression during adulthood. The *Tcf4*$^{-/-}$ mouse does not express *Cdx1* in the small intestinal epithelium and thus the Wnt/β-catenin pathway appears to induce Cdx1 expression with Tcf4 during the development of intestinal crypts (165). Mice heterozygous for a *Cdx2* mutation develop colonic polyps comprised squamous, body, and antral gastric mucosa with small intestinal tissue. These region-specific homeobox genes appear to help define the morphological features of differential regions of the intestine and regulate the proliferation and differentiation of the stem cells [for review, see Ref. 166].

The winged helix–forkhead family of transcription factors are vital components of the development of the ectodermal and endodermal regions of the gut. *Fkh6* is expressed in gastrointestinal mesenchymal cells (167,168) and *Fkh61*$^{-/-}$ null mice have elongated villi and goblet cell hyperplasia (169). They show up-regulated levels of heparan sulfate proteoglycans which increase Wnt-binding efficacy to the frizzled receptors on epithelial cells. This results in the overactivation of the Wnt pathways and increased nuclear β-catenin, the downstream effects of which have been summarized above.

A further potential regulator is the gene encoding the human BRCA2 tumor suppressor, which is mutated in a number of different tumor types, most notably inherited breast cancers. The primary role of BRCA2 is thought to lie in the maintenance of genomic stability via its role in the homologous recombination pathway. In a recent elegant study using generated mice in which *Brca2* was deleted from virtually all cells within the adult small intestine, using a CYP1A1-driven Cre-Lox approach, Clarke's group (170) noted a significant p53-dependent increase in the levels of spontaneous apoptosis which persisted for several months after removal of the gene. This study went on to show that *Brca2* deficiency results in the spontaneous deletion of stem cells, thereby protecting the small intestine against tumorigenesis.

Morphological development of the small intestinal mucosa involves the stepwise remodeling of a smooth-surfaced endodermal tube to form finger-like luminal projections (villi) and flask-shaped invaginations (crypts). These remodeling processes are orchestrated by instructive signals that pass bidirectionally between the epithelium and underlying mesenchyme [for review, see Ref. 17]. Sonic (Shh) and Indian (Ihh) hedgehogs are expressed in the epithelium throughout these morphogenic events and mice lacking either factor exhibit intestinal abnormalities (171). Hedgehog (Hh) signaling in the mouse neonatal intestine is paracrine, from epithelium to Ptch1-expressing ISEMFs and smooth muscle cells. Strong inhibition of this signal compromises epithelial remodeling and villus formation. Surprisingly, modest attenuation of Hh also perturbs villus patterning. Desmin-positive smooth muscle progenitors are expanded and ISEMFs are mislocated. This mesenchymal change secondarily affects the epithelium—Tcf4/β-catenin target gene activity is enhanced, proliferation is increased, and ectopic precrypt structures form on villus tips.

The Stem-Cell Molecular Signature

Gordon's group (172–174) have profiled the gene expression of murine intestinal stem cells and demonstrated that they are richly equipped with genes involved in c-MYC pathways. This laboratory developed approaches to overcome the problems of physically retrieving intestinal stem cells using germ-free transgenic mice lacking Paneth cells and harvesting a consolidated population of stem cells using laser-capture microdissection.

Table 2 c-MYC-Related Genes in the Gastrointestinal Stem Cells

Gene	SI crypt	Colon crypt	Stomach crypt	c-MYC-related action
Pituitary tumor-transforming factor (Pttg1)	2.3	2.0	3.7	Erk1 (extracellular-regulated kinase 1) phosphorylates Pttg1 allowing nuclear translocation and c-MYC transactivation
Ubiquitin-conjugating enzyme (Ube2ve1)	1.8	2.1	1.2	Ubiquitin degrades c-MYC, a process prevented by CKII phosphorylation
Macrophage migration-inhibitory factor (secreted) (Mif)	11.5	4.3	2.6	Erk1 stimulates cell proliferation in response to a variety of factors including Mif
BRG1/brm-associated factor 53A (Baf53a)	4.2	2.4	2.5	c-MYC together with the co-factor Baf53 forms a complex that functions in nucleosome remodeling during transcription
Histone acetyltransferase, type B subunit 2 (Rbbp7)	13.8	10.5	3.3	c-MYC binding to chromatin induces histone acetylation by histone acetyltransferase (Rbbp7)
Casein kinase II, beta subunit (CKII) (Csnk2b)	1.7	3.8	1.2	c-MYC is phosphorylated by CKII in response to polyamines, known as intestinal stem-cell mitogens
Protein phosphatase 2A catalytic subunit (Ppp2cb)	2.4	15.7	0.8	CKII substrates are dephosphorylated by Pp2A (Ppp2cb is the β-isoform), which promotes ubiquitin degradation

Note: Values refer to x-fold increases in gene expression compared with non–stem cells.
Abbreviation: SI, small intestine; c-MYC, the human homologue of an oncogene carried by an acutely transforming retrovirus known as Avian **myelocytomatos** virus.
Source: From Ref. 173.

Expression profiling identified 15 predominantly expressed genes, of which, seven are intimately involved in c-MYC related actions (Table 2).

EARLY MOLECULAR EVENTS IN COLORECTAL TUMORIGENESIS

Colorectal cancers are believed to originate from a mutated intestinal stem cell (detailed subsequently) and progress through a series of well-characterized histological changes—the *adenoma–carcinoma sequence* (175)—with corresponding accumulation of genetic changes (176–179). The paradigm early step in this pathway is mutation of tumor suppressor gene *APC* [for review, see Ref. 180]. Genetic changes subsequent to *APC* mutation commonly include mutational activation of the *K-ras* oncogene, inactivation of the *p53* tumor suppressor, and deletion of material on the long arm of chromosome 18, occasionally accompanied by mutation of the tumor suppressor *DPC4/SMAD4/MADH4*. APC mutations are generally sufficient for colorectal tumors to grow to about 1 cm diameter (181). Several other genes have been reported to undergo activating or inactivating mutation at low frequencies and a large panel of genes show evidence of promoter methylation and consequent transcriptional silencing (182). In addition to mutations that directly promote tumor growth, there exist (epi)mutations that lead to various forms of genomic instability in colorectal cancers. The best-characterized form of genomic instability is defective mismatch repair (MMR) which usually results from transcriptional silencing of the *MLH1* gene. These cancers comprise 15% of all colorectal malignancies and follow a different—though overlapping—genetic pathway from the classical pathway delineated above, which is followed by many MMR-proficient tumors. MMR-deficient tumors have few karyotypic changes, but are prone to frameshift mutations and tend to acquire fewer mutations of *APC*, mutation of *BRAF* rather than *K-ras*, inactivation of *BAX* rather than *p53*, and mutation of *TGFB2R* rather than *SMAD4*. Hereditary non-polyposis colorectal cancer (HNPCC) is the hereditary form of the MMR-deficient pathway [for review, see Ref. 183] and results from germline mutations in not only *MLH1*, but also the related MMR genes *MSH2* and *MSH6* mechanism. Given that they act as tumor suppressors, one intriguing possibility is that mutations in the MMR genes may affect not only DNA repair, but also cell proliferation. Additional pathways of color-ectal tumorigenesis exist, including one driven by germline defects in the base excision repair gene, *MYH*, leading to a failure to repair oxidative damage and hypermutation of *APC* and *K-ras* [for review, see Ref. 184].

Although there is no conclusive evidence of genomic instability in all colorectal tumors, especially early lesions, the increased genetic instability associated with colorectal cancer development is generally considered to be of the MMR-deficient (or *microsatellite instability*, MIN) type or of the *chromosomal instability* (CIN) type (185,186). Intriguingly, it has been suggested that mutation of *APC* itself can cause chromosomal mis-segregation, although this has to date only been demonstrated in in vitro models.

Mutation of *APC* yields mainly truncated forms of the protein that lack all of the axin-binding and most of the β-catenin-binding sites (187). In turn, the efficient degradation of β-catenin is blocked (188), leading to increased associations with TCFs. Accordingly, immunohistochemical studies demonstrate nuclear accumulation of nuclear β-catenin in early lesions such as microadenomas (189–192). Other Wnt pathway molecules such as conductin (193) and axin (194) may also be perturbed. Interestingly, from the perspective of crypt homeostasis, *APC* mutations do not simply inactivate the protein. Instead, there are tight constraints on the levels of APC activity of the proteins encoded by the two mutant alleles (195). Specifically, most colorectal adenomas have mutations that truncate the APC protein so as to leave a total of one or two of the 20 amino acid repeats involved in β-catenin degradation. The consequence of this may be to set an optimal level of Wnt signaling for colorectal tumorigenesis that is different from the norm, but is neither too weak not too strong. The pattern of *APC* mutations shows that this level is different between adenomas from the large and small bowels in FAP.

On the basis of these observations, various models of molecular and genetic changes in colorectal tumorigenesis have emerged (185). One view proposes that the multifunctional nature of APC confers one of the rate-limiting steps in tumor initiation and progression. Loss of β-catenin regulation by APC provides the intestinal cell, perhaps a mutated daughter cell, with a selective advantage and allows the initial clonal expansion (196). At this stage, CIN caused by loss of the C-terminal functional motifs of APC is latent due to surveillance by the cell cycle and mitotic checkpoint machinery (197). The early activation of the oncogenes *K-ras* (by point mutation) and MYC (as a downstream target of the Wnt pathway) will synergize with *APC* in triggering CIN and the subsequent allelic imbalances at chromosomal positions 17p and 18q (Fig. 8). Additional synergisms between *APC* and other tumor suppressor genes in eliciting aneuploidy and CIN will progressively lead to malignant transformation and metastasis (198).

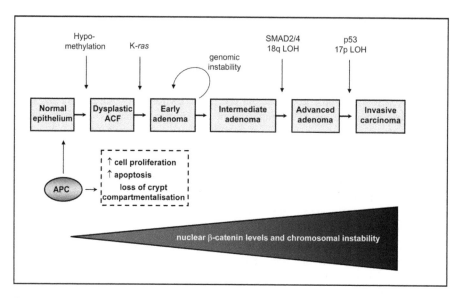

Figure 8 The accumulation of genetic alterations in the adenoma–carcinoma sequence. Sequential genetic changes leading to the evolution of colorectal cancer. Diagram mainly representative of the gatekeeper pathway. (*See color insert.*)

Altered maturation is increasingly recognized as an early event in tumorigenesis. Loss of imprinting (LOI) of the insulin-like growth factor II gene (IGF-2) is an epigenetic alteration that results in a modest increase in IGF-2 expression, which in turn may drive adenoma growth in a paracrine manner (199). Mice with LOI developed twice as many intestinal tumors as did control littermates (200). These mice also showed a shift toward a less differentiated normal intestinal epithelium, reflected by an increase in crypt length and increased staining with progenitor cell markers. A similar shift in differentiation was seen in the normal colonic mucosa of humans with LOI.

INTESTINAL STEM CELLS AND THE ORIGIN OF COLORECTAL CANCER

It is a well-held view (although not proven) that colorectal carcinomas are derived from the clonal expansion of a single intestinal stem cell (5,201,202). The evidence in support of this postulate is as follows:

- For the most part (caveat below), human tumors are monoclonal, suggesting that they arise from a single transformed cell. Examples supporting monoclonality include immunoglobulins and their heavy- and light-chain components secreted by malignant plasma cells in multiple myeloma and some B-cell lymphomas, and the expression of a single isoenzyme of *G6pd* (an X chromosome gene) in tumors (203,204).
- Given the evidence for monoclonal origin of most human tumors, there are two candidate target cells in the process of carcinogenesis: "proliferating" differentiated cells and stem cells. The acquisition of proliferative features in a differentiated cell necessitates the concept of de-differentiation and a cessation of migration. These scenarios are arguably unlikely. In the intestinal crypt, the

proliferating cells or *transit* cells are short lived (five to seven days) and rapidly migrating, whereas the natural history of carcinoma development is of many years or decades. Stem cells are the only cells with such a lifespan. Transit cells could conceivably be transformed but would *ceteris paribus* be lost with the upward drift of cells in the crypt—the so-called "escalator" effect.

- Most human tumors contain cell types consistent with an origin from the stem cells of that tissue. Thus, a cloned colorectal adenocarcinoma cell line, subcutaneously xenografted in immunodeficient mice, gives rise to four single-cell-type tumors (91).

- Changes in stem-cell-regulation mechanisms during early tumorigenesis is a further line of evidence (205). A number of studies (195,206) have been able to demonstrate that by *APC* mutation in embryonic stem cells, and it is possible to alter the level of cell differentiation through modulation of β-catenin dosage in these pluripotent cells (207).

Bearing in mind the fact that cancer development requires at least four mutational-type changes that span a long period of time and hence can only accumulate in the long-lived stem cell, there will be a point in time when a stem cell will have accumulated all but the last of these changes. This may be expressed by distortional and/or proliferative changes in the crypt. However, at this point, any cell in the crypt could sustain the final mutational change and develop into a neoplastic lesion, but this still requires changes to have occurred in the stem cell.

Caveat to the Monoclonal Theory of Human Colorectal Tumors

An unusual (and very rare) patient found both to be a constitutional XO/XY mosaic and to have FAP was used to explore the relationship between adenoma evolution and XO or XY karyotype utilizing the technique of in situ hybridization with a Y chromosome probe (79). The intestinal crypts of normal tissue were indeed monoclonal, as demonstrated in previous studies (described earlier), but 76% of microadenomas, although not the larger adenomas, were found to be polyclonal. The clonality of adenomas has also been assessed in a chimeric *Min*/ROSA26 mouse model (208). This study found 79% of adenomas to be polyclonal. It is not known precisely how these polyclonal tumors arise and what, if any interaction occurs between the genetically distinct cell types during the early stages of tumor evolution (209). The most likely explanations include synergy and/or fusion between adjacent early lesions and entrapment of genetically normal tissue within growing adenomas.

COLORECTAL TUMOR MORPHOGENESIS

On the basis of the histological distribution of proliferating cells within the crypt (described earlier), colonic adenoma formation is conventionally attributed to an upward extension of proliferating epithelial cells from the crypt base toward the colonic lumen, followed by an *outward* extension beyond the crypt surface into the lumen, ultimately forming a polyp (210,211). In this theory, adenomas form when the rate of cell proliferation exceeds that of the adjacent normal mucosa (212). Apoptosis is also important in determining tissue mass, and it is increasingly appreciated that its impairment may be an early event in the neoplastic process (38). Indeed, mean proliferation and apoptosis rates increase from early to advanced adenomas, although the rate of change of the former is

Table 3 Proliferation and Apoptotic Characteristics in the Adenoma–Carcinoma Sequence

	Early adenoma	Advanced adenoma	Adeno-carcinoma	P-value[a]
	Median (range)			
N	38	23	67	
Proliferation index (%)	16.0 (1.0–95.0)	20.5 (0–95.0)	49.4 (2.0–98.0)	<0.0001
Apoptotic index (%)	0.3 (0–7.9)	1.0 (0–7.3)	1.9 (0.1–12.4)	<0.0001
	Percentages (%)			
N	21	16	27	
p53 immunopositivity (%)	28	38	70	0.009
bcl2 immunopositivity (%)	81	88	22	<0.001

Notes: Unpublished data from Potten laboratory. Early and advanced adenomas defined by clinicopathological criteria in accordance with the Flexi-Scope trial. Advanced adenoma: ≥1 cm, >20% villous component on histology, and/or severe dysplasia.
[a]Percentages compared using χ^2-tests: median compared using Kruskal–Wallis nonparametric tests.

greater than that of the latter (Table 3). Abnormal expression of oncoproteins and apoptosis-related proteins is also observed in microadenomas. Thus, for example, p21$^{WAF1/CIP1}$ protein expression is reduced early in the progression of both sporadic and FAP adenomas (213), but not those arising in HNPCC patients (214). Bcl2 (an anti-apoptotic peptide), but not p53 protein expression, is also increased in early adenoma formation (38,215,216).

Top-Down Morphogenesis

Some investigators have challenged the outward theory of adenoma formation. Building on the earlier observations of Maskens (217) and Nigro and Bull (218) that proliferative activity is increased on the villous table in mucosa adjacent to colon tumors, two of the studies demonstrated that the distribution of proliferative cells and apoptotic cells is strikingly reversed in adenomas compared with normal colonic mucosa (216,219). This was referred to as loss of crypt compartmentalization and led the authors to conclude that cell migration in adenomas is not toward the lumen but rather *inward* toward the polyp base. Extending this concept further, Vogelstein's group have recently coined the term *Top-down morphogenesis* for adenoma development, demonstrating that early molecular markers of neoplastic change (*APC* gene mutation and β-catenin) are first noted in cells in the superficial portions of mucosa, and migrate laterally and downwards to form new crypts (190). The top-down hypothesis holds favor in some contemporary medical literature (220). A modification of this proposal is that a mutant cell in the crypt base, classically the site of the stem-cell compartment, migrates to the crypt apex, where it expands as above (221).

Bottom-Up Morphogenesis

Although there is undoubtedly loss of crypt compartmentalization and early molecular markers of neoplastic changes, such as *APC* mutation and β-catenin nuclear accumulation, in cells in the upper aspects of the crypt, there are several strong arguments against the "top-down" morphogenesis hypothesis. First, the evidence that stem cells originate from the base of the crypt and form lineages that spread to the surface is considerable.

Second, for clonal expansion of a mutated stem cell, there must be proliferation. Yet, in the intestinal crypt, the proliferating cells are short lived (five to seven days) and migrating, whereas the natural history of carcinoma development is of many years. Third, in the upper aspects of the crypt, the molecular signals are those of cell growth arrest. For expansion of a stem cell in this crypt position, there is a need for a change in the molecular environment (the "niche") and/or de-differentiation of cells, scenarios that seem unlikely.

An alternative hypothesis—*bottom-up histogenesis*—involves the recognition of the earliest lesion, the *unicryptal* or *monocryptal* adenoma (Fig. 9A), where the dysplastic epithelium occupies an entire single crypt (222). These lesions are very common in FAP, and although they are rare in non-FAP patients, they have certainly been described (223). Here a stem cell apparently acquires a second hit and expands (either stochastically or more probably because of a selective advantage, to colonize the entire crypt). Such monocryptal lesions thus should be clonal (79). Similar crypt-restricted expansion of mutated stem cells has been well documented in mice after ENU treatment (70) and also in humans heterozygous for the *OAT* gene, where, after LOH, initially half and then the whole crypt is colonized by the progeny of the mutated stem cell (78). Interestingly, OAT^+/OAT^- individuals with FAP show increased rates of stem-cell mutation with clustering of mutated crypts (224). In this scenario, in sharp contrast to the "top-down" model, the mutated clone further expands, not by lateral migration but by crypt fission, where the crypt divides, usually symmetrically at the base, or by budding (Fig. 9B–D).

In several studies, fission of adenomatous crypts is regarded as the main mode of adenoma progression, certainly in FAP, where such events are readily evaluated

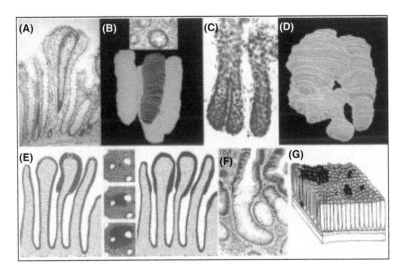

Figure 9 Adenoma morphogenesis. (**A**) A monocryptal or unicryptal adenoma. (**B**) A three-dimensional reconstruction of a unicryptal adenoma (inset) from serial sections, showing the adenoma in blue. Note that the adenomatous epithelium extends to the base of the crypt. (**C**) The mechanism of crypt fission in the normal colon whereby a crypt divides into two by this fission process. (**D**) A larger adenoma, from a three-dimensional reconstruction, showing expansion by basal fission and budding. (**E**) Lateral migration at the margins of an adenoma, with adenomatous epithelium invading crypt territories. (**F**) "Top-down" morphogenesis where a single cell incurs APC activation, passes to the top of the crypt, and proliferates or transforms in situ at the top of the crypt. Both concepts lead to expansion of the clone in the intercrypt zone. (**G**) Representation of how mutated clones expand in the colorectal epithelium by crypt fission. *Source*: From Ref. 231. (*See color insert.*)

(225,226), but also in sporadic adenomas (201). In fact, the non-adenomatous mucosa of FAP, with only one *APC* mutation, shows a large increase in the incidence of crypts in fission (226). Aberrant crypt foci, murine lesions that are putative precursors of adenomas and which can show *K-ras* and *APC* mutations (227), grow by crypt fission (228,229), as do hyperplastic polyps (230), but this concept does not exclude the possibility that the clone later expands by lateral migration and spreads downward into adjacent crypts, with the initial lesion being the monocryptal adenoma. This model of morphogenesis is conceptually very different from that proposed by Shih and co-workers (190) (Fig. 9E) (231).

On the basis of their observations, Boman et al. (232) hypothesize that tumor initiation in the colon is caused by crypt stem-cell overproduction. Using a combination of cell proliferation data in crypts from FAP patients and mathematical simulation, they show that only an increase in stem-cell numbers can explain the observed altered crypt labeling pattern in the premalignant crypt. This postulate is consistent with the previously mentioned "threshold" number of stem cells to trigger crypt fission and also the concept of hierarchical proliferation as described earlier. One possibility is that stem-cell overproduction increases the number of susceptible cells, leading to an increased likelihood of tumor initiation. Further studies suggest that mutated *APC* may confer enhanced stem-cell survival either directly (233) or indirectly through increased surviving expression (234). In addition, there are instances where mutated clones expand and remain cohesive, often involving a large area of tissue. The main example is the movement of mutated clonal crypts through the colorectal epithelium, again, by the process of crypt fission (235) (Fig. 9G).

The most persuasive evidence for the bottom-up hypothesis has recently come from the Wright laboratory. Preston et al. (236) examined 10 sporadic adenomas, the flat mucosa of three FAP patients, and specimens from the XO/XY individual with FAP. In the earliest sporadic adenomas, there were crypts entirely filled with the adenomatous epithelium, which showed proliferative activity and nuclear localization of β-catenin. There was a sharp cut-off between crypt epithelial cells showing nuclear β-catenin and surface cells with membrane staining. In contrast, in the larger lesions, downward adenomatous spread spilling over from an adjacent crypt was seen. Microdissected adenomas showed multiple fission events, with proliferation distributed equally throughout. In FAP tissues, numerous isolated monocryptal adenomas, which were clonal in origin, were seen, and examination of adenomas in the XO/XY individual also showed no instances of XY or XO adenomatous epithelium growing down into crypts of the other genotype. These data, together with an earlier study (237), provide the strongest evidence to date that both sporadic and FAP adenomas start as a unicryptal adenomas and grow initially by crypt fission, that is, bottom-up histogenesis. Later, in sporadic adenomas, there is evidence of growth down into adjacent crypts (top-down).

CLINICAL IMPLICATIONS AND FUTURE DIRECTIONS

The study of stem cells, with specific reference to intestinal stem cells, is of medical importance for a number of reasons.

- This chapter has shown that mechanisms of normal intestinal stem-cell homeostasis are at least some of the same processes that become dysregulated in carcinogenesis. Discovery of these pathways, therefore, brings us a step closer to treating uncontrolled clonal expansion and providing us with targets for future cancer treatments (238), lifestyle and dietary changes (239), and gene therapy (240).

- Molecular targeting of the Wnt/β-catenin pathway for cancer prevention and anti-cancer therapy is already a reality and is summarized in Figure 10 [for review, see Refs. 241,242].

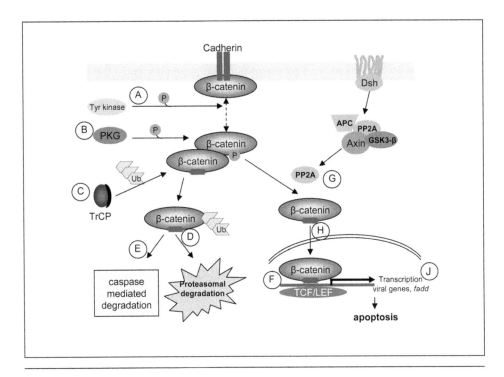

Site	Agent (class)	Mechanism of action
A	Glivec (tyrosine kinase inhibitor) Celecoxib (selective COX-2 inhibitor)	Inhibition of tyrosine phosphorylation/ re-localisation of β-catenin to the plasma membrane
B	Sulindac sulphide (NSAID)	Induction of β-catenin degradation by the proteasome
C	F-Box chimera (artificial biologically based protein)	Induction of β-catenin degradation by the proteasome
D	Endostatin (endogenous fragment of Collagen XVIII)	Induction of β-catenin degradation by the proteasome
E	Sulindac sulphide (NSAID)	Induction of β-catenin degradation by the proteasome
F	NO-aspirin (NSAID), PFK115-584 (small molecular inhibitor)	Disruption of β-catenin/TCF interaction
G	Aspirin (NSAID)	Stabilisation of the serine/threonine phosphorylated inactive β-catenin
H	Indomethacin (NSAID)	Decreased expression of Wnt/β-catenin target genes
J	TCF restrictive gene therapy	Interruption of TCF transcription

Figure 10 Different therapeutic agents block constitutive β-catenin/TCF signaling activity. (*See color insert.*)

- The rapidly dividing tissues of bone marrow, gut, and skin are the first to be affected by cancer treatment. Toxicity in these tissues is dose limiting for many chemotherapeutic agents or radiotherapeutic practices (243). Isolation of viable stem cells could be used as a therapeutic maneuver to repopulate such tissues following cancer therapy (244). Alternatively, growth factor manipulation to alter their sensitivity to treatment or improve their regenerative potential could also have benefits.
- The culture of intestinal stem cells may eventually facilitate tissue engineering. Already skin can be manufactured from its constitutive elements to provide cover following ulceration or burns. Hopefully, this tissue and others may eventually be derived from clones of our own progenitor cells (245).
- The identification of stem cells by novel markers may aid the development of stem-cell gene therapy for single gene conditions, such as *APC*, similar to examples in cystic fibrosis (246).

The importance of LOI of the *IGF-2* gene as an epigenetic alteration in early colorectal carcinogenesis is already mentioned. Importantly, IGF-2 LOI may be determined in human leucocytes and this is a potential powerful marker of the presence of colorectal adenomas (247,248). The *IGF-2* gene is one of the most commonly up-regulated genes in colorectal neoplasia (249)—there is concomitant overexpression of IGF-2 protein, which may spill over into the circulation, and corresponding levels of serum IGF-2 are elevated (250). This may serve as an inexpensive tumor marker.

Bone Marrow–Derived Cells and Intestinal Cancer Development

The evidence and potential clinical implications of plasticity of adult human stem cells (the ability to generate cells of different lineage from their organ of origin) have been discussed elsewhere (Chaps. 2, 9, and 14). Studies in several species have demonstrated that bone marrow-derived cells (BMDCs) are not simply stromal or hemopoietic stem cells, but they are precursors for many peripheral tissues, and may have potential implications in the clinical management of common diseases of the heart and central nervous system (Chap. 14).

In the intestinal tract, BMDCs are frequently recruited to sites of tissue injury and inflammation (251–253), and thus, may also be a potential source of malignancy. In a recent study, using a Helicobacter-gastric cancer model, Wang's laboratory showed that although acute injury, acute inflammation, or transient parietal cell loss within the stomach do not lead to BMDC recruitment, chronic infection of C57BL/6 mice with Helicobacter (a known gastric cancer carcinogen) induces repopulation with BMDCs (254). These cells go on to progress through metaplasia and dysplasia to intraepithelial cancer. However, BMDCs have a tendency to fuse with other cells, a trait not demonstrated in this study. Although these experiments have yet to be reported for intestinal cancer development, nonetheless they suggest for the first time that intestinal cancers can originate from BMDCs. As current hypotheses are generally based on the tenet that epithelial cancers originate from transformation of tissue-specific stem cells, these observations challenge the multi-step model of cancer progression discussed earlier and open a whole new field of intestinal carcinogenesis (255).

SUMMARY

The main issues of this chapter are summarized in Table 4. Over the past three decades, we have gained huge insight into the origin and characteristics of intestinal stem cells. Along

Table 4 Summary of Chapter

Indirect evidence from many lines of investigation shows that intestinal stem cells are located at or near the base of the intestinal crypt.

Novel markers of stem cells have been identified, for example, Musashi-1, allowing direct visualization, and in the future, isolation of intestinal stem cells.

The adult crypt has developed a stem-cell hierarchical system with four to six ultimate stem cells and approximately 20 to 30 potential stem cells.

In a simple model, intestinal crypts are monoclonal but probably arise through competition between multiple stem cells with the eventual dominance of a single stem cell following successive rounds of cell division, that is, initial polyclonality; this single stem cell gives rise to a few divisions to attain the adult crypt complement of four to six ultimate stem cells.

There are four main cell lineages in the intestinal epithelium—columnar, mucosecreting, enteroendocrine, and Paneth cells—a single stem cell may give rise to all four lineages and thus demonstrate pluripotentiality.

Crypts grow by fission, an important mechanism for determining crypt number.

The ultimate intestinal stem cells selectively retain old "immortal" DNA strands while donating newly synthesized strands to their descendents; this process is not in operation in the other levels of the stem-cell hierarchy.

The canonical Wnt/β-catenin signaling pathway is a key regulator of intestinal stem-cell homeostasis including a proliferation/differentiation "master switch," cell migration, and cell fate specification.

Many early colorectal neoplasms are characterized by *APC* mutation and nuclear accumulation of β-catenin.

It is generally held that most colorectal cancers arise from a mutated stem cell and that the mutation is in the *APC* gene.

Very early adenoma formation is monoclonal and later the adenoma becomes polyclonal.

There are two hypotheses of colon tumor morphogenesis—"top-down" versus "bottom-up"; emerging evidence suggests that the bottom-up pattern occurs in the earliest recognizable lesion— the monocryptal adenoma—and that the top-down pattern may prevail as the lesion enlarges.

---

the route, we have learned that the stem-cell systems differ between the small and large intestines, an observation that may explain the vast difference in cancer incidences at these respective cancer sites. More recently, scientists have identified the individual molecules that regulate proliferation, migration, and fate of stem-cell daughter cells within the crypt and identified that dysregulation of these processes are common features of early tumorigenesis. These findings strike at the very origin of colorectal carcinogenesis and offer untapped opportunities for primary prevention of this common cancer.

ACKNOWLEDGMENTS

We thank Dr. Catherine Booth and Dr. Ged Brady for their helpful advice during the preparation of this chapter. This work was in part supported by the Christie Hospital Endowment Fund and Royal College of Surgeons of Edinburgh.

REFERENCES

1. Potten CS. Intestinal Stem Cells. London: Academic Press, 1997.
2. Wright NA. Stem cell repertoire in the intestine. In: Potten CS, ed. Stem Cells. London: Academic Press, 1997:315–330.

3. Pinto D, Gregorieff A, Begthel H, Clevers H. Canonical Wnt signals are essential for homeostasis of the intestinal epithelium. Genes Dev 2003; 17(14):1709–1713.

4. Parkin DM. Global cancer statistics in the year 2000. Lancet Oncol 2001; 2(9):533–543.

5. Bach SP, Renehan AG, Potten CS. Stem cells: the intestinal stem cell as a paradigm. Carcinogenesis 2000; 21(3):469–476.

6. Booth C, Potten CS. Gut instincts: thoughts on intestinal epithelial stem cells. J Clin Invest 2000; 105(11):1493–1499.

7. Brittan M, Wright NA. Gastrointestinal stem cells. J Pathol 2002; 197(4):492–509.

8. Potten CS, Booth C, Pritchard DM. The intestinal epithelial stem cell: the mucosal governor. Int J Exp Pathol 1997; 78(4):219–243.

9. Potten CS. Stem cells in gastrointestinal epithelium: numbers, characteristics and death. Philos Trans R Soc Lond B Biol Sci 1998; 353(1370):821–830.

10. Potten CS, Loeffler M. Stem cells: attributes, cycles, spirals, pitfalls and uncertainties. Lessons for and from the crypt. Development 1990; 110(4):1001–1020.

11. Wright NA. Epithelial stem cell repertoire in the gut: clues to the origin of cell lineages, proliferative units and cancer. Int J Exp Pathol 2000; 81(2):117–143.

12. Renehan AG, Potten CS. Stem cells and cancer. In: Schwab M, ed. Encyclopaedia of Cancer. Berlin: Springer, 2001:846–853.

13. Preston SL, Alison MR, Forbes SJ, Direkze NC, Poulsom R, Wright NA. The new stem cell biology: something for everyone. Mol Pathol 2003; 56(2):86–96.

14. Renehan AG, O'Dwyer ST, Haboubi NJ, Potten CS. Early cellular events in colorectal carcinogenesis. Colorectal Dis 2002; 4(2):76–89.

15. Marshman E, Booth C, Potten CS. The intestinal epithelial stem cell. Bioessays 2002; 24(1):91–98.

16. Leedham SJ, Brittan M, McDonald SA, Wright NA. Intestinal stem cells. J Cell Mol Med 2005; 9(1):11–24.

17. Clatworthy JP, Subramanian V. Stem cells and the regulation of proliferation, differentiation and patterning in the intestinal epithelium: emerging insights from gene expression patterns, transgenic and gene ablation studies. Mech Dev 2001; 101(1–2):3–9.

18. de Santa Barbara P, van den Brink GR, Roberts DJ. Development and differentiation of the intestinal epithelium. Cell Mol Life Sci 2003; 60(7):1322–1332.

19. Sancho E, Batlle E, Clevers H. Live and let die in the intestinal epithelium. Curr Opin Cell Biol 2003; 15(6):763–770.

20. Schonhoff SE, Giel-Moloney M, Leiter AB. Minireview: development and differentiation of gut endocrine cells. Endocrinology 2004; 145(6):2639–2644.

21. Potten CS. Structure, function and proliferative organisation of mammalian gut. In: Potten CS, Hendry JH, eds. Radiation and Gut. New York: Elsevier, 1995:1–31.

22. Potten CS, Roberts SA, Chwalinski S, Loeffler M, Paulus U. Scoring mitotic activity in longitudinal sections of crypts of the small intestine. Cell Tissue Kinet 1988; 21(4):231–246.

23. Ijiri K, Potten CS. Further studies on the response of intestinal crypt cells of different hierarchical status to eighteen different cytotoxic agents. Br J Cancer 1987; 55(2):113–123.

24. Potten CS, Owen G, Roberts SA. The temporal and spatial changes in cell proliferation within the irradiated crypts of the murine small intestine. Int J Radiat Biol 1990; 57(1):185–199.

25. Totafurno J, Bjerknes M, Cheng H. Variation in crypt size and its influence on the analysis of epithelial cell proliferation in the intestinal crypt. Biophys J 1988; 54(5):845–858.

26. Kellett M, Potten CS, Rew DA. A comparison of in vivo cell proliferation measurements in the intestine of mouse and man. Epithelial Cell Biol 1992; 1(4):147–155.

27. Potten CS, Kellett M, Roberts SA, Rew DA, Wilson GD. Measurement of in vivo proliferation in human colorectal mucosa using bromodeoxyuridine. Gut 1992; 33(1):71–78.

28. Potten CS, Booth C. The role of radiation-induced and spontaneous apoptosis in the homeostasis of the gastrointestinal epithelium: a brief review. Comp Biochem Physiol B Biochem Mol Biol 1997; 118(3):473–478.

29. Potten CS. What is an apoptotic index measuring? A commentary. Br J Cancer 1996; 74(11):1743–1748.

30. Potten CS. The significance of spontaneous and induced apoptosis in the gastrointestinal tract of mice. Cancer Metastasis Rev 1992; 11(2):179–195.

31. Renehan AG, Bach SP, Potten CS. The relevance of apoptosis for cellular homeostasis and tumorigenesis in the intestine. Can J Gastroenterol 2001; 15(3):166–176.

32. Potten CS, Wilson JW, Booth C. Regulation and significance of apoptosis in the stem cells of the gastrointestinal epithelium. Stem Cells 1997; 15(2):82–93.

33. Merritt AJ, Potten CS, Kemp CJ, Hickman JA, Balmain A, Lane DP, et al. The role of p53 in spontaneous and radiation-induced apoptosis in the gastrointestinal tract of normal and p53-deficient mice. Cancer Res 1994; 54(3):614–617.

34. Merritt AJ, Allen TD, Potten CS, Hickman JA. Apoptosis in small intestinal epithelial from p53-null mice: evidence for a delayed, p53-independent G2/M-associated cell death after gamma-irradiation. Oncogene 1997; 14(23):2759–2766.

35. Hickman JA. Apoptosis as a therapeutic target. In: Yarnold JR, Stratton MR, McMillan TJ, eds. Molecular Biology for Oncologists. London: Chapman & Hall, 1996:230–239.

36. Potten CS. Radiation, the ideal cytotoxic agent for studying the cell biology of tissues such as the small intestine. Radiat Res 2004; 161(2):123–136.

37. Merritt AJ, Potten CS, Watson AJ, Loh DY, Nakayama K, Nakayama K, et al. Differential expression of bcl-2 in intestinal epithelia: correlation with attenuation of apoptosis in colonic crypts and the incidence of colonic neoplasia. J Cell Sci 1995; 108(Pt 6):2261–2271.

38. Watson AJ, Merritt AJ, Jones LS, Askew JN, Anderson E, Becciolini A, et al. Evidence of reciprocity of bcl-2 and p53 expression in human colorectal adenomas and carcinomas. Br J Cancer 1996; 73(8):889–895.

39. Kitada S, Krajewska M, Zhang X, Scudiero D, Zapata JM, Wang HG, et al. Expression and location of pro-apoptotic Bcl-2 family protein BAD in normal human tissues and tumor cell lines. Am J Pathol 1998; 152(1):51–61.

40. Sherr CJ. The Pezcoller lecture: cancer cell cycles revisited. Cancer Res 2000; 60(14):3689–3695.

41. Fero ML, Rivkin M, Tasch M, Porter P, Carow CE, Firpo E, et al. A syndrome of multiorgan hyperplasia with features of gigantism, tumorigenesis, and female sterility in p27(Kip1)-deficient mice. Cell 1996; 85(5):733–744.

42. Wilson JW, Pritchard DM, Hickman JA, Potten CS. Radiation-induced p53 and p21WAF-1/CIP1 expression in the murine intestinal epithelium: apoptosis and cell cycle arrest. Am J Pathol 1998; 153(3):899–909.

43. el-Deiry WS, Tokino T, Waldman T, Oliner JD, Velculescu VE, Burrell M, et al. Topological control of p21WAF1/CIP1 expression in normal and neoplastic tissues. Cancer Res 1995; 55(13):2910–2919.

44. Chandrasekaran C, Coopersmith CM, Gordon JI. Use of normal and transgenic mice to examine the relationship between terminal differentiation of intestinal epithelial cells and accumulation of their cell cycle regulators. J Biol Chem 1996; 271(45):28414–28421.

45. Kaur P, Potten CS. Cell migration velocities in the crypts of the small intestine after cytotoxic insult are not dependent on mitotic activity. Cell Tissue Kinet 1986; 19(6):601–610.

46. Qiu JM, Roberts SA, Potten CS. Cell migration in the small and large bowel shows a strong circadian rhythm. Epithelial Cell Biol 1994; 3(4):137–148.

47. Al-Dewachi HS, Wright NA, Appleton DR, Watson AJ. The effect of a single injection of hydroxyurea on cell population kinetics in the small bowel mucosa of the rat. Cell Tissue Kinet 1977; 10(3):203–213.

48. Potten CS, Loeffler M. A comprehensive model of the crypts of the small intestine of the mouse provides insight into the mechanisms of cell migration and the proliferation hierarchy. J Theor Biol 1987; 127(4):381–391.

49. Loeffler M, Birke A, Winton D, Potten C. Somatic mutation, monoclonality and stochastic models of stem cell organization in the intestinal crypt. J Theor Biol 1993; 160(4): 471–491.

50. Spradling A, Drummond-Barbosa D, Kai T. Stem cells find their niche. Nature 2001; 414(6859):98–104.

51. Kai T, Spradling A. An empty *Drosophila* stem cell niche reactivates the proliferation of ectopic cells. Proc Natl Acad Sci USA 2003; 100(8):4633–4638.

52. Plateroti M, Rubin DC, Duluc I, Singh R, Foltzer-Jourdainne C, Freund JN, et al. Subepithelial fibroblast cell lines from different levels of gut axis display regional characteristics. Am J Physiol 1998; 274(5 Pt 1):G945–G954.

53. Powell DW, Mifflin RC, Valentich JD, Crowe SE, Saada JI, West AB. Myofibroblasts. II. Intestinal subepithelial myofibroblasts. Am J Physiol 1999; 277(2 Pt 1):C183–C201.

54. Potten CS, Booth C, Tudor GL, Booth D, Brady G, Hurley P et al. Identification of a putative intestinal stem cell and early lineage marker; musashi-1. Differentiation 2003; 71(1):28–41.

55. Nakamura M, Okano H, Blendy JA, Montell C. Musashi, a neural RNA-binding protein required for *Drosophila* adult external sensory organ development. Neuron 1994; 13(1):67–81.

56. Moser AR, Luongo C, Gould KA, McNeley MK, Shoemaker AR, Dove WF. ApcMin: a mouse model for intestinal and mammary tumorigenesis. Eur J Cancer 1995; 31A (7–8):1061–1064.

57. Kayahara T, Sawada M, Takaishi S, Fukui H, Seno H, Fukuzawa H, et al. Candidate markers for stem and early progenitor cells, Musashi-1 and Hes1, are expressed in crypt base columnar cells of mouse small intestine. FEBS Lett 2003; 535(1–3):131–135.

58. Nishimura S, Wakabayashi N, Toyoda K, Kashima K, Mitsufuji S. Expression of Musashi-1 in human normal colon crypt cells: a possible stem cell marker of human colon epithelium. Dig Dis Sci 2003; 48(8):1523–1529.

59. Withers HR, Elkind MM. Microcolony survival assay for cells of mouse intestinal mucosa exposed to radiation. Int J Radiat Biol Relat Stud Phys Chem Med 1970; 17(3):261–267.

60. Hendry JH, Roberts SA, Potten CS. The clonogen content of murine intestinal crypts: dependence on radiation dose used in its determination. Radiat Res 1992; 132(1):115–119.

61. Cai WB, Roberts SA, Potten CS. The number of clonogenic cells in crypts in three regions of murine large intestine. Int J Radiat Biol 1997; 71(5):573–579.

62. Potten CS. Extreme sensitivity of some intestinal crypt cells to X and gamma irradiation. Nature 1977; 269(5628):518–521.

63. Potten CS, Hendry JH, Moore JV. Estimates of the number of clonogenic cells in crypts of murine small intestine. Virchows Arch B Cell Pathol Incl Mol Pathol 1987; 53(4):227–234.

64. Loeffler M, Bratke T, Paulus U, Li YQ, Potten CS. Clonality and life cycles of intestinal crypts explained by a state-dependent stochastic model of epithelial stem cell organization. J Theor Biol 1997; 186(1):41–54.

65. Mills JC, Gordon JI. The intestinal stem cell niche: there grows the neighborhood. Proc Natl Acad Sci USA 2001; 98(22):12334–12336.

66. Winton DJ, Blount MA, Ponder BA. A clonal marker induced by mutation in mouse intestinal epithelium. Nature 1988; 333(6172):463–466.

67. Bjerknes M, Cheng H. Clonal analysis of mouse intestinal epithelial progenitors. Gastroenterology 1999; 116(1):7–14.

68. Ponder BA, Schmidt GH, Wilkinson MM, Wood MJ, Monk M, Reid A. Derivation of mouse intestinal crypts from single progenitor cells. Nature 1985; 313(6004):689–691.

69. Griffiths DF, Davies SJ, Williams D, Williams GT, Williams ED. Demonstration of somatic mutation and colonic crypt clonality by X-linked enzyme histochemistry. Nature 1988; 333(6172):461–463.

70. Park HS, Goodlad RA, Wright NA. Crypt fission in the small intestine and colon: a mechanism for the emergence of G6PD locus-mutated crypts after treatment with mutagens. Am J Pathol 1995; 147(5):1416–1427.

71. Bjerknes M, Cheng H. The stem-cell zone of the small intestinal epithelium. I. Evidence from Paneth cells in the adult mouse. Am J Anat 1981; 160(1):51–63.

72. Bjerknes M, Cheng H. The stem-cell zone of the small intestinal epithelium. II. Evidence from Paneth cells in the newborn mouse. Am J Anat 1981; 160(1):65–75.

73. Bjerknes M, Cheng H. The stem-cell zone of the small intestinal epithelium. III. Evidence from columnar, enteroendocrine, and mucous cells in the adult mouse. Am J Anat 1981; 160(1):77–91.

74. Bjerknes M, Cheng H. The stem-cell zone of the small intestinal epithelium. IV. Effects of resecting 30% of the small intestine. Am J Anat 1981; 160(1):93–103.

75. Bjerknes M, Cheng H. The stem-cell zone of the small intestinal epithelium. V. Evidence for controls over orientation of boundaries between the stem-cell zone, proliferative zone, and the maturation zone. Am J Anat 1981; 160(1):105–112.

76. Jass JR, Roberton AM. Colorectal mucin histochemistry in health and disease: a critical review. Pathol Int 1994; 44(7):487–504.

77. Fuller CE, Davies RP, Williams GT, Williams ED. Crypt restricted heterogeneity of goblet cell mucus glycoprotein in histologically normal human colonic mucosa: a potential marker of somatic mutation. Br J Cancer 1990; 61(3):382–384.

78. Campbell F, Williams GT, Appleton MA, Dixon MF, Harris M, Williams ED. Post-irradiation somatic mutation and clonal stabilisation time in the human colon. Gut 1996; 39(4):569–573.

79. Novelli MR, Williamson JA, Tomlinson IP, Elia G, Hodgson SV, Talbot IC, et al. Polyclonal origin of colonic adenomas in an XO/XY patient with FAP. Science 1996; 272(5265): 1187–1190.

80. Novelli M, Cossu A, Oukrif D, Quaglia A, Lakhani S, Poulsom R, et al. X-inactivation patch size in human female tissue confounds the assessment of tumor clonality. Proc Natl Acad Sci USA 2003; 100(6):3311–3314.

81. Brierley EJ, Johnson MA, Lightowlers RN, James OF, Turnbull DM. Role of mitochondrial DNA mutations in human aging: implications for the central nervous system and muscle. Ann Neurol 1998; 43(2):217–223.

82. Michikawa Y, Mazzucchelli F, Bresolin N, Scarlato G, Attardi G. Aging-dependent large accumulation of point mutations in the human mtDNA control region for replication. Science 1999; 286(5440):774–779.

83. Carew JS, Huang P. Mitochondrial defects in cancer. Mol Cancer 2002; 1(1):9.

84. Taylor RW, Barron MJ, Borthwick GM, Gospel A, Chinnery PF, Samuels DC et al. Mito-chondrial DNA mutations in human colonic crypt stem cells. J Clin Invest 2003; 112(9):1351–1360.

85. Pearse AG, Takor TT. Neuroendocrine embryology and the APUD concept. Clin Endocrinol (Oxf) 1976; 5(suppl):229S–244S.

86. Le Douarin N, Le Lievre C. Demonstration of neural origin of calcitonin cells of ultimobran-chial body of chick embryo. CR Acad Sci Hebd Seances Acad Sci D 1970; 270(23):2857–2860.

87. Holzer H. Cell lineages, stem cells and the "quantal cell cycle concept". In: Potten CS, Lord BI, Schofield R, eds. Stem Cells and Tissue Homeostasis. Cambridge: Cambridge University Press, 1978:1–27.

88. Cheng H, Leblond CP. Origin, differentiation and renewal of the four main epithelial cell types in the mouse small intestine. V. Unitarian theory of the origin of the four epithelial cell types. Am J Anat 1974; 141(4):537–561.

89. Potten CS, Schofield R, Lajtha LG. A comparison of cell replacement in bone marrow, testis and three regions of surface epithelium. Biochim Biophys Acta 1979; 560(2):281–299.

90. Karam SM, Leblond CP. Dynamics of epithelial cells in the corpus of the mouse stomach. I. Identification of proliferative cell types and pinpointing of the stem cell. Anat Rec 1993; 236(2):259–279.

91. Kirkland SC. Clonal origin of columnar, mucous, and endocrine cell lineages in human color-ectal epithelium. Cancer 1988; 61(7):1359–1363.

92. Cox WF Jr, Pierce GB. The endodermal origin of the endocrine cells of an adenocarcinoma of the colon of the rat. Cancer 1982; 50(8):1530–1538.

93. Henderson K, Kirkland SC. Multilineage differentiation of cloned HRA-19 cells in serum-free medium: a model of human colorectal epithelial differentiation. Differentiation 1996; 60(4):259–268.

94. Wong WM, Wright NA. Cell proliferation in gastrointestinal mucosa. J Clin Pathol 1999; 52(5):321–333.

95. St Clair WH, Osborne JW. Crypt fission and crypt number in the small and large bowel of postnatal rats. Cell Tissue Kinet 1985; 18(3):255–262.

96. Maskens AP, Dujardin-Loits RM. Kinetics of tissue proliferation in colorectal mucosa during postnatal growth. Cell Tissue Kinet 1981; 14(5):467–477.

97. Cairnie AB, Millen BH. Fission of crypts in the small intestine of the irradiated mouse. Cell Tissue Kinet 1975; 8(2):189–196.

98. Wright NA, Al-Nafussi A. The kinetics of villus cell populations in the mouse small intestine. II. Studies on growth control after death of proliferative cells induced by cytosine arabinoside, with special reference to negative feedback mechanisms. Cell Tissue Kinet 1982; 15(6): 611–621.

99. Totafurno J, Bjerknes M, Cheng H. The crypt cycle: crypt and villus production in the adult intestinal epithelium. Biophys J 1987; 52(2):279–294.

100. Bjerknes M. The crypt cycle and the asymptotic dynamics of the proportion of differently sized mutant crypt clones in the mouse intestine. Proc R Soc Lond B Biol Sci 1995; 260(1357):1–6.

101. Bjerknes M. Expansion of mutant stem cell populations in the human colon. J Theor Biol 1996; 178(4):381–385.

102. Kim KM, Shibata D. Tracing ancestry with methylation patterns: most crypts appear distantly related in normal adult human colon. BMC Gastroenterol 2004; 4(1):8.

103. Potten CS, Hume WJ, Reid P, Cairns J. The segregation of DNA in epithelial stem cells. Cell 1978; 15(3):899–906.

104. Cairns J. Mutation selection and the natural history of cancer. Nature 1975; 255(5505): 197–200.

105. Potten CS, Owen G, Booth D. Intestinal stem cells protect their genome by selective segregation of template DNA strands. J Cell Sci 2002; 115(Pt 11):2381–2388.

106. Merok JR, Lansita JA, Tunstead JR, Sherley JL. Cosegregation of chromosomes containing immortal DNA strands in cells that cycle with asymmetric stem cell kinetics. Cancer Res 2002; 62(23):6791–6795.

107. Smith GH. Label-retaining epithelial cells in mouse mammary gland divide asymmetrically and retain their template DNA strands. Development 2005; 132(4):681–687.

108. Cairns J. Somatic stem cells and the kinetics of mutagenesis and carcinogenesis. Proc Natl Acad Sci USA 2002; 99(16):10567–10570.

109. Heddle JA, Cosentino L, Dawod G, Swiger RR, Paashuis-Lew Y. Why do stem cells exist? Environ Mol Mutagen 1996; 28(4):334–341.

110. Schmidt GH, Winton DJ, Ponder BA. Development of the pattern of cell renewal in the crypt-villus unit of chimaeric mouse small intestine. Development 1988; 103(4):785–790.

111. Yatabe Y, Tavare S, Shibata D. Investigating stem cells in human colon by using methylation patterns. Proc Natl Acad Sci USA 2001; 98(19):10839–10844.

112. Vidrich A, Buzan JM, Cohn SM. Intestinal stem cells and mucosal gut development. Curr Opin Gastroenterol 2003; 19(6):583–590.

113. Jensen J, Pedersen EE, Galante P, Hald J, Heller RS, Ishibashi M, et al. Control of endodermal endocrine development by Hes-1. Nat Genet 2000; 24(1):36–44.

114. Yang Q, Bermingham NA, Finegold MJ, Zoghbi HY. Requirement of Math1 for secretory cell lineage commitment in the mouse intestine. Science 2001; 294(5549):2155–2158.

115. Stappenbeck TS, Gordon JI. Rac1 mutations produce aberrant epithelial differentiation in the developing and adult mouse small intestine. Development 2000; 127(12):2629–2642.

116. Korinek V, Barker N, Moerer P, van Donselaar E, Huls G, Peters PJ, et al. Depletion of epithelial stem-cell compartments in the small intestine of mice lacking Tcf-4. Nat Genet 1998; 19(4):379–383.

117. Bienz M, Clevers H. Linking colorectal cancer to Wnt signaling. Cell 2000; 103(2):311–320.

118. Bienz M. APC. Curr Biol 2003; 13(6):R215–R216.

119. Karim R, Tse G, Putti T, Scolyer R, Lee S. The significance of the Wnt pathway in the pathology of human cancers. Pathology 2004; 36(2):120–128.

120. Reya T, Clevers H. Wnt signalling in stem cells and cancer. Nature 2005; 434(7035): 843–850.
121. Radtke F, Clevers H. Self-renewal and cancer of the gut: two sides of a coin. Science 2005; 307(5717):1904–1909.
122. Pinto D, Clevers H. Wnt, stem cells and cancer in the intestine. Biol Cell 2005; 97(3): 185–196.
123. Sancho E, Batlle E, Clevers H. Signaling pathways in intestinal development and cancer. Annu Rev Cell Dev Biol 2004; 20:695–723.
124. Booth C, Brady G, Potten CS. Crowd control in the crypt. Nat Med 2002; 8(12):1360–1361.
125. Wong MH, Stappenbeck TS, Gordon JI. Living and commuting in intestinal crypts. Gastroenterology 1999; 116(1):208–210.
126. Staal FJ, van Noort M, Strous GJ, Clevers HC. Wnt signals are transmitted through N-terminally dephosphorylated beta-catenin. EMBO Rep 2002; 3(1):63–68.
127. de Lau W, Clevers H. LEF1 turns over a new leaf. Nat Genet 2001; 28(1):3–4.
128. Brantjes H, Barker N, van Es J, Clevers H. TCF: Lady Justice casting the final verdict on the outcome of Wnt signalling. Biol Chem 2002; 383(2):255–261.
129. Pelengaris S, Khan M, Evan G. c-MYC: more than just a matter of life and death. Nat Rev Cancer 2002; 2(10):764–776.
130. Issack PS, Ziff EB. Altered expression of helix–loop–helix transcriptional regulators and cyclin D1 in Wnt-1-transformed PC12 cells. Cell Growth Differ 1998; 9(10):837–845.
131. Blache P, van de Wetering M, Duluc I, Domon C, Berta P, Freund JN, et al. SOX9 is an intestine crypt transcription factor, is regulated by the Wnt pathway, and represses the CDX2 and MUC2 genes. J Cell Biol 2004; 166(1):37–47.
132. van Noort M, Meeldijk J, van der Zee R, Destree O, Clevers H. Wnt signaling controls the phosphorylation status of beta-catenin. J Biol Chem 2002; 277(20):17901–17905.
133. van de Wetering M, Sancho E, Verweij C, de Lau W, Oving I, Hurlstone A, et al. The beta-catenin/TCF-4 complex imposes a crypt progenitor phenotype on colorectal cancer cells. Cell 2002; 111(2):241–250.
134. Kim KM, Shibata D. Methylation reveals a niche: stem cell succession in human colon crypts. Oncogene 2002; 21(35):5441–5449.
135. Heinen CD, Goss KH, Cornelius JR, Babcock GF, Knudsen ES, Kowalik T, et al. The APC tumor suppressor controls entry into S-phase through its ability to regulate the cyclin D/RB pathway. Gastroenterology 2002; 123(3):751–763.
136. Glinka A, Wu W, Delius H, Monaghan AP, Blumenstock C, Niehrs C. Dickkopf-1 is a member of a new family of secreted proteins and functions in head induction. Nature 1998; 391(6665):357–362.
137. Lickert H, Kispert A, Kutsch S, Kemler R. Expression patterns of Wnt genes in mouse gut development. Mech Dev 2001; 105(1–2):181–184.
138. McBride HJ, Fatke B, Fraser SE. Wnt signaling components in the chicken intestinal tract. Dev Biol 2003; 256(1):18–33.
139. Batlle E, Henderson JT, Beghtel H, van den Born MM, Sancho E, Huls G, et al. Beta-catenin and TCF mediate cell positioning in the intestinal epithelium by controlling the expression of EphB/ephrinB. Cell 2002; 111(2):251–263.
140. Frisen J, Holmberg J, Barbacid M. Ephrins and their Eph receptors: multitalented directors of embryonic development. EMBO J 1999; 18(19):5159–5165.
141. Stephenson SA, Slomka S, Douglas EL, Hewett PJ, Hardingham JE. Receptor protein tyrosine kinase EphB4 is up-regulated in colon cancer. BMC Mol Biol 2001; 2(1):15.
142. Jenny M, Uhl C, Roche C, Duluc I, Guillermin V, Guillemot F, et al. Neurogenin3 is differentially required for endocrine cell fate specification in the intestinal and gastric epithelium. EMBO J 2002; 21(23):6338–6347.
143. Mao B, Wu W, Davidson G, Marhold J, Li M, Mechler BM, et al. Kremen proteins are Dickkopf receptors that regulate Wnt/beta-catenin signalling. Nature 2002; 417(6889):664–667.
144. Zheng JL, Shou J, Guillemot F, Kageyama R, Gao WQ. Hes1 is a negative regulator of inner ear hair cell differentiation. Development 2000; 127(21):4551–4560.

145. Katz JP, Perreault N, Goldstein BG, Lee CS, Labosky PA, Yang VW, et al. The zinc-finger transcription factor Klf4 is required for terminal differentiation of goblet cells in the colon. Development 2002; 129(11):2619–2628.

146. Ng AY, Waring P, Ristevski S, Wang C, Wilson T, Pritchard M, et al. Inactivation of the transcription factor Elf3 in mice results in dysmorphogenesis and altered differentiation of intestinal epithelium. Gastroenterology 2002; 122(5):1455–1466.

147. Zhu L, Skoultchi AI. Coordinating cell proliferation and differentiation. Curr Opin Genet Dev 2001; 11(1):91–97.

148. Deschenes C, Vezina A, Beaulieu JF, Rivard N. Role of p27(Kip1) in human intestinal cell differentiation. Gastroenterology 2001; 120(2):423–438.

149. Quaroni A, Tian JQ, Seth P, Ap Rhys C. p27(Kip1) is an inducer of intestinal epithelial cell differentiation. Am J Physiol Cell Physiol 2000; 279(4):C1045–C1057.

150. Yang W, Velcich A, Mariadason J, Nicholas C, Corner G, Houston M, et al. p21(WAF1/cip1) is an important determinant of intestinal cell response to sulindac in vitro and in vivo. Cancer Res 2001; 61(16):6297–6302.

151. Wong MH, Huelsken J, Birchmeier W, Gordon JI. Selection of multipotent stem cells during morphogenesis of small intestinal crypts of Lieberkühn is perturbed by stimulation of Lef-1/beta-catenin signaling. J Biol Chem 2002; 277(18):15843–15850.

152. Ambrosini G, Adida C, Altieri DC. A novel anti-apoptosis gene, survivin, expressed in cancer and lymphoma. Nat Med 1997; 3(8):917–921.

153. Ambrosini G, Adida C, Sirugo G, Altieri DC. Induction of apoptosis and inhibition of cell proliferation by survivin gene targeting. J Biol Chem 1998; 273(18):11177–11182.

154. Kim PJ, Plescia J, Clevers H, Fearon ER, Altieri DC. Survivin and molecular pathogenesis of colorectal cancer. Lancet 2003; 362(9379):205–209.

155. Nohe A, Keating E, Knaus P, Petersen NO. Signal transduction of bone morphogenetic protein receptors. Cell Signal 2004; 16(3):291–299.

156. Hardwick JC, van den Brink GR, Bleuming SA, Ballester I, van den Brande JM, Keller JJ, et al. Bone morphogenetic protein 2 is expressed by, and acts upon, mature epithelial cells in the colon. Gastroenterology 2004; 126(1):111–121.

157. Sayed MG, Ahmed AF, Ringold JR, Anderson ME, Bair JL, Mitros FA, et al. Germline SMAD4 or BMPR1A mutations and phenotype of juvenile polyposis. Ann Surg Oncol 2002; 9(9):901–906.

158. He XC, Zhang J, Tong WG, Tawfik O, Ross J, Scoville DH, et al. BMP signaling inhibits intestinal stem cell self-renewal through suppression of Wnt-beta-catenin signaling. Nat Genet 2004; 36(10):1117–1121.

159. Waite KA, Eng C. Protean PTEN: form and function. Am J Hum Genet 2002; 70(4):829–844.

160. Tian Q, He XC, Hood L, Li L. Bridging the BMP and Wnt pathways by PI3 kinase/Akt and 14-3-3zeta. Cell Cycle 2005; 4(2):215–216.

161. Wilson JW, Nostro MC, Balzi M, Faraoni P, Cianchi F, Becciolini A, et al. Bcl-w expression in colorectal adenocarcinoma. Br J Cancer 2000; 82(1):178–185.

162. Potten CS, Owen G, Hewitt D, Chadwick CA, Hendry H, Lord BI, et al. Stimulation and inhibition of proliferation in the small intestinal crypts of the mouse after in vivo administration of growth factors. Gut 1995; 36(6):864–873.

163. Booth D, Haley JD, Bruskin AM, Potten CS. Transforming growth factor-B3 protects murine small intestinal crypt stem cells and animal survival after irradiation, possibly by reducing stem-cell cycling. Int J Cancer 2000; 86(1):53–59.

164. Cui W, Fowlis DJ, Bryson S, Duffie E, Ireland H, Balmain A, et al. TGFbeta1 inhibits the formation of benign skin tumors, but enhances progression to invasive spindle carcinomas in transgenic mice. Cell 1996; 86(4):531–542.

165. Lickert H, Domon C, Huls G, Wehrle C, Duluc I, Clevers H, et al. Wnt/(beta)-catenin signaling regulates the expression of the homeobox gene Cdx1 in embryonic intestine. Development 2000; 127(17):3805–3813.

166. Walters JR. Cell and molecular biology of the small intestine: new insights into differentiation, growth and repair. Curr Opin Gastroenterol 2004; 20(2):70–76.

167. Kaestner KH, Bleckmann SC, Monaghan AP, Schlondorff J, Mincheva A, Lichter P, et al. Clustered arrangement of winged helix genes fkh-6 and MFH-1: possible implications for mesoderm development. Development 1996; 122(6):1751–1758.

168. Kaestner KH, Knochel W, Martinez DE. Unified nomenclature for the winged helix/forkhead transcription factors. Genes Dev 2000; 14(2):142–146.

169. Kaestner KH, Silberg DG, Traber PG, Schutz G. The mesenchymal winged helix transcription factor Fkh6 is required for the control of gastrointestinal proliferation and differentiation. Genes Dev 1997; 11(12):1583–1595.

170. Hay T, Patrick T, Winton D, Sansom OJ, Clarke AR. Brca2 deficiency in the murine small intestine sensitizes to p53-dependent apoptosis and leads to the spontaneous deletion of stem cells. Oncogene 2005; 28:28.

171. Madison BB, Braunstein K, Kuizon E, Portman K, Qiao XT, Gumucio DL. Epithelial hedgehog signals pattern the intestinal crypt-villus axis. Development 2005; 132(2): 279–289.

172. Mills JC, Andersson N, Hong CV, Stappenbeck TS, Gordon JI. Molecular characterization of mouse gastric epithelial progenitor cells. Proc Natl Acad Sci USA 2002; 99(23):14819–14824.

173. Stappenbeck TS, Mills JC, Gordon JI. Molecular features of adult mouse small intestinal epithelial progenitors. Proc Natl Acad Sci USA 2003; 100(3):1004–1009.

174. Mariadason JM, Nicholas C, L'Italien KE, Zhuang M, Smartt HJ, Heerdt BG, et al. Gene expression profiling of intestinal epithelial cell maturation along the crypt-villus axis. Gastroenterology 2005; 128(4):1081–1088.

175. Hill MJ, Morson BC, Bussey HJ. Aetiology of adenoma—carcinoma sequence in large bowel. Lancet 1978; 1(8058):245–247.

176. Vogelstein B, Fearon ER, Hamilton SR, Kern SE, Preisinger AC, Leppert M, et al. Genetic alterations during colorectal-tumor development. N Engl J Med 1988; 319(9):525–532.

177. Fearon ER, Vogelstein B. A genetic model for colorectal tumorigenesis. Cell 1990; 61(5):759–767.

178. Kinzler KW, Vogelstein B. Cancer-susceptibility genes: gatekeepers and caretakers. Nature 1997; 386(6627):761, 763.

179. Vogelstein B, Kinzler KW. Cancer genes and the pathways they control. Nat Med 2004; 10(8):789–799.

180. Fodde R. The APC gene in colorectal cancer. Eur J Cancer 2002; 38(7):867–871.

181. Lamlum H, Papadopoulou A, Ilyas M, Rowan A, Gillet C, Hanby A, et al. APC mutations are sufficient for the growth of early colorectal adenomas. Proc Natl Acad Sci USA 2000; 97(5):2225–2228.

182. Kondo Y, Issa JP. Epigenetic changes in colorectal cancer. Cancer Metastasis Rev 2004; 23(1–2):29–39.

183. Lynch HT, de la Chapelle A. Hereditary colorectal cancer. N Engl J Med 2003; 348(10): 919–932.

184. Lipton L, Tomlinson I. The multiple colorectal adenoma phenotype and MYH, a base excision repair gene. Clin Gastroenterol Hepatol 2004; 2(8):633–638.

185. Fodde R, Smits R, Clevers H. APC, signal transduction and genetic instability in colorectal cancer. Nat Rev Cancer 2001; 1(1):55–67.

186. Nowak MA, Komarova NL, Sengupta A, Jallepalli PV, Shih IM, Vogelstein B, et al. The role of chromosomal instability in tumor initiation. Proc Natl Acad Sci USA 2002; 99(25):16226–16231.

187. Morin PJ, Sparks AB, Korinek V, Barker N, Clevers H, Vogelstein B, et al. Activation of beta-catenin-Tcf signaling in colon cancer by mutations in beta-catenin or APC. Science 1997; 275(5307):1787–1790.

188. Hao X, Frayling IM, Willcocks TC, Han W, Tomlinson IP, Pignatelli MN, et al. Beta-catenin expression and allelic loss at APC in sporadic colorectal carcinogenesis. Virchows Arch 2002; 440(4):362–366.

189. Brabletz T, Herrmann K, Jung A, Faller G, Kirchner T. Expression of nuclear beta-catenin and c-myc is correlated with tumor size but not with proliferative activity of colorectal adenomas. Am J Pathol 2000; 156(3):865–870.

190. Shih IM, Wang TL, Traverso G, Romans K, Hamilton SR, Ben-Sasson S, et al. Top-down morphogenesis of colorectal tumors. Proc Natl Acad Sci 2001; 98(5):2640–2645.

191. Bird RP, McLellan EA, Bruce WR. Aberrant crypts, putative precancerous lesions, in the study of the role of diet in the aetiology of colon cancer. Cancer Surv 1989; 8(1):189–200.

192. Roncucci L, Pedroni M, Vaccina F, Benatti P, Marzona L, De Pol A. Aberrant crypt foci in colorectal carcinogenesis: cell and crypt dynamics. Cell Prolif 2000; 33(1):1–18.

193. Liu W, Dong X, Mai M, Seelan RS, Taniguchi K, Krishnadath KK, et al. Mutations in AXIN2 cause colorectal cancer with defective mismatch repair by activating beta-catenin/TCF signalling. Nat Genet 2000; 26(2):146–147.

194. Satoh S, Daigo Y, Furukawa Y, Kato T, Miwa N, Nishiwaki T, et al. AXIN1 mutations in hepatocellular carcinomas, and growth suppression in cancer cells by virus-mediated transfer of AXIN1. Nat Genet 2000; 24(3):245–250.

195. Lamlum H, Ilyas M, Rowan A, Clark S, Johnson V, Bell J, et al. The type of somatic mutation at APC in familial adenomatous polyposis is determined by the site of the germline mutation: a new facet to Knudson's 'two-hit' hypothesis. Nat Med 1999; 5(9):1071–1075.

196. Sansom OJ, Reed KR, Hayes AJ, Ireland H, Brinkmann H, Newton IP, et al. Loss of Apc in vivo immediately perturbs Wnt signaling, differentiation, and migration. Genes Dev 2004; 18(12):1385–1390.

197. Sieber OM, Heinimann K, Gorman P, Lamlum H, Crabtree M, Simpson CA, et al. Analysis of chromosomal instability in human colorectal adenomas with two mutational hits at APC. Proc Natl Acad Sci USA 2002; 99(26):16910–16915.

198. Fodde R, Kuipers J, Rosenberg C, Smits R, Kielman M, Gaspar C, et al. Mutations in the APC tumour suppressor gene cause chromosomal instability. Nat Cell Biol 2001; 3(4): 433–438.

199. Hassan AB, Howell JA. Insulin-like growth factor II supply modifies growth of intestinal adenoma in Apc(Min/+) mice. Cancer Res 2000; 60(4):1070–1076.

200. Sakatani T, Kaneda A, Iacobuzio-Donahue CA, Carter MG, de Boom Witzel S, Okano H, et al. Loss of imprinting of Igf2 alters intestinal maturation and tumorigenesis in mice. Science 2005; 307(5717):1976–1978.

201. Wong WM, Garcia SB, Wright NA. Origins and morphogenesis of colorectal neoplasms. Apmis 1999; 107(6):535–544.

202. Fearon ER, Hamilton SR, Vogelstein B. Clonal analysis of human colorectal tumors. Science 1987; 238(4824):193–197.

203. Fialkow PJ. Clonal origin of human tumors. Biochim Biophys Acta 1976; 458(3):283–321.

204. Garcia SB, Novelli M, Wright NA. The clonal origin and clonal evolution of epithelial tumours. Int J Exp Pathol 2000; 81(2):89–116.

205. Brickman JM, Burdon TG. Pluripotency and tumorigenicity. Nat Genet 2002; 32(4): 557–558.

206. Kielman MF, Rindapaa M, Gaspar C, Van Poppel N, Breukel C, Van Leeuwen S, et al. Apc modulates embryonic stem-cell differentiation by controlling the dosage of beta-catenin signaling. Nat Genet 2002; 32(4):594–605.

207. Gaspar C, Fodde R. APC dosage effects in tumorigenesis and stem cell differentiation. Int J Dev Biol 2004; 48(5–6):377–386.

208. Merritt AJ, Gould KA, Dove WF. Polyclonal structure of intestinal adenomas in ApcMin/+ mice with concomitant loss of Apc+ from all tumor lineages. Proc Natl Acad Sci USA 1997; 94(25):13927–13931.

209. Playford RJ. Tales from the human crypt—intestinal stem cell repertoire and the origins of human cancer. J Pathol 1998; 185(2):119–122.

210. Deschner EE. Early proliferative changes in gastrointestinal neoplasia. Am J Gastroenterol 1982; 77(4):207–211.

211. Itzkowitz SH, Kim YS. Polyp and benign neoplasms of the colon. In: Sleisenger MH, Fordtran JS, Scharschmidt BF, Feldman M, eds. Gastrointestinal Disease: Pathophysiology/ Diagnosis/Management. Philadelphia: Saunders WB, 1993:1402–1430.

212. Risio M. Cell proliferation in colorectal tumor progression: an immunohistochemical approach to intermediate biomarkers. J Cell Biochem Suppl 1992; 16g:79–87.

213. Polyak K, Hamilton SR, Vogelstein B, Kinzler KW. Early alteration of cell-cycle-regulated gene expression in colorectal neoplasia. Am J Pathol 1996; 149(2):381–387.

214. Sinicrope FA, Roddey G, Lemoine M, Ruan S, Stephens LC, Frazier ML, et al. Loss of p21WAF1/Cip1 protein expression accompanies progression of sporadic colorectal neoplasms but not hereditary nonpolyposis colorectal cancers. Clin Cancer Res 1998; 4(5):1251–1261.

215. Sinicrope FA, Ruan SB, Cleary KR, Stephens LC, Lee JJ, Levin B. bcl-2 and p53 oncoprotein expression during colorectal tumorigenesis. Cancer Res 1995; 55(2):237–241.

216. Sinicrope FA, Roddey G, McDonnell TJ, Shen Y, Cleary KR, Stephens LC. Increased apoptosis accompanies neoplastic development in the human colorectum. Clin Cancer Res 1996; 2(12):1999–2006.

217. Maskens AP. Histogenesis of adenomatous polyps in the human large intestine. Gastroenterology 1979; 77(6):1245–1251.

218. Nigro ND, Bull AW. Experimental intestinal carcinogenesis. Br J Surg 1985; 72(suppl): S36–S37.

219. Moss SF, Liu TC, Petrotos A, Hsu TM, Gold LI, Holt PR. Inward growth of colonic adenomatous polyps. Gastroenterology 1996; 111(6):1425–1432.

220. Shiff SJ, Rigas B. Colon adenomatous polyps—do they grow inward? Lancet 1997; 349(9069):1853–1854.

221. Lamprecht SA, Lipkin M. Migrating colonic crypt epithelial cells: primary targets for transformation. Carcinogenesis 2002; 23(11):1777–1780.

222. Nakamura S, Kino I. Morphogenesis of minute adenomas in familial polyposis coli. J Natl Cancer Inst 1984; 73(1):41–49.

223. Woda BA, Forde K, Lane N. A unicryptal colonic adenoma, the smallest colonic neoplasm yet observed in a non-polyposis individual. Am J Clin Pathol 1977; 68(5):631–632.

224. Campbell F, Geraghty JM, Appleton MA, Williams ED, Williams GT. Increased stem cell somatic mutation in the non-neoplastic colorectal mucosa of patients with familial adenomatous polyposis. Hum Pathol 1998; 29(12):1531–1535.

225. Chang WW, Whitener CJ. Histogenesis of tubular adenomas in hereditary colonic adenomatous polyposis. Arch Pathol Lab Med 1989; 113(9):1042–1049.

226. Wasan HS, Park HS, Liu KC, Mandir NK, Winnett A, Sasieni P, et al. APC in the regulation of intestinal crypt fission. J Pathol 1998; 185(3):246–255.

227. Smith AJ, Stern HS, Penner M, Hay K, Mitri A, Bapat BV, et al. Somatic APC and K-ras codon 12 mutations in aberrant crypt foci from human colons. Cancer Res 1994; 54(21):5527–5530.

228. Fujimitsu Y, Nakanishi H, Inada K, Yamachika T, Ichinose M, Fukami H, et al. Development of aberrant crypt foci involves a fission mechanism as revealed by isolation of aberrant crypts. Jpn J Cancer Res 1996; 87(12):1199–1203.

229. Siu IM, Robinson DR, Schwartz S, Kung HJ, Pretlow TG, Petersen RB, et al. The identification of monoclonality in human aberrant crypt foci. Cancer Res 1999; 59(1):63–66.

230. Araki K, Ogata T, Kobayashi M, Yatani R. A morphological study on the histogenesis of human colorectal hyperplastic polyps. Gastroenterology 1995; 109(5):1468–1474.

231. Wright NA, Poulsom R. Top-down or bottom-up? Competing management structures in the morphogenesis of colorectal neoplasms. Gut 2002; 51(3):306–308.

232. Boman BM, Fields JZ, Bonham Carter O, Runquist OA. Computer modeling implicates stem cell overproduction in colon cancer initiation. Cancer Res 2001; 61(23):8408–8411.

233. Kim KM, Calabrese P, Tavare S, Shibata D. Enhanced stem cell survival in familial adenomatous polyposis. Am J Pathol 2004; 164(4):1369–1377.

234. Zhang T, Otevrel T, Gao Z, Ehrlich SM, Fields JZ, Boman BM. Evidence that APC regulates survivin expression: a possible mechanism contributing to the stem cell origin of colon cancer. Cancer Res 2001; 61(24):8664–8667.

235. Garcia SB, Park HS, Novelli M, Wright NA. Field cancerization, clonality, and epithelial stem cells: the spread of mutated clones in epithelial sheets. J Pathol 1999; 187(1):61–81.

236. Preston SL, Wong WM, Chan AO, Poulsom R, Jeffery R, Goodlad RA, et al. Bottom-up histogenesis of colorectal adenomas: origin in the monocryptal adenoma and initial expansion by crypt fission. Cancer Res 2003; 63(13):3819–3825.

237. Wong WM, Mandir N, Goodlad RA, Wong BC, Garcia SB, Lam SK, et al. Histogenesis of human colorectal adenomas and hyperplastic polyps: the role of cell proliferation and crypt fission. Gut 2002; 50(2):212–217.

238. Renehan AG, Booth C, Potten CS. What is apoptosis, and why is it important? BMJ 2001; 322(7301):1536–1538.

239. Yang WC, Mathew J, Velcich A, Edelmann W, Kucherlapati R, Lipkin M, et al. Targeted inactivation of the p21(WAF1/cip1) gene enhances Apc-initiated tumor formation and the tumor-promoting activity of a Western-style high-risk diet by altering cell maturation in the intestinal mucosal. Cancer Res 2001; 61(2):565–569.

240. Reya T, Morrison SJ, Clarke MF, Weissman IL. Stem cells, cancer, and cancer stem cells. Nature 2001; 414(6859):105–111.

241. Luu HH, Zhang R, Haydon RC, Rayburn E, Kang Q, Si W, et al. Wnt/beta-catenin signaling pathway as a novel cancer drug target. Curr Cancer Drug Targets 2004; 4(8):653–671.

242. Dihlmann S, von Knebel Doeberitz M. Wnt/beta-catenin-pathway as a molecular target for future anti-cancer therapeutics. Int J Cancer 2005; 113(4):515–524.

243. Potten CS, Booth C, Hargreaves D. The small intestine as a model for evaluating adult tissue stem cell drug targets. Cell Prolif 2003; 36(3):115–129.

244. Prockop DJ, Gregory CA, Spees JL. One strategy for cell and gene therapy: harnessing the power of adult stem cells to repair tissues. Proc Natl Acad Sci USA 2003; 100(suppl 1): 11917–11923.

245. Kawaguchi AL, Dunn JC, Fonkalsrud EW. In vivo growth of transplanted genetically altered intestinal stem cells. J Pediatr Surg 1998; 33(4):559–563.

246. Wang G, Bunnell BA, Painter RG, Quiniones BC, Tom S, Lanson NA Jr, et al. Adult stem cells from bone marrow stroma differentiate into airway epithelial cells: potential therapy for cystic fibrosis. Proc Natl Acad Sci USA 2005; 102(1):186–191.

247. Cui H, Horon IL, Ohlsson R, Hamilton SR, Feinberg AP. Loss of imprinting in normal tissue of colorectal cancer patients with microsatellite instability. Nat Med 1998; 4(11): 1276–1280.

248. Cui H, Cruz-Correa M, Giardiello FM, Hutcheon DF, Kafonek DR, Brandenburg S, et al. Loss of IGF2 imprinting: a potential marker of colorectal cancer risk. Science 2003; 299(5613):1753–1755.

249. St Croix B, Rago C, Velculescu V, Traverso G, Romans KE, Montgomery E, et al. Genes expressed in human tumor endothelium. Science 2000; 289(5482):1197–1202.

250. Renehan AG, Painter JE, O'Halloran D, Atkin WS, Potten CS, O'Dwyer ST, et al. Circulating insulin-like growth factor II and colorectal adenomas. J Clin Endocrinol Metab 2000; 85(9):3402–3408.

251. Krause DS, Theise ND, Collector MI, Henegariu O, Hwang S, Gardner R, et al. Multi-organ, multi-lineage engraftment by a single bone marrow-derived stem cell. Cell 2001; 105(3):369–377.

252. Jiang Y, Jahagirdar BN, Reinhardt RL, Schwartz RE, Keene CD, Ortiz-Gonzalez XR, et al. Pluripotency of mesenchymal stem cells derived from adult marrow. Nature 2002; 418(6893):41–49.

253. Matsumoto T, Okamoto R, Yajima T, Mori T, Okamoto S, Ikeda Y, et al. Increase of bone marrow-derived secretory lineage epithelial cells during regeneration in the human intestine. Gastroenterology 2005; 128(7):1851–1867.

254. Houghton J, Stoicov C, Nomura S, Rogers AB, Carlson J, Li H, et al. Gastric cancer originating from bone marrow-derived cells. Science 2004; 306(5701):1568–1571.

255. Marx J. Medicine. Bone marrow cells: the source of gastric cancer? Science 2004; 306(5701):1455–1457.

12

Stem Cells in Neurodegeneration and Injury

Reaz Vawda, Nigel L. Kennea, and Huseyin Mehmet
Institute of Reproductive and Developmental Biology, Imperial College, London, London, U.K.

INTRODUCTION

One in four people worldwide suffer some form of neurodegenerative disorder. The World Health Organization estimates that there are currently four million people worldwide with Parkinson's disease (PD), 37 million with Alzheimer's disease (AD) (4.5 million in the United States), and 5.5 million die each year as a result of cerebrovascular events. In the United States, more than 50 million people are affected by various central nervous system (CNS) diseases. Each year, 11,000 people sustain spinal cord injury, adding to the 400,000 or so already affected. Two million people have been disabled by head injuries with 1.5 million people a year suffering traumatic brain injury (TBI), adding to the 5.3 million already living with disabilities resulting from TBI (http://www.who.int). With an aging global population, the number of people with neurodegenerative and cerebrovascular conditions continues to grow, as does the cost to the health service.

The adult CNS has only a limited capacity for self-repair, although varying degrees of functional recovery are achievable, often with little or no clinical intervention. As the brain matures, it loses its ability to support the growth of axons and consists of a cellular environment largely dominated by growth-inhibitory molecules, including myelin (1,2). Because neuronal loss is at the core of most neurodegenerative conditions, clinical therapy has until recently been limited to their symptomatic relief. Cellular replacement therapy is aimed at restoring neural circuits damaged either by trauma or neurodegeneration. Different strategies have been attempted both in the laboratory and at the clinical level, especially for PD (given its localized and well-characterized etiology), which has served as a platform for testing cell-based restorative therapies. These include the implantation of adrenal cells, Sertoli cells, and fetal mesencephalic cells (3). Unlike their less developed counterparts, mature or differentiated neural cells are mitotically quiescent and do not survive well after intracerebral implantation. Therefore, alternative sources of expandable cells are needed for regenerative therapy. The use of human fetal mesencephalic tissue, despite providing some promising early results, is constrained by ethical and practical concerns surrounding the availability and survival of CNS tissue from elective

271

abortions. The use of fetal mesencephalic tissue of porcine origin (xenotransplantation) has been attempted in some patients with late-stage PD (4), but the immunological constraints and unsatisfactory benefit have prompted a radical rethink of this approach. Stem cells from a range of sources have the potential to overcome these obstacles.

STEM CELLS

Although a universal definition of stem cells has yet to be agreed, the following criteria represent the current consensus: they are karyotypically normal, undifferentiated cells (lacking a specific morphology and not expressing antigens of mature cells) with extensive proliferative capacity, long-term self-renewal, and pluripotency (that is, capable of giving rise to multiple types of cell lineages). It is not always possible to assess all these parameters simultaneously, although a number of antigenic markers have been used to characterize and detect these cells (Table 1) (5).

Pluripotent stem cells can be derived from embryonic, fetal, and adult tissues (Fig. 1). At least three different types of mammalian stem cells have been identified in the embryo: embryonal carcinoma cells, embryonic stem (ES) cells derived from the inner cell mass of blastocysts, and embryonic germ cells obtained from postimplantation embryos. In the early 1990s, several groups (13,14) reported the existence of a subset of stem cells in the CNS. However, they were more restricted in their differentiation potential than ES cells, giving rise predominantly to the three major cell types of the CNS: neurones, astrocytes, and oligodendrocytes and were therefore named neural stem cells (NSCs) (15). These can be isolated from both fetal and adult CNS.

More recently, stem cells referred to as "mesenchymal stem cells" (MSCs) with neurogenic potential have been derived from non-neural tissues such as blood, bone marrow (BM), umbilical cord matrix Wharton's jelly (WJ), liver, skin, muscle, and adipose tissue. MSCs are derived from the developing mesoderm of the embryo and give rise to connective tissue in the adult, which retains a population of MSCs throughout life (16).

Table 1 Major Antigenic Markers of Undifferentiated Stem Cells—ES cells, NSCs, and MSCs

Marker type	Designation	References
ES cells	Oct-4	(6,7)
	SSEA-1, SSEA-3, SSEA-4	(8,9)
	TRA-1-60, TRA-1-81	(10,11)
	nanog	(12)
NSCs	Nestin,	(5)
	PSA-NCAM	
	Sox1, Sox2	
	Bcrp1	
MSCs	CD105 (SH2 antibody), CD73 (SH3 and SH4 antibodies)	(5)
	STRO-1	
	α-smooth muscle actin	
	prolyl-4 hydroxylase	

Note: Neural induction results in a down-regulation of these markers and an up-regulation of neural markers.
Abbreviations: ES, embryonic stem; NSCs, neural stem cells; MSCs, mesenchymal stem cells.
Source: From Ref. 5.

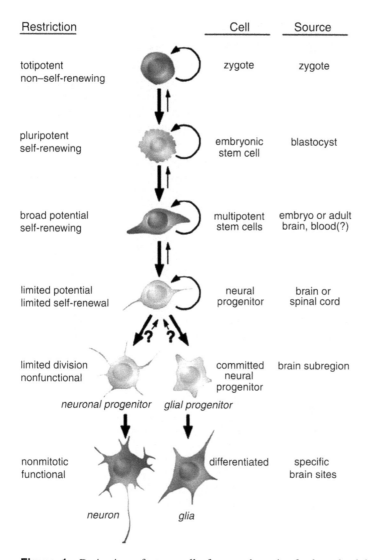

Restriction		Cell	Source

totipotent non–self-renewing — zygote — zygote

pluripotent self-renewing — embryonic stem cell — blastocyst

broad potential self-renewing — multipotent stem cells — embryo or adult brain, blood(?)

limited potential limited self-renewal — neural progenitor — brain or spinal cord

limited division nonfunctional — committed neural progenitor — brain subregion

neuronal progenitor *glial progenitor*

nonmitotic functional — differentiated — specific brain sites

neuron *glia*

Figure 1 Derivation of stem cells from embryonic, fetal, and adult tissues. *Source*: From Ref. 16.

Embryonic Stem Cells

ES cells are totipotent (can give rise to all tissues in the body, including those of the nervous system) (17) and, as such, are a promising source of material for therapeutic applications. They can be propagated in vitro and can be engineered to express therapeutic genes. ES cells can be cultured as floating aggregates called embryoid bodies and retain their ability to differentiate into cell types of all three germ layers. The first demonstration that mouse ES cells can be differentiated into multiple neural phenotypes in culture was reported by Bain and colleagues (18), using retinoic acid. The newly formed neurones not only expressed lineage-specific markers, but were also capable of generating action potentials. Several groups have now enriched neural progenitors from murine and human ES cells (19,20).

Neural Stem Cells—Fetal and Adult

NSCs are found in both the developing (embryonic and neonatal) and adult mammalian CNS mostly in two main active neurogenic germinal zones: the subgranular zone of the dentate gyrus (which generates hippocampal interneurones) and the subventricular zone (SVZ) (Fig. 2). The main source of mature neurones and glia in the CNS seems to be the undifferentiated precursor cell population of the embryonic germinal periventricular neuroepithelium (21,22). Adult mammalian CNS non-neurogenic regions have also been reported to contain small numbers of stem-like cells, including the spinal cord (which has important clinical implications for spinal cord repair), where neurogenesis has not been described in the adult (Fig. 2), the septum, and striatal parenchyma (23). NSCs from different CNS regions have different growth factor requirements for the maintenance of their undifferentiated and proliferative state. The diagram below summarizes the various locations in the rodent CNS from which NSCs have been isolated.

Although the bulk of experimental data has been obtained using rodent NSCs, similar multipotent cells have been identified in the human, with the antigenic phenotype $CD133^+$, $CD34^-$, $CD45^-$, $CD24^{-/lo}$, and $5E12^+$. These have been shown to generate neurospheres, to self-renew, and to differentiate into neurones and glia (24,25). When these cells are injected into the lateral ventricles of immunodeficient newborn NOD/SCID mice, they show engraftment, migration, and region-specific neuronal differentiation up to seven months later (26).

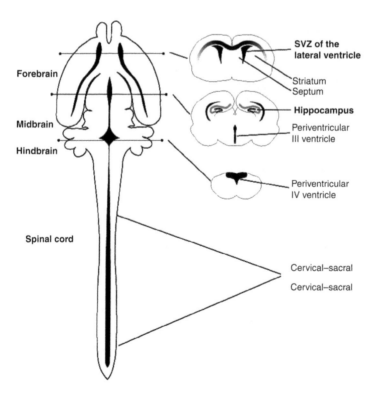

Figure 2   Anatomical distribution of NSCs in the rodent CNS, and their growth factor responsiveness. *Abbreviations*: NSCs, neural stem cells; CNS, central nervous system; SVZ, subventricular zone. *Source*: From Ref. 23.

There are important similarities between the in vivo and culture properties of rodent and human NSCs, but there are also notable differences (27). One of the most evident is that differentiation of rat neurospheres generates large proportions of oligodendrocytes, whereas human NSCs generate relatively few. The reason for this is not clear and more work is needed to determine the signals that influence the cell phenotype, and indeed what the specific cues are for differentiation and maturation.

Another potential source of human NSCs is the adult brain, and stem cells have even been cultured from human cadavers up to five days after death (28). At present, NSCs derived from adults are difficult to isolate, found in smaller numbers, have a more limited proliferative capacity, and seem to have a more restricted differentiation potential than their fetal counterparts. In the light of current knowledge, it appears that fetal NSCs are likely to be more useful for cell therapy. However, the highly controversial nature of this source of neural tissue has provided an impetus for understanding adult NSCs in more detail. Whether fetal or adult NSCs are used in cerebral transplants, little is known about the cell-intrinsic and external factors that influence proliferative capacity and fate choice. By careful examination of the effects of defined neurotrophic factors and accurate definition of the factors in the transplant microenvironment, it may well be possible to improve the potential of NSC therapy in the future.

Although their discovery and isolation is useful for the understanding of brain development and repair, the basic biology of NSCs is still not well understood. Indeed, the unequivocal identification of true NSC remains difficult, as there are no specific markers for this cell type. Nestin, for example, is a widely used marker and is highly expressed in the developing neuroepithelium and NSCs (29,30), but is also found in other cell types such as endothelial cells, developing myoblasts, and reactive astrocytes (31). Other markers used to define NSCs include Musashi, *tai-ji*, and notch-1 (32–34). However, none of these markers is exclusive or definitive. Moreover, the apparent continuum of stem, precursor, and progenitor cell subtypes hinders a straightforward classification of cell lineages (35). Therefore, the identification of neural stem cells in culture relies on their functional attributes, including their growth factor responsiveness, their multilineage differentiation potential (ability to generate one or more cell types), and their capacity for proliferative self-renewal (36). In vivo, the survival and migration of implanted cells can be measured using magnetic resonance microscopy, complemented by the identification of cells at matched time points by conventional immunohistochemistry (37).

When considering NSCs for replacement therapies, it is important to recognize that cells from different gestational ages and anatomical sites are not identical, displaying different growth characteristics, trophic factor requirements, and specific patterns of differentiation (38–42). The gene expression of isolated neural stem cells seems unaltered within the neurospheres they generate in culture (43–46). Rodent neural stem cells express developmentally regulated genes in vitro, such as the *Pax* family, which is influenced by the ECM molecules to which they are exposed (43–46). The spatio-temporal variations in NSC growth factor receptor expression (especially epidermal growth factor receptor and fibroblast growth factor receptor) also seem to influence NSC fate, as might do differences in receptor expression levels (47,48).

Stem Cells from Non-neural Tissues

Recent studies have suggested that MSCs from certain adult and fetal tissues have the potential both in vitro and in vivo to exhibit phenotypic characteristics of cells not expected within the tissue of origin (49,50), including neural phenotypes. These tissues include BM (51), peripheral blood (PB) (52–54), umbilical cord blood (55–57), umbilical cord matrix (WJ) cells (58,59), amniotic fluid (60), amniotic epithelium (61), endothelium (62,63), adipose tissue (64,65), muscle (66,67), liver (68), dental pulp (69) and skin (70). As well as representing a plentiful, ethically acceptable, and easily accessible source of neural tissue for therapeutic brain and spinal cord repair and regeneration, these cells could potentially be obtained from the very patient receiving the intracerebral cell graft, thereby reducing the risk of tissue rejection. To date, however, little is known about the sources, frequency, and characteristics of cells with the potential to adopt neural lineages and the mechanisms by which they are generated outside the nervous system. This chapter will focus on MSCs isolated from BM and WJ.

Bone Marrow

Adult hematopoiesis (the generation of blood cells) takes place in the BM (located within the vascular sinuses of flat and short bones), which consists of two stem-cell types: hematopoietic (HSCs) and mesenchymal (MSCs) (71). HSCs self-renew and continually generate blood cells throughout life. They express CD34, CD45, c-kit/CD117, and HLA-DR; are typically non-adherent in cell culture; and can be maintained in long-term culture using a bone marrow stromal cell (BMSC) feeder layer. Until recently, they were thought of as tissue-specific stem cells able to differentiate into blood-lineage cells only. However, recent evidence suggests that HSCs may have greater differentiation potential (72,73).

    BMSCs are non-hematopoietic cells, also referred to as colony forming unit fibroblasts, which provide the structural and functional support for the generation of blood lineages from HSCs (51). They also have the potential to differentiate into a range of morphologically and biochemically distinct cell types from all three germ layers (ectoderm, mesoderm, and endoderm), including adipocytes, osteoblasts, macrophages, chondrocytes, tendon, hepatocytes, muscle, cardiac myocytes, endothelial cells, and neural cells (51,73). BMSCs are present within the BM at a very low frequency (less than one per million mononuclear cells) (74,75), hence the need for expanding them in culture before phenotypic characterization and implantation. The different enrichment and expansion methods used have been shown not to influence the immunophenotype and proliferation rate of BMSCs (76). Due to their expression of several adhesion-related antigens, including the integrin subunits $\alpha4$, $\alpha5$, $\beta1$, integrins $\alpha v \beta3$ and $\alpha v \beta5$, ICAM-1, and CD44H, proliferative BMSCs can be enriched or purified by spontaneous adherence to tissue culture plastic, but only in the presence of serum. This forms the basis on which they were first isolated by Friedenstein et al. in the early 1970s (77–85). There is as yet no single specific antigenic marker for BMSCs. However, they express CD29, CD44, and CD166, but lack HSC-associated markers. They do not express antigenic markers of mature blood lineages either: CD14 (monocytes and macrophages), CD31 (endothelial cells), and CD11a (lymphocytes). The replacement of serum with a defined formulation for BMSC expansion in vitro is an important challenge if hBMSCs (human BMSCs) are to be used in clinical transplantation, as serum of bovine origin may contain as yet unidentified xenogenic pathogens potentially harmful to man. Even serum of human origin may transmit prions, which would escape standard, routine pathogen screening (86,87).

    There is still debate as to whether BMSCs are bona fide stem cells. In addition to displaying the characteristics described earlier, are they capable of self-renewing in long-term culture and pluripotent or multipotent lineage specification and differentiation? In this respect, some groups have successfully passaged BMSCs for 60 to 120 population doublings [as in the case of multipotent adult progenitor cells (MAPCs)], with little or no apparent change in their multipotency (88). The differentiation potential of BMSCs has not been extensively studied at the clonal level, given the difficulty in generating clonal BMSC lines. This is important because colonies of BMSCs derived from several cells may consist of a number of clones each capable of differentiating into specific lineages. Although the antigenic enrichment of neurones and their progenitors from ES cells has been reported (20), the same has not yet been attempted for BM cells after the induction of differentiation into a specific lineage. This would allow the subsequent comparison of the in vivo efficacy of positively-selected, negatively-selected, and unselected hBMSCs. The superiority of positively-labeled cells in effecting CNS repair will rely on their hypothetical ability either to re-establish functional synaptic connections and restore lost or damaged neural circuitry, or to release specific, diffusible, trophic factors capable of enhancing CNS repair.

The demonstration of the functionality of hBMSCs is based primarily on their ability to function electrophysiologically in vitro, as this would suggest that they might establish appropriate synaptic connections with depleted host neurones after implantation in vivo. Few studies have examined the electrophysiological properties of BM-derived neurones using whole-cell patch clamp recording (88–90), although in one study, a resting membrane potential similar to that of neurones and a rectifying ionic current typical of voltage-dependent potassium ion (K^+) channels were recorded (90). A rapid and reversible rise in calcium ion (Ca^{2+}) levels in response to acetylcholine that is characteristic of neurones was also measured. In another study (89), voltage-sensitive ionic currents were also detected, as well as intracellular Ca^{2+} concentrations, which could be elevated by high K^+ and glutamate, in β-mercaptoethanol-treated size-sieved hBMSCs, and not in untreated cells.

Wharton's Jelly Cells of the Umbilical Cord

The human umbilical cord (Fig. 3) consists of an outer layer of amniotic epithelial cells (5) enclosing a gelatinous matrix referred to as WJ, first described by Thomas Wharton (91). The latter encases a single vein (1) and two arteries (2 and 3). WJ is the gelatinous connective tissue (4) that constitutes the umbilical cord and is composed of myofibroblast-like stromal WJ cells, collagen fibers, and proteoglycans (92). WJ cells are highly proliferative and can be propagated for over 80 population doublings, while maintaining high levels of telomerase activity. They reportedly express several stem cells markers, including c-kit and Oct-4, as well as telomerase, an enzyme that inhibits cell senescence by maintaining telomere length. They also seem to have neurogenic potential (58). Their osteogenic, chondrogenic, and adipogenic differentiation potential is currently being assessed in order to test whether they are genuine MSCs similar to those isolated from other sources. WJ cells have been shown to survive for at least six weeks following intracerebral transplantation or systemic infusion without the need for immunosuppression of the host rat. The enhanced green fluorescent protein (eGFP)-labeled cells migrate extensively following implantation and co-express neuronal filament 70 (59). To date, no electrophysiological confirmation of neuronal differentiation has been reported for WJ cells and, similarly, no behavioral assessment of animals transplanted with WJ cells has yet been published, as they have not yet been used in any disease model.

Differentiation Potential of Stem Cells

Multipotent stem cells undergo a progressive restriction of their lineage potential as development or differentiation proceeds, until the terminal final fate is specified. This "developmentally driven lineage restriction" operates within all stem-cell systems

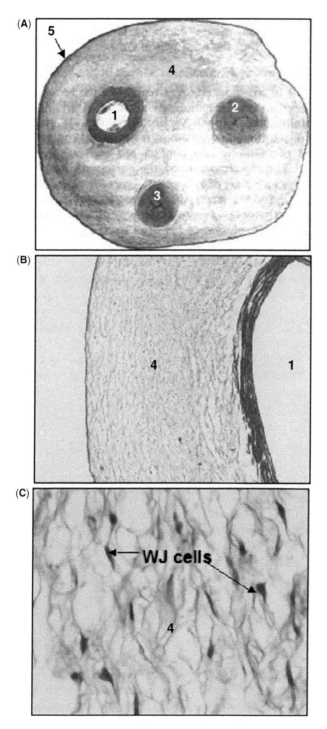

Figure 3   H&E of human umbilical cord in transverse section: (**A**) magnification (×2) 1, umbilical vein; 2 and 3, umbilical arteries; 4, matrix or WJ; 5, amniotic epithelium. (**B**) low magnification (×4) of transverse section through the umbilical vein and WJ. (**C**) Higher magnification (×40) of WJ and the cells making up the matrix. *Abbreviation*: WJ, Wharton's Jelly. *Source*: From Ref. 278. (*See color insert.*)

(93,94), and occurs alongside a reduction in proliferative capacity (36). The stage of development at which fate specification occurs and the regulatory mechanisms involved are poorly understood, although both extrinsic or epigenetic (extracellular matrix, growth factor availability, and cell–cell contact) and intrinsic genetic influences are likely to operate (48). Differentiation involves the sequential expression of specific genes leading to a mature phenotype in a temporally defined manner. The maturation of stem cells involves a continuous loss of pluripotency and increased phenotypic commitment, until terminal differentiation into a specific cell type or subtype (95).

Transdifferentiation (Fig. 4) is the genetic reprogramming of a differentiated cell into a pluripotent one (51,88), as occurs in reproductive cloning (when a somatic nucleus is transferred into an enucleated egg). It has also been used to describe a switch in cellular phenotype or lineage fate without genetic reprogramming, as occurs in the conversion of cells of mesodermal lineage (such as osteoblasts) into cells of ectodermal origin, including neural cells (90). Another example of transdifferentiation is the expression of neurogenic phenotypes (including neurone-specific enolase, neurofilament, and neurotrophic growth factor receptor) in bone-derived "Ewing" sarcomas, which are rare neoplastic growths of bone and extra-osseous tissue (96–98). This has important implications for brain regeneration and repair after traumatic injury and degenerative disease, as it suggests that bone-derived cells are capable of generating neural cells under certain circumstances, and may therefore represent an alternative, more easily accessible source of neural tissue for therapeutic implantation than neural or ES cells.

The reverse may also be true and so NSCs are not restricted to a neural fate. They can apparently generate a variety of blood cell types including myeloid, lymphoid, and

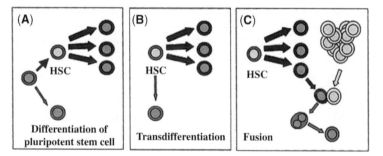

Figure 4 Different mechanisms by which observed "transdifferentiation" could occur. The three models shown represent mechanisms of differentiation from BM-derived cells into an alternate phenotype. (**A**) Consistent with a paradigm that cells always travel from a less differentiated to a more differentiated state, this model predicts that there is a highly pluripotent cell that has not yet committed to the hematopoietic lineage and maintains the ability to differentiate into multiple diverse cells types. (**B**) In direct transdifferentiation, an HSC may be able to directly change its gene expression pattern from that of an HSC into an alternate cell type. (**C**) If cell fusion is the mechanism by which HSCs acquire a non-hematopoietic phenotype, a marrow-derived cell, perhaps a macrophage, fuses with a non-hematopoietic cell and the nucleus of the marrow-derived cell assumes the gene expression pattern of the non-hematopoietic cell type. The two nuclei do not necessarily need to fuse. Note that these models are not mutually exclusive and may all reflect the in vivo mechanisms involved. These models apply equally well to MSCs, which may be highly pluripotent stem cells with the ability to differentiate directly into multiple cell types, transdifferentiate into non-mesenchymal cell lineages, or have the ability to fuse with different cell types. *Source*: From Ref. 130. (*See color insert.*)

early hematopoietic cells when transplanted into the marrow of irradiated rodent hosts (99). Similarly, co-culture of NSCs with muscle cells can induce their differentiation into myocytes (100,101). Although one should not underestimate the controversial nature of these findings, they suggest a possible lineage relationship between stem cells in unrelated parts of the body. Such studies also illustrate the deterministic influence of environmental cues on the fate of NSCs.

This phenotypic plasticity is not restricted to NSCs (102). There are several reports of non-neural stem cells undergoing transdifferentiation to a proneural form. Some of these have demonstrated the presence of cells expressing markers of differentiated neural lineages within BM (51,103) and WJ cell cultures (58), although only few have shown the presence of putative NSCs. Furthermore, none has isolated, purified, or enriched those neural-like cells from BM and WJ, and many have used non-physiological (toxic and carcinogenic) stimuli to induce or promote the emergence of neural-like cells, which would limit their clinical application. The use of substances toxic to cells can cause them to react non-specifically with a range of antigenic neural markers (104).

The early in vivo experimental data indicative of transdifferentiation is based on bone marrow transplantation (BMT) studies in lethally irradiated mice. Transplanted cells were tracked by genetic differences (e.g., mouse strain, sex, expression of green fluorescent protein (GFP)) between the injected cells and the host. In one case of female mice, which were rescued by male BMT, up to 2.4% of neurones were male (carrying the Y chromosome) (103). Further studies have demonstrated the presence of donor markers in neurones of the olfactory bulb (105). Similar work has also been undertaken in humans (106,107). Postmortem studies of females who had BMT from male donors revealed that 0.1% of Purkinje cells carried the Y chromosome and were presumed to be of donor origin (108,109). However, no account was taken of previous pregnancies with male offspring and the possibility of microchimerism (110).

A range of physiological and non-physiological agents, which have direct or indirect effects on gene transcription, have been used to induce neural specification of BMSCs in vitro (111). Furthermore, co-culture with neural tissue or with differentiated neural cells (astrocytes, oligodendrocytes, or neurones) has been used in an attempt to increase the proportions of BMSC- (51,88) and UCB (umbilical and cord blood) cell-derived (112) neural antigen-expressing cells. No study has yet examined the effect of co-culturing NSCs with BMSCs in order to generate neural-like cells from the latter. Similarly, WJ cells have not been co-cultured with other cell types for the purposes of directed differentiation.

The precise identity of the cell type or subpopulation of BM and WJ cells with the potential to generate neural-like cells remains unknown, as most studies have used heterogeneous or poorly characterized starting populations of cells (113). For instance, it is not yet certain whether only BMSCs (putatively $CD34^-$) have the ability to express neural antigenic markers, or whether HSCs (putatively $CD34^+$ or c-kit/$CD117^+$) also have this potential. This is complicated by the finding that HSCs can be both $CD34^+$ or $CD34^-$ (114). This uncertainty also applies to studies on UCB. For instance, two recent studies on stem cells from cord blood have published conflicting data with regard to their expression of CD133 (55,57). This controversy may in part be the result of different isolation and culture protocols prior to characterization. Nevertheless, a number of BM-derived HSC (115,116) and non-HSC populations (including cells expressing SSEA1, a marker of ES cells) [Bonnet, pers. commun., Nov. 2003] have been shown to express neural antigens both in vitro and in vivo. The in vitro expression of neural antigens by cells derived from HSCs has only very recently gained attention (52,116,117), although in vivo observations of their transdifferentiation potential have previously been reported (115,118). Hematopoietic c-kit/$CD117^+$ cells from murine BM were shown to generate oligodendrocytes in vivo after transplantation (115). However, no account was taken of the possibility of spontaneous donor–host cell fusion (119,120), and, therefore, confirmatory tests remain outstanding.

However, the concept of stem-cell transdifferentiation is not universally accepted (121). In one study, Bjornson et al. intravenously injected clonally derived NSCs into the tail veins of irradiated mice and found that they generated significant numbers of hematopoietic cells (99). However, this could not be reproduced by Morshead et al. (122), who failed to detect significant numbers of hematopoietic cells arising from the injection of labeled neurospheres into the tail vein. The long-term culture of stem cells in the earlier experiments might account for this apparent discrepancy.

More recently, two independent studies have provided an alternative explanation for transdifferentiation. In the first, by co-culturing GFP-labeled neurones with hygromy-cin-resistant ES cells, undifferentiated hybrid cells with molecular characteristics of both neurones and ES cells were obtained (120). In the second study, co-culture of BM cells (labeled with GFP) with ES cells resulted in fused cells that were GFP-positive and expressed a number of ES cell genes. Significantly, the hybrid cells were predominantly tetraploid, demonstrating that BM cells can fuse spontaneously with other cell types, adopting their phenotype (119,120).

Cell fusion can also influence in vivo observations. It has been claimed that the apparent donor-derived cells do not represent "transdifferentiation" of donor cells, but are the result of cell fusion. In the case of female patients who had had BMT from male donors and revealed a small proportion (0.1%) of Purkinje cells carrying the Y chromosome, this fusion would be between endogenous Purkinje cells and donor cells forming stable heterokaryons that express markers of both recipient and donor tissue. This possibility has been confirmed in murine models, and it is thought that the fusion event reprogrammes the donor cell (123).

Fusion of donor cells with host brain cells may explain the apparent transdifferentia-tion of blood to brain in vivo, but cannot explain all of the in vitro data. There are now several published protocols for directing BMSC to neural lineages (51,86,88,111, 124,125). The approaches used are different and range from using chemical demethylating or reducing agents to more physiological growth factors. Although controversial, much of the transdifferentiation data is tantalizing and is not easily explained by cell fusion.

In these experiments, the generation of neural cells from blood could be a result of either:

1. Reprogramming of the gene expression profile of a tissue-specific committed progenitor or fully differentiated cell into that of a pluripotent or multipotent cell. This could operate through a process of transdifferentiation in response to local microenvironmental cues or through cell fusion, although some authors choose to consider the latter simply as a possible mechanism of plas-ticity, rather than a distinct phenomenon (Fig. 4A–C); or

2. Proliferation or differentiation of a pluripotent progenitor cell residing within an adult tissue, either after migrating into the tissue early in development, or at a later stage from a different tissue (126) (Fig. 4A). Verfaillie's group has described the multipotent adult progenitor cell as a BM-derived cell that has multi-tissue differentiation potential, including neural lineages. When transplanted, these cells have been shown to ameliorate neurological deficits in a rat model of cerebral ischemia (127).

Therefore, fusion, although unable to increase the number of cells in a damaged CNS, may contribute to replacing damaged cells. If the mechanisms of reprogramming could be understood in detail and the frequency of fusion increased, this approach could still be useful for therapy, although the possible genomic instability of hetero-karyotic cells would have to be taken into consideration.

Ethically, the transdifferentiation debate is of particular interest as, if adult stem cells can be induced to differentiate into any cell type, the need for ES cell research is reduced (128). However, a large body of research still suggests that fetal stem cells are significantly more plastic than their adult counterparts (129).

Embryonic, Fetal, or Adult Stem Cells?

A number of factors determine the usefulness of specific cell populations in clinical transplantation. However, if stem-cell replacement is necessary, which source of stem cells is the best for which conditions?

Embryonic Stem Cells

Several groups have found that enriched neural precursors from human ES cells can incorporate into brain tissue and differentiate in vivo (131). Zhang et al. (132) transplanted neural precursors enriched from ES cells into the lateral ventricles of newborn mice, and observed migration to multiple brain regions, followed by differentiation into cells with mature neuronal and astrocytic phenotypes, although, interestingly, no mature oligodendrocytes were identified. Such in vivo studies are important in two respects. First, they suggest that transplanted cells do have the potential to populate the brain and, second, they highlight the fact that more manipulation may be necessary before the required neural cell types are efficiently generated (21,133). Early successes in neural differentiation of ES cell grafts in vivo have led to further work in injury models to demonstrate that transplanted ES cells can integrate and functionally improve outcome following CNS injury (134,135). However, it is clear that there is still a significant gap in our knowledge of how to direct the appropriate differentiation of ES cells in vivo.

Advantages of ES Cells. ES cells provide the most promising alternative source of cells for therapeutic transfer into neural tissue. They are multipotent, can be propagated in vitro and can be engineered to express therapeutic genes. They migrate and differentiate into regionally appropriate cell types and do not appear to interfere with normal brain development (133). ES cells can also be differentiated in vitro into oligodendrocyte precursors that effectively myelinate host axons in animal models of human demyelinating disease (136,137).

Disadvantages of ES Cells. In addition to the ethical considerations restricting the therapeutic use of ES cells, their capacity for unlimited growth in culture reflects their tendency to form teratomas after implantation. Until a reliable means of completely eliminating undifferentiated ES cells from populations intended for implantation are developed and tested, ES cells remain an experimental tool with which to explore proof-of-principle therapies for neurodegenerative conditions. Thus, there are more than just ethical reasons for using adult stem cells rather than those derived from embryos (129).

Neural Stem Cells

Cortical neurones undergoing injury-induced apoptosis can be replaced by transplanted fetal NSCs that can also differentiate in situ into region-specific neuronal and glial subtypes when implanted into the lesioned hippocampus, neocortex, or striatum in adult rodents (138). In the developing brain, stable clones of NSCs can participate in aspects of normal brain development when injected into the germinal zones of newborn mice (133).

Advantages of Fetal NSCs.   Comparisons of fetal stem cells with those derived from defined developmental niches in the mature brain seem to indicate that NSC plasticity decreases with developmental age. Although the precise mechanisms for this are not fully understood, there is evidence that many of the fate decisions may be cell intrinsic (95). For example, EGF-R expression increases in cortical stem cells with increased passage number, and this may result in an increased responsiveness to this ligand (47).

Human fetal NSCs have been successfully isolated from fresh post-mortem brain tissue by exploiting the unique expression of the cell surface marker CD133 together with the absence of CD34 and CD45. Not only can single cells with this antigenic phenotype form neurosphere cultures that differentiate into both neurones and glia, they can also be successfully transplanted into brains of immune-deficient neonatal mice (26). Importantly, such studies have been extrapolated to non-human primates, using BrdU-labeled human NSCs. Ourednik et al. observed that transplantation into the developing primate forebrain resulted in the integration of the donor cells into both the mature cerebral cortex and the SVZ where, presumably, they remained until needed for further neurogenesis or post-injury repair (139). One further advantage of fetal NSCs is that they can be rapidly propagated in vitro with little or no apparent change in their plasticity. In one study, human neural progenitors isolated from embryonic forebrain were expanded for up to a year in culture using EGF, FGF and leukemia inhibitory factor. Subsequent injection of these cell lines into the developing rat brain showed extensive migration and integration (140,141).

Disadvantages of Fetal NSCs.   Clinical use of fetal tissue for stem-cell transplantation is made difficult by ethical constraints. Confronted with the spectre of couples conceiving for the sole purpose of obtaining aborted brain tissue for the treatment of either parent or afflicted siblings, scientists have turned to the most unlikely sources for NSCs. Indeed, investigators have claimed to isolate functional NSCs from adult post-mortem brain tissue as late as five days after death (28). Although it is suspected that adult NSCs have a more limited ability to form all the neural subtypes, they may have an even broader potential than first thought. Using a chick-mouse chimera approach, adult NSCs were reported to readily give rise to cells in all the germ layers, demonstrating a high degree of plasticity and indicating that their application may go beyond the treatment of CNS disorders (142). Although it is known that stem cells in the SVZ of the adult mammalian brain can proliferate, migrate to the olfactory bulb, and ultimately differentiate into mature neurones, are these cells functional? In an elegant study from Frisen's lab, labeled stem cells were shown to respond to a specific odor-induced signal in the olfactory bulb by upregulating expression of the c-Fos proto-oncogene. These results indicate that newly formed adult neurones can functionally integrate into the synaptic circuits of the mature brain (143).

Mesenchymal Stem Cells

Advantages of MSCs.   MSCs offer a number of advantages over NSCs and ES cells for clinical implantation (Table 2). They may:

1. be more easily and ethically isolated than NSCs,
2. have a greater ability to "home in" on the brain than NSCs after intravenous infusion, although no systematic comparison has yet been carried out (103,105,144,145),
3. negate the need for immunosuppression in the case of autologous transplants and possibly even in the case of heterologous transplants (59,146,147). The same may not be true of NSCs (148,149). MSCs have already been used in

Table 2   Factors Affecting Clinical Use of WJ, thMSC, ES, NSC, BM, PB and UCB Cells

	WJ	fhMSC	ES	NSC	BM	PB	UCB
Ease of collection/Availability	++	+/−	−	−	+	+++	++
Ethical constraints	+++	+/−	−	−	++	+++	+++
Risk of contamination with tumorigenic cells	−	−	+++	−	++	+++	−
Relative stem-cell complement	?		+++	+++	++	+	+++
Proliferative potential	+++	+++	+++	++	++	+	++
Differentiation potential	Neural	All	All	Neural	All	All	All
Immunogenicity	?	−	?	−/?	−	?	?
Homing capacity	?	?	?	+	+++	+++	+++
Possibility of autologous transplantation	+	−/+	−	−	++	+++	+
Gene therapy (ex vivo)	?			+			
Integration into host CNS parenchyma; migration and growth	?			+++			

Notes: −, none or worst; +, good; ++, better or more; +++, best or most; ?, unknown.
Abbreviations: WJ, Whartons Jelly; thMSCs, fetal human mesenchymal stem cells; ES, embryonic stem cells; NSC, neural stem cells; BM, bone marrow; UCB, umbilical cord blood; CNS, central nervous system.

several clinical trials of autologous transplantation for a wide range of conditions and were found to be well tolerated with minimal side-effects (86),
4. present fewer ethical constraints, unlike NSCs isolated from human fetal CNS tissue and human ES cells,
5. be confronted with fewer regulatory obstacles. Autologous transplantations of BMSCs are already possible and such cells from postnatal tissue would open up the possibility of using autologous transplants to treat neurodegenerative conditions (51),
6. have a greater differentiation potential than NSCs, which may be restricted to neural fates (71,130,150–153).

In addition, there are some possible advantages to the use of UCB and WJ cells over that of BM-derived and PB-derived cells:

1. There is negligible risk of contamination of UCB or WJ cells by leukemic cells.
2. They may have a greater self-renewal capacity, proliferation rate, and differentiation potential than adult BM and PB cells (154), although no systematic comparison of different non-neural sources of neural cells has yet been published.

Disadvantages of MSCs.   There are a number of possible drawbacks of using non-neural sources of neural-like cells for intracerebral implantation. A report was made recently of tumor formation after BM cell intracerebral implantation in rats (155). So far there has been only one report to date of tumor formation following NSC implantation (156). Furthermore, neural-like cells derived from non-neural tissue might not be able to respond appropriately to positional signals within the recipient brain, as indicated by their presence in inappropriate areas (157). This latter observation contrasts with published observations of the fate of donor-derived BM cells in the human CNS (107) and that of BM-derived MAPCs implanted into the blastocyst-stage mouse embryos (158), which have indicated that they might respond to local positional and migrational signals within the recipient brain, but nevertheless highlights the need for caution during the design of transplantation studies and the subsequent interpretation of results. A number of other factors have to be taken into account when selecting the most appropriate cell types for transplantation, and these are listed in Table 2.

THE USE OF STEM CELLS IN BRAIN DISEASE AND INJURY

While the therapeutic replacement of entire tissues with stem cells is some way off, BMT are routinely used to treat immune-deficient patients and similarly, tentative steps are being taken to repair PD with neural cell grafts. It should be borne in mind that the treatment of localized diseases [such as PD, Huntington's disease (HD), stroke, and trauma] and global or disseminated diseases [such as multiple sclerosis (MS), AD, and metastatic brain tumors] require different approaches (159). The latter do not preclude the use of cellular therapy, as stem cells are able to migrate extensively within the mature brain and even home to sites of injury and degeneration. Furthermore, it has been estimated in PD, for instance, that more than 80% of striatal neurones and more than 50% of nigral neurones have died by the time the first observable symptoms of neurodegeneration appear (160, 161), and a relatively small degree of repair or trophic compensation is likely to lead to a dramatic clinical improvement.

Cells of non-neural derivation, such as MSCs have also been used for clinical transplantation. At least four experimental models have been used to assess the therapeutic effectiveness of MSCs from BM, UCB, and WJ in neurodegeneration and injury: rodents with induced parkinsonism, rodents and marmosets with ischemic infarcts, and rodents with traumatic brain or spinal cord injury. Similarly, three methods of implantation have been used in most in vivo studies on rodents: direct implantation into the brain parenchyma, injection into the cerebral ventricles, usually the lateral, systemic infusion via the tail vein, and rodents with induced demyelination (162).

The precise mechanism of recovery afforded by the transplantation of MSCs from BM and UCB is unclear, although there is widespread acceptance that humoral factors secreted by the grafted cells play a significant role (51,163). For instance, hBMSCs release factors capable of supporting the prolonged expansion of human ES cells in vitro (164). These diffusible factors might enhance the endogenous repair systems by the provision of trophic support.

Parkinson's Disease

PD is attributed to a selective loss of dopaminergic neurones in the substantia nigra and is typified by motor symptoms including tremor, rigidity, and bradykinesia (slowness of gait). Symptomatic relief is provided by pharmacological dopamine stimulation in the form of levodopa, which despite being highly effective for the first few years of treatment, causes disabling side-effects, including tachykinesia (fast involuntary movements), dyskinesia (abnormal movements), and tremors (165).

From a cell therapy perspective, experience already exists of fetal neural cell transfer in humans suffering neurodegenerative disorders. Over 300 patients with PD have now received grafts of fetal mesencephalic cells into the striatum. These grafts are spontaneously active and can restore dopamine release to near-normal levels with symptomatic improvement. There is a downside, however; in a clinical trial from Denver and New York, 15% of grafted patients developed unacceptable dyskinesias (166).

Furthermore, the supply of fetal neural tissue is limited and consequently only small numbers of neurones are available. This could be partially overcome by in vitro expansion, but fetal tissue contains a heterogenous population of cells, many of which are inappropriate for use in clinical transplantation. The limitation of poor tissue supply might be overcome by generating dopaminergic neurones from NSCs or ES cells. Indeed, rat NSCs can be propagated in culture while retaining the capacity to differentiate into dopaminergic neurones and to improve outcome in a rat model of PD (167,168), although different studies have obtained conflicting results (169). Human fetal neural stem cells can also be expanded in vitro and their progeny transplanted into the injured rodent brain, where they migrate and integrate into numerous regions of the brain, suggesting that they can respond to local cues (170,171). Furthermore, EGF- and FGF-2-expanded human neural precursors have been shown to repopulate the dopamine-depleted striatum, with some observable differentiation into tyrosine hydroxylase (TH)-positive cells (169). Perhaps surprisingly, NSCs from adult human mesencephalon labeled with a nestin-GFP transgene have also been shown to have the potential to generate dopaminergic neurones (172). Recently, Nurr1-overexpressing murine ES cells were shown to generate a highly enriched population of dopaminergic neurones, which integrated (without forming teratomas) into the brains of 6-hydroxydopamine (6-OHDA)-lesioned rats and exhibited electrophysiological properties of mesencephalic neurones (173). Investigations are ongoing as to whether human ES cells can generate such neurones with similar efficiency. In spite of these promising results, the transplantation of NSCs into PD models have shown only limited success. For instance, even though solid grafts can be found two weeks after transplanting expanded human NSCs into lesioned animals, these often decrease in size due to cells migrating out and differentiating into other cell types, such as astrocytes, while only a small percentage become neurones (169). Interestingly, in two animals, a significant number of these neurones became dopaminergic, and rotational deficits associated with the lesion were reversed. These rather disappointing results might be explained by variations in major histocompatibility antigen status within the cohort of wild-type rats used (169) and transplantation into immunosuppressed rats are now being assessed (21). Transplantation of human fetal mesencephalic tissue is a clinically promising experimental treatment for PD. However, ethical and technical issues surrounding the limited supply of donor tissue are obstacles to its transfer to widespread clinical practice. This might be overcome, in part, by the in vitro expansion of primary CNS precursor cells, along with region-specific differentiation in vitro prior to implantation (168). So, although these preliminary studies suggest that NSCs may provide a highly proliferative pool of cells with considerable advantages over fetal tissue grafts, further studies of phenotypic specification, controlled growth, and functional integration into the injured adult brain are needed before full clinical trials of cell therapy.

Huntington's Disease

HD is a result of a very similar type of cell loss as PD, except that in the former, inhibitory GABAergic striatal neurones are damaged and degenerate. This has a direct consequence on motor and cognitive functions (174,175). In contrast to the availability of pharmacological agents for PD, there is currently no effective therapy for HD. Clinical trials of fetal human and porcine striatal cell implantation into the striata of patients with HD have yielded encouraging but limited success (176,177). More dramatic improvements in motor function have been observed by several groups at the experimental level using MSCs (163) and NSCs (178).

Ischemic Stroke

Ischemic stroke is an acute onset cerebral deficit caused by a vascular event such as a thromboembolism. Current pharmacological interventions include the administration of aspirin and other anticoagulants that are designed primarily to disperse blood clots and minimize the risks of a recurrence (179). Despite the established knowledge that wide-spread cell death follows such cerebrovascular incidents, pharmacological interventions to minimize this (using anti-apoptotic agents) are not common practice. Anti-apoptotic agents cannot address the necrotic cell death that occurs immediately after an ischemic stroke, although they can decrease the amount of delayed cell death in the subsequent hours, days, and weeks. In severe cases, however, the amount of damage caused by the ischemic event can be so extensive that a lasting motor or cognitive deficit is sustained. In such cases, cell replacement would be an ideal way to restore lost cells and function. Clinical trials have been undertaken to evaluate the efficacy of whole body hypothermia as a neuroprotective treatment, following a successful pilot study (180), but have yielded mixed results (181,182).

Hypoxic ischemic encephalopathy (HIE) is an important cause of newborn brain injury, and follows a very similar etiology and progression to ischemic stroke in adults. Therefore, similar factors would need to be taken into consideration when applying restorative therapy to cases of HIE. In this respect, fetal neural cell transplantation has been successfully applied to several experimental models of adult brain injury. Fetal cortical grafts survive in the infarct area following focal forebrain ischemic injury in adult rats and appear to receive connections from the surrounding brain with a resulting improvement in motor function (183), spatial learning, and memory (184). Furthermore, trans-formed NSCs have been used to replace cortical neurones undergoing photolytic injury-induced apoptosis. Significantly, these cells demonstrated appropriate differentiation in situ into region-specific neuronal and glial subtypes determined by the site of injection (138). Similarly encouraging observations have been made with MSCs from BM and UCB (185–197).

Traumatic Brain Injury

TBI or spinal cord injury (SCI) is a multifactorial condition in that several mechanisms of injury and cellular damage contribute to cell death following the initial insult. These include inflammation, excitotoxicity, demyelination, and ischemia. In this respect, when considering cases of TBI or SCI for restorative cellular therapy, several factors need to be taken into account in order to address the various levels and modes of damage, and to set realistic targets of clinical improvement (198). Early results obtained with embryonic murine NSCs implanted into the striata of adult mice following cortical impact injury are encouraging: behavioral (motor and cognitive) recovery was sustained for at least a year post-implantation. Surprisingly, most of the grafted cells expressed the NG2 antigen, a marker of oligodendrocytic progenitors, and not neuronal, astrocytic, or micro-glial markers. The implanted cells also migrated extensively towards the injury site (199). MSCs implanted into the experimentally injured CNS have been found to behave in a way similar to NSCs (200,201).

Multiple Sclerosis

MS is an acute autoimmune inflammatory demyelinating condition that leads to axonal loss and the formation of chronic, multifocal, sclerotic plaques in the CNS. With a lifetime risk of 1 in 400, it is one of the major causes of neurological disability in young adults

(202). Current treatment is pharmacological, involving the administration of β-interferons for the functional antagonism of proinflammatory cytokines and down-regulation of major histocompatibility class II (MHC II) antigen expression (203). A major drawback of β-interferons is their limited ability to suppress relapses and the ensuing neural damage. Furthermore, they are of limited effectiveness once axonal degeneration has reached a critical threshold and clinical progression is under way (202). Endogenous remyelination can partially restore axonal conduction and motor function, but is limited to acute inflammatory lesions within which oligodendrocytic progenitors are found but are unable to remyelinate stripped axons (204–206).

Cell restoration for MS is still at the experimental stage, using cells from a range of sources, including peripheral nerve Schwann cells (207), olfactory bulb ensheathing cells (208), as well as NSCs and MSCs. One should consider the extra demands placed on the implanted cells compared to focal neurodegenerative conditions—they will need to migrate extensively through the adult brain towards lesion sites, survive, proliferate, and then remyelinate bare axons. Nevertheless, rodent ES cell-derived oligodendrocytes transplanted into an adult rat model of MS have shown efficient remyelination (136). Similarly, the implantation of an immortalized cerebellar stem-cell line into newborn *shiverer* mice brains (which produce oligodendrocytes but not myelin) leads to the replacement of dysfunctional oligodendrocytes and myelination (209). Rodent NSCs can generate large numbers of oligodendrocytes at all passages, which can remyelinate experimental MS-like spinal cord lesions (210). Canine and rodent "oligospheres" can be generated by supplementing the culture medium with neuroblastoma B104 conditioned medium (B104CM) (211,212). Similarly, neonatal rat NSCs exposed in vitro to EGF and B104CM to induce their differentiation into oligodendrocytes have been shown to produce myelin when transplanted into the myelin-deficient rat (213).

An important aspect of cell therapy is the developmental stage at which cells are transplanted. More committed oligodendrocyte progenitor cells (OPCs) from 21–23 week human fetal brains have been isolated, purified, and cultured. These cells were xenografted into *shiverer* brains and developed into oligodendrocytes that myelinated host axons (214). Interestingly, in this study, OPCs from adults generated oligodendrocytes more efficiently than fetal OPCs. Although this indicates that both NSC and more committed progenitors can replace a single cell type, other studies in different models are less encouraging. Thus, NSC transplantation into *Twitcher* mice (a model of Krabbe's disease, where there is an absence of galactocerebroside resulting in the accumulation of the toxic lipid, psychosine) resulted in no improvement in disease symptoms or survival in spite of extensive differentiation and myelination (215). It is possible that the transplanted cells could not sufficiently overcome the toxic environment of the endogenous cells. Recently, CD117$^+$ HSCs from adult BM have been shown to differentiate into oligodendrocytes after intracerebral transplantation into experimental mice (115). This represents an important development in the search for more effective cell types for cell replacement in MS. However, further analyses are required in order to confirm the neurogenic potential of this abundant source of stem cells and establish the safety of the procedure. MSCs have also been used successfully in experimental models of CNS demyelination. For example, MSCs implanted into an ethidium bromide-mediated lesion of the adult rat spinal cord extensively remyelinated damaged axons (216).

Alzheimer's Disease

Alzheimer's disease (AD) is a progressive degenerative brain disease and is the commonest form of dementia. The characteristic histopathological features of the condition are senile plaques and neurofibrillary tangles associated with a neurochemical abnormality resulting in a cholinergic deficit. Although there is no cure, some of the symptoms such as decline can be partially relieved with anticholinesterases (217). This aside, current pharmacological approaches are unable to halt or delay, let alone reverse, the progression of the disease. A promising development was made recently when murine NSCs were differentiated into high proportions of cholinergic neurones, which could potentially be used for neuronal replacement in AD and even motor neurone replacement in amyotrophic lateral sclerosis/Lou Gehrig's syndrome (218,219).

Brain Tumors

Gliomas are a highly invasive form of CNS neoplasms with a poor survival rate. Due to the highly motile phenotype of glioma cells, tumor margins tend to be diffuse and complete resection is almost impossible. In this context, NSCs are also highly motile and can migrate through the brain parenchyma towards the area of pathology before integrating into the host cytoarchitecture (220). They also exhibit a striking degree of tropism for glioma cells (221,222), and when genetically engineered to secrete the proinflammatory cytokine interleukin-12 (IL-12), have been shown to prolong the survival of tumor-bearing mice by promoting a targeted T cell-mediated immune response against the glioma cells (223,224). The relative ease with which NSCs can be genetically manipulated makes them an attractive option for the treatment of otherwise intractable malignancies (209). A recent study found that MSCs also have the same effect on glioma cells in vitro and in vivo after transplantation into experimental tumors (225).

Pediatric CNS Disorders

Until recently, the application of cell therapies was only considered for focal brain diseases or insults. This was based on the assumption that it would not be practicable to deliver cells to multiple sites in the brain. The fact that the majority of pediatric neurological diseases are global in nature, often affecting widespread areas of the CNS, and that their pathogenesis is poorly understood at the molecular and cellular levels suggest that they might be refractile to stem-cell therapy. These diseases include genetic abnormalities such as inborn errors of metabolism (226,227), lysosomal storage diseases (220), and leukodystrophies, as well as the widespread degeneration that can follow acute brain injury after asphyxia (HIE), or the more subtle white-matter abnormality that occurs in the majority of extremely preterm infants (228–230). Perhaps an early indication of the potential for cell replacement in perinatal therapy is the observation that stable clones of NSCs can contribute to normal brain development when injected into the germinal zones of fetal [Kennea et al., submitted] or newborn mice (133).

Other Considerations

Endogenous Repair

Is stem-cell replacement really necessary? There is increasing evidence supporting the existence of endogenous compensatory mechanisms that are activated in response to injury and disease (231–233). For example, a low level of ongoing neurogenesis has recently been shown to occur in the adult mammalian striatum (234). Similarly, targeted apoptotic degeneration of murine cortical neurones has been shown to trigger the formation of new cortical neurones, whose axons extend into the thalamus (235), and a similar process has been observed following ischemia, which promotes neurogenesis in the rat SVZ, with newly generated neurones migrating into the striatum where they mature into spinal striatal neurones (236,237). Factors that promote neurogenesis might also encourage endogenous repair. Thus, infusion of EGF after experimental ischemia in mice triggers an enhanced neurogenic response with subsequent partial replacement of parvalbumin-expressing striatal interneurones (238), while a combination of EGF and fibroblast growth factor-2 (FGF-2) after global forebrain ischemia in rats results in the partial regeneration of the hippocampal CA1 pyramidal neurones, which subsequently form afferent and efferent connections and partially reverse functional deficits (239). Studies using BrdU labeling to identify proliferating cells have demonstrated the expansion and subsequent differentiation of endogenous neural precursors following experimental stroke (240). Similarly, NSC proliferation has been found to increase tenfold in the subgranular zone of the dentate gyrus after global ischemia in the gerbil (241). Endogenous repair in response to stroke can also involve the proliferation of neural progenitor cells in the SVZ. Following middle cerebral artery occlusion, injection of BrdU specifically labeled astrocytes in the ependymal and subependymal layers that later acquired the characteristic antigenic markers of neurones after injury (242). In a separate model employing chemically induced seizures in the rodent, a pronounced increase in the generation of new neuronal precursors in the SVZ and their subsequent migration and integration towards the olfactory bulb were reported (243). While it has been proposed that ischemia-induced neurogenesis might contribute to the specific recovery of memory function lost following injury, a high proportion of the dividing cells are lost over the weeks following injury. Adult NSCs might function after injury to maintain or increase levels of trophic factors so as to promote neural cell survival. This hypothesis is supported by the increased survival of mature neurones when co-transplanted with NSCs (241).

Injection of sonic hedgehog into the Parkinsonian brain stimulates the proliferation and subsequent differentiation into TH$^+$ (dopamine (DA)-producing) and gamma-aminobutyric acid (GABA) neurones of endogenous adult NSCs. The latter might serve to protect the dividing stem cells and promote recovery through their inhibitory signals (244). The injection of certain growth factors into the lateral ventricles can expand the sub-ependymal zone (SEZ) population of neuropoietic cells and trigger their migration into adjacent neural structures, including the striatum (245, 246). Localized injection of specific growth factors into defined locations within the brain, such as the striatum in PD, may promote the ingrowth of axons from endogenous neuronally-committed progenitors. However, this strategy might be limited by the accessibility of certain areas of the brain in other pathologies, the more diffuse nature of many of the latter (209,247), and also the possibility of epileptogenesis as a result of such treatment, as demonstrated for brain-derived neurotrophic factor (BDNF) (248). The role of the intrinsic receptor competence of endogenous NSC populations has also been shown to affect the outcome of such treatment (249).

Similar observations have been made in demyelinating diseases, such as MS. In chronic MS lesions, the presence of NG2$^+$ premyelinating oligodendrocytic progenitors has been reported (206,250), although the relationship between endogenous gliogenesis and remission is still unclear. In a broader sense, it remains to be seen whether such responses are patient- or disease-specific or represent a generic global response that occurs in areas that already have ongoing adult neurogenesis. However, the demonstration of the continued production and survival of neural cell types following injury has led to renewed interest in mechanisms of the endogenous cell response and whether this could be exploited further in order to instruct repair following injury.

Differentiation, Migration, and Integration of Donor Stem Cells

The stage of differentiation of transplanted cells may have a profound effect on outcome following cell transplantation. For demyelinating diseases, multipotential neural precursors rather than more restricted oligodendrocytic precursors may be more useful (209,251,252). Although cells committed to a defined lineage before injection may generate a larger proportion of a given cell type, these will not have the advantage of cell plasticity and may display reduced proliferative potential.

Stem-cell migration is also likely to influence the success of neural cell grafts. Like NSCs, BM-derived cells may also exhibit tropism for sites of pathology after transplantation (221). This has been verified in vitro using ischemic brain tissue and chemotaxis assays (253,254). Any preferential migration of implanted cells towards lesion sites in vivo, however, will have to be interpreted due to possible confounding factors, any of which might create a false impression of directed cellular migration. These include changes in vascularity (neoangiogenesis) or the permeability of the vasculature in and around a lesion site, which might increase the likelihood of systemically infused cells being present in and around the lesion. This might lead to a rise in the extracellular concentration of endogenous diffusible trophic factors, which in turn, might chemotactically attract implanted cells to the area. Similarly, enhanced trophic support from endogenous cells could lead to an increased survival of donor cells in the area compared to those further away from the lesion site.

Graft Survival

A number of other factors can influence the success of cell-based therapies, including the tissue source of stem cells, their developmental stage, and the receptiveness of the host environment. In this regard, one area that is largely neglected by current studies of stem-cell biology is the poor survival of grafted cells. It has been estimated that as many as 80% to 97% of transplanted cells die by apoptosis (255,256). Minimizing cell death both during in vitro expansion and postoperatively is vital for the success of intracerebral cell transplantation (255–260). Among the causes of cell death are immune rejection, hypoxia, hypoglycemia, mechanical trauma, free radicals, growth factor deprivation, and exposure to excitatory amino acids within the host brain (256,261). The immunogenicity of grafted cells is a major obstacle in xenotransplantation, and consequently the use of non-human tissue in clinical trials has relied heavily on long-term immunosuppression of the recipients (262). There are a number of avenues (genetic, epigenetic, and pharmacological) via which cell survival has been increased in vitro and in vivo, including the overexpression of genes such as *Akt*, which encode prosurvival proteins (263,264) and pharmacological caspase inhibition (255,265). Exposure to epigenetic signals (growth factors) such as FGF-2 can also reduce graft cell death (256). Thus, the survival of dopaminergic grafts into the striatum of parkinsonian rats can also be significantly increased if the transplants are "spiked" with a small population of fibroblasts expressing FGF, which acts both as a survival signal and enhances neuronal differentiation (266). In the neural differentiation of mouse ES cells, increased numbers of dopaminergic neurones were obtained in the presence of survival factors, including interleukin-1β and glial cell-derived neurotrophic factor (267). In parallel experiments, the mRNA level of the anti-apoptotic gene *bcl-2* was also increased in these cultures. NSCs isolated from transgenic mice overexpressing this anti-apoptotic gene display improved fiber outgrowth (268). Although these preliminary studies are encouraging, it should be noted that engineering stem cells to express prosurvival genes could increase the risk of tumorigenesis. Clearly, more work is needed to determine the role of apoptosis in the survival of stem-cell transplants.

It is often overlooked that, in the context of neural transplantation, stem cells will often be introduced into refractory host environments (e.g., where activated macroglia and pro-inflammatory cytokines are present). This might be overcome by employing genetically modified donor cells equipped to counteract this hostile environment (269–271). For instance, ectopic expression of the neural cell adhesion molecule, L1, in astrocytes can increase the speed and efficiency of innervation of branching axons, thus improving the transplant success of grafted NSCs (272). In this respect, a major advantage of stem cells is that they can be readily modified using cloning technology to express the patient's own genotype or a transgene. This technology could potentially be used to provide a source of immune-compatible cells for transplantation or even to transfer a gene product.

A further problem when considering many of the neurodegenerative conditions of childhood is that the process is ongoing and the environment inherently toxic and changeable. Unless the transplanted cells or the local environment can be manipulated, the graft might itself be vulnerable and ultimately lost. Some groups have found that implanted fetal cells are not significantly affected by disease progression (273).

In addition to cell–cell interactions, the importance of external cues from the environment has been demonstrated both in vitro and in vivo in determining the correct terminal differentiation of NSCs. In culture, embryonic precursors or adult subependymal cells in the presence of FGF-2 yielded only small numbers of striatal neurones, while the inclusion of conditioned medium from glial cell cultures increased the yield more than 17-fold (274). The embryonic striatal precursors were significantly more responsive to the differentiation environment than their adult counterparts, further indicating that stem cells from earlier developmental sources may provide more successful transplants. The importance of the host environment has also been demonstrated by transplantation into the cerebral ventricles of embryonic hosts in utero. Not only do donor cells differentiate, but they acquire the specific phenotype of the surrounding cells. McKay and coworkers found that cells that had incorporated into the host hippocampus assumed morphologies resembling granule and pyramidal neurones, whereas those that integrated into the inferior colliculus resembled tectal neurones that reside in this region (136). Although there are encouraging data suggesting that pluripotent cells can respond appropriately to developmental cues from the brain, more research needs to be centered on the extrinsic signals and molecular events that direct this process.

It seems that the brain can detect and respond to even small changes in cell number or subtle perturbations in normal function by providing the appropriate cues for stem cells to differentiate and repair the damage (138). At the other extreme, what would be the outcome of grafts in a situation where cell loss was so extensive that tissue structure was significantly disrupted? (136). In an important development, transplantation of a polymer scaffold seeded with NSCs was found to offer a significant improvement in motor function in a severe traumatic spinal cord injury (SCI) model in rats (275).

Possible Mechanisms of Benefit from Stem-Cell Therapy

There is little direct evidence to suggest that implanted stem cells participate in the structural reconstruction of neural circuits damaged or lost as a result of injury or disease. Indeed, there are a growing number of transplantation studies where functional improvements have been observed in lesioned animals after stem-cell therapy even though the graft does not appear to have integrated. It has been speculated in such cases that the behavioral improvement may be due to trophic signals from donor cells promoting survival and repair of endogenous tissue. In support of this, the intravenous injection of MSCs into ischemic rats reduces cell death, enhances endogenous FGF-2 synthesis and host-cell proliferation, and promotes functional recovery (185). Similarly, it was recently shown that MSCs expressing the prosurvival gene *Akt* improved outcome in a model of myocardial ischemia by producing trophic factors for endogenous repair (276).

FUTURE PERSPECTIVES

There are many issues that remain to be clarified about stem-cell transplantation into injured or diseased brains, including the fundamental one as to which cell sources are best suited for therapy (277). The pathogenesis of many CNS disorders is not fully understood and, in many cases, this precludes the directed use of stem cells for restorative therapy. In an ideal world, one would be able to stimulate the proliferation and appropriate differentiation of endogenous stem cells. Indeed, a number of gene delivery growth factor-based therapies may work, at least in part, through this mechanism. Early experiments in stem-cell transplantation suggested that embryonic tissue is significantly more plastic than that derived from the adult. Although subsequent research has indicated that adult NSCs possess a broader developmental potential than was first thought, they have a more limited lifespan compared to ES- or fetal-derived cells. Any research that relies on fetal tissues (especially when derived by therapeutic cloning) will be ethically controversial. Consequently, efforts should also focus on adult sources of stem cells for neural cell replacement. Whether the starting material is embryonic, fetal or adult-derived, cell replacement strategies must also contend with the influence of environmental signals. In several models of adult brain repair, transplants are prone to apoptosis for prolonged periods after transfer and so clinical improvement may only be temporary (134). Considerable work is therefore needed to identify the triggers for specific neural cell survival and integration and to further determine how the environment of the injured brain may be manipulated to become more permissive for effective repair.

REFERENCES

1. Steindler DA. Glial boundaries in the developing nervous system. Annu Rev Neurosci 1993; 16:445–470.
2. Schwab ME. Kapfhammer JP, Bandtlow CE. Inhibitors of neurite growth. Annu Rev Neurosci 1993; 16:565–595.
3. Lindvall O et al. Grafts of Fetal Dopamine Neurones Survive & Improve Motor Function in Parkinson's Disease. Science 1990; 247:574–577.
4. Deacon T et al. Histological evidence of fetal pig neural cell survival after transplantation into a patient with Parkinson's disease. Nat Med 1997; 3(3):350–353.
5. Minguell JJ, Erices A, Conget P. Mesenchymal stem cells. Exp Biol Med (Maywood) 2001; 226(6):507–520.
6. Pesce M, Gross MK, Scholer HR. In line with our ancestors: Oct-4 and the mammalian germ. Bioessays 1998; 20(9):722–732.
7. Pesce, M et al. Differential expression of the Oct-4 transcription factor during mouse germ cell differentiation. Mech Dev 1998; 71(1–2):89–98.
8. Solter D, Knowles BB. Monoclonal antibody defining a stage-specific mouse embryonic antigen (SSEA-1). Proc Natl Acad Sci USA 1978; 75(11):5565–5569.
9. Gooi HC et al. Stage-specific embryonic antigen involves alpha 1 goes to 3 fucosylated type 2 blood group chains. Nature 1981; 292(5819):156–158.

10. Andrews PW et al. Two monoclonal antibodies recognizing determinants on human embryonal carcinoma cells react specifically with the liver isozyme of human alkaline phosphatase. Hybridoma 1984; 3(1):33–39.
11. Andrews PW et al. Three monoclonal antibodies defining distinct differentiation antigens associated with different high molecular weight polypeptides on the surface of human embryonal carcinoma cells. Hybridoma 1984; 3(4):347–361.
12. Zhou S et al. The ABC transporter Bcrp1/ABCG2 is expressed in a wide variety of stem cells and is a molecular determinant of the side-population phenotype. Nat Med 2001; 7(9):1028–1034.
13. Murphy M, Drago J, Bartlett PF. Fibroblast growth factor stimulates the proliferation and differentiation of neural precursor cells in vitro. J Neurosci Res 1990; 25(4):463–475.
14. Reynolds BA, Weiss S. Generation of neurones and astrocytes from isolated cells of the adult mammalian central nervous system. Science 1992; 255:1707–1710.
15. Gage FH. Mammalian neural stem cells. Science 2000; 287(5457):1433–1438.
16. Roufosse CA et al. Circulating mesenchymal stem cells. Int J Biochem Cell Biol 2004; 36(4):585–597.
17. Thomson JA et al. Embryonic stem cell lines derived from human blastocysts. Science 1998; 282(5391):1145–1147.
18. Bain G et al. Embryonic stem cells express neuronal properties in vitro. Dev Biol 1995; 168(2):342–357.
19. Brustle O et al. In vitro-generated neural precursors participate in mammalian brain development. Proc Natl Acad Sci USA 1997; 94(26):14809–14814.
20. Carpenter MK et al. Enrichment of neurones and neural precursors from human embryonic stem cells. Exp Neurol 2001; 172(2):383–397.
21. Svendsen CN, Caldwell MA, Ostenfeld T. Human neural stem cells: isolation, expansion and transplantation. Brain Pathol 1999; 9(3):499–513.
22. McKay R. Stem cells in the central nervous system. Science 1997; 276:66–71.
23. Temple S, Alvarez-Buylla A. Stem cells in the adult mammalian central nervous system. Curr Opin Neurobiol 1999; 9(1):135–141.
24. Carpenter MK et al. In vitro expansion of a multipotent population of human neural progenitor cells. Exp Neurol 1999; 158(2):265–278.
25. Piper DR et al. Identification and characterization of neuronal precursors and their progeny from human fetal tissue. J Neurosci Res 2001; 66(3):356–358.
26. Uchida N et al. Direct isolation of human central nervous system stem cells. Proc Natl Acad Sci USA 2000; 97(26):14720–14725.
27. Ginis I, Rao MS. Toward cell replacement therapy: promises and caveats. Exp Neurol 2003; 184(1):61–77.
28. Palmer TD et al. Cell culture. Progenitor cells from human brain after death. Nature 2001; 411(6833):42–43.
29. Tohyama T et al. Nestin expression in embryonic human neuroepithelium and in human neuroepithelial tumor cells. Lab Invest 1992; 66(3):303–313.
30. Lendahl U, Zimmerman LB, McKay RD. CNS stem cells express a new class of intermediate filament protein. Cell 1990; 60(4):585–595.
31. Lin RC et al. Re-expression of the intermediate filament nestin in reactive astrocytes. Neurobiol Dis 1995; 2(2):79–85.
32. Sakakibara S et al. Mouse-Musashi-1, a neural RNA-Binding protein highly enriched in the mammalian CNS stem cell. Dev Biol 1996; 176(2):230–242.
33. Okano H, Imai T, Okabe M. Musashi: a translational regulator of cell fate. J Cell Sci 2002: 115(Pt 7):1355–1359.
34. Allen T, Lobe CG. A comparison of Notch, Hes and Grg expression during murine embryonic and post-natal development. Cell Mol Biol 1999; 45(5):687–708.
35. Graham GJ, Wright EG. Haemopoietic stem cells: their heterogeneity and regulation. Int J Exp Pathol 1997; 78(4):197–218.

36. Vescovi AL, Snyder EY. Establishment and properties of neural stem cell clones: plasticity in vitro and in vivo. Brain Pathol 1999; 9(3):569–598.
37. Modo M et al. Tracking transplanted stem cell migration using bifunctional, contrast agent-enhanced, magnetic resonance imaging. Neuroimage 2002; 17(2):803–811.
38. Ostenfeld T et al. Regional specification of rodent and human neurospheres. Brain Res Dev Brain Res 2002; 134(1–2):43–55.
39. Morrison SJ. Neuronal potential and lineage determination by neural stem cells. Curr Opin Cell Biol 2001; 13(6):666–672.
40. Morrison SJ. Neuronal differentiation: proneural genes inhibit gliogenesis. Curr Biol 2001; 11(9):R349–R351.
41. MayerProschel M et al. Isolation of lineage-restricted neuronal precursors from multipotent neuroepithelial stem cells. Neuron 1997; 19(4):773–785.
42. Rao MS. Multipotent and restricted precursors in the central nervous system. Anat Rec 1999; 257(4):137–148.
43. Cameron HA, McKay R. Discussion point—stem cells and neurogenesis in the adult brain. Curr Opin Neurobiol 1998; 8(5):677–680.
44. Kukekov VG et al. A nestin-negative precursor cell from the adult mouse brain gives rise to neurones and glia. Glia 1997; 21(4):399–407.
45. Lillien L. Neural progenitors and stem cells: mechanisms of progenitor heterogeneity. Curr Opin Neurobiol 1998; 8(1):37–44.
46. Lillien L, Wancio D. Changes in epidermal growth factor receptor expression and competence to generate glia regulate timing and choice of differentiation in the retina. Mol Cell Neurosci 1998; 10(5–6):296–308.
47. Burrows RC et al. Response, diversity and the timing of progenitor cell maturation are regulated by developmental changes in EGFR expression in the cortex. Neuron 1997; 19(2):251–267.
48. Tropepe V et al. Distinct neural stem cells proliferate in response to EGF and FGF in the developing mouse telencephalon. Dev Biol 1999; 208(1):166–188.
49. Meletis K, Frisen J. Blood on the tracks: a simple twist of fate? Trends Neurosci 2003; 26(6):292–296.
50. Howell JC et al. Pluripotent stem cells identified in multiple murine tissues. Ann N Y Acad Sci 2003; 996:158–173.
51. Sanchez-Ramos JR. Neural cells derived from adult bone marrow and umbilical cord blood. J Neurosci Res 2002; 69(6):880–893.
52. Abuljadayel IS. Induction of stem cell-like plasticity in mononuclear cells derived from unmobilised adult human peripheral blood. Curr Med Res Opin 2003; 19(5):355–375.
53. Milhaud, G. A human pluripotent stem cell in the blood of adults: towards a new cellular therapy for tissue repair. Bull Acad Natl Med 2001; 185(3):567–577; discussion 577–582.
54. Zhao Y, Glesne D, Huberman E. A human peripheral blood monocyte-derived subset acts as pluripotent stem cells. Proc Natl Acad Sci USA 2003; 100(5):2426–2431.
55. Ha Y et al. Intermediate filament nestin expressions in human cord blood monocytes (HCMNCs). Acta Neurochir (Wien) 2003; 145(6):483–487.
56. Ha Y et al. Neural phenotype expression of cultured human cord blood cells in vitro. Neuroreport 2001; 12(16):3523–3527.
57. Lee OK et al. Isolation of multi-potent mesenchymal stem cells from umbilical cord blood. Blood 2003; 103(5):1669–1675.
58. Mitchell KE et al. Matrix cells from Wharton's jelly form neurones and glia. Stem Cells 2003; 21(1):50–60.
59. Weiss ML et al. Transplantation of porcine umbilical cord matrix cells into the rat brain. Exp Neurol 2003; 182(2):288–299.
60. In't Anker PS et al. Amniotic fluid as a novel source of mesenchymal stem cells for therapeutic transplantation. Blood 2003; 102(4):1548–1549.
61. In't Anker PS et al. Isolation of mesenchymal stem cells of fetal or maternal origin from human placenta. Stem Cells 2004; 22(7):1338–1345.

62. Condorelli G et al. Cardiomyocytes induce endothelial cells to trans-differentiate into cardiac muscle: implications for myocardium regeneration. Proc Natl Acad Sci USA 2001; 98(19):10733–10738.

63. Frid MG, Kale VA, Stenmark KR. Mature vascular endothelium can give rise to smooth muscle cells via endothelial-mesenchymal transdifferentiation: in vitro analysis. Circ Res 2002; 90(11):1189–1196.

64. Ashjian PH et al. In vitro differentiation of human processed lipoaspirate cells into early neural progenitors. Plast Reconstr Surg 2003; 111(6):1922–1931.

65. Tholpady SS, Katz AJ, Ogle RC. Mesenchymal stem cells from rat visceral fat exhibit multi-potential differentiation in vitro. Anat Rec 2003; 272A(1):398–402.

66. Jay KE, Gallacher L, Bhatia M. Emergence of muscle and neural hematopoiesis in humans. Blood 2002; 100(9):3193–3202.

67. Mahmud N et al. Primate skeletal muscle contains cells capable of sustaining in vitro hematopoiesis. Exp Hematol 2002; 30(8):925–936.

68. Deng J et al. Neural trans-differentiation potential of hepatic oval cells in the neonatal mouse brain. Exp Neurol 2003; 182(2):373–382.

69. Gronthos S et al. Stem cell properties of human dental pulp stem cells. J Dent Res 2002; 81(8):531–535.

70. Liang L, Bickenbach JR. Somatic epidermal stem cells can produce multiple cell lineages during development. Stem Cells 2002; 20(1):21–31.

71. Prockop DJ. Marrow stromal cells as stem cells for nonhematopoietic tissues. Science 1997; 276(5309):71–74.

72. Scheffler B et al. Marrow-mindedness: a perspective on neuropoiesis. Trends Neurosci 1999; 22(8):348–357.

73. Tao H, Ma DD. Evidence for transdifferentiation of human bone marrow-derived stem cells: recent progress and controversies. Pathology 2003; 35(1):6–13.

74. Bruder SP, Fink DJ, Caplan AI. Mesenchymal stem cells in bone development, bone repair, and skeletal regeneration therapy. J Cell Biochem 1994; 56(3):283–294.

75. Caplan AI. The mesengenic process. Clin Plast Surg 1994; 21(3):429–435.

76. Lodie TA et al. Systematic analysis of reportedly distinct populations of multipotent bone marrow-derived stem cells reveals a lack of distinction. Tissue Eng 2002; 8(5):739–754.

77. Friedenstein AJ, Chailakhjan RK, Lalykina KS. The development of fibroblast colonies in monolayer cultures of guinea-pig bone marrow and spleen cells. Cell Tissue Kinet 1970; 3(4):393–403.

78. Friedenstein AJ et al. Stromal cells responsible for transferring the microenvironment of the hemopoietic tissues. Cloning in vitro and retransplantation in vivo. Transplantation 1974; 17(4):331–340.

79. Friedenstein AJ et al. Precursors for fibroblasts in different populations of hematopoietic cells as detected by the in vitro colony assay method. Exp Hematol 1974; 2(2):83–92.

80. Friedenstein AJ, Gorskaja JF, Kulagina NN. Fibroblast precursors in normal and irradiated mouse hematopoietic organs. Exp Hematol 1976; 4(5):267–274.

81. Friedenstein AJ. Stromal mechanisms of bone marrow: cloning in vitro and retransplantation in vivo. Hamatol Bluttransfus 1980; 25:19–29.

82. Friedenstein AJ et al. Marrow microenvironment transfer by heterotopic transplantation of freshly isolated and cultured cells in porous sponges. Exp Hematol 1982; 10(2):217–227.

83. Friedenstein AJ, Chailakhyan RK, Gerasimov UV. Bone marrow osteogenic stem cells: in vitro cultivation and transplantation in diffusion chambers. Cell Tissue Kinet 1987; 20(3):263–272.

84. Kuznetsov SA, Friedenstein AJ, Robey PG. Factors required for bone marrow stromal fibroblast colony formation in vitro. Br J Haematol 1997; 97(3):561–570.

85. Luria EA, Panasyuk AF, Friedenstein AY. Fibroblast colony formation from monolayer cultures of blood cells. Transfusion 1971; 11(6):345–349.

86. Kassem M, Kristiansen M, Abdallah BM. Mesenchymal stem cells: cell biology and potential use in therapy. Basic Clin Pharmacol Toxicol 2004; 95(5):209–214.

87. Krebsbach PH et al. Bone marrow stromal cells: characterization and clinical application. Crit Rev Oral Biol Med 1999; 10(2):165–181.

88. Jiang Y et al. Neuroectodermal differentiation from mouse multipotent adult progenitor cells. Proc Natl Acad Sci USA 2003; 100(1):11854–11860.

89. Hung SC et al. In vitro differentiation of size-sieved stem cells into electrically active neural cells. Stem Cells 2002; 20(6):522–529.

90. Kohyama J et al. Brain from bone: efficient "meta-differentiation" of marrow stroma-derived mature osteoblasts to neurones with Noggin or a demethylating agent. Differentiation 2001; 68(4–5):235–244.

91. Speert H. Obstetric-gynecologic eponyms; Thomas Wharton and the jelly of the umbilical cord. Obstet Gynecol 1956; 8(3):380–382.

92. Kobayashi K, Kubota T, Aso T. Study on myofibroblast differentiation in the stromal cells of Wharton's jelly: expression and localization of alpha-smooth muscle actin. Early Hum Dev 1998; 51(3):223–233.

93. Quesenberry PJ et al. Correlates between hematopoiesis and neuropoiesis: neural stem cells. J Neurotrauma 1999; 16(8):661–666.

94. Weissman IL. Stem cells—lessons from hematopoieses. In Gage F, Christen Y, eds. Isolation, Characterisation & Utilisation of CNS Stem Cells. Berlin, Heidelberg: Spring Verlag, 1997:1–8.

95. Morrison SJ. The last shall not be first: the ordered generation of progeny from stem cells. Neuron 2000; 28(1):1–3.

96. Gardner LJ et al. Identification of CD56 and CD57 by flow cytometry in Ewing's sarcoma or primitive neuroectodermal tumor. Virchows Arch 1998; 433(1):35–40.

97. Chung DH et al. ILK (beta1-integrin-linked protein kinase): a novel immunohistochemical marker for Ewing's sarcoma and primitive neuroectodermal tumour. Virchows Arch 1998; 433(2):113–117.

98. Sugimoto T, Umezawa A, Hata J. Neurogenic potential of Ewing's sarcoma cells. Virchows Arch 1997; 430(1):41–46.

99. Bjornson CR et al. Turning brain into blood: a hematopoietic fate adopted by adult neural stem cells in vivo. Science 1999; 283(5401):534–537.

100. Galli R et al. Skeletal myogenic potential of human and mouse neural stem cells. Nat Neurosci 2000; 3(10):986–991.

101. Rietze RL et al. Purification of a pluripotent neural stem cell from the adult mouse brain. Nature 2001; 412(6848):736–739.

102. Vescovi A et al. Neural stem cells: plasticity and their transdifferentiation potential. Cells Tissues Organs 2002; 171(1):64–76.

103. Mezey E, Chandross KJ. Bone marrow: a possible alternative source of cells in the adult nervous system. Eur J Pharmacol 2000; 405(1–3):297–302.

104. Lu P, Blesch A, Tuszynski MH. Induction of bone marrow stromal cells to neurones: differentiation, transdifferentiation, or artifact? J Neurosci Res 2004; 77(2):174–191.

105. Brazelton TR et al. From marrow to brain: expression of neuronal phenotypes in adult mice. Science 2000; 290(5497):1775–1779.

106. Mezey E et al. Comment on failure of bone marrow cells to transdifferentiate into neural cells in vivo. Science 2003; 299(5610):1184; author reply 1184.

107. Mezey E et al. Transplanted bone marrow generates new neurones in human brains. Proc Natl Acad Sci USA 2003; 100(3):1364–1369.

108. Weimann JM et al. Contribution of transplanted bone marrow cells to Purkinje neurones in human adult brains. Proc Natl Acad Sci USA 2003; 100(4):2088–2093.

109. Cogle CR et al. Bone marrow transdifferentiation in brain after transplantation: a retrospective study. Lancet 2004; 363(9419):1432–1437.

110. Khosrotehrani K, Bianchi DW. Multi-lineage potential of fetal cells in maternal tissue: a legacy in reverse. J Cell Sci 2005; 118(Pt 8):1559–1563.

111. Woodbury D et al. Adult rat and human bone marrow stromal cells differentiate into neurones. J Neurosci Res 2000; 61(4):364–370.

112. Buzanska L et al. Human cord blood-derived cells attain neuronal and glial features in vitro. J Cell Sci 2002; 115(Pt 10):2131–2138.

113. Vogel W et al. Heterogeneity among human bone marrow-derived mesenchymal stem cells and neural progenitor cells. Haematologica 2003; 88(2):126–133.

114. Donnelly DS, Krause DS. Hematopoietic stem cells can be CD34+ or CD34−. Leuk Lymphoma 2001; 40(3–4):221–234.

115. Bonilla S et al. Haematopoietic progenitor cells from adult bone marrow differentiate into cells that express oligodendroglial antigens in the neonatal mouse brain. Eur J Neurosci 2002; 15(3):575–582.

116. Hao HN et al. Fetal human hematopoietic stem cells can differentiate sequentially into neural stem cells and then astrocytes in vitro. J Hematother Stem Cell Res 2003; 12(1):23–32.

117. Goolsby J et al. Hematopoietic progenitors express neural genes. Proc Natl Acad Sci USA 2003; 100(25):14926–14931.

118. Krause DS et al. Multi-organ, multi-lineage engraftment by a single bone marrow-derived stem cell. Cell 2001; 105(3):369–377.

119. Terada N et al. Bone marrow cells adopt the phenotype of other cells by spontaneous cell fusion. Nature 2002; 416(6880):542–545.

120. Ying QL et al. Changing potency by spontaneous fusion. Nature 2002; 416(6880):545–548.

121. Anderson DJ, Gage FH, Weissman IL. Can stem cells cross lineage boundaries? Nat Med 2001; 7(4):393–395.

122. Morshead CM et al. Hematopoietic competence is a rare property of neural stem cells that may depend on genetic and epigenetic alterations. Nat Med 2002; 8(3):268–273.

123. Alvarez-Dolado M et al. Fusion of bone-marrow-derived cells with Purkinje neurones, cardiomyocytes and hepatocytes. Nature 2003; 425(6961):968–973.

124. Black IB, Woodbury D. Adult rat and human bone marrow stromal stem cells differentiate into neurones. Blood Cells Mol Dis 2001; 27(3):632–636.

125. Barry FP, Murphy JM. Mesenchymal stem cells: clinical applications and biological characterization. Int J Biochem Cell Biol 2004; 36(4):568–584.

126. McKinney-Freeman SL et al. Muscle-derived hematopoietic stem cells are hematopoietic in origin. Proc Natl Acad Sci USA 2002; 99(3):1341–1346.

127. Zhao LR et al. Human bone marrow stem cells exhibit neural phenotypes and ameliorate neurological deficits after grafting into the ischemic brain of rats. Exp Neurol 2002; 174(1):11–20.

128. Wells WA. Is transdifferentiation in trouble? J Cell Biol 2002; 157(1):15–18.

129. Labat ML. Stem cells and the promise of eternal youth: embryonic versus adult stem cells. Biomed Pharmacother 2001; 55(4):179–185.

130. Herzog EL, Chai L, Krause DS. Plasticity of marrow derived stem cells. Blood 2003; 102(10): 3483–3493.

131. Reubinoff BE et al. Neural progenitors from human embryonic stem cells. Nat Biotechnol 2001; 19(12):1134–1140.

132. Zhang SC et al. In vitro differentiation of transplantable neural precursors from human embryonic stem cells. Nat Biotechnol 2001; 19(12):1129–1133.

133. Flax JD et al. Engraftable human neural stem cells respond to developmental cues, replace neurones, and express foreign genes. Nat Biotechnol 1998; 16(11):1033–1039.

134. Svendsen CN et al. Survival and differentiation of rat and human epidermal growth factor-responsive precursor cells following grafting into the lesioned adult central nervous system. Exp Neurol 1996; 137(2):376–388.

135. McDonald JW et al. Transplanted embryonic stem cells survive, differentiate and promote recovery in injured rat spinal cord. Nat Med 1999; 5(12):1410–1412.

136. Brustle O et al. Embryonic stem cell-derived glial precursors: a source of myelinating transplants. Science 1999; 285(5428):754–756.

137. Liu S et al. Embryonic stem cells differentiate into oligodendrocytes and myelinate in culture and after spinal cord transplantation. Proc Natl Acad Sci USA 2000; 97(11):6126–6131.

138. Snyder EY et al. Multipotent neural precursors can differentiate toward replacement of neurones undergoing targeted apoptotic degeneration in adult mouse neocortex. Proc Nat Acad Sci USA 1997; 94(21):11663–11668.

139. Ourednik V et al. Segregation of human neural stem cells in the developing primate forebrain. Science 2001; 293(5536):1820–1824.

140. Englund U et al. Transplantation of human neural progenitor cells into the neonatal rat brain: extensive migration and differentiation with long-distance axonal projections. Exp Neurol 2002; 173(1):1–21.

141. Englund U, Bjorklund A, Wictorin K. Migration patterns and phenotypic differentiation of long-term expanded human neural progenitor cells after transplantation into the adult rat brain. Brain Res Dev Brain Res 2002; 134(1–2):123–141.

142. Clarke DL et al. Generalized potential of adult neural stem cells. Science 2000; 288(5471):1660–1663.

143. Carlen M et al. Functional integration of adult-born neurones. Curr Biol 2002; 12(7):606–608.

144. Chu K et al. Human neural stem cells can migrate, differentiate, and integrate after intravenous transplantation in adult rats with transient forebrain ischemia. Neurosci Lett 2003; 343(2):129–133.

145. Eglitis MA, Mezey E. Hematopoietic cells differentiate into both microglia and macroglia in the brains of adult mice. Proc Natl Acad Sci USA 1997; 94(8):4080–4085.

146. Le Blanc K. Immunomodulatory effects of fetal and adult mesenchymal stem cells. Cytotherapy 2003; 5(6):485–489.

147. Le Blanc K et al. HLA expression and immunologic properties of differentiated and undifferentiated mesenchymal stem cells. Exp Hematol 2003; 31(10):890–896.

148. Hori J et al. Neural progenitor cells lack immunogenicity and resist destruction as allografts. Stem Cells 2003; 21(4):405–416.

149. Armstrong RJ et al. Transplantation of expanded neural precursor cells from the developing pig ventral mesencephalon in a rat model of Parkinson's disease. Exp Brain Res 2003; 151(2):204–217.

150. Ortiz LA et al. Mesenchymal stem cell engraftment in lung is enhanced in response to bleomycin exposure and ameliorates its fibrotic effects. Proc Natl Acad Sci USA 2003; 100(14):8407–8411.

151. Magrassi L et al. Freshly dissociated fetal neural stem/progenitor cells do not turn into blood. Mol Cell Neurosci 2003; 22(2):179–187.

152. Magrassi L. Differences and similarities among phenotypes of mesenchymal and neural stem cells. Haematologica 2003; 88(2):121.

153. Kicic A et al. Differentiation of marrow stromal cells into photoreceptors in the rat eye. J Neurosci 2003; 23(21):7742–7749.

154. Lewis ID, Verfaillie CM. Multi-lineage expansion potential of primitive hematopoietic progenitors: superiority of umbilical cord blood compared to mobilized peripheral blood. Exp Hematol 2000; 28(9):1087–1095.

155. Bonnet, pers. commun., Nov. 2003.

156. Zheng T, Steindler DA, Laywell ED. Transplantation of an indigenous neural stem cell population leading to hyperplasia and atypical integration. Cloning Stem Cells 2002; 4(1):3–8.

157. Mehmet, pers. commun., Nov. 2003

158. Keene CD et al. Neural differentiation and incorporation of bone marrow-derived multipotent adult progenitor cells after single cell transplantation into blastocyst stage mouse embryos. Cell Transplant 2003; 12(3):201–213.

159. Rossi F, Cattaneo E. Opinion: neural stem cell therapy for neurological diseases: dreams and reality. Nat Rev Neurosci 2002; 3(5):401–409.

160. Fearnley JM, Lees AJ. Ageing and Parkinson's disease: substantia nigra regional selectivity. Brain 1991; 114(Pt 5):2283–2301.

161. Kish SJ, Shannak K, Hornykiewicz O. Uneven pattern of dopamine loss in the striatum of patients with idiopathic Parkinson's disease. Pathophysiologic and clinical implications. N Engl J Med 1988; 318(14):876–880.

162. Zhang J et al. Human bone marrow stromal cell treatment improves neurological functional recovery in EAE mice. Exp Neurol 2005; 195(1):16–26.

163. Lescaudron L, Unni D, Dunbar GL. Autologous adult bone marrow stem cell transplantation in an animal model of huntington's disease: behavioral and morphological outcomes. Int J Neurosci 2003; 113(7):945–956.

164. Cheng L et al. Human adult marrow cells support prolonged expansion of human embryonic stem cells in culture. Stem Cells 2003; 21(2):131–142.

165. Samii A, Nutt JG, Ransom BR. Parkinson's disease. Lancet 2004; 363(9423):1783–1793.

166. Freed CR et al. Transplantation of embryonic dopamine neurones for severe Parkinson's disease. N Engl J Med 2001; 344(10):710–719.

167. Studer L, Tabar V, McKay RDG. Transplantation of expanded mesencephalic precursors leads to recovery in parkinsonian rats. Nat Neurosci 1998; 1(4):290–295.

168. Laywell ED, Kukekov VG, Steindler DA. Multipotent neurospheres can be derived from forebrain subependymal zone and spinal cord of adult mice after protracted postmortem intervals. Experimental Neurology 1999; 156(2):430–433.

169. Svendsen CN et al. Long-term survival of human central nervous system progenitor cells transplanted into a rat model of Parkinson's disease. Experimental Neurology 1997; 148(1):135–146.

170. Weiss S. Pathways for Neural Stem Cell Biology & Repair., in Nature. 1999:850–851.

171. Fricker RA et al. Site-specific migration and neuronal differentiation of human neural progenitor cells after transplantation in the adult rat brain. J Neurosci 1999; 19(14):5990–6005.

172. Sawamoto K et al. Generation of dopaminergic neurones in the adult brain from mesencephalic precursor cells labeled with a nestin-GFP transgene. J Neurosci 2001; 21(11):3895–3903.

173. Kim JH et al. Dopamine neurones derived from embryonic stem cells function in an animal model of Parkinson's disease. Nature 2002; 418(6893):50–56.

174. Crossman AR. Functional anatomy of movement disorders. J Anat 2000; 196(Pt 4): 519–525.

175. Crossman AR. Primate models of dyskinesia: the experimental approach to the study of basal ganglia-related involuntary movement disorders. Neuroscience 1987; 21(1):1–40.

176. Isacson O, Breakefield XO. Benefits and risks of hosting animal cells in the human brain. Nat Med 1997; 3(9):964–969.

177. Philpott LM et al. Neuropsychological functioning following fetal striatal transplantation in Huntington's chorea: three case presentations. Cell Transplant 1997; 6(3):203–212.

178. Armstrong RJ et al. Survival, neuronal differentiation, and fiber outgrowth of propagated human neural precursor grafts in an animal model of Huntington's disease. Cell Transplant 2000; 9(1):55–64.

179. Warlow C et al. Stroke. Lancet 2003; 362(9391):1211–1224.

180. Azzopardi D et al. Pilot study of treatment with whole body hypothermia for neonatal encephalopathy. Pediatrics 2000; 106(4):684–694.

181. Hill MD, Hachinski V. Stroke treatment: time is brain. Lancet 1998; 352(Suppl 3): SIII10–SIII14.

182. Gluckman PD et al. Selective head cooling with mild systemic hypothermia after neonatal encephalopathy: multicentre randomised trial. Lancet 2005; 365(9460):663–670.

183. Sorensen JC et al. Fetal neocortical tissue blocks implanted in brain infarcts of adult rats interconnect with the host brain. Exp Neurol 1996; 138(2):227–235.

184. Hodges H et al. Contrasting effects of fetal CA1 and CA3 hippocampal grafts on deficits in spatial learning and working memory induced by global cerebral ischaemia in rats. Neuroscience 1996; 72(4):959–988.

185. Chopp M, Li Y. Treatment of neural injury with marrow stromal cells. Lancet Neurol 2002; 1(2):92–100.

186. Chopp M, Li Y. Treatment of stroke with marrow stromal cells. International Congress Series 2003; 1252:465–470.

187. Chen J et al. Intravenous bone marrow stromal cell therapy reduces apoptosis and promotes endogenous cell proliferation after stroke in female rat. J Neurosci Res 2003; 73(6):778–786.

188. Chen J et al. Therapeutic benefit of intracerebral transplantation of bone marrow stromal cells after cerebral ischemia in rats. J Neurol Sci 2001; 189(1–2):49–57.

189. Chen J et al. Therapeutic benefit of intravenous administration of bone marrow stromal cells after cerebral ischemia in rats. Stroke 2001; 32(4):1005–1011.

190. Chen J et al. Combination therapy of stroke in rats with a nitric oxide donor and human bone marrow stromal cells enhances angiogenesis and neurogenesis. Brain Res 2004; 1005(1–2):21–28.

191. Chen J et al. Intravenous administration of human umbilical cord blood reduces behavioral deficits after stroke in rats. Stroke 2001; 32(11):2682–2688.

192. Chen J et al. Intravenous administration of human bone marrow stromal cells induces angiogenesis in the ischemic boundary zone after stroke in rats. Circ Res 2003; 92(6):692–699.

193. Li Y et al. Human marrow stromal cell therapy for stroke in rat: neurotrophins and functional recovery. Neurology 2002; 59(4):514–523.

194. Li Y, Chen, J, Chopp, M. Adult bone marrow transplantation after stroke in adult rats. Cell Transplant 2001; 10(1):31–40.

195. Li Y et al. Treatment of stroke in rat with intracarotid administration of marrow stromal cells. Neurology 2001; 56(12):1666–1672.

196. Li Y et al. Intrastriatal transplantation of bone marrow nonhematopoietic cells improves functional recovery after stroke in adult mice. J Cereb Blood Flow Metab 2000; 20(9):1311–1319.

197. Li Y et al. Gliosis and brain remodeling after treatment of stroke in rats with marrow stromal cells. Glia 2004; 49(3):407–417.

198. Okano H et al. Transplantation of neural stem cells into the spinal cord after injury. Semin Cell Dev Biol 2003; 14(3):191–198.

199. Shear DA et al. Neural progenitor cell transplants promote long-term functional recovery after traumatic brain injury. Brain Res 2004; 1026(1):11–22.

200. Lu D et al. Adult bone marrow stromal cells administered intravenously to rats after traumatic brain injury migrate into brain and improve neurological outcome. Neuroreport 2001; 12(3):559–563.

201. Chopp M et al. Spinal cord injury in rat: treatment with bone marrow stromal cell transplantation. Neuroreport 2000; 11(13):3001–3005.

202. Compston A, Coles A. Multiple sclerosis. Lancet 2002; 359(9313):1221–1231.

203. Hall GL, Compston A, Scolding NJ. Beta-interferon and multiple sclerosis. Trends Neurosci 1997; 20(2):63–67.

204. Scolding N et al. Oligodendrocyte progenitors are present in the normal adult human CNS and in the lesions of multiple sclerosis. Brain 1998; 121(Pt 12):2221–2228.

205. Wolswijk G. Oligodendrocyte regeneration in the adult rodent CNS and the failure of this process in multiple sclerosis. Prog Brain Res 1998; 117:233–247.

206. Chang A et al. Premyelinating oligodendrocytes in chronic lesions of multiple sclerosis. N Engl J Med 2002; 346(3):165–173.

207. Kohama I et al. Transplantation of cryopreserved adult human Schwann cells enhances axonal conduction in demyelinated spinal cord. J Neurosci 2001; 21(3):944–950.

208. Barnett SC et al. Identification of a human olfactory ensheathing cell that can effect transplant-mediated remyelination of demyelinated CNS axons. Brain 2000; 123(Pt 8): 1581–1588.

209. Yandava BD, Billinghurst LL, Snyder EY. "Global" cell replacement is feasible via neural stem cell transplantation: evidence from the dysmyelinated shiverer mouse brain. Proc Nat Acad Sci USA 1999; 96(12):7029–7034.

210. Hammang JP, Archer DR, Duncan ID. Myelination following transplantation of EGF-responsive neural stem cells into a myelin-deficient environment. Exp Neurol 1997; 147(1):84–95.

211. Avellana-Adalid V et al. Expansion of rat oligodendrocyte progenitors into proliferative "oligospheres" that retain differentiation potential. J Neurosci Res 1996; 45(5):558–570.

212. Zhang SC, Lipsitz D, Duncan ID. Self-renewing canine oligodendroglial progenitor expanded as oligospheres. J Neurosci Res 1998; 54(2):181–190.

213. Zhang SC et al. Generation of oligodendroglial progenitors from neural stem cells. J Neurocytol 1998; 27(7):475–489.

214. Windrem MS et al. Fetal and adult human oligodendrocyte progenitor cell isolates myelinate the congenitally dysmyelinated brain. Nat Med 2004; 10(1):93–97.

215. Park KI et al. Global gene and cell replacement strategies via stem cells. Gene Ther 2002; 9(10):613–624.

216. Bizen A et al. Transplantation of mesenchymal stem cells derived from the bone marrow into the demyelinated spinal cord. Int Cong Series 2003; 1252:471–475.

217. Burns A, Byrne EJ, Maurer K. Alzheimer's disease. Lancet 2002; 360(9327):163–165.

218. Silani V et al. Stem-cell therapy for amyotrophic lateral sclerosis. Lancet 2004; 364(9429):200–202.

219. Wu P et al. Region-specific generation of cholinergic neurones from fetal human neural stem cells grafted in adult rat. Nat Neurosci 2002; 5(12):1271–1278.

220. Snyder EY, Taylor RM, Wolfe JH. Neural progenitor cell engraftment corrects lysosomal storage throughout the MPS VII mouse brain. Nature 1995; 374(6520):367–370.

221. Aboody KS et al. From the cover: neural stem cells display extensive tropism for pathology in adult brain: evidence from intracranial gliomas. Proc Natl Acad Sci USA 2000; 97(23):12846–12851.

222. Benedetti S et al. Gene therapy of experimental brain tumors using neural progenitor cells. Nat Med 2000; 6(4):447–450.

223. Ehtesham M et al. The use of interleukin 12-secreting neural stem cells for the treatment of intracranial glioma. Cancer Res 2002; 62(20):5657–5663.

224. Ehtesham M et al. Induction of glioblastoma apoptosis using neural stem cell-mediated delivery of tumor necrosis factor-related apoptosis-inducing ligand. Cancer Res 2002; 62(24):7170–7174.

225. Nakamura K et al. Antitumor effect of genetically engineered mesenchymal stem cells in a rat glioma model. Gene Ther 2004; 11(14):1155–1164.

226. Gieselmann V et al. Gene therapy: prospects for glycolipid storage diseases. Philos Trans R Soc Lond B Biol Sci 2003; 358(1433):921–925.

227. Meng XL et al. Brain transplantation of genetically engineered human neural stem cells globally corrects brain lesions in the mucopolysaccharidosis type VII mouse. J Neurosci Res 2003; 74(2):266–277.

228. Maalouf EF et al. Magnetic resonance imaging of the brain in a cohort of extremely preterm infants. J Pediatr 1999; 135(3):351–357.

229. Counsell SJ et al. Diffusion-weighted imaging of the brain in preterm infants with focal and diffuse white matter abnormality. Pediatrics 2003; 112(1 Pt 1):1–7.

230. Ness JK et al. Perinatal hypoxia-ischemia induces apoptotic and excitotoxic death of periventricular white matter oligodendrocyte progenitors. Dev Neurosci 2001; 23(3):203–208.

231. Li Y et al. Apoptosis and protein expression after focal cerebral ischemia in rat. Brain Res 1997; 765(2):301–312.

232. Li Y, Chopp M. Temporal profile of nestin expression after focal cerebral ischemia in adult rat. Brain Res 1999; 838(1–2):1–10.

233. Lindvall O, McKay R. Brain repair by cell replacement and regeneration. Proc Natl Acad Sci USA 2003; 100(13):7430–7431.

234. Zhao M et al. Evidence for neurogenesis in the adult mammalian substantia nigra. Proc Natl Acad Sci USA 2003; 100(13):7925–7930.

235. Magavi SS, Leavitt BR, Macklis JD. Induction of neurogenesis in the neocortex of adult mice. Nature 2000; 405(6789):951–955.

236. Arvidsson A et al. Neuronal replacement from endogenous precursors in the adult brain after stroke. Nat Med 2002; 8(9):963–970.

237. Parent JM et al. Rat forebrain neurogenesis and striatal neuron replacement after focal stroke. Ann Neurol 2002; 52(6):802–813.

238. Teramoto T et al. EGF amplifies the replacement of parvalbumin-expressing striatal interneurones after ischemia. J Clin Invest 2003; 111(8):1125–1132.

239. Nakatomi H et al. Regeneration of hippocampal pyramidal neurones after ischemic brain injury by recruitment of endogenous neural progenitors. Cell 2002; 110(4):429–441.

240. Abe K. Therapeutic potential of neurotrophic factors and neural stem cells against ischemic brain injury. J Cereb Blood Flow Metab 2000; 20(10):1393–1408.
241. Liu J et al. Increased neurogenesis in the dentate gyrus after transient global ischemia in gerbils. J Neurosci 1998; 18(19):7768–7778.
242. Li Y, Chen J, Chopp M. Cell proliferation and differentiation from ependymal, subependymal and choroid plexus cells in response to stroke in rats. J Neurol Sci 2002; 193(2):137–146.
243. Parent JM, Valentin VV, Lowenstein DH. Prolonged seizures increase proliferating neuro-blasts in the adult rat subventricular zone-olfactory bulb pathway. J Neurosci 2002; 22(8):3174–3188.
244. Miao N et al. Sonic hedgehog promotes the survival of specific CNS neuron populations and protects these cells from toxic insult In vitro. J Neurosci 1997; 17(15):5891–5899.
245. Craig CG et al. In Vivo Growth Factor Expansion of Endogenous Subependymal Neural Precursor Cell Populations in the Adult Mouse Brain. J Neurosci 1996; 16(8):2649–2658.
246. Kuhn HG et al. Epidermal growth factor and fibroblast growth factor-2 have different effects on neural progenitors in the adult rat brain. J Neurosci 1997; 17(15):5820–5829.
247. Taylor RM, Snyder EY. Widespread engraftment of neural progenitor and stem-like cells throughout the mouse brain. Transplant Proc 1997; 29(1–2):845–847.
248. Kokaia M et al. Suppressed epileptogenesis in BDNF mutant mice. Exp Neurol 1995; 133(2):215–224.
249. Sheen VL et al. Neural precursor differentiation following transplantation into neocortex is dependent on intrinsic developmental state and receptor competence. Exp Neurol 1999; 158(1):47–62.
250. Chang A et al. NG2-positive oligodendrocyte progenitor cells in adult human brain and mul-tiple sclerosis lesions. J Neurosci 2000; 20(17):6404–6412.
251. Smith PM, Blakemore WF. Porcine neural progenitors require commitment to the oligoden-drocyte lineage prior to transplantation in order to achieve significant remyelination of demyelinated lesions in the adult CNS. Eur J Neurosci 2000; 12(7):2414–2424.
252. Zhang SC, Ge B, Duncan ID. Adult brain retains the potential to generate oligodendroglial progenitors with extensive myelination capacity. Proc Nat Acad Sci USA 1999; 96(7):4089–4094.
253. Wang L et al. Ischemic cerebral tissue and MCP-1 enhance rat bone marrow stromal cell migration in interface culture. Exp Hematol 2002; 30(7):831–836.
254. Wang L et al. MCP-1, MIP-1, IL-8 and ischemic cerebral tissue enhance human bone marrow stromal cell migration in interface culture. Hematology 2002; 7(2):113–117.
255. Schierle GS et al. Caspase inhibition reduces apoptosis and increases survival of nigral trans-plants. Nat Med 1999; 5(1):97–100.
256. Brundin P et al. Improving the survival of grafted dopaminergic neurones: a review over current approaches. Cell Transplant 2000; 9(2):179–195.
257. Emgard M, Blomgren K, Brundin P. Characterisation of cell damage and death in embryonic mesencephalic tissue: a study on ultrastructure, vital stains and protease activity. Neuro-science 2002; 115(4):1177–1187.
258. Emgard M et al. Both apoptosis and necrosis occur early after intracerebral grafting of ventral mesencephalic tissue: a role for protease activation. J Neurochem 2003; 86(5):1223–1232.
259. Hagell P, Brundin P. Cell survival and clinical outcome following intrastriatal transplantation in Parkinson disease. J Neuropathol Exp Neurol 2001; 60(8):741–752.
260. Hansson O et al. Additive effects of caspase inhibitor and lazaroid on the survival of transplanted rat and human embryonic dopamine neurones. Exp Neurol 2000; 164(1):102–111.
261. Schierle GS, Brundin P. Excitotoxicity plays a role in the death of tyrosine hydroxylase-immunopositive nigral neurones cultured in serum-free medium. Exp Neurol 1999; 157(2):338–348.
262. Lindvall O. Parkinson disease. Stem cell transplantation. Lancet, 2001; 358 Suppl:S48.
263. Mangi AA et al. Mesenchymal stem cells modified with Akt prevent remodeling and restore performance of infarcted hearts. Nat Med 2003; 9(9):1195–1201.

264. Koc, O.N. and Gerson SL. Akt helps stem cells heal the heart. Nat Med 2003; 9(9): 1109–1110.

265. Chen J et al. Caspase inhibition by Z-VAD increases the survival of grafted bone marrow cells and improves functional outcome after MCAo in rats. J Neurol Sci, 2002. 199(1–2):17–24.

266. Takayama H et al. Basic fibroblast growth factor increases dopaminergic graft survival and function in a rat model of Parkinson's disease. Nat Med 1995; 1(1):53–58.

267. Rolletschek A et al. Differentiation of embryonic stem cell-derived dopaminergic neurones is enhanced by survival-promoting factors. Mech Dev 2001; 105(1–2):93–104.

268. Schierle GS et al. Differential effects of Bcl-2 overexpression on fibre outgrowth and survival of embryonic dopaminergic neurones in intracerebral transplants. Eur J Neurosci 1999; 11(9):3073–3081.

269. Ourednik V et al. Neural stem cells—a versatile tool for cell replacement and gene therapy in the central nervous system. Clin Genet 1999; 56(4):267–278.

270. Sabate O et al. Transplantation to the rat brain of human neural progenitors that were genetically modified using adenoviruses. Nat Genet 1995; 9(3):256–260.

271. Sabate O et al. Adenovirus for neurodegenerative diseases: in vivo strategies and ex vivo gene therapy using human neural progenitors. Clin Neurosci 1995; 3(5):317–321.

272. Ourednik J et al. Ectopic expression of the neural cell adhesion molecule L1 in astrocytes leads to changes in the development of the corticospinal tract. Eur J Neurosci 2001; 14(9):1464–1474.

273. Tabbal S, Fahn S, Frucht S. Fetal tissue transplantation in Parkinson's disease. Curr Opin Neurol 1998; 11(4):341–349.

274. Daadi MM, Weiss S. Generation of tyrosine hydroxylase-producing neurones from precursors of the embryonic and adult forebrain. J Neurosci 1999; 19(11):4484–4497.

275. Teng YD et al. Functional recovery following traumatic spinal cord injury mediated by a unique polymer scaffold seeded with neural stem cells. Proc Natl Acad Sci USA 2002; 99(5):3024–3029.

276. Gnecchi M et al. Paracrine action accounts for marked protection of ischemic heart by Akt-modified mesenchymal stem cells. Nat Med 2005; 11(4):367–368.

277. Svendsen, C., Adult versus embryonic stem cells: which is the way forward? Trends Neurosci 2000; 23(10):450.

278. The University of Western Australia, School of Anatomy and Human Biology, Teaching Website (http://www.lab.anhb.uwa.edu.au/mb140/CorePages/Connective/Connect.htm #labmucous).

13
Adult Stem Cells and Gene Therapy

Ilaria Bellantuono
Academic Unit of Bone Biology, The Sheffield Medical School, Sheffield, U.K.

Leslie J. Fairbairn†
Cancer Research U.K. Gene Therapy Group, Paterson Institute for Cancer Research, Manchester, U.K.

INTRODUCTION

Stem cells are viewed as an important target for gene therapy because of their ability to self-renew at least for the lifetime of the individual and to give rise to a large number of differentiated progenies. Thus, the transfer of a therapeutic gene to stem cells has the potential to provide long-lasting correction of a number of acquired or inherited disorders. Most research in stem cells has focused on either embryonic or adult stem cells. Embryonic stem cells are isolated from early embryos and generally possess enormous proliferative capacity and the potential to differentiate into multiple lineages (1,2). They are more versatile when used for the regeneration or repair of different tissues because of their multipotentiality. However, ethical and technical concerns limit the use of those cells due to the need to use human embryos for their derivation and in the form of an allograft, unless therapeutic cloning techniques are utilized to generate autologous embryonic stem cells (3).

Adult stem cells can be isolated from several tissues of individuals and until recently their capacity for multilineage differentiation has been considered limited. Recently, the potential of stem cells from some tissues has been revisited, and the results have generated some controversy (4,5). Some studies suggested that adult stem cells are not restricted to generating progenies identical to their tissue of origin but instead exhibit plasticity, which can be harnessed to generate progenies of all germ layers (6). Thus, unfractionated bone marrow or bone marrow cells enriched by various methods for stem-cell activity have apparently contributed to multiple nonhematopoietic tissues following administration to lethally irradiated or injured recipient mice and humans (7–15). Similarly, studies have suggested that brain or muscle-derived stem cells harbored hematopoietic potential (14,16,17). However, caution is warranted given that a number of studies have failed to reproduce such results (18–22). In green fluorescent protein (GFP+ :GFP−) parabiotic

---

†This work is dedicated to Lez Fairbairn, who unexpectedly passed away just after this chapter was written. His mentoring and Friendship are greatly missed.

mice, substantial chimerism of hematopoietic but not nonhematopoietic cells was found, indicating that "transdifferentiation" of circulating hemopoietic stem cells (HSCs) and/or their progenies was an extremely rare event, if it occurred at all (22). Other studies have shown that stem-cell plasticity may have been mistaken for cell fusion (23,24).

In view of these contradictory results, assessment of plasticity needs to be more rigorous. Reconstitution of multiple tissues should come from a single stem cell, well characterized as derived from a different tissue by reliable markers. The putative cells so derived should be differentiated and functional, and fusion should be excluded. Such criteria are yet to be met and consequently "transdifferentiation" remains controversial. For this reason, we will consider in this chapter, only those applications derived from the use of genetically modified adult stem cells in their tissue of origin.

ADULT STEM CELLS FOR GENE THERAPY

Regardless of how restricted the potential of adult stem cells may or may not be, they represent an autologous source of cells with enough proliferative and differentiative capacity to provide a wealth of opportunities for the treatment or prevention of disease. So far, stem cells have been derived from bone marrow (25), skin (26), gut (27,28), muscle (29), brain (30), liver (31), and pancreas (32). As most gene therapy protocols involve ex vivo transduction of the cells, when considering whether a stem-cell type is a suitable target for gene therapy, reliable methods of isolation, identification, and in vitro culture are major requirements. Furthermore, as gene transfer and expression are likely to be required over an extended period of time, it is crucial that in vitro and in vivo functional assays are available to test that long-term repopulating stem cells have been transduced and that transduction is polyclonal. Even with the most rigorous isolation protocol, only a small percentage of enriched cells are bona fide stem cells, which replicate infrequently but have the capacity of extended growth and when transplanted are responsible for long-term engraftment (33–35). The majority of the cells are progenitor (or transient amplifying) cells, which replicate frequently but cycle only a limited number of times before undergoing terminal differentiation and on transplantation, provide only short-term engraftment (33–36). This latter cell type is easier to transduce with current vector systems and this creates an initial, false impression of efficient transduction, whereas disappointing levels of gene-modified long-term repopulating cells are more normally seen (37–39).

Stem cells that meet those requirements listed above are mainly HSCs and one type of skin stem cell, the keratinocyte. Stem cells from other tissues such as liver and pancreas are currently poorly characterized (40–44). Endothelial progenitor cells (EPCs) have been isolated from peripheral blood, cord blood, and bone marrow (45–47) and it has been shown that in adults these can home to sites of neovascularization and differentiate into endothelial cells, suggesting that they could be used as a gene/protein delivery vehicle (48). However, their origin, self-renewal, and differentiation potential are still being explored. Precursors of muscle cells have shown a low survival rate following transplantation (49,50). As muscle fibers have a slow turnover and gene expression could, in principle, be maintained for a long period of time even if the transgene is not integrated, direct in vivo delivery to differentiated muscle fibers is currently believed to be a more suitable strategy for the correction of severe muscle disorders such as muscular dystrophy (51–53). However, a stem-cell-like subpopulation able to survive transplantation into irradiated host muscles has been recently described (54). In the future, isolation of those cells may be invaluable for gene therapy of muscle disorders. Recently, postmortem adult brain has been shown to represent an important source of human neuronal stem cells (NSCs) even 20 hours postmortem (55,56). In the past, fetal NSCs have been shown to be able to grow in culture, expand as cell aggregates called neurospheres, and differentiate along multiple lineages. Both rodent and human fetal NSCs are capable of generating neuronal grafts in an animal model (57–59). Although adult NSCs have been shown to have a similar potential to their fetal counterparts in vitro, further studies are required to determine whether human adult neural progenitor cells will be useful in transplantation, and enormous challenges are to be faced before the NSC-mediated therapy becomes feasible (60).

Hematopoietic Stem Cells

The predominant target for genetic intervention has been the HSC, mainly because of accessibility for in vitro manipulation, a greater understanding of its biology, and the number of clinical settings of primary medical and scientific relevance that have presented. These include inherited disorders, such as hemoglobinopathies or metabolic storage disorders, as well as acquired disorders such as cancer. The latter is an area that is receiving much attention. Delivery of drug-resistant genes such as MDR1, dihydrofolate reductase, and O^6-methylguanine-DNA-methyltransferase may be used to reduce toxicity to bone marrow, which is often dose limiting due to myelosuppression. Thus transfer of a drug-resistant gene to HSC can protect the bone marrow from toxicity and allow dose intensification (61). Increased immune responses to tumor cells can be achieved by generation of bone-marrow-derived dendritic cells expressing tumor-associated antigens capable of inducing a specific cytotoxic T-cell response (62). HSCs can also be genetically modified to generate T-cells expressing a T-cell receptor specific for a tumor antigen (63). Such strategies have been tested in mouse models (in some cases in larger animal models) and await clinical trial.

For HSCs, a potential stem-cell marker (CD34) is available for isolation and is currently the only such marker used clinically. CD34 is expressed on 0.5% to 5% of human bone marrow cells and is found on early hematopoietic progenitor cells but not on their mature counterparts and has been used to provide cells that achieve clinical engraftment following transplantation (64,65). Other surface markers have been used in the laboratory in association with CD34 to identify more primitive populations of HSCs, such as in the case of $CD34^+$ $CD38^{low}$ (66,67). However, it is unclear whether further selection would represent an advantage as the elimination of the more committed progenitor cell population could compromise short-term engraftment capability leaving patients exposed to risks related to prolonged cytopenia. Moreover, such isolation protocols, based on flow cytometry, may be more at risk of microbial contamination.

A range of in vivo and in vitro assays has been developed over the last 20 to 30 years in an attempt to define HSCs. With the development of gene-marking protocols, views on the most appropriate assays to test HSC function have changed. Initially, in vitro assays, for example, long-term culture-initiating cell (LTC-IC) (68), and high proliferative potential cell assays (69) were postulated to test the ability to generate additional stem and progenitor cells for variable periods of time and to differentiate in at least one type of highly differentiated descendant. However, the generation of gene-marked LTC-IC proved very efficient, whereas transplantation of these same cells in an ablated host resulted in very low (<1%) levels of circulating, gene-marked cells. This suggested that those assays detected progenitor cells or transient amplifying cells, responsible at most for short-term engraftment (70,71).

It became clear that the only conclusive assay for HSCs was to assess their ability to give rise to cells of the lymphoid and myeloerythroid lineages in a potentially, lethally irradiated host following transplantation. Most studies have been carried out in mice where HSCs were assessed by in vivo competitive repopulation assays and long-term engraftment of HSCs could be assessed by secondary and tertiary transplants. Assays for the detection of human HSCs consist in the engraftment of human cells in a range of xenogenic hosts, particularly the immunodeficient non-obese diabetic/severe combined immunodeficiency (NOD/SCID) mouse model. Transplantation of HSCs in these mice results in terminally differentiated human cells from multiple hematopoietic lineages including B-cells, immature progenitor cells, mature erythrocytes, and all lineages of myeloid cells (72,73). In these systems, multilineage engraftment of human cells could be achieved for up to four months in optimized systems (71,74,75).

Gene transfer and expression in repopulating mouse HSCs and in human HSCs, transplanted into NOD/SCID mice, have proved successful (76–81). In contrast, with the exception of gene therapy of inherited immunodeficiencies, the use of gene-modified human HSCs in humans has remained difficult, with poor transduction efficiencies, often less than 1%, evident in cells repopulating patient hematopoiesis (37,39,82). This discrepancy questions the utility of murine models in predicting outcomes in terms of the efficacy of gene transfer in human long-term repopulating cells. Human HSCs exhibit differences in stem cell kinetics, cytokine responsiveness, and retroviral receptor levels when compared with their murine counterparts. It seems that, although NOD/SCID repopulating cells have been used as a measure of long-term repopulating cells, they probably more closely resemble short-term repopulating cells. Engraftment levels of baboon hematopoietic cells in NOD/SCID mice were different to engraftment levels of the same cells in baboons (83). In baboons, clones responsible for long-term engraftment and contributing to all lineages for nearly two years appeared only six to eight weeks post-transplantation (84). This time is usually beyond the follow-up time used in most studies where NOD/SCID mice have been used as a model. The use of larger animal models, including dogs, cats, or non-human primates such as baboons or rhesus monkeys, has stronger relevance to human gene therapy. These yield similar gene transfer efficiencies to those seen in clinical trial, and the increased life expectancy of larger animals allows for more extended follow-up.

Skin Stem Cells

Skin keratinocytes have proved to be a useful source of skin stem cells because of their accessibility and capacity for growth in culture (85). No specific marker is available and usually a combination of markers such as size, integrin expression levels, and DNA content is suggestive of keratinocytes stem cells (86–88). Enrichment relies on culture conditions that are permissive and selective for the growth of stem and progenitor cells (34). Cultured keratinocytes may be used to form confluent epithelial sheets that can be gently removed from the cultured dish and applied to reconstitute portions of the damaged epithelium (89,90). This system has the advantage of using the patient's own skin and allows covering of a larger surface area starting from a relatively small amount of unaffected skin. At present, this approach is limited by the slow growth of keratinocytes in culture, which leaves the patient prone to infection where the skin has been removed, and by the fragility and poor adhesion properties of the cultured epithelial sheets when returned to the patient (91,92). However, skin substitutes are under development to temporarily cover wounds and to improve adhesiveness of the expanded keratinocytes (93). Using this system, keratinocytes were harvested and genetically modified by retroviral or lentiviral vectors and were shown to correct single gene defects in epidermolysis bullosa and lamellar ichthyosis (94–97). The corrected epithelial sheets were engrafted onto nude mice skin to produce healthy epithelia. Gene expression was detected for extended periods in grafted keratinocytes although in some studies transgene expression has been reported to decrease with time (98,99).

The major problem in assessing the efficiency and long-lasting effects of gene transfer in skin stem cells is still the limited knowledge about keratinocyte stem cells. It is not clear which particular population of cells is responsible for long-term engraftment and whether any of the assays available is suitable for their measurement. So far, the presence of the transgene has been examined for the ability of the transduced cells to form holoclones in vitro or to be able to engraft in immunodeficient xenograft murine models (34). Holoclones are believed to have properties suggestive of stem cells in that they have a high colony-forming efficiency and give rise to meroclones and paraclones, which have progressively less colony-forming efficiency. Moreover, holoclones can produce mature epithelium in vivo when transplanted in immunodeficient mice (98). However, Li et al. (100) have shown that slowly dividing putative stem cells, rapidly dividing transient amplifying cells, and differentiated keratinocytes were all capable of prolonged tissue regeneration in SCID mice, revealing that either a greater capacity for cell renewal than predicted of the more committed progeny of stem cells or the short follow-up available with murine xenotransplantation models does not allow a proper assessment of long-term repopulating skin stem cells. This latter seems most likely given the experience developed in assessing HSC engraftment using immunodeficient murine models for xenografts. No studies in larger animals are available and a phase I/II clinical trial to determine safety and efficacy in the long term of this approach in epidermolysis bullosa is ongoing (101). If this approach was to be successful, genetic modification of keratinocytes could be used not only in inherited disorders but also to accelerate the healing rate or to inhibit post-ulcer complications such as keloid formation or scarring by overexpression of mitogenic factors such as platelet-derived growth factor–aplastic anemia that participate in wound healing (102). The skin may also be amenable as a system for protein delivery to the blood stream as it is accessible and rich in vascularization. Engineered keratinocytes could thus be used to deliver an appropriate effector molecule in the treatment of, for example, growth hormone deficiency, or hemophilia (103,104).

Marrow Stromal Cells

Bone marrow stromal cells or mesenchymal stem cells (MSCs), which are the postnatal precursors of osteogenic cells and retain the capacity to differentiate into chondrocytes, adipocytes, and myelosupportive cells, have gained attention for their potential in tissue correction and regeneration especially with regard to inherited skeletal disorders (105,106). At present, no specific markers are available for the isolation of these cells. Isolation occurs mostly by plastic adherence in permissive culture conditions. They are usually recognized as $CD45^-$, $CD34^-$, $Stro-I^+$, $SH2^+$, or $SH3^+$ following isolation in culture (107,108). They are also recognized by their clonogenic capacity [defined as colony-forming unit-fibroblast (CFU-F)] and their ability to differentiate into a broad spectrum of fully differentiated connective tissues, including cartilage, bone, adipose tissue, and myelosupportive stroma (109,110). However, remarkable differences are observed among the different CFU-Fs in terms of cell morphology, rate of replication, expression of markers for osteoblastic, chondroblastic, adipogenic phenotypes, and the number of differentiated progeny they can give rise to, with some CFU-Fs capable of multipotent differentiation, others only capable of forming bone, and yet others capable of giving origin to myelosupportive stroma (111–114). The frequency of each type of CFU-Fs in a culture depends on many factors, including methods of isolation, seeding density, serum used for cell culture, species from which the cells are derived, donor age, and time in culture.

In vitro transduction efficiencies are very high, with the majority of studies reporting over 90% using a variety of viral vectors (115–119). However, demonstration of gene transfer in long-term, multipotential CFU-Fs capable of contributing to bone remodeling has not been achieved at present. Transplantation of MSCs by intravenous infusion has shown disappointing results. The only example where infusion of MSCs has been claimed to be able to temporarily correct a clinical phenotype is in osteogenesis imperfecta (120). However, the authors did not reconcile the important effect seen on bone growth with the poor engraftment rates observed (only 1–2% of the cells was found to be of donor origin). The clinical assessment of the disease lacked appropriate controls, and the study did not provide convincing histological data. The majority of studies indicate that MSCs are very poorly transplanted during this procedure (121,122). Intrafemoral injection of a subset of murine marrow stromal cells shows limited engraftment at the site of injection at four to six weeks post-transplantation with a few cells expressing osteoblastic markers (123). This lack of engraftment could be due to current isolation protocols, which may select a progenitor cell population with limited lifespan rather than a long-term engrafting stem-cell population. Alternatively, primitive cells may be selected but the current culture conditions fail to maintain self-renewal of these, in much the same way as early HSC treatment protocols led to dramatic losses in hematopoietic repopulating capacity. Certainly, most MSC cultures have limited proliferative capacity undergoing at most 40 population doublings with loss of multipotential capacity with time in culture (124). However, even the more primitive multipotent adult progenitor cells (125,126), a subpopulation of MSCs that have extended proliferative capacity independent of donor age and that at the single cell level can differentiate into MSCs and cells of mesodermal origin such as endothelial cells, do not engraft efficiently. This suggests that other factors, such as lack of homing and poor migrative capacity, may contribute to poor engraftment. This may correlate to the role of MSCs as resident, support cells rather than circulating progenitors. Certainly, MSCs have been shown to lack chemokines receptors such as CXCR4 (127) necessary for the migration and homing of HSCs to bone and bone marrow.

At present, evaluation of the relevance of gene therapy in MSCs rests primarily in the transplantation of such cells in open systems, reproducing experiments performed 30 years ago by Friedenstein et al. (128). This consisted of MSC transplantation under the kidney capsule with the formation of a chimeric ossicle with a structure replicating the histology and architecture of a miniature bone (129). This can now be achieved more simply by implanting MSCs in a scaffold (130–133). In such systems, it is possible to monitor survival, proliferation, and differentiation of transduced MSCs for long period of time. MSCs transduced with an erythropoietin cDNA and transplanted (along with hydroxyapatite particles) under the skin of nude mice were able to differentiate into osteoblasts and the animals maintained increased hematocrit levels for 10 to 12 weeks (130). Re-isolation of MSCs from the implant and their use in secondary transplantation showed a sustained increase of hematocrit at levels similar to those obtained in primary transplantation. These data suggest that in principle MSCs could be used for long-term delivery of proteins and possibly for bone regeneration provided homing and migration properties are improved. However, even if cells were capable of homing to bone marrow, a reliable quantitative estimate of engrafted progeny as well as evaluation of cell identity and function within the host environment would require long-term follow-up, due to the slow turnover of this tissue. This may be only achieved in large animals.

GENE DELIVERY

There are a number of requirements for successful gene transfer to stem cells. The gene must be transferred in such a way that it can be passed to all progenies of stem cells.

Currently, that is best achieved via stable integration within the genome of the target cells. It would be desirable for such integration to occur at a specific chromosomal locus in order to avoid insertional mutagenesis. The gene transfer procedure should not compromise the repopulating capacity of the stem-cell pool and should be efficient enough to have an impact on the disease phenotype. The gene product should be stably expressed from the vector and therapeutically useful levels should be obtained in the corrected cells.

Delivery Systems

As DNA molecules are charged and are too large to readily transit the cell membrane, various methods have been developed to facilitate the delivery of genes into cells. Some of these methods involve either physically delivering DNA via electrical impulse, with cationic lipids, DNA polymer complexes (non-viral systems), or by use of viral vectors.

Nonviral Systems

Nonviral systems have several advantages as they are easy to prepare and scale up, they are more flexible with regard to the size of the DNA being transferred, and may be safer in vivo (131,134), They might be particularly useful for topical application, for example, in the skin (135). However, non-viral systems developed to date are rather inefficient, producing a low number of integration events and resulting in permanent genetic correction of only about one in 10^4 cells. Obstacles to effective non-viral gene therapy are many: First, in manufacturing and formulation of DNA complexes to confer stability and low toxicity. The use of lipid/liposome DNA complexes has been shown to confer pulmonary inflammation, production of reactive oxygen species, and inhibition of proliferation (136). Second, the interaction with the cell membrane and entry, which occurs primarily via the endocytic route of internalization, leads to degradation of most of the delivered DNA by lysosomal nucleases (137). Third, the small amount of DNA remaining has to be transported intracellularly and translocated to the nucleus. Only small molecules of 200 to 310 bp can enter the nucleus by passive diffusion, whereas larger macromolecules only enter via highly regulated processes, which often involve either cell division with disassembly of the nuclear envelope or the presence of signals with a nuclear targeting component (138).

More recent developments have exploited specific viral proteins, which are responsible for the superior efficiency of viral vectors, to boost delivery of non-viral plasmid DNA into the cytosol, along with elements that facilitate nuclear localization and integration. Proteins such as the transactivating transcriptional activator protein from human immunodeficiency virus (HIV) have been shown to increase transduction efficiencies and decrease the toxicity of the liposome complexes (136,139,140). Nuclear localization signal motifs within the vector can bind small peptides, which in turn interact with molecules of the nuclear pore complexes mediating nuclear import, leading to improved nuclear transfer and better transfection efficiencies (141). The use of elements such as retrotransposons or enzymes such as phage integrases, which mediate unidirectional site-specific recombination between two DNA recognition sequences, has been shown to increase the efficiency of transduction up to 55-fold due to increased integration into the genome (96,142). Although these results are encouraging, non-viral systems are still at an early stage of development. Success is still limited due to instability of the linkage of DNA with the different elements, modification of the DNA, and by the increase in size of the DNA/polymer complexes, which hampers DNA entry into the cells.

Viral Systems

The most efficient method of gene transfer uses viruses as delivery vectors. Some viruses insert their genetic information into the chromosomal DNA of the cells as part of their life cycle. They can be manipulated, whereby some of the structural genes important for viral replication are substituted by the insertion of genes of interest. Usually, sequences that are required in *cis* for functions such as packaging the vector genome into the virus capsid or the integration of vector DNA into the host chromatin are left intact. The deleted genes, encoding proteins that are involved in viral replication, or elements of the capsid/envelope, are included in one or more separate plasmid constructs, the packaging constructs (Fig. 1A). These are co-transfected with the vector genome into packaging cells to provide helper functions in *trans* and then produce the recombinant vector particle (Fig. 1B).

A wide variety of viruses have been considered as vehicles for gene transfer. Viruses such as adenoviruses that do not result in chromosomal integration are of limited use in stem cells. The use of elements (e.g., from Epstein-Barr or *Herpes simplex* virus) that allow efficient episomal replication may surmount the need for integration but these elements are not sufficiently developed to be considered for immediate clinical application (143,144). An alternative viral vector, which to date has not lived up to expectation, is the adeno-associated virus (AAV)-derived vector (145). It was thought that AAV vectors would retain some characteristics of the parental virus in terms of stability, wide host range, lack of pathogenicity, site-specific integration site, and ability to infect quiescent cells. However, AAV vectors exhibit different characteristics from the wild-type virus probably as a consequence of removal of the AAV structural genes. Thus, AAV vectors are not as efficient at genomic integration as the wild-type virus and the specificity of integration is lost (146,147). Moreover, AAV can accommodate only small inserts (up to 5 kb), viral production is very labor-intensive with the possibility of contamination with helper virus, and expression in stem cells such as HSCs has been questioned (148,149).

The majority of clinical trials use retroviral vectors, particularly type-C oncoviruses of murine origin. A number of characteristics make recombinant retrovirus vectors particularly suitable for gene transfer in stem cells, and they are still the best system available to date. They are small, simple, and well characterized, thus allowing manipulation of the genome; they have a wide host range and are able to efficiently and stably integrate into the chromosome of the target cell. Targeting of primitive hematopoietic cells has proved most successful in murine systems (79,150–152). However, when similar approaches were taken in clinical trials or with non-human primate stem cells, transduction efficiencies were very low (37,153–157). The barriers to stem-cell transduction by oncoretrovirus are best considered in relation to the retroviral life cycle. Although capable of efficiently transducing dividing cells, they are largely unable to transduce quiescent cells (158). This is limiting because most stem cells are quiescent and therefore require stimulation into cycle for efficient transduction (159).

Recent developments have put much effort into another type of retroviral-based vector, based on lentiviruses. These can accommodate fairly large inserts with complex transgene cassettes, which are useful for genes that require tight regulation such as β-globin (160–162). Moreover, they provide long-term expression thanks to their ability of chromosomal integration. Most importantly, as they can naturally penetrate an intact nuclear membrane without in principle requiring cellular division (163,164), they were thought to represent a great improvement over the oncoretrovirus. However, transduction efficiencies by lentiviruses have been shown to vary with cell-cycle status, with highest efficiency in M and lowest in G_0 (165,166). Lentiviruses have been shown to transduce

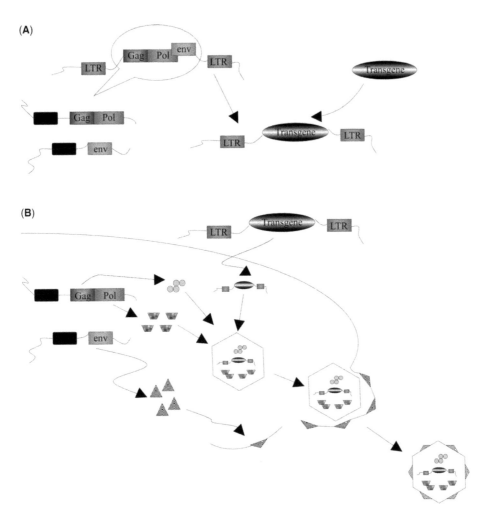

Figure 1   (**A**) A typical type-C retroviral vector comprises three structural genes organized in the sequence gag-pol-env driven by the viral promoter, the LTR. They encode, respectively, for components of the nucleoprotein core, reverse transcriptase, and components of the envelope of the viral particle. Retroviral vectors retain only the *cis*-acting sequences from the original viral genome. These sequences allow the recombinant structure to be transcribed (LTR) and the RNA to be processed and packaged into a virion particle. The other viral functions (encoded by the structural genes) are placed on a different plasmid and (**B**) introduced into cells that express those genes stably (packaging cells). In this way, they can complement a replicative defective vector construct introduced into the packaging cells by transfection or transduction leading to mature pseudovirus particle production. *Abbreviation*: LTR, long terminal repeat.

human repopulating cells at higher levels than oncoretroviruses in immunodeficient murine models (167,168), but their ability to transduce HSCs in non-human primate models has been disappointingly similar to that of oncoretrovirus vectors (169,170) with less than 5% long-term transduction in most studies and only one report of approximately 10%, using a simian immunodeficiency virus-based vector with a follow-up to one year (171). Similarly, no improvement in transduction efficiency over that obtained with oncoretroviruses was seen in keratinocytes engrafted in NOD/SCID mice after four weeks in cultures (172). Moreover, there are still safety concerns in using lentiviral vectors in clinical trials as most are based on the HIV. There are also technical barriers to lentiviral gene therapy due to a lack of methods to produce high-titer virus stocks. Transient transfection systems, where plasmids encoding the vector genome and the viral structural components are co-transfected into producer cells with vector, are not suitable for large-scale production and do not allow the thorough characterization of vector producer cells, something required by regulatory agencies prior to clinical use. Thus, although lentiviral vectors continue to be developed, and these will most likely be used clinically in time, there is still much to be done to bring them forward (173).

Modality of Delivery—In Vivo Versus Ex Vivo

Gene transfer may be approached using either ex vivo or in vivo gene delivery. In vivo gene transfer consists of direct delivery of the vector to the tissue of interest. In vivo gene transfer is attractive due to the potential ease of application and for some tissues (e.g., brain and lung) may be necessary. However, current approaches fall short of efficiency and safety as transduction and expression may not be limited only to target cells and vector may be disseminated. For example, an AAV-factor IX vector was found to be present in the semen of one patient following hepatic artery infusion (174). Profound immune responses to transgene have been seen probably due to transduction of antigen-presenting cells (175). This was especially true when high-titer retrovirus was directly applied to the skin, a tissue with unique immunological features (97,176). A further problem for in vivo gene delivery is the presence of extracellular barriers. In tissues such as central nervous system or muscle, direct injection of viral vectors has resulted in limited diffusion of the virus and in efficient transduction of cells only within a few millimeters from the site of injection (51,177). Intravenous injection of viral particles has mainly led to expression of the transgene in the liver with low gene expression levels in other cell types and no, or partial, correction of disease phenotype (178–180). This is due to antibody and serum proteins that can directly inactivate the vector or direct it for rapid clearance to the liver (181–183). In the last 10 years, numerous attempts have been undertaken to develop strategies that improve the efficacy and safety of in vivo delivery. Injection of viral particles in neonates has shown promising results probably due to a muted immune response and perhaps to a decreased density of tissues leading to decreased extracellular barriers (184). Viruses have been pseudotyped with modified envelopes that have shown higher resistance to serum clearance and may improve transduction efficiency in future applications (182,185,186). Retargeting viral entry through specific cell surface molecules expressed on human cells has been attempted by expressing single-chain antibodies (scFv) directed against the extracellular domain of cell surface receptors and natural ligands of surface molecules or by a protease-activable chimeric receptor (187,188). The limited success of this approach reflected the fact that attention was focused on receptor binding and not on efficient viral entry and the fusogenic activity of the chimeric receptor was compromised or lacked specificity (189,190). A more promising approach may be to use elements that limit gene expression to the cell of interest. Keratinocyte-specific enhancer elements have been used in retroviral vectors to restrict expression, thus avoiding activation of dendritic cells, and these have shown maintenance of transgene expression in the epidermis up to 20 weeks (97). High levels of human factor IX were reached and maintained long term in immunocompetent mice, without inducing antibodies, following in vivo infusion of a lentiviral vector that incorporated a hepatocyte-specific expression cassette (175).

Despite promising advancements in in vivo delivery, thus far the major experimental focus has been on ex vivo gene therapy using viral delivery systems, particularly retroviral and lately lentiviral systems. The ex vivo approach relies on the isolation of stem cells from a tissue of interest, expansion, genetic modification in culture, and subsequently reinfusion to a patient. Although this approach is cumbersome and costly, it ensures that appropriate target cells are exposed to the vector and with high efficiency of gene transfer prior to reinfusion to the patient. However, as this system requires isolation and culture of the stem cells, problems with loss of multipotentiality and engraftment limit the success of this strategy.

HURDLES TO STEM-CELL GENE THERAPY

Regardless of the modality of delivery, studies so far have highlighted some common problems. (*i*) Transduction levels can be low, particularly in cells capable of conferring long-term engraftment. (*ii*) Retroviral vectors, most commonly used in stem-cell gene therapy, bear the risk of multiple integration and insertional mutagenesis. (*iii*) Transgene expression is variable and can be silenced. (*iv*) Depending on the type of transgene expressed and the vector used, immune responses can compromise the outcome of therapy.

Inefficient Stem-Cell Transduction

Most experience in stem-cell transduction comes from work with HSCs. Targeting of HSCs has proved most successful in murine systems. It has been shown that genes can be introduced into pluripotent stem cells with production of genetically modified progeny in both lymphoid and myeloid lineages (191). In murine models of immunodeficiencies and chronic granulomatous disease (77–80,192–194), therapeutic sequences have been successfully transferred at high levels with correction of the disease phenotype. Clinically, however, the much lower efficiency of gene transfer in human HSCs has hampered success. In general, a high transduction frequency in hematopoietic cells transduced in vitro has not been reflected by high levels of gene-modified hematopoiesis at recovery post-transplant. This has been attributed to the inability to infect sufficiently large numbers of cells with marrow-reconstituting capacity. There are several reasons behind the low levels of gene-modified stem-cell engraftment in human HSCs.

First, the ability of a virus to bind to and to enter target cells is determined by the viral envelope protein and its correspondent receptor on the target cell. Murine HSCs express high levels of the receptor for the ecotropic envelope of the Moloney Murine Leukemia Virus and are consequently easily transduced by an ecotropic retrovirus. Human HSCs do not express the ecotropic receptor, and most studies in humans have targeted the amphotropic retrovirus receptor, ram-1. However, it has subsequently been determined that ram-1 is expressed only at low levels on human HSCs (195,196). To overcome this problem, alternative envelopes such as Gibbon Ape Leukemia Virus (GALV) (197,198), Feline Leukemia Virus envelope RD114 (199–201), and Vesicular Stomatitis Virus-G (VSV-G) (202,203) have been tested. Receptors for these were thought to be expressed at higher levels on HSCs, and VSV-G in particular was thought to represent an improvement as viral entry is based on fusion, not requiring a specific protein receptor. Although initial studies suggested an improved performance of viruses pseudotyped with such alternative envelopes (204,205), subsequent studies in NOD/SCID models and large animals have failed to show a clear superiority of any of these pseudotyped vectors in consistently transferring genes to human or non-human primate long-term repopulating cells, with the exception of GALV, which was found to be superior to the amphotropic receptor in baboon (206–209). These studies cannot be considered conclusive and more thorough testing is required as the studies in large animals have been very limited in numbers, with transduction levels low and variable and no investigation on possible immune response to the transgene carried out. Reproducibility among the different studies has been scant as different source of stem cells and different protocols of transduction have been used. Levels of receptors on the different subsets of stem cells have not been properly assessed and it is possible that, although no difference is seen in the overall number of mature blood cells that harbor the transgene, the proportions of transduced cells in different compartments may change with the different envelopes. Amphotropic virus was found to be able to infect mostly the lymphoid compartment. A more homogenous distribution between the lymphoid and myeloid compartments was seen in cells infected with the RD114 viral envelope, suggesting that RD114 and amphotropic receptor may target different subsets of CD34$^+$ cells (210).

A second, and serious, barrier to transduction of HSCs reflects the life cycle of the viruses used. So far, most systems have used gamma-retrovirus vectors that require target cell division for vector integration into the genome (211), and HSC cycle slowly. Even lentiviral vectors, which were supposed to enter the nucleus of target cells in the absence of cellular division, have been found to transduce quiescent cells very poorly (165). In murine studies, treatment of donor animals with 5-fluorouracil can be used to kill committed progenitors and to induce cycling of stem cells in vivo prior to harvest. Such an approach would not be appropriate in human subjects. The use of factors such as granulocyte colony-stimulating factor (G-CSF), stem cell factor (SCF), Flt-3 to induce cycling of stem cells before harvest has been tested. In mice and non-human primates, pretreatment of donor animals with a combination of G-CSF and SCF can prime HSCs for efficient transduction (154,212–214). G-CSF mobilization of human CD34 cells, however, does not significantly increase transduction levels in repopulating cells, and SCF has not been proscribed in human trials in the United States due to adverse effects. Efforts have focused on improving in vitro conditions for gene transfer in human HSCs. To facilitate cell cycling and to potentially retain self-renewal capacity of HSCs, hematopoietic growth factors have been employed. Again this has proved more successful in murine studies, whereas large animals and human studies have been characterized by a pronounced loss of repopulating HSCs (153,215,216). Currently preferred conditions for HSC transduction incorporate early cytokines such as Flt-3 and thrombopoietin in place of interleukin-3 (which has been associated with impaired self-renewal of human HSCs) and with some protocols additionally incorporating G-CSF (81,217–222). A further improvement in the transduction of human HSCs has been achieved with the use of fibronectin fragments, which serve to mimic stromal layers by engaging integrins on the HSC surface and facilitate co-localization of viral vector and target cell (217,223–225). Although such modifications have undoubtedly improved transduction frequencies, optimal in vitro transduction conditions remain to be defined.

The greatest success in clinical gene therapy has been seen in the treatment of severe combined immune deficiency (SCID). Patients with a deficiency in the common γ-chain of the lymphoid cytokine receptors (X-SCID) or in adenosine deaminase (SCID-ADA) have been corrected with long-term clinical benefit (226,227). All other clinical trials to date have fallen short of engraftment of therapeutically useful level transduced cells. Analysis of the elements that led to success of the SCID trials point to a strong selective repopulating advantage of the gene-corrected cells in comparison to the uncorrected cells. In X-SCID, the presence of a functional common γ-chain confers to lymphoid cells the ability to respond to appropriate cytokines and leads to a strong selective advantage for survival, growth, and differentiation of gene-corrected cells. In the SCID-ADA patients, an initial developmental advantage to transduced HSCs was provided by creating a space in the bone marrow of the patients with a low-intensity myeloablative conditioning regimen and by selecting a progenitor cell population. Moreover, ADA-corrected T-cells also exhibit a survival advantage over their uncorrected counterparts. The importance of a selective advantage is further underlined when the non-lymphoid population is examined in patients from these clinical trials. In all cases, engraftment within the non-lymphoid compartment (where no advantage of gene-correction is anticipated) is significantly lower than that in the lymphoid compartment.

Other than in SCID, the most successful clinical trial is one where patients undergoing autologous bone marrow transplantation as part of their treatment for germ cell tumors were infused with CD34+ cells modified to express MDR. At one month post-transplant, most patients had vector-containing cells but became negative in the following months, consistent with the effects of in vivo transduction on long-term repopulating cells. After completion of the chemotherapy, however, four out of 11 patients exhibited 5% to 15% transgene positivity in assayable progenitors, and those levels were maintained for up to a year (228).

The data from the SCID and MDR trials imply that, post-engraftment, engineered cells constitute a minority of the total HSC pool of a transplanted individual because the in vitro manipulation associated with the transduction process compromises the fitness of the gene-modified HSC to compete with their endogenous counterparts. Unless the modification offers a competitive advantage (survival, proliferative, or both), gene-modified cells tend to disappear with time.

Therefore, for disorders for which there is no inherent selective advantage of gene correction, it may be desirable to adopt strategies to provide an in vivo selection. One strategy is to use vectors that co-express a drug-resistant gene with the gene of interest. After transplantation and hematopoietic reconstitution, the selective agent can be administered, thus conferring a selectable phenotype to the transduced cells and increasing their proportion in comparison with the untransduced population (61). However, there are concerns in the use of this strategy. HSCs have high endogenous levels of some of these proteins that may lead to lack of killing of the unmodified HSCs, with reduced clinical efficacy. Higher levels of selective agents might be used to overcome this. However, this might prove a dangerous strategy were the number of gene-modified HSCs to be below a certain threshold. In such a case, hematopoietic reconstitution might be impaired with consequent hematopoietic failure. This scenario was seen recently in two dogs infused with transduced MDR1 HSC, when challenged with high levels of paclitaxel (229). As well as selection in the hematopoietic system, MDR has been used to achieve the in vivo selection of genetically modified keratinocytes following topical administration of colchicines (99).

An alternative to cytotoxic drug selection might be the enforced self-renewal of transduced stem cells. One approach has used a "selective amplifier gene" encoding a modified membrane receptor that can be activated by a specific chemical agent. Perhaps, the most successful example of this is the production of fusion proteins containing the signaling domain of a cytokine receptor and a binding site for a drug known as a chemical inducer of dimerization. In the absence of drugs, the modified cytokine receptor remains monomeric and does not transmit a signal (230,231). On binding drugs, however, the chimeric receptor dimerizes and provides a proliferative stimulus. Some encouraging results have been achieved, most particularly with the signaling domain from the thrombopoietin receptor (232–234). However, there are a number of issues to be addressed, most notably, the transient nature of the selective advantage, which suggests that it is operating at the level of the committed progenitor, and an apparent lineage bias, with better selection in erythroid and platelet compartments seen at the expense of other lineages when using the thrombopoietin receptor (235).

Another option might be the use of genes encoding factors, which may play a role in self-renewal of stem cells. The primary example of this is HoxB4, which is up-regulated in primitive hematopoietic cells and down-regulated with maturation (236). Retroviral-mediated transfer of HoxB4 to murine or human bone marrow cells leads to in vitro expansion of repopulating cells (237,238). In competitive repopulating assays in mice, HoxB4 overexpression confers an engraftment advantage over untransduced or control-vector-transduced cells (239,240). One study, using human CD34+ cells, however, indicated that enforced HoxB4 expression impaired myeloerythroid differentiation and reduced B-cell output (241). Impaired differentiation can be one characteristic of tumor cells. Thus, although the engraftment advantage offered by HoxB4 is attractive for gene therapy purposes, any potential for adverse outcomes in terms of cellular transformation will have to be carefully analyzed. Recent studies have highlighted the potential of recombinant HOXB4 protein in maintaining and expanding HSCs ex vivo (242,243). This approach may be useful in maximizing the numbers of HSCs transplanted following ex vivo manipulation, but will not discriminate between transduced and untransduced cells. Nor will it provide a selection strategy in vivo. In view of this, further work with HOXB4 is merited.

Insertional Mutagenesis

Until recently, it was thought that the risks related to random insertion of a vector in the genome was very low if the multiplicity of infection was adjusted so that only one proviral insertion occurred per transduced cell. A number of factors were thought to contribute to this low risk: (*i*) most of the genome is inactive and therefore the likelihood of random integration in a transcriptionally active area of the genome would be small (244), (*ii*) malignant transformation tends to be a result of multiple genetic events and thus a single insertion would be unlikely to be sufficient to cause acute transformation (245), (*iii*) retroviral insertion is monoallelic, thus reducing the chance of mutagenic events, such as loss of tumor suppressor function (211), and (*iv*) acute up-regulation of oncogenes in the absence of other enabling mutations can lead to apoptosis or cytostasis, rather than transformation (246).

This view was supported indirectly by the fact that few side effects related to insertional mutagenesis by a replication-incompetent vector had been reported in any preclinical studies or clinical trials (247). The discovery that two of 11 patients treated during the successful X-linked severe combined immunodeficiency (X-SCID) trial developed leukemia led the gene therapy community to reconsider the risks (248). It became clear that in both cases the leukemic event was associated with retroviral vector integration in or near the oncogene LM02 (249). These results prompted a series of studies which concluded that the risk of retroviral insertion in transcriptionally active regions was much higher than predicted (250), as viral integration occurred preferentially in euchromatin, possibly because of its accessibility (251). Moreover, there was a propensity for retroviral vectors to integrate in the vicinity of cellular promoters (252).

The frequency of leukemic induction in the X-SCID clinical trial is obviously at odds with the data from previous preclinical and clinical studies. A number of factors may account for this. First, in most other clinical trials of HSC gene therapy, the total number of transduced repopulating cells infused has been low, and thus the chances of transplantation of a leukemic clone may have been low. Most preclinical studies have been carried out in murine models with a follow up of three to six months, whereas in the X-SCID patients, leukemia became evident only two years after transplant. One recent study has assessed the long-term sequelae of gene therapy in large animals. This involved analysis of rhesus macaques ($n = 42$) and dogs ($n = 17$) that had been transplanted with HSCs transduced with a variety of retroviral vectors and mostly via a protocol similar to that used in the X-SCID clinical trial (253). The minimum follow-up was one year post-transplant and in some cases extended up to six years, with all animals, exhibiting at least 1% vector positivity in white blood cells. No animal showed any evidence of neoplastic disease or, indeed, any form of clonal amplification. This suggests that the high incidence of leukemia seen in the X-SCID trial may not be a general effect of retroviral random insertion but may depend on disease/transgene-specific factors. These may include age of transplantation, dose of cells administered, the immunosuppressed phenotype of the patients, a proleukemic or co-operating effect of the transgene, and combinations of some, or all, of these.

The patients that developed disease were the youngest to be treated (one and three months old)—HSCs may have different characteristics in young patients. Indeed, a higher number of CD34+ cells were found in these patients and a higher number of cells/kg was infused in the two patients that developed leukemia. As insertional mutagenesis is a stochastic process, this may have increased the mutagenic risk. X-SCID is caused by the absence of the common γ-chain of a number of lymphoid cytokine receptors. This leads to a maturation block in the lymphoid compartment, thus likely expanding the pool of immature T-cells available for transduction and expansion. Certainly, post-transplant, a higher transduction frequency and expansion of the lymphoid compartment was seen in the face of very poor transduction frequencies in the myeloid compartment. Correction of the γ-chain defect probably confers a profound selection advantage to gene-corrected cells. Moreover, it may be that overexpression of γ-chain co-operates in some way with the LM02 oncogene to further facilitate clonal expansion. It is not certain yet which, if any, of these factors are important. However, it is clear that, given the effects seen in the X-SCID study, the risks of neoplastic transformation will need to be assessed specifically for each disease/transgene combination and tested in animal models that allow for long-term follow-up.

Concomitant with efforts to more accurately assess the risks associated with any therapeutic strategy, the development of safer vectors is also a priority. Central to this will be efforts to reduce the risks of insertional mutagenesis. The retroviral long terminal repeat (LTR) is the major contributor to the activation of neighboring cellular sequences. This is due to strong enhancer functions that regulate initiation of transcription of the LTR, but can also affect cellular promoters. Transcriptional read-through into cellular sequences may occur, because the viral polyadenylation signal, which should terminate transcription, is often very weak (254). Where this occurs, the retroviral splice acceptor site may interact with a splice donor site of a cellular gene positioned downstream to provide a novel transcript. To improve vector design and obviate these problems, a number of strategies may be employed. Thus, a stronger polyadenylation signal and a strong internal splice acceptor may be incorporated into the vector, to reduce problems associated with read-through (254,255). The enhancer/promoter that drives expression of the transgene may be positioned internally, in a self-inactivating vector, to reduce interaction with the neighboring sequences—this may be reinforced by the inclusion of dominant DNA insulator regions, which can block enhancer effects on distant promoters (256). Alternatively, integration could be targeted to specific sites within the genome. Recently, the site-specific integration machinery of the bacteriophage ΦC31 has been exploited in non viral delivery approaches to achieve the targeted integration of a transgene in murine and human cells in vitro (96,257,258). All these modifications to vectors will take some time to develop and to test. In the short term, vectors could be made safer by including suicide genes, which could provide a self-destructive mechanisms if needed. Such an approach has been tested clinically with some success, to reduce graft-versus-host complications following the transfusion of activated T-cells for anti-cancer immunotherapy (259).

Transgene Expression

Effective gene therapy needs not only the transfer of a gene to an appropriate target cell, but also the expression of that gene to a level that is therapeutically useful. Most studies to date have employed viral sequences to achieve transcription of transgenes, generally with a view to maximizing expression. However, many viral promoters and enhancers are targets for cellular mechanisms of gene-silencing, and thus transgene expression can be completely inactivated in specific cell types, particularly in primitive cells responsible for repopulation in vivo (260–262). In HSCs and other stem-cell types, epigenetic events such as DNA methylation, histone modification, chromatin remodeling, and post-transcriptional gene silencing have all been reported to attenuate transgene expression (263–266). Attempts to increase expression from retrovirally transduced genes focus on the removal of silencer sequences from the LTR, and the use of alternative promoters to drive expression, as well as the inclusion of cellular elements such as locus control region elements, and chromatin insulators (267).

However, as previously discussed, due to the perils of insertional mutagenesis, uncontrolled, high-level expression of transgenes may be undesirable in some cases. This is exemplified by HOXB4, which may be used to enhance the engraftment potential of HSCs. The effects of this transcription factor are clearly dose-dependent. Thus, although expression from a relatively strong retroviral LTR conferred an engraftment advantage, with self-renewal and a mild perturbation in differentiation (241), higher levels of expression from an adenoviral vector were associated with a loss of self-renewal capacity and a push toward differentiation (268). The ability to control transgene expression, in terms of levels, tissue specificity, and temporal expression is likely to be central to many therapeutic applications. For example, delivery of trophic factors to the brain may require tight regulation as the unregulated, inappropriate, excessive, or ectopic release of effector molecules such as neural growth factor or tyrosine hydroxylase may prove harmful (269–272). Similarly, regulation of insulin expression is likely to be a desirable characteristic of gene therapy vectors for the treatment of diabetes (273).

Immune Response

Many of the immunological defense systems that tackle wild-type viral infections are also activated against viral vectors. Moreover, new transgene products may also stimulate an immune response, and this may lead to deletion of transduced cells. Vectors based on retroviruses and lentiviruses, which are those most commonly used for stem cells, are not particularly immunogenic and no side effects due to immunological response to the vector have been observed so far. Immune responses mounted against transgene antigens may develop with some latency. This is particularly true if artificial sequences are introduced or, in the case of inherited disorders, where the genetic defect leads to the absence of all or part of a gene product. Immune-mediated rejection of hematopoietic cells expressing green fluorescent protein (GFP), neomycine, hygromicine thymidine kinase (HyTK) gene, or α-L-iduronidase has been seen in preclinical animal models and gene therapy trials (274–278). However, long-term persistence of gene expression in the hematopoietic system was reported by several groups, implying that tolerance to the transgene could be established although no detailed analysis of immune response was carried out and levels of transgene were very low (217,279–281).

Heim et al. (282) showed that tolerance to the transgene can occur following transplantation of *GFP* gene-modified HSCs but not in all animals. One of the factors contributing to establishment of tolerance is likely to be related to the levels of transgene expression and its time in circulation. Indeed, in this study, a reduced immune response was seen when a higher number of transduced CD34+ cells engrafted, suggesting that a threshold level of transgene expression is required for a sustained period of time in order to induce tolerance and thus prevent clearance of genetic-modified cells. This is also observed in the study by Mingozzi et al. (283) where higher levels of transgene expression promoted a shift from a Th1-driven antibody response to a Th2 response, typical of a state of tolerance. The degree of immunosuppression by the conditioning regimen can also contribute to prolonged transgene expression although on its own is unlikely to be sufficient to prevent an immune response. Treatment with a non-myeloablative regimen interfered with the induction of transgene-specific cytotoxic T lymphocytes (CTL) and facilitated in vivo persistence of gene-modified cells in a non-primed host. However, in one study sustained tolerance was not achieved despite myeloablation, and a complete loss of genetically modified cells after transplantation in baboons was seen (284). Other factors such as immunogenicity of the protein transduced, source of stem cells, treatment with cytokine, major histocompatibility complex tissue type of the individual, and the route of introduction of the antigen are likely to play a role in the induction of tolerance. A greatly reduced risk of response against factor IX was associated with specific induction of tolerance due to the introduction of the transgene via the liver (283). Liver has been previously seen as an organ capable of inducing immunological tolerance (285). The tissue to which genetically-modified cells are returned is also important in this regard. Tissues such as skin have an important immune-associated function, serving as a primary barrier against foreign antigens and contain large number of antigen-presenting cells and keratinocytes, which are capable of secreting numerous inflammatory cytokines.

Regardless of the mechanism, immune response to neoantigen may prove to be a serious limitation to gene therapy. Most studies are carried out in immunocompromised hosts and have concentrated on increasing the number of gene-modified cells. Now that this goal is within reach, the immunological barriers to successful gene therapy are becoming more apparent. Thus, a major challenge in the coming years will be a systematic approach to understanding these barriers and to overcome them.

CONCLUSION

The major problems in the design and implementation of therapeutic strategies using gene modification arise from the significant deficits in our understanding of the mechanisms regulating adult stem cells development, effective gene expression and modification, and immunological responses to neoantigens. Initial successes in murine models have delayed awareness of the problems to be faced and have led to numerous clinical trials with very poor outcome and, in a few cases, with adverse effects. There is a pressing need to develop adequate animal models, which allow testing of new protocols of cell isolation, expansion, transduction, engraftment, and the side effects of new vector systems with long-term follow-up. Murine models may not be fully predictive for long-term effects due to differences in stem-cell turnover and to the shorter lifespan of mice. Therefore, the efficiency of engraftment of transduced cells should be tested in larger animal models. However, serious side effects such as tumor induction may take a long time to develop in such systems. A more rapid screen may be developed by the use of modified murine models engineered to be ultra-sensitive to adverse effects, for example, by introduction of co-operating proto-oncogene genetic lesions or by inducing replicative stress using serial transplantation.

Alongside the development of more appropriate animal models, further improvements in vector design are required to allow more control over gene expression and over integration. Additionally, improved transplantation protocols will be required to reduce immune responses to neoantigens.

Notwithstanding these considerations, recent experience has shown that clinical benefit can ensue from therapeutic gene transfer to stem cells. Thus, although technological advances are required, and will be obtained with time, stem-cell gene therapy should be pursued wherever the benefit to risk ratio is significantly weighted in favor of benefit.

REFERENCES

1. Thomson JA, Itskovitz-Eldor J, Shapiro SS, Waknitz MA, Swiergiel JJ, Marshall VS, Jones JM. Embryonic stem cell lines derived from human blastocysts. Science 1998; 282:1145–1147.
2. Reubinoff BE, Pera MF, Fong CY, Trounson A, Bongso A. Embryonic stem cell lines from human blastocysts: somatic differentiation in vitro. Nat Biotechnol 2000; 18:399–404.
3. Donovan PJ, Gearhart J. The end of the beginning for pluripotent stem cells. Nature 2001; 414:92–97.
4. Wagers AJ, Weissman IL. Plasticity of adult stem cells. Cell 2004; 116:639–648.
5. Goodell MA. Stem-cell "plasticity": befuddled by the muddle. Curr Opin Hematol 2003; 10:208–213.
6. Herzog EL, Chai L, Krause DS. Plasticity of marrow-derived stem cells. Blood 2003; 102:3483–3493.
7. Krause DS, Theise ND, Collector MI, Henegariu O, Hwang S, Gardner R, Neutzel S, Sharkis SJ. Multi-organ, multi-lineage engraftment by a single bone marrow-derived stem cell. Cell 2001; 105:369–377.
8. Lagasse E, Connors H, Al-Dhalimy M, Reitsma M, Dohse M, Osborne L, Wang X, Finegold M, Weissman IL, Grompe M. Purified hematopoietic stem cells can differentiate into hepatocytes in vivo. Nat Med 2000; 6:1229–1234.
9. Petersen BE, Bowen WC, Patrene KD, Mars WM, Sullivan AK, Murase N, Boggs SS, Greenberger JS, Goff JP. Bone marrow as a potential source of hepatic oval cells. Science 1999; 284:1168–1170.
10. Theise ND, Badve S, Saxena R, Henegariu O, Sell S, Crawford JM, Krause DS. Derivation of hepatocytes from bone marrow cells in mice after radiation-induced myeloablation. Hepatology 2000; 31:235–240.
11. Theise ND, Henegariu O, Grove J, Jagirdar J, Kao PN, Crawford JM, Badve S, Saxena R, Krause DS. Radiation pneumonitis in mice: a severe injury model for pneumocyte engraftment from bone marrow. Exp Hematol 2002; 30:1333–1338.
12. Brazelton TR, Nystrom M, Blau HM. Significant differences among skeletal muscles in the incorporation of bone marrow-derived cells. Dev Biol 2003; 262:64–74.
13. Ferrari G, Cusella-De Angelis G, Coletta M, Paolucci E, Stornaiuolo A, Cossu G, Mavilio F. Muscle regeneration by bone marrow-derived myogenic progenitors. Science 1998; 279:1528–1530.
14. Gussoni E, Soneoka Y, Strickland CD, Buzney EA, Khan MK, Flint AF, Kunkel LM, Mulligan RC. Dystrophin expression in the mdx mouse restored by stem cell transplantation. Nature 1999; 401:390–394.

15. Jackson KA, Majka SM, Wang H, Pocius J, Hartley CJ, Majesky MW, Entman ML, Michael LH, Hirschi KK, Goodell MA. Regeneration of ischemic cardiac muscle and vascular endothelium by adult stem cells. J Clin Invest 2001; 107:1395–1402.

16. Bjornson CR, Rietze RL, Reynolds BA, Magli MC, Vescovi AL. Turning brain into blood: a hematopoietic fate adopted by adult neural stem cells in vivo. Science 1999; 283:534–537.

17. Morshead CM, Benveniste P, Iscove NN, van der Kooy D. Hematopoietic competence is a rare property of neural stem cells that may depend on genetic and epigenetic alterations. Nat Med 2002; 8:268–273.

18. Castro RF, Jackson KA, Goodell MA, Robertson CS, Liu H, Shine HD. Failure of bone marrow cells to transdifferentiate into neural cells in vivo. Science 2002; 297:1299.

19. Choi JB, Uchino H, Azuma K, Iwashita N, Tanaka Y, Mochizuki H, Migita M, Shimada T, Kawamori R, Watada H. Little evidence of transdifferentiation of bone marrow-derived cells into pancreatic beta cells. Diabetologia 2003; 46:1366–1374.

20. Ono K, Yoshihara K, Suzuki H, Tanaka KF, Takii T, Onozaki K, Sawada M. Preservation of hematopoietic properties in transplanted bone marrow cells in the brain. J Neurosci Res 2003; 72:503–507.

21. Vallieres L, Sawchenko PE. Bone marrow-derived cells that populate the adult mouse brain preserve their hematopoietic identity. J Neurosci 2003; 23:5197–5207.

22. Wagers AJ, Sherwood RI, Christensen JL, Weissman IL. Little evidence for developmental plasticity of adult hematopoietic stem cells. Science 2002; 297:2256–2259.

23. Wang X, Willenbring H, Akkari Y, Torimaru Y, Foster M, Al-Dhalimy M, Lagasse E, Finegold M, Olson S, Grompe M. Cell fusion is the principal source of bone-marrow-derived hepatocytes. Nature 2003; 422:897–901.

24. Vassilopoulos G, Wang PR, Russell DW. Transplanted bone marrow regenerates liver by cell fusion. Nature 2003; 422:901–904.

25. Weissman IL. Stem cells: units of development, units of regeneration, and units in evolution. Cell 2000; 100:157–168.

26. Watt FM. Epidermal stem cells as targets for gene transfer. Hum Gene Ther 2000; 11:2261–2266.

27. Whitehead RH, Demmler K, Rockman SP, Watson NK. Clonogenic growth of epithelial cells from normal colonic mucosa from both mice and humans. Gastroenterology 1999; 117:858–865.

28. Potten CS, Booth C, Tudor GL, Booth D, Brady G, Hurley P, Ashton G, Clarke R, Sakakibara S, Okano H. Identification of a putative intestinal stem cell and early lineage marker: musashi-1. Differentiation 2003; 71:28–41.

29. Seale P, Rudnicki MA. A new look at the origin, function, and "stem-cell" status of muscle satellite cells. Dev Biol 2000; 218:115–124.

30. Gage FH. Mammalian neural stem cells. Science 2000; 287:1433–1438.

31. Alison M. Hepatic stem cells. Transplant Proc 2002; 34:2702–2705.

32. Lechner A, Habener JF. Stem/progenitor cells derived from adult tissues: potential for the treatment of diabetes mellitus. Am J Physiol Endocrinol Metab 2003; 284:E259–E266.

33. Lajtha LG. Stem cell concepts. Differentiation 1979; 14:23–34.

34. Barrandon Y, Green H. Three clonal types of keratinocyte with different capacities for multiplication. Proc Natl Acad Sci USA 1987; 84:2302–2306.

35. Morrison SJ, Wandycz AM, Hemmati HD, Wright DE, Weissman IL. Identification of a lineage of multipotent hematopoietic progenitors. Development 1997; 124:1929–1939.

36. Lavker RM, Miller S, Wilson C, Cotsarelis G, Wei ZG, Yang JS, Sun TT. Hair follicle stem cells: their location, role in hair cycle, and involvement in skin tumor formation. J Invest Dermatol 1993; 101:16S–26S.

37. Dunbar CE, Cottler-Fox M, O'Shaughnessy JA, Doren S, Carter C, Berenson R, Brown S, Moen RC, Greenblatt J, Stewart FM, et al. Retrovirally marked CD34-enriched peripheral blood and bone marrow cells contribute to long-term engraftment after autologous transplantation. Blood 1995; 85:3048–3057.

38. Hanania EG, Giles RE, Kavanagh J, Fu SQ, Ellerson D, Zu Z, Wang T, Su Y, Kudelka A, Rahman Z, et al. Results of MDR-1 vector modification trial indicate that granulocyte/macrophage colony-forming unit cells do not contribute to post-transplant hematopoietic recovery following intensive systemic therapy. Proc Natl Acad Sci USA 1996; 93:15346–15351.

39. Kohn DB, Weinberg KI, Nolta JA, Heiss LN, Lenarsky C, Crooks GM, Hanley ME, Annett G, Brooks JS, el-Khoureiy A, et al. Engraftment of gene-modified umbilical cord blood cells in neonates with adenosine deaminase deficiency. Nat Med 1995; 1:1017–1023.

40. Fausto N. Liver regeneration and repair: hepatocytes, progenitor cells, and stem cells. Hepatology 2004; 39:1477–1487.

41. Suzuki A, Nakauchi H. Identification and propagation of liver stem cells. Semin Cell Dev Biol 2002; 13:455–461.

42. Zulewski H, Abraham EJ, Gerlach MJ, Daniel PB, Moritz W, Muller B, Vallejo M, Thomas MK, Habener JF. Multipotential nestin-positive stem cells isolated from adult pancreatic islets differentiate ex vivo into pancreatic endocrine, exocrine, and hepatic phenotypes. Diabetes 2001; 50:521–533.

43. Holland AM, Gonez LJ, Harrison LC. Progenitor cells in the adult pancreas. Diabetes Metab Res Rev 2004; 20:13–27.

44. Zhang YQ, Sarvetnick N. Development of cell markers for the identification and expansion of islet progenitor cells. Diabetes Metab Res Rev 2003; 19:363–374.

45. Asahara T, Murohara T, Sullivan A, Silver M, van der Zee R, Li T, Witzenbichler B, Schatteman G, Isner JM. Isolation of putative progenitor endothelial cells for angiogenesis. Science 1997; 275:964–967.

46. Murohara T. Therapeutic vasculogenesis using human cord blood-derived endothelial progenitors. Trends Cardiovasc Med 2001; 11:303–307.

47. Quirici N, Soligo D, Caneva L, Servida F, Bossolasco P, Deliliers GL. Differentiation and expansion of endothelial cells from human bone marrow CD133(+) cells. Br J Haematol 2001; 115:186–194.

48. Masuda H, Asahara T. Postnatal endothelial progenitor cells for neovascularization in tissue regeneration. Cardiovasc Res 2003; 58:390–398.

49. Beauchamp JR, Morgan JE, Pagel CN, Partridge TA. Dynamics of myoblast transplantation reveal a discrete minority of precursors with stem cell-like properties as the myogenic source. J Cell Biol 1999; 144:1113–1122.

50. Hodgetts SI, Beilharz MW, Scalzo AA, Grounds MD. Why do cultured transplanted myoblasts die in vivo? DNA quantification shows enhanced survival of donor male myoblasts in host mice depleted of CD4+ and CD8+ cells or Nk1.1+ cells. Cell Transplant 2000; 9:489–502.

51. Xiao X, Li J, Samulski RJ. Efficient long-term gene transfer into muscle tissue of immunocompetent mice by adeno-associated virus vector. J Virol 1996; 70:8098–8108.

52. Kessler PD, Podsakoff GM, Chen X, McQuiston SA, Colosi PC, Matelis LA, Kurtzman GJ, Byrne BJ. Gene delivery to skeletal muscle results in sustained expression and systemic delivery of a therapeutic protein. Proc Natl Acad Sci USA 1996; 93:14082–14087.

53. Hagstrom JN, Couto LB, Scallan C, Burton M, McCleland ML, Fields PA, Arruda VR, Herzog RW, High KA. Improved muscle-derived expression of human coagulation factor IX from a skeletal actin/CMV hybrid enhancer/promoter. Blood 2000; 95:2536–2542.

54. Beauchamp JR, Heslop L, Yu DS, Tajbakhsh S, Kelly RG, Wernig A, Buckingham ME, Partridge TA, Zammit PS. Expression of CD34 and Myf5 defines the majority of quiescent adult skeletal muscle satellite cells. J Cell Biol 2000; 151:1221–1234.

55. Palmer TD, Schwartz PH, Taupin P, Kaspar B, Stein SA, Gage FH. Cell culture: progenitor cells from human brain after death. Nature 2001; 411:42–43.

56. Sanai N, Tramontin AD, Quinones-Hinojosa A, Barbaro NM, Gupta N, Kunwar S, Lawton MT, McDermott MW, Parsa AT, Manuel-Garcia Verdugo J, Berger MS, Alvarez-Buylla A. Unique astrocyte ribbon in adult human brain contains neural stem cells but lacks chain migration. Nature 2004; 427:740–744.

57. Ostenfeld T, Caldwell MA, Prowse KR, Linskens MH, Jauniaux E, Svendsen CN. Human neural precursor cells express low levels of telomerase in vitro and show diminishing cell proliferation with extensive axonal outgrowth following transplantation. Exp Neurol 2000; 164:215–226.

58. Svendsen CN, ter Borg MG, Armstrong RJ, Rosser AE, Chandran S, Ostenfeld T, Caldwell MA. A new method for the rapid and long-term growth of human neural precursor cells. J Neurosci Meth 1998; 85:141–152.

59. Snyder EY, Taylor RM, Wolfe JH. Neural progenitor cell engraftment corrects lysosomal storage throughout the MPS VII mouse brain. Nature 1995; 374:367–370.

60. Snyder EY, Park KL. Limitations in brain repair. Nat Med 2002; 8:928–930.

61. Milsom MD, Fairbairn LJ. Protection and selection for gene therapy in the hematopoietic system. J Gene Med 2004; 6:133–146.

62. Paczesny S, Ueno H, Fay J, Banchereau J, Palucka AK. Dendritic cells as vectors for immunotherapy of cancer. Semin Cancer Biol 2003; 13:439–447.

63. Yang L, Qin XF, Baltimore D, Van Parijs L. Generation of functional antigen-specific T cells in defined genetic backgrounds by retrovirus-mediated expression of TCR cDNAs in hematopoietic precursor cells. Proc Natl Acad Sci USA 2002; 99:6204–6209.

64. Civin CI, Strauss LC, Fackler MI, Trischmann TM, Wiley JM, Loken MR. Positive stem cell selection—basic science. Prog Clin Biol Res 1990; 333:387–401; discussion 402.

65. Brugger W, Henschler R, Heimfeld S, Berenson RJ, Mertelsmann R, Kanz L. Positively selected autologous blood CD34+ cells and unseparated peripheral blood progenitor cells mediate identical hematopoietic engraftment after high-dose VP16, ifosfamide, carboplatin, and epirubicin. Blood 1994; 84:1421–1426.

66. Bhatia M, Wang JC, Kapp U, Bonnet D, Dick JE. Purification of primitive human hematopoietic cells capable of repopulating immune-deficient mice. Proc Natl Acad Sci USA 1997; 94:5320–5325.

67. Glimm H, Eisterer W, Lee K, Cashman J, Holyoake TL, Nicolini F, Shultz LD, von Kalle C, Eaves CJ. Previously undetected human hematopoietic cell populations with short-term repopulating activity selectively engraft NOD/SCID-beta2 microglobulin-null mice. J Clin Invest 2001; 107:199–206.

68. Szilvassy SJ, Humphries RK, Lansdorp PM, Eaves AC, Eaves CJ. Quantitative assay for totipotent reconstituting hematopoietic stem cells by a competitive repopulation strategy. Proc Natl Acad Sci USA 1990; 87:8736–8740.

69. McNiece IK, Bertoncello I, Kriegler AB, Quesenberry PJ. Colony-forming cells with high proliferative potential (HPP-CFC). Int J Cell Cloning 1990; 8:146–160.

70. Bhatia M, Bonnet D, Murdoch B, Gan OI, Dick JE. A newly discovered class of human hematopoietic cells with SCID-repopulating activity. Nat Med 1998; 4:1038–1045.

71. Larochelle A, Vormoor J, Hanenberg H, Wang JC, Bhatia M, Lapidot T, Moritz T, Murdoch B, Xiao XL, Kato I, Williams DA, Dick JE. Identification of primitive human hematopoietic cells capable of repopulating NOD/SCID mouse bone marrow: implications for gene therapy. Nat Med 1996; 2:1329–1337.

72. Gan OI, Murdoch B, Larochelle A, Dick JE. Differential maintenance of primitive human SCID-repopulating cells, clonogenic progenitors, and long-term culture-initiating cells after incubation on human bone marrow stromal cells. Blood 1997; 90:641–650.

73. Conneally E, Cashman J, Petzer A, Eaves C. Expansion in vitro of transplantable human cord blood stem cells demonstrated using a quantitative assay of their lympho-myeloid repopulating activity in nonobese diabetic-SCID/SCID mice. Proc Natl Acad Sci USA 1997; 94:9836–9841.

74. Larochelle A, Vormoor J, Lapidot T, Sher G, Furukawa T, Li Q, Shultz LD, Olivieri NF, Stamatoyannopoulos G, Dick JE. Engraftment of immune-deficient mice with primitive hematopoietic cells from beta-thalassemia and sickle cell anemia patients: implications for evaluating human gene therapy protocols. Hum Mol Genet 1995; 4:163–172.

75. Cashman J, Bockhold K, Hogge DE, Eaves AC, Eaves CJ. Sustained proliferation, multi-lineage differentiation and maintenance of primitive human haemopoietic cells in NOD/SCID mice transplanted with human cord blood. Br J Haematol 1997; 98:1026–1036.

76. Wolfe JH, Sands MS, Barker JE, Gwynn B, Rowe LB, Vogler CA, Birkenmeier EH. Reversal of pathology in murine mucopolysaccharidosis type VII by somatic cell gene transfer. Nature 1992; 360:749–753.

77. Strom TS, Turner SJ, Andreansky S, Liu H, Doherry PC, Srivastava DK, Cunningham JM, Nienhuis AW. Defects in T-cell-mediated immunity to influenza virus in murine Wiskott-Aldrich syndrome are corrected by oncoretroviral vector-mediated gene transfer into repopulating hematopoietic cells. Blood 2003; 102:3108–3116.

78. Otsu M, Anderson SM, Bodine DM, Puck JM, O'Shea JJ, Candotti F. Lymphoid development and function in X-linked severe combined immunodeficiency mice after stem cell gene therapy. Mol Ther 2000; 1:145–153.

79. Bjorgvinsdottir H, Ding C, Pech N, Gifford MA, Li LL, Dinauer MC. Retroviral-mediated gene transfer of gp91phox into bone marrow cells rescues defect in host defense against *Aspergillus fumigatus* in murine X-linked chronic granulomatous disease. Blood 1997; 89:41–48.

80. Bunting KD, Sangster MY, Ihle JN, Sorrentino BP. Restoration of lymphocyte function in Janus kinase 3-deficient mice by retroviral-mediated gene transfer. Nat Med 1998; 4:58–64.

81. Hennemann B, Conneally E, Pawliuk R, Leboulch P, Rose-John S, Reid D, Chuo JY, Humphries RK, Eaves CJ. Optimization of retroviral-mediated gene transfer to human NOD/SCID mouse repopulating cord blood cells through a systematic analysis of protocol variables. Exp Hematol 1999; 27:817–825.

82. Malech HL, Maples PB, Whiting-Theobald N, Linton GF, Sekhsaria S, Vowells SJ, Li F, Miller JA, DeCarlo E, Holland SM, et al. Prolonged production of NADPH oxidase-corrected granulocytes after gene therapy of chronic granulomatous disease. Proc Natl Acad Sci USA 1997; 94:12133–12138.

83. Horn PA, Thomasson BM, Wood BL, Andrews RG, Morris JC, Kiem HP. Distinct hematopoietic stem/progenitor cell populations are responsible for repopulating NOD/SCID mice compared with nonhuman primates. Blood 2003; 102:4329–4335.

84. Schmidt M, Zickler P, Hoffmann G, Haas S, Wissler M, Muessig A, Tisdale JF, Kuramoto K, Andrews RG, Wu T, Kiem HP, Dunbar CE, von Kalle C. Polyclonal long-term repopulating stem cell clones in a primate model. Blood 2002; 100:2737–2743.

85. Rheinwald JG, Green H. Epidermal growth factor and the multiplication of cultured human epidermal keratinocytes. Nature 1977; 265:421–424.

86. Potten CS, Kovacs L, Hamilton E. Continuous labelling studies on mouse skin and intestine. Cell Tissue Kinet 1974; 7:271–283.

87. Li A, Simmons PJ, Kaur P. Identification and isolation of candidate human keratinocyte stem cells based on cell surface phenotype. Proc Natl Acad Sci USA 1998;95:3902–3907.

88. Bickenbach JR, Mackenzie IC. Identification and localization of label-retaining cells in hamster epithelia. J Invest Dermatol 1984; 82:618–622.

89. Compton CC, Gill JM, Bradford DA, Regauer S, Gallico GG, O'Connor NE. Skin regenerated from cultured epithelial autografts on full-thickness burn wounds from 6 days to 5 years after grafting: a light, electron microscopic and immunohistochemical study. Lab Invest 1989; 60:600–612.

90. Gallico GG III, O'Connor NE, Compton CC, Kehinde O, Green H. Permanent coverage of large burn wounds with autologous cultured human epithelium. N Engl J Med 1984; 311:448–451.

91. Sheridan RL, Hurley J, Smith MA, Ryan CM, Bondoc CC, Quinby WC Jr, Tompkins RG, Burke JF. The acutely burned hand: management and outcome based on a ten-year experience with 1047 acute hand burns. J Trauma 1995; 38:406–411.

92. Harris PA, Leigh IM, Navsaria HA. Pre-confluent keratinocyte grafting: the future for cultured skin replacements? Burns 1998; 24:591–593.

93. Ehrlich HP. Understanding experimental biology of skin equivalent: from laboratory to clinical use in patients with bums and chronic wounds. Am J Surg 2004; 187:29S–33S.

94. Chen M, Kasahara N, Keene DR, Chan L, Hoeffler WK, Finlay D, Barcova M, Cannon PM, Mazurek C, Woodley DT. Restoration of type VII collagen expression and function in dystrophic epidermolysis bullosa. Nat Genet 2002; 32:670–675.

95. Freiberg RA, Choate KA, Deng H, Alperin ES, Shapiro LJ, Khavari PA. A model of corrective gene transfer in X-linked ichthyosis. Hum Mol Genet 1997; 6:927–933.

96. Ortiz-Urda S, Thyagarajan B, Keene DR, Lin Q, Calos MP, Khavari PA. PhiC31 integrase-mediated nonviral genetic correction of junctional epidermolysis bullosa. Hum Gene Ther 2003; 14:923–928.

97. Ghazizadeh S, Doumeng C, Taichman LB. Durable and stratum-specific gene expression in epidermis. Gene Ther 2002; 9:1278–1285.

98. Mathor MB, Ferrari G, Dellambra E, Cilli M, Mavilio F, Cancedda R, De Luca M. Clonal analysis of stably transduced human epidermal stem cells in culture. Proc Natl Acad Sci USA 1996; 93:10371–10376.

99. Pfutzner W, Terunuma A, Tock CL, Snead EK, Kolodka TM, Gottesman MM, Taichman L, Vogel JC. Topical colchicine selection of keratinocytes transduced with the multidrug resistance gene (MDR1) can sustain and enhance transgene expression in vivo. Proc Natl Acad Sci USA 2002; 99:13096–13101.

100. Li A, Pouliot N, Redvers R, Kaur P. Extensive tissue-regenerative capacity of neonatal human keratinocyte stem cells and their progeny. J Clin Invest 2004; 113:390–400.

101. Dellambra E, Pellegrini G, Guerra L, Ferrari G, Zambruno G, Mavilio F, De Luca M. Toward epidermal stem cell-mediated ex vivo gene therapy of junctional epidermolysis bullosa. Hum Gene Ther 2000; 11:2283–2287.

102. Tyrone JW, Mogford IE, Chandler LA, Ma C, Xia Y, Pierce GF, Mustoe TA. Collagen-embedded platelet-derived growth factor DNA plasmid promotes wound healing in a dermal ulcer model. J Surg Res 2000; 93:230–236.

103. Wang X, Zinkel S, Polonsky K, Fuchs E. Transgenic studies with a keratin promoter-driven growth hormone transgene: prospects for gene therapy. Proc Natl Acad Sci USA 1997; 94:219–226.

104. Siprashvili Z, Khavari PA. Lentivectors for regulated and reversible cutaneous gene delivery. Mol Ther 2004; 9:93–100.

105. Prockop DJ. Marrow stromal cells as stem cells for nonhematopoietic tissues. Science 1997; 276:71–74.

106. Caplan AI. Mesenchymal stem cells. J Orthop Res 1991; 9:641–650.

107. Haynesworth SE, Baber MA, Caplan AL. Cell surface antigens on human marrow-derived mesenchymal cells are detected by monoclonal antibodies. Bone 1992; 13:69–80.

108. Simmons PJ, Torok-Storb B. Identification of stromal cell precursors in human bone marrow by a novel monoclonal antibody, STRO-1. Blood 1991; 78:55–62.

109. Pittenger MF, Mackay AM, Beck SC, Jaiswal RK, Douglas R, Mosca JD, Moorman MA, Simonetti DW, Craig S, Marshak DR. Multilineage potential of adult human mesenchymal stem cells. Science 1999; 284:143–147.

110. Bianco P, Gehron Robey P. Marrow stromal stem cells. J Clin Invest 2000; 105:1663–1668.

111. Muraglia A, Cancedda R, Quarto R. Clonal mesenchymal progenitors from human bone marrow differentiate in vitro according to a hierarchical model. J Cell Sci 2000; 113(Pt 7):1161–1166.

112. Castro-Malaspina H, Gay RE, Resnick G, Kapoor N, Meyers P, Chiarieri D, McKenzie S, Broxmeyer HE, Moore MA. Characterization of human bone marrow fibroblast colony-forming cells (CFU-F) and their progeny. Blood 1980; 56:289–301.

113. Banfi A, Muraglia A, Dozin B, Mastrogiacomo M, Cancedda R, Quarto R. Proliferation kinetics and differentiation potential of ex vivo expanded human bone marrow stromal cells: implications for their use in cell therapy. Exp Hematol 2000; 28:707–715.

114. Digirolamo CM, Stokes D, Colter D, Phinney DG, Class R, Prockop DJ. Propagation and senescence of human marrow stromal cells in culture: a simple colony-forming assay identifies samples with the greatest potential to propagate and differentiate. Br J Haematol 1999; 107:275–281.

115. Campagnoli C, Bellantuono I, Kumar S, Fairbairn LJ, Roberts I, Fisk NM. High transduction efficiency of circulating first trimester fetal mesenchymal stem cells: potential targets for in utero ex vivo gene therapy. BJOG 2002; 109:952–954.

116. Olmsted-Davis EA, Gugala Z, Gannon FH, Yotnda P, McAlhany RE, Lindsey RW, Davis AR. Use of a chimeric adenovirus vector enhances BMP2 production and bone formation. Hum Gene Ther 2002; 13:1337–1347.

117. Marx JC, Allay JA, Persons DA, Nooner SA, Hargrove PW, Kelly PF, Vanin EF, Horwitz EM. High-efficiency transduction and long-term gene expression with a murine stem cell retroviral vector encoding the green fluorescent protein in human marrow stromal cells. Hum Gene Ther 1999; 10:1163–1173.

118. Bartholomew A, Patil S, Mackay A, Nelson M, Buyaner D, Hardy W, Mosca J, Sturgeon C, Siatskas M, Mahmud N, et al. Baboon mesenchymal stem cells can be genetically modified to secrete human erythropoietin in vivo. Hum Gene Ther 2001; 12:1527–1541.

119. Baxter MA, Wynn RF, Deakin JA, Bellantuono I, Edington KG, Cooper A, Besley GT, Church HJ, Wraith JE, Carr TF, Fairbairn LJ. Retrovirally mediated correction of bone marrow-derived mesenchymal stem cells from patients with mucopolysaccharidosis type I. Blood 2002; 99:1857–1859.

120. Horwitz EM, Gordon PL, Koo WK, Marx JC, Neel MD, McNall RY, Muul L, Hofmann T. Isolated allogeneic bone marrow-derived mesenchymal cells engraft and stimulate growth in children with osteogenesis imperfecta: implications for cell therapy of bone. Proc Natl Acad Sci USA 2002; 99:8932–8937.

121. Devine SM, Cobbs C, Jennings M, Bartholomew A, Hoffman R. Mesenchymal stem cells distribute to a wide range of tissues following systemic infusion into nonhuman primates. Blood 2003; 101:2999–3001.

122. Gao J, Dennis JE, Muzic RF, Liradberg M, Caplan AI. The dynamic in vivo distribution of bone marrow-derived mesenchymal stem cells after infusion. Cells Tissues Organs 2001; 169:12–20.

123. Short B, Brouard N, Simmons PJ. Prospective isolation of mesenchymal progenitor cells from mouse compact bone. ISSCR 2nd Annual Meeting, Boston, 2004:134.

124. Baxter MA, Wynn RF, Jowitt SN, Wraith JE, Fairbairn LJ, Bellantuono I. Study of telomere length reveals rapid aging of human marrow stromal cells following in vitro expansion. Stem Cells 2004; 22(5):675–682.

125. Reyes M, Lund T, Lenvik T, Aguiar D, Koodie L, Verfaillie CM. Purification and ex vivo expansion of postnatal human marrow mesodermal progenitor cells. Blood 2001; 98:2615–2625.

126. Jiang Y, Jahagirdar BN, Reinhardt RL, Schwartz RE, Keene CD, Ortiz-Gonzalez XR, Reyes M, Lenvik T, Lund T, Blackstad M, et al. Pluripotency of mesenchymal stem cells derived from adult marrow. Nature 2002; 418:41–49.

127. Wynn RF, Hart CA, Corradi-Perini C, O'Neill L, Evans CA, Wraith JE, Fairbairn LJ, Bellantuono I. A small proportion of mesenchymal stem cells strongly express functionally active CXCR4 receptor capable of promoting migration to bone marrow. Blood 2004; 104:2643–2645.

128. Friedenstein AJ, Deriglasova UF, Kulagina NN, Panasuk AF, Rudakowa SF, Luria EA, Ruadkow IA. Precursors for fibroblasts in different populations of hematopoietic cells as detected by the in vitro colony assay method. Exp Hematol 1974; 2:83–92.

129. Friedenstein AJ, Latzinik NW, Grosheva AG, Gorskaya UF. Marrow microenvironment transfer by heterotopic transplantation of freshly isolated and cultured cells in porous sponges. Exp Hematol 1982; 10:217–227.

130. Daga A, Muraglia A, Quarto R, Cancedda R, Corte G. Enhanced engraftment of EPO-transduced human bone marrow stromal cells transplanted in a 3D matrix in non-conditioned NOD/SCID mice. Gene Ther 2002; 9:915–921.

131. Felgner PL. Nonviral strategies for gene therapy. Sci Am 1997; 276:102–106.

132. Bruder SP, Kraus KH, Goldberg VM, Kadiyala S. The effect of implants loaded with autologous mesenchymal stem cells on the healing of canine segmental bone defects. J Bone Joint Surg Am 1998; 80:985–996.

133. Bruder SP, Kurth AA, Shea M, Hayes WC, Jaiswal N, Kadiyala S. Bone regeneration by implantation of purified, culture-expanded human mesenchymal stem cells. J Orthop Res 1998; 16:155–162.

134. Lollo CP, Banaszczyk MG, Chiou HC. Obstacles and advances in non-viral gene delivery. Curr Opin Mol Ther 2000; 2:136–142.

135. Vogel JC. Nonviral skin gene therapy. Hum Gene Ther 2000; 11:2253–2259.

136. Torchilin VP, Lukyanov AN, Gao Z, Papahadjopoulos-Sternberg B. Immunomicelles: targeted pharmaceutical carriers for poorly soluble drugs. Proc Natl Acad Sci USA 2003; 100:6039–6044.

137. Lechardeur D, Lukacs GL. Intracellular barriers to non-viral gene transfer. Curr Gene Ther 2002; 2:183–194.

138. Ludtke JJ, Zhang G, Sebestyen MG, Wolff JA. A nuclear localization signal can enhance both the nuclear transport and expression of 1 kb DNA. J Cell Sci 1999; 112(Pt 12):2033–2041.

139. Frankel AD, Pabo CO. Cellular uptake of the tat protein from human immunodeficiency virus. Cell 1988; 55:1189–1193.

140. Vocero-Akbani A, Lissy NA, Dowdy SF. Transduction of full-length Tat fusion proteins directly into mammalian cells: analysis of T cell receptor activation-induced cell death. Meth Enzymol 2000; 322:508–521.

141. Munkonge FM, Dean DA, Hillery E, Griesenbach U, Alton EW. Emerging significance of plasmid DNA nuclear import in gene therapy. Adv Drug Deliv Rev 2003; 55:749–760.

142. Ortiz-Urda S, Lin Q, Yant SR, Keene D, Kay MA, Khavari PA. Sustainable correction of junctional epidermolysis bullosa via transposon-mediated nonviral gene transfer. Gene Ther 2003; 10:1099–1104.

143. Delecluse HJ, Pich D, Hilsendegen T, Baum C, Hammerschmidt W. A first-generation packaging cell line for Epstein-Barr virus-derived vectors. Proc Natl Acad Sci USA 1999; 96:5188–5193.

144. Bowers WJ, Olschowka JA, Federoff HJ. Immune responses to replication-defective HSV-1 type vectors within the CNS: implications for gene therapy. Gene Ther 2003; 10:941–945.

145. Flotte TR. Gene therapy progress and prospects: recombinant adeno-associated virus (rAAV) vectors. Gene Ther 2004; 11:805–810.

146. Flotte TR, Afione SA, Zeitlin PL. Adeno-associated virus vector gene expression occurs in nondividing cells in the absence of vector DNA integration. Am J Respir Cell Mol Biol 1994; 11:517–521.

147. Flotte TR, Carter BJ. Adeno-associated virus vectors for gene therapy. Gene Ther 1995; 2:357–362.

148. Nathwani AC, Hanawa H, Vandergriff J, Kelly P, Vanin EF, Nienhuis AW. Efficient gene transfer into human cord blood CD34+ cells and the CD34 + CD38-subset using highly purified recombinant adeno-associated viral vector preparations that are free of helper virus and wild-type AAV. Gene Ther 2000; 7:183–195.

149. Snyder RO, Flotte TR. Production of clinical-grade recombinant adeno-associated virus vectors. Curr Opin Biotechnol 2002; 13:418–423.

150. Correll PH, Colilla S, Karlsson S. Retroviral vector design for long-term expression in murine hematopoietic cells in vivo. Blood 1994; 84:1812–1822.

151. Ohashi T, Boggs S, Robbins P, Bahnson A, Patrene K, Wei FS, Wei JF, Li J, Lucht L, Fei Y, et al. Efficient transfer and sustained high expression of the human glucocerebrosidase gene in mice and their functional macrophages following transplantation of bone marrow transduced by a retroviral vector. Proc Natl Acad Sci USA 1992; 89:11332–11336.

152. Bunting KD, Flynn KJ, Riberdy JM, Doherty PC, Sorrentino BP. Virus-specific immunity after gene therapy in a murine model of severe combined immunodeficiency. Proc Natl Acad Sci USA 1999; 96:232–237.

153. Tisdale JF, Hanazono Y, Sellers SE, Agricola BA, Metzger ME, Donahue RE, Dunbar CE. Ex vivo expansion of genetically marked rhesus peripheral blood progenitor cells results in diminished long-term repopulating ability. Blood 1998; 92:1131–1141.

154. Dunbar CE, Seidel NE, Doren S, Sellers S, Cline AP, Metzger ME, Agricola BA, Donahue RE, Bodine DM. Improved retroviral gene transfer into murine and rhesus peripheral blood or bone marrow repopulating cells primed in vivo with stem cell factor and granulocyte colony-stimulating factor. Proc Natl Acad Sci USA 1996; 93:11871–11876.

155. Kiem HP, Darovsky B, Von Kalle C, Goehle S, Graham T, Miller AD, Storb R, Schuening FG. Long-term persistence of canine hematopoietic cells genetically marked by retrovirus vectors. Hum Gene Ther 1996; 7:89–96.

156. Kiem HP, Heyward S, Winkler A, Potter J, Allen JM, Miller AD, Andrews RG. Gene transfer into marrow repopulating cells: comparison between amphotropic and gibbon ape leukemia virus pseudotyped retroviral vectors in a competitive repopulation assay in baboons. Blood 1997; 90:4638–4645.

157. Hesdorffer C, Ayello J, Ward M, Kaubisch A, Vahdat L, Balmaceda C, Garrett T, Fetell M, Reiss R, Bank A, Antman K. Phase I trial of retroviral-mediated transfer of the human MDR1 gene as marrow chemoprotection in patients undergoing high-dose chemotherapy and autologous stem-cell transplantation. J Clin Oncol 1998; 16:165–172.

158. Miller DG, Adam MA, Miller AD. Gene transfer by retrovirus vectors occurs only in cells that are actively replicating at the time of infection. Mol Cell Biol 1990; 10:4239–4242.

159. Cheshier SH, Morrison SJ, Liao X, Weissman IL. In vivo proliferation and cell cycle kinetics of long-term self-renewing hematopoietic stem cells. Proc Natl Acad Sci USA 1999; 96:3120–3125.

160. Hanawa H, Hargrove PW, Kepes S, Srivastava DK, Nienhuis AW, Persons DA. Extended {beta}-globin locus control region elements promote consistent therapeutic expression of a {gamma}-globin lentiviral vector in murine {beta}-thalassemia. Blood 2004; 104:2281–2290.

161. Rivella S, Sadelain M. Therapeutic globin gene delivery using lentiviral vectors. Curr Opin Mol Ther 2002; 4:505–514.

162. Pawliuk R, Westerman KA, Fabry ME, Payen E, Tighe R, Bouhassira EE, Acharya SA, Ellis J, London IM, Eaves CJ, Humphries RK, Beuzard Y, Nagel RL, Leboulch P. Correction of sickle cell disease in transgenic mouse models by gene therapy. Science 2001; 294:2368–2371.

163. Uchida N, Sutton RE, Friera AM, He D, Reitsma MJ, Chang WC, Veres G, Scollay R, Weissman IL. HIV, but not murine leukemia virus, vectors mediate high efficiency gene transfer into freshly isolated G0/G1 human hematopoietic stem cells. Proc Natl Acad Sci USA 1998; 95:11939–11944.

164. Naldini L, Blomer U, Gallay P, Ory D, Mulligan R, Gage FH, Verma IM, Trono D. In vivo gene delivery and stable transduction of nondividing cells by a lentiviral vector. Science 1996; 272:263–267.

165. Sutton RE, Reitsma MJ, Uchida N, Brown PO. Transduction of human progenitor hematopoietic stem cells by human immunodeficiency virus type 1-based vectors is cell-cycle-dependent. J Virol 1999; 73:3649–3660.

166. Korin YD, Zack JA. Progression to the G1b phase of the cell cycle is required for completion of human immunodeficiency virus type 1 reverse transcription in T cells. J Virol 1998; 72:3161–3168.

167. Guenechea G, Gan OI, Inamitsu T, Dorrell C, Pereira DS, Kelly M, Naldini L, Dick JE. Transduction of human CD34 + CD38 − bone marrow and cord blood-derived SCID-repopulating cells with third-generation lentiviral vectors. Mol Ther 2000; 1:566–573.

168. Miyoshi H, Smith KA, Mosier DE, Verma IM, Torbett BE. Transduction of human CD34+ cells that mediate long-term engraftment of NOD/SCID mice by HIV vectors. Science 1999; 283:682–686.

169. Horn PA, Morris JC, Bukovsky AA, Andrews RG, Naldini L, Kurre P, Kiem HP. Lentivirus-mediated gene transfer into hematopoietic repopulating cells in baboons. Gene Ther 2002; 9:1464–1471.

170. An DS, Rung SK, Bonifacino A, Wersto RP, Metzger ME, Agricola BA, Mao SH, Chen IS, Donahue RE. Lentivirus vector-mediated hematopoietic stem cell gene transfer of common gamma-chain cytokine receptor in rhesus macaques. J Virol 2001; 75:3547–3555.

171. Hanawa H, Hematti P, Keyvanfar K, Metzger ME, Krouse A, Donahue RE, Kepes S, Gray J, Dunbar CE, Persons DA, Nienhuis AW. Efficient gene transfer into rhesus repopulating hematopoietic stem cells using a simian immunodeficiency virus-based lentiviral vector system. Blood 2004; 103:4062–4069.

172. Kuhn U, Terunuma A, Pfutzner W, Foster RA, Vogel JC. In vivo assessment of gene delivery to keratinocytes by lentiviral vectors. J Virol 2002; 76:1496–1504.

173. Farson D, Witt R, McGuinness R, Dull T, Kelly M, Song J, Radeke R, Bukovsky A, Consiglio A, Naldini L. A new-generation stable inducible packaging cell line for lentiviral vectors. Hum Gene Ther 2001; 12:981–997.

174. Marshall E. Gene therapy: panel reviews risks of germ line changes. Science 2001; 294:2268–2269.

175. Follenzi A, Battaglia M, Lombardo A, Annoni A, Roncarolo MG, Naldini L. Targeting lentiviral vector expression to hepatocytes limits transgene-specific immune response and establishes long-term expression of human antihemophilic factor IX in mice. Blood 2004; 103:3700–3709.

176. Ghazizadeh S, Kalish RS, Taichman LB. Immune-mediated loss of transgene expression in skin: implications for cutaneous gene therapy. Mol Ther 2003; 7:296–303.

177. Davidson BL, Breakefield XO. Viral vectors for gene delivery to the nervous system. Nat Rev Neurosci 2003; 4:353–364.

178. Xu L, Haskins ME, Melniczek JR, Gao C, Weil MA, O'Malley TM, ODonnell PA, Mazrier B, Ellinwood NM, Zweigle J, Wolfe JH, Ponder KP. Transduction of hepatocytes after neonatal delivery of a Moloney murine leukemia virus based retroviral vector results in long-term expression of beta-glucuronidase in mucopolysaccharidosis VII dogs. Mol Ther 2002; 5:141–153.

179. Xu L, Gao C, Sands MS, Cai SR, Nichols TC, Bellinger DA, Raymer RA, McCorquodale S, Ponder KP. Neonatal or hepatocyte growth factor-potentiated adult gene therapy with a retroviral vector results in therapeutic levels of canine factor IX for hemophilia B. Blood 2003; 101:3924–3932.

180. McCormack JE, Edwards W, Sensintaffer J, Lillegren L, Kozloski M, Brumm D, Karavodin L, Jolly DJ, Greengard J. Factors affecting long-term expression of a secreted transgene product after intravenous administration of a retroviral vector. Mol Ther 2001; 3:516–525.

181. DePolo NT, Reed JD, Sheridan PL, Townsend K, Sauter SL, Jolly DJ, Dubensky TW Jr. VSV-G pseudotyped lentiviral vector particles produced in human cells are inactivated by human serum. Mol Ther 2000; 2:218–222.

182. Takeuchi Y, Cosset FL, Lachmann PJ, Okada H, Weiss RA, Collins MK. Type C retrovirus inactivation by human complement is determined by both the viral genome and the producer cell. J Virol 1994; 68:8001–8007.

183. Takeuchi Y, Porter CD, Strahan KM, Preece AF, Gustafsson K, Cosset FL, Weiss RA, Collins MK. Sensitization of cells and retroviruses to human serum by (alpha 1–3) galactosyltransferase. Nature 1996; 379:85–88.

184. Ponder KP, Melniczek JR, Xu L, Weil MA, O'Malley TM, O'Donnell PA, Knox VW, Aguirre GD, Mazrier H, Ellinwood NM, et al. Therapeutic neonatal hepatic gene therapy in mucopolysaccharidosis VH dogs. Proc Natl Acad Sci USA 2002; 99:13102–13107.

185. Sandrin V, Boson B, Salmon P, Gay W, Negre D, Le Grand R, Trono D, Cosset FL. Lentiviral vectors pseudotyped with a modified RD114 envelope glycoprotein show increased stability in sera and augmented transduction of primary lymphocytes and CD34+ cells derived from human and nonhuman primates. Blood 2002; 100:823–832.

186. Pensiero MN, Wysocki CA, Nader K, Kikuchi GE. Development of amphotropic murine retrovirus vectors resistant to inactivation by human serum. Hum Gene Ther 1996; 7:1095–1101.

187. Valsesia-Wittmann S, Morling FJ, Nilson BH, Takeuchi Y, Russell SJ, Cosset FL. Improvement of retroviral retargeting by using amino acid spacers between an additional binding domain and the N terminus of Moloney murine leukemia virus SU. J Virol 1996; 70:2059–2064.

188. Cosset FL, Morling FJ, Takeucbi Y, Weiss RA, Collins MK, Russell SJ. Retroviral retargeting by envelopes expressing an N-terminal binding domain. J Virol 1995; 69:6314–6322.

189. Morling FJ, Peng KW, Cosset FL, Russell SJ. Masking of retroviral envelope functions by oligomerizing polypeptide adaptors. Virology 1997; 234:51–61.

190. Peng KW, Morling FJ, Cosset FL, Murphy G, Russell SJ. A gene delivery system activatable by disease-associated matrix metalloproteinases. Hum Gene Ther 1997; 8:729–738.

191. Dick JE, Magli MC, Huszar D, Phillips RA, Bernstein A. Introduction of a selectable gene into primitive stem cells capable of long-term reconstitution of the hemopoietic system of W/Wv mice. Cell 1985; 42:71–79.

192. Bunting KD, Galipeau J, Topham D, Benaim E, Sorrentino BP. Transduction of murine bone marrow cells with an MDR1 vector enables ex vivo stem cell expansion, but these expanded grafts cause a myeloproliferative syndrome in transplanted mice. Blood 1998; 92:2269–2279.

193. Mardiney M III, Jackson SH, Spratt SK, Li F, Holland SM, Malech HL. Enhanced host defense after gene transfer in the murine p47phox-deficient model of chronic granulomatous disease. Blood 1997; 89:2268–2275.

194. Soudais C, Shiho T, Sharara LI, Guy-Grand D, Taniguchi T, Fischer A, Di Santo JP. Stable and functional lymphoid reconstitution of common cytokine receptor gamma chain deficient mice by retroviral-mediated gene transfer. Blood 2000; 95:3071–3077.

195. Crooks GM, Kohn DB. Growth factors increase amphotropic retrovirus binding to human CD34+ bone marrow progenitor cells. Blood 1993; 82:3290–3297.

196. Orlic D, Girard LI, Jordan CT, Anderson SM, Cline AP, Bodine DM. The level of mRNA encoding the amphotropic retrovirus receptor in mouse and human hematopoietic stem cells is low and correlates with the efficiency of retrovirus transduction. Proc Natl Acad Sci USA 1996; 93:11097–11102.

197. Kavanaugh MP, Miller DG, Zhang W, Law W, Kozak SL, Kabat D, Miller AD. Cell-surface receptors for gibbon ape leukemia virus and amphotropic murine retrovirus are inducible sodium-dependent phosphate symporters. Proc Natl Acad Sci USA 1994; 91:7071–7075.

198. Pedersen L, van Zeijl M, Johann SV, O'Hara B. Fungal phosphate transporter serves as a receptor backbone for gibbon ape leukemia virus. J Virol 1997; 71:7619–7622.

199. Kelly PF, Vandergriff J, Nathwani A, Nienhuis AW, Vanin EF. Highly efficient gene transfer into cord blood nonobese diabetic/severe combined immunodeficiency repopulating cells by oncoretroviral vector particles pseudotyped with the feline endogenous retrovirus (RD114) envelope protein. Blood 2000; 96:1206–1214.

200. Kelly PF, Carrington J, Nathwani A, Vanirt EF. RD114-pseudotyped oncoretroviral vectors: biological and physical properties. Ann NY Acad Sci 2001; 938:262–276; discussion 276–267.

201. Gatlin J, Melkus MW, Padgett A, Kelly PF, Garcia JV. Engraftment of NOD/SCID mice with human CD34(+) cells transduced by concentrated oncoretroviral vector particles pseudotyped with the feline endogenous retrovirus (RD114) envelope protein. J Virol 2001; 75:9995–9999.

202. Emi N, Friedmann T, Yee JK. Pseudotype formation of murine leukemia virus with the G protein of vesicular stomatitis virus. J Virol 1991; 65:1202–1207.

203. Burns JC, Friedmann T, Driever W, Burrascano M, Yee JK. Vesicular stomatitis virus G glycoprotein pseudotyped retroviral vectors: concentration to very high titer and efficient gene transfer into mammalian and nonmammalian cells. Proc Natl Acad Sci USA 1993; 90:8033–8037.

204. Brenner S, Whiting-Theobald NL, Linton GF, Holmes KL, Anderson-Cohen M, Kelly PF, Vanin EF, Pilon AM, Bodine DM, Horwitz ME, Malech HL. Concentrated RD114-pseudotyped MFGS-gp91phox vector achieves high levels of functional correction of the chronic granulomatous disease oxidase defect in NOD/SCID/beta− microglobulin−/− repopulating mobilized human peripheral blood CD34+ cells. Blood 2003; 102:2789–2797.

205. von Kalle C, Kiem HP, Goehle S, Darovsky B, Heimfeld S, Torok-Storb B, Storb R, Schuening FG. Increased gene transfer into human hematopoietic progenitor cells by extended in vitro exposure to a pseudotyped retroviral vector. Blood 1994; 84:2890–2897.

206. Relander T, Brun AC, Olsson K, Pedersen L, Richter J. Overexpression of gibbon ape leukemia virus (GALV) receptor (GLVR1) on human CD34(+) cells increases gene transfer mediated by GALV pseudotyped vectors. Mol Ther 2002; 6:400–406.

207. Shi PA, Angioletti MD, Donahue RE, Notaro R, Luzzatto L, Dunbar CE. In vivo gene marking of rhesus macaque long-term repopulating hematopoietic cells using a VSV-G pseudotyped versus amphotropic oncoretroviral vector. J Gene Med 2004; 6:367–373.

208. van der Loo JC, Liu BL, Goldman AI, Buckley SM, Chrudimsky KS. Optimization of gene transfer into primitive human hematopoietic cells of granulocyte-colony stimulating factor-mobilized peripheral blood using low-dose cytokines and comparison of a gibbon ape leukemia virus versus an RD114-pseudotyped retroviral vector. Hum Gene Ther 2002; 13:1317–1330.

209. Hanawa H, Kelly PF, Nathwani AC, Persons DA, Vandergriff JA, Hargrove P, Vanin EF, Nienhuis AW. Comparison of various envelope proteins for their ability to pseudotype lentiviral vectors and transduce primitive hematopoietic cells from human blood. Mol Ther 2002; 5:242–251.

210. Hu J, Kelly P, Bonifacino A, Agricola B, Donahue R, Vanin E, Dunbar CE. Direct comparison of RD1 14-pseudotyped versus amphotropic-pseudotyped retroviral vectors for transduction of rhesus macaque long-term repopulating cells. Mol Ther 2003; 8:611–617.

211. Roe T, Reynolds TC, Yu G, Brown PO. Integration of murine leukemia virus DNA depends on mitosis. EMBO J 1993; 12:2099–2108.

212. Thomasson B, Peterson L, Thompson J, Goerner M, Kiem HP. Direct comparison of steady-state marrow, primed marrow, and mobilized peripheral blood for transduction of hematopoietic stem cells in dogs. Hum Gene Ther 2003; 14:1683–1686.

213. Bodine DM, Seidel HE, Gale MS, Nienhuis AW, Orlic D. Efficient retrovirus transduction of mouse pluripotent hematopoietic stem cells mobilized into the peripheral blood by treatment with granulocyte colony-stimulating factor and stem cell factor. Blood 1994; 84:1482–1491.

214. Hematti P, Sellers SE, Agricola BA, Metzger ME, Donahue RE, Dunbar CE. Retroviral transduction efficiency of G-CSF + SCF− mobilized peripheral blood CD34+ cells is superior to G-CSF or G-CSF + Flt3-L-mobilized cells in nonhuman primates. Blood 2003; 101:2199–2205.

215. Dorrell C, Gan OI, Pereira DS, Hawley RG, Dick JE. Expansion of human cord blood CD34(+)CD38(−) cells in ex vivo culture during retroviral transduction without a corresponding increase in SCID repopulating cell (SRC) frequency: dissociation of SRC phenotype and function. Blood 2000; 95:102–110.

216. Kittler EL, Peters SO, Crittenden RB, Debatis ME, Ramshaw HS, Stewart FM, Quesenberry PJ. Cytokine-facilitated transduction leads to low-level engraftment in nonablated hosts. Blood 1997; 90:865–872.

217. Kiem HP, Andrews RG, Morris J, Peterson L, Heyward S, Allen JM, Rasko JE, Potter J, Miller AD. Improved gene transfer into baboon marrow repopulating cells using recombinant human fibronectin fragment CH-296 in combination with interleukin-6, stem cell factor, FLT-3 ligand, and megakaryocyte growth and development factor. Blood 1998; 92:1878–1886.

218. Yonemura Y, Ku H, Hirayama F, Souza LM, Ogawa M. Interleukin 3 or interleukin 1 abrogates the reconstituting ability of hematopoietic stem cells. Proc Natl Acad Sci USA 1996; 93:4040–4044.

219. Wognum AW, Visser TP, Peters K, Bierhuizen MF, Wagemaker G. Stimulation of mouse bone marrow cells with kit ligand, FLT3 ligand, and thrombopoietin leads to efficient retrovirus-mediated gene transfer to stem cells, whereas interleukin 3 and interleukin 11 reduce transduction of short- and long-term repopulating cells. Hum Gene Ther 2000; 11:2129–2141.

220. Ku H, Yonemura Y, Kaushansky K, Ogawa M. Thrombopoietin, the ligand for the Mpl receptor, synergizes with steel factor and other early acting cytokines in supporting proliferation of primitive hematopoietic progenitors of mice. Blood 1996; 87:4544–4551.

221. Kurre P, Morris J, Horn PA, Harkey MA, Andrews RG, Kiem HP. Gene transfer into baboon repopulating cells: a comparison of Flt-3 ligand and megakaryocyte growth and development factor versus IL-3 during ex vivo transduction. Mol Ther 2001; 3:920–927.

222. Dunbar CE, Takatoku M, Donahue RE. The impact of ex vivo cytokine stimulation on engraftment of primitive hematopoietic cells in a non-human primate model. Ann NY Acad Sci 2001; 938:236–244; discussion: 244–235.

223. Hanenberg H, Hashino K, Konishi H, Hock RA, Kato I, Williams DA. Optimization of fibro-nectin-assisted retroviral gene transfer into human CD34+ hematopoietic cells. Hum Gene Ther 1997; 8:2193–2206.

224. Xu LC, Kluepfel-Stahl S, Blanco M, Schiffmann R, Dunbar C, Karlsson S. Growth factors and stromal support generate very efficient retroviral transduction of peripheral blood CD34+ cells from Gaucher patients. Blood 1995; 86:141–146.

225. Sellers SE, Tisdale IF, Agricola BA, Donahue RE, Dunbar CE. The presence of the carboxy-terminal fragment of fibronectin allows maintenance of non-human primate long-term hema-topoietic repopulating cells during extended ex vivo culture and transduction. Exp Hematol 2004; 32:163–170.

226. Aiuti A, Slavin S, Aker M, Ficara F, Deola S, Mortellaro A, Morecki S, Andolfi G, Tabucchi A, Carlucci F, et al. Correction of ADA-SCID by stem cell gene therapy combined with nonmye-loablative conditioning. Science 2002; 296:2410–2413.

227. Cavazzana-Calvo M, Hacein-Bey S, de Saint Basile G, Gross F, Yvon E, Nusbaum P, Selz F, Hue C, Certain S, Casanova JL, Bousso P, Deist FL, Fischer A. Gene therapy of human severe combined immunodeficiency (SCID)-X1 disease. Science 2000; 288:669–672.

228. Abonour R, Williams DA, Einhorn L, Hall KM, Chen J, Coffman J, Traycoff CM, Bank A, Kato I, Ward M, et al. Efficient retrovirus-mediated transfer of the multidrug resistance 1 gene into autologous human long-term repopulating hematopoietic stem cells. Nat Med 2000; 6:652–658.

229. Licht T, Haskins M, Henthom P, Kleiman SE, Bodine DM, Whitwam T, Puck JM, Gottesman MM, Melniczek JR. Drag selection with paclitaxel restores expression of linked TE-2 receptor gamma-chain and multidrug resistance (MDR1) transgenes in canine bone marrow. Proc Natl Acad Sci USA 2002; 99:3123–3128.

230. Jin L, Zeng H, Chien S, Otto KG, Richard RE, Emery DW, Blau CA. In vivo selection using a cell-growth switch. Nat Genet 2000; 26:64–66.

231. Neff T, Blau CA. Pharmacologically regulated cell therapy. Blood 2001; 97:2535–2540.

232. Jin L, Siritanaratkul N, Emery DW, Richard RE, Kaushansky K, Papayannopoulou T, Blau CA. Targeted expansion of genetically modified bone marrow cells. Proc Natl Acad Sci USA 1998; 95:8093–8097.

233. Richard RE, Blau CA. Small-molecule-directed mp1 signaling can complement growth factors to selectively expand genetically modified cord blood cells. Stem Cells 2003; 21:71–78.

234. Richard RE, Wood B, Zeng H, Jin L, Papayannopoulou T, Blau CA. Expansion of genetically modified primary human hemopoietic cells using chemical inducers of dimerization. Blood 2000; 95:430–436.

235. Richard RE, Weinreich M, Chang KH, Jeremia J, Stevenson MM, Blau CA. Modulating erythrocyte chimerism in a mouse model of pyruvate kinase deficiency. Blood 2004; 103:4432–4439.

236. Lawrence HJ, Sauvageau G, Humphries RK, Largman C. The role of HOX homeobox genes in normal and leukemic hematopoiesis. Stem Cells 1996; 14:281–291.

237. Sauvageau G, Thorsteinsdottir U, Eaves CJ, Lawrence HJ, Largman C, Lansdorp PM, Humphries RK. Overexpression of HOXB4 in hematopoietic cells causes the selective expan-sion of more primitive populations in vitro and in vivo. Genes Dev 1995; 9:1753–1765.

238. Thorsteinsdottir U, Sauvageau G, Humphries RK. Enhanced in vivo regenerative potential of HOXB4-transduced hematopoietic stem cells with regulation of their pool size. Blood 1999; 94:2605–2612.

239. Antonchuk J, Sauvageau G, Humphries RK. HOXB4 overexpression mediates very rapid stem cell regeneration and competitive hematopoietic repopulation. Exp Hematol 2001; 29:1125–1134.

240. Antonchuk J, Sauvageau G, Humphries RK. HOXB4-induced expansion of adult hematopoie-tic stem cells ex vivo. Cell 2002; 109:39–45.

241. Schiedlmeier B, Klump H, Will E, Arman-Kalcek G, Li Z, Wang Z, Rimek A, Friel J, Baum C, Ostertag W. High-level ectopic HOXB4 expression confers a profound in vivo competitive

growth advantage on human cord blood CD34+ cells, but impairs lymphomyeloid differentiation. Blood 2003; 101:1759–1768.

242. Amsellem S, Pflumio F, Bardinet D, Izac B, Charneau P, Romeo PH, Dubart-Kupperschmitt A, Fichelson S. Ex vivo expansion of human hematopoietic stem cells by direct delivery of the HOXB4 homeoprotein. Nat Med 2003; 9:1423–1427.

243. Krosl J, Austin P, Beslu N, Kroon E, Humphries RK, Sauvageau G. In vitro expansion of hematopoietic stem cells by recombinant TAT-HOXB4 protein. Nat Med 2003; 9:1428–1432.

244. King W, Patel MD, Lobel LI, Goff SP, Nguyen-Huu MC. Insertion mutagenesis of embryonal carcinoma cells by retroviruses. Science 1985; 228:554–558.

245. Hahn WC, Weinberg RA. Modelling the molecular circuitry of cancer. Nat Rev Cancer 2002; 2:331–341.

246. Evan GI, Vousden KH. Proliferation, cell cycle and apoptosis in cancer. Nature 2001; 411:342–348.

247. Kohn DB, Sadelain M, Dunbar C, Bodine D, Kiem HP, Candotti F, Tisdale J, Riviere I, Blau CA, Richard RE, Sorrentino B, et al. American Society of Gene Therapy (ASGT) ad hoc subcommittee on retroviral-mediated gene transfer to hematopoietic stem cells. Mol Ther 2003; 8:180–187.

248. Hacein-Bey-Abina S, von Kalle C, Schmidt M, Le Deist F, Wulffraat N, McIntyre E, Radford L, Villeval JL, Fraser CC, Cavazzana-Calvo M, Fischer A. A serious adverse event after successful gene therapy for X-linked severe combined immunodeficiency. N Engl J Med 2003; 348:255–256.

249. Hacein-Bey-Abina S, Von Kalle C, Schmidt M, McCormack MP, Wulffraat N, Leboulch P, Lim A, Osborne CS, Pawliuk R, Morillon E, et al. LMO2-associated clonal T cell proliferation in two patients after gene therapy for SCID-X1. Science 2003; 302:415–419.

250. Schroder AR, Shinn P, Chen H, Berry C, Ecker JR, Bushman F. HIV-1 integration in the human genome favors active genes and local hotspots. Cell 2002; 110:521–529.

251. Rohdewohld H, Weiher H, Reik W, Jaenisch R, Breindl M. Retrovirus integration and chromatin structure: Moloney murine leukemia proviral integration sites map near DNase I-hypersensitive sites. J Virol 1987; 61:336–343.

252. Wu X, Li Y, Crise B, Burgess SM. Transcription start regions in the human genome are favored targets for MLV integration. Science 2003; 300:1749–1751.

253. Kiem HP, Sellers S, Thomasson B, Morris JC, Tisdale JF, Horn PA, Hematti P, Adler R, Kuramoto K, Calmels B, et al. Long-term clinical and molecular follow-up of large animals receiving retrovirally transduced stem and progenitor cells: no progression to clonal hematopoiesis or leukemia. Mol Ther 2004; 9:389–395.

254. Furger A, Monks J, Proudfoot KT. The retroviruses human immunodeficiency virus type 1 and Moloney murine leukemia virus adopt radically different strategies to regulate promoter-proximal polyadenylation. J Virol 2001; 75:11735–11746.

255. Zaiss AK, Son S, Chang LJ. RNA3′ read-through of oncoretrovirus and lentivirus: implications for vector safety and efficacy. J Virol 2002; 76:7209–7219.

256. Emery DW, Yannaki E, Tubb J, Nishino T, Li Q, Stamatoyannopoulos G. Development of virus vectors for gene therapy of beta chain hemoglobinopathies: flanking with a chromatin insulator reduces gamma-globin gene silencing in vivo. Blood 2002; 100:2012–2019.

257. Olivares EC, Hollis RP, Chalberg TW, Meuse L, Kay MA, Calos MP. Site-specific genomic integration produces therapeutic Factor DC levels in mice. Nat Biotechnol 2002; 20:1124–1128.

258. Groth AC, Olivares EC, Thyagarajan B, Calos MP. A phage integrase directs efficient site-specific integration in human cells. Proc Natl Acad Sci USA 2000; 97:5995–6000.

259. De Palma M, Venneri MA, Roca C, Naldini L. Targeting exogenous genes to tumor angiogenesis by transplantation of genetically modified hematopoietic stem cells. Nat Med 2003; 9:789–795.

260. Jaenisch R, Schnieke A, Harbers K. Treatment of mice with 5-azacytidine efficiently activates silent retroviral genomes in different tissues. Proc Natl Acad Sci USA 1985; 82:1451–1455.

261. Challita PM, Kohn DB. Lack of expression from a retroviral vector after transduction of murine hematopoietic stem cells is associated with methylation in vivo. Proc Natl Acad Sci USA 1994; 91:2567–2571.

262. Pannell D, Osborne CS, Yao S, Sukonnik T, Pasceri P, Karaiskakis A, Okano M, Li E, Lipshitz HD, Ellis J. Retrovirus vector silencing is de novo methylase independent and marked by a repressive histone code. EMBO J 2000; 19:5884–5894.

263. Lorincz MC, Schubeler D, Groudine M. Methylation-mediated proviral silencing is associated with MeCP2 recruitment and localized histone H3 deacetylation. Mol Cell Biol 2001; 21:7913–7922.

264. Hoeben RC, Migchielsen AA, van der Jagt RC, van Ormondt H, van der Eb AJ. Inactivation of the Moloney murine leukemia virus long terminal repeat in murine fibroblast cell lines is associated with methylation and dependent on its chromosomal position. J Virol 1991; 65:904–912.

265. Cherry SR, Biniszkiewicz D, van Parijs L, Baltimore D, Jaenisch R. Retroviral expression in embryonic stem cells and hematopoietic stem cells. Mol Cell Biol 2000; 20:7419–7426.

266. Bednarik DP, Cook JA, Pitha PM. Inactivation of the HIV LTR by DNA CpG methylation: evidence for a role in latency. EMBO J 1990; 9:1157–1164.

267. Lipps HJ, Jenke AC, Nehlsen K, Scinteie MF, Stehle IM, Bode J. Chromosome-based vectors for gene therapy. Gene 2003; 304:23–33.

268. Bran AC, Fan X, Bjornsson JM, Humphries RK, Karlsson S. Enforced adenoviral vector-mediated expression of HOXB4 in human umbilical cord blood CD34+ cells promotes myeloid differentiation but not proliferation. Mol Ther 2003; 8:618–628.

269. Chen KS, Gage FH. Somatic gene transfer of NGF to the aged brain: behavioral and morphological amelioration. J Neurosci 1995; 15:2819–2825.

270. Liste I, Navarro B, Johansen J, Bueno C, Villa A, Johansen TE, Martinez-Serrano A. Low-level tyrosine hydroxylase (TH) expression allows for the generation of stable TH+ cell lines of human neural stem cells. Hum Gene Ther 2004; 15:13–20.

271. Markowska AL, Koliatsos VE, Breckler SI, Price DL, Olton DS. Human nerve growth factor improves spatial memory in aged but not in young rats. J Neurosci 1994; 14:4815–4824.

272. Pallage V, Toniolo G, Will B, Hefti F. Long-term effects of nerve growth factor and neural transplants on behavior of rats with medial septal lesions. Brain Res 1986; 386:197–208.

273. Gros L, Riu E, Montoliu L, Ontiveros M, Lebrigand L, Bosch F. Insulin production by engineered muscle cells. Hum Gene Ther 1999; 10:1207–1217.

274. Rosenzweig M, Connole M, Glickman R, Yue SP, Noren B, DeMaria M, Johnson RP. Induction of cytotoxic T lymphocyte and antibody responses to enhanced green fluorescent protein following transplantation of transduced CD34(+) hematopoietic cells. Blood 2001; 97:1951–1959.

275. Riddell SR, Elliott M, Lewinsohn DA, Gilbert MJ, Wilson L, Manley SA, Lupton SD, Overell RW, Reynolds TC, Corey L, Greenberg PD. T-cell mediated rejection of gene-modified HIV-specific cytotoxic T lymphocytes in HIV-infected patients. Nat Med 1996; 2:216–223.

276. Shull R, Lu X, Dube I, Lutzko C, Kruth S, Abrams-Ogg A, Kiem HP, Goehle S, Schuening F, Millan C, Carter R. Humoral immune response limits gene therapy in canine MPS I. Blood 1996; 88:377–379.

277. Stripecke R, Carmen Villacres M, Skelton D, Satake N, Halene S, Kohn D. Immune response to green fluorescent protein: implications for gene therapy. Gene Ther 1999; 6:1305–1312.

278. Lutzko C, Kruth S, Abrams-Ogg AC, Lau K, Li L, Clark BR, Ruedy C, Nanji S, Foster R, Kohn D, Shull R, Dube ID. Genetically corrected autologous stem cells engraft, but host immune responses limit their utility in canine alpha-L-iduronidase deficiency. Blood 1999; 93:1895–1905.

279. Goerner M, Bruno B, McSweeney PA, Buron G, Storb R, Kiem HP. The use of granulocyte colony-stimulating factor during retroviral transduction on fibronectin fragment CH-296 enhances gene transfer into hematopoietic repopulating cells in dogs. Blood 1999; 94:2287–2292.

280. Goemer M, Horn PA, Peterson L, Kurre P, Storb R, Rasko JE, Kiem HP. Sustained multilineage gene persistence and expression in dogs transplanted with CD34(+) marrow cells transduced by RD114-pseudotype oncoretrovirus vectors. Blood 2001; 98:2065–2070.
281. Takatoku M, Sellers S, Agricola BA, Metzger ME, Kato L, Donahue RE, Dunbar CE. Avoidance of stimulation improves engraftment of cultured and retrovirally transduced hematopoietic cells in primates. J Clin Invest 2001; 108:447–455.
282. Heira DA, Hanazono Y, Giri N, Wu T, Childs R, Sellers SE, Muul L, Agricola BA, Metzger ME, Donahue RE, Tisdale IF, Dunbar CE. Introduction of a xenogeneic gene via hematopoietic stem cells leads to specific tolerance in a rhesus monkey model. Mol Ther 2000; 1:533–544.
283. Mingozzi F, Liu YL, Dobrzynski E, Kaufhold A, Liu JH, Wang Y, Arruda VR, High KA, Herzog RW. Induction of immune tolerance to coagulation factor IX antigen by in vivo hepatic gene transfer. J Clin Invest 2003; 111:1347–1356.
284. Morris JC, Conerly M, Thomasson B, Storek J, Riddell SR, Kiem HP. Induction of cytotoxic T-lymphocyte responses to enhanced green and yellow fluorescent proteins after myeloablative conditioning. Blood 2004; 103:492–499.
285. Limmer A, Ohl J, Kurts C, Ljunggren HG, Reiss Y, Groettrup M, Momburg F, Arnold B, Knolle PA. Efficient presentation of exogenous antigen by liver endothelial cells to CD8+ T cells results in antigen-specific T-cell tolerance. Nat Med 2000; 6:1348–1354.

14

Clinical Applications of Hematopoietic Stem Cells

Joanne Ewing
Birmingham Heartlands Hospital, Bordesley Green East, Birmingham, U.K.

Yvonne Summers
Northern Ireland Cancer Centre, Belfast City Hospital, Belfast, U.K.

INTRODUCTION

Hematopoietic stem-cell (HSC) therapy is well established in clinical practice. Both autologous (ASCT) and allogeneic stem-cell transplantations (alloSCT) have been critical in the progress toward potential cure of leukemia, lymphoma, and other malignancies and the ability to transplant allogeneic stem cells has not only contributed in the field of malignant hematology but also been a key approach toward the cure of some inherited hemoglobinopathy and bone marrow (BM) failure conditions. Stem cells have the potential to regenerate a variety of tissues, as indicated by a number of groundbreaking but preliminary reports on stem-cell plasticity in vitro, but ethical issues and safety considerations preclude the use of human embryonic stem cells in the clinical setting at present. The exploitation of adult stem cells might circumvent the controversial issues posed by stem cells derived from embryonal tissue, although the true potential for plasticity of adult stem cells remains under scrutiny, with many contradictory reports in the field of stem-cell research.

In this chapter, we present the current indications for both ASCT and alloSCT in clinical practice, rationale for transplant strategies, and evaluate some of the evidence to support this approach (Fig. 1). We explore some of the innovative and developmental approaches emerging from translational research programs for the clinical use of adult HSCs, including cord blood transplantation and use of mesenchymal stem cells in myocardial repair and in inherited bone disorders such as osteogenesis imperfecta (OI). We also examine the current status of clinical gene therapy studies using HSC.

CLINICAL USE OF HSCs IN TRANSPLANTATION

The clinical decisions surrounding the optimal use, including type and timing, of adult hematopoietic stem-cell transplantation (HSCT), are frequently very difficult. The optimal use of transplantation within the total therapeutic strategy for an individual patient with

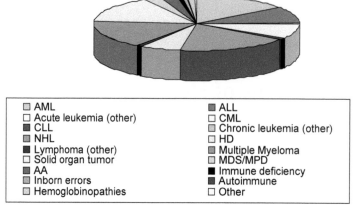

Figure 1   EBMT 2003–2004 indications for transplant. *Abbreviation*: EBMT, European Group for Blood and Marrow Transplantation. (*See color insert*.)

hematopoietic malignancy is contingent upon a number of key factors that must be weighted to allow risk:benefit analysis. These factors are determined not only by factors relating to the patient status but also those of the disease itself. With respect to biological characteristics of the disease, an assessment of the expected natural course of the disease, alternative therapeutic options, disease chemosensitivity, and likely outcome of salvage therapy must be undertaken. From the perspective of the patient, the likely procedural mortality must be appraised along with a consideration of the acceptable level of morbidity likely to be experienced. This evaluation will include issues such as patient age and co-morbidity, risk of infection, disease characteristics that correlate with predicted transplant outcome as well as the ability to mobilize peripheral blood stem cells (PBSCs) in the case of ASCT or donor availability, and degree of histocompatibility [human leukocyte antigen (HLA)-match] in the case of alloSCT.

Comparison of transplant outcome data must always be made with best available contemporary chemotherapy protocols rather than historical comparisons. The role of transplantation is changing partly because of the continued evolution and development of alternative therapies; for example, the introduction of the tyrosine kinase inhibitor, *imatinib*, for chronic myeloid leukemia (CML). The advent of this novel therapy has meant that during the past four years, centers have seen a 50% to 75% reduction in the number of transplants for chronic-phase (CP) CML because patients have opted for imatinib as first-line treatment rather than alloSCT. The introduction of targeted and monoclonal antibody therapies may drive a further sea change in the indications for transplant. Conversely, the role of transplantation has increased for some indications as we have gained a greater understanding of stem-cell mobilization, developed safer conditioning protocols, and improved graft-versus-host disease (GVHD) management permitting safer delivery of these intensive schedules, reducing transplant-related mortality.

The ability to risk-stratify patients with hematologic malignancy is improving rapidly and the assessment of outcome from large multi-center clinical trial data has imparted a greater ability to target those patients that are likely to benefit from stem-cell transplantation as part of a first-line treatment strategy, refining our approach. We are starting to define risk groups using features of disease at presentation such as cytogenetic abnormalities and the presence of mutations such as tyrosine kinase receptor for vascular endothelial growth factor (Flt-3) in acute myeloid leukemia (AML), as well as molecular minimal residual disease (MRD) detection following initial treatment in the case of leukemia. The use of technology such as 2-fluoro-deoxyglucose-positron-emission tomography in combination with CT scanning (FDG-PET-CT) allows assessment of disease activity following first-line treatment in the case of Hodgkin's disease and non-Hodgkin's lymphoma (NHL) so that treatment intensification can be directed toward those at risk of relapse. The rational use of this information in the clinical decision-making process will be critical and the indications for transplantation are a continually changing and challenging field.

AUTOLOGOUS STEM-CELL TRANSPLANTATION

General Issues in Clinical Application of Autologous Transplantation

The rationale for the use of ASCT in the treatment of hematological malignancy is based on the effectiveness of dose-intensified therapy to achieve a log-scale reduction in tumor burden. The transplantation of cryopreserved autologous HSCs then acts to rescue the hematopoietic system following high-dose chemotherapy and/or radiotherapy. The disadvantages include potential for contamination with occult tumor cells and the lack of any immunological anti-tumor effect (Table 1).

Stem-Cell Source

There has been an almost universal shift toward the use of cytokine-primed granulocyte colony-stimulating factor (G-CSF) mobilized PBSCs for ASCT. The first report of PBSC mobilized into the peripheral blood with hematopoietic growth factors for transplantation was from Juttner et al. in Australia (1), confirming that PBSCs collected following induction regimens are capable of hematopoietic reconstitution in AML. The major benefit seen in early A-PBSC trials was a considerably greater speed of hematopoietic recovery with a median time to attain neutrophils $>0.5 \times 10^9 \, \text{L}^{-1}$ of 10 to 15 days and platelets $>50 \times 10^9 \, \text{L}^{-1}$ of 19 days, impacting favorably on procedure-related mortality, morbidity, cost, and earlier discharge from hospital (2–4). The reconstitutive capacity of HSCs is predictable and related to the progenitor cell dose infused and analysis of CD34+ cells by flow cytometry has allowed rapid quantitative analysis of harvest inoculum and standardization to improve safety (5).

Initially, it was hypothesized that the use of PBSCs may lead to a reduction in graft contamination due to the differences in the mobilization profile of normal cells compared with blasts, although others disagreed with this suggesting that the overall higher cell numbers infused with the PBSCs may actually increase the dose of re-infused occult

Table 1 Advantages and Disadvantages of ASCT

Advantages	Disadvantages
No requirement for matched donor	No GVL effect
No GVHD	Risk of leukemic cell contamination
Shorter recovery period	Requires ability to mobilize stem cells
More complete immune reconstitution	
Able to offer up to 65 yrs and above	

Abbreviations: ASCT, autologous hematopoietic stem-cell transplantation; GVL, graft-versus-leukemia; GVHD, graft-versus-host disease.

leukemia cells (6,7). However, this has never been demonstrated to be the case. Paired analysis of marrow and leucopheresis products with karyotypic markers has suggested that PBSCs actually contain fewer contaminating cells (8,9).

In Vitro Purging of Autologous HSCs

One of the prerequisites determining the success of ASCT for leukemia is that re-infused stem cells should not contribute to relapse, however, the origin of cells contributing to relapse remains controversial and makes assessment of purging effectiveness difficult. Results of syngeneic (identical twin) transplantation, when compared with those from ASCT, show relapse rates of similar magnitude. This led to an assumption that many patients relapse following ASCT owing to an inability of the preparative conditioning regimen to fully ablate the tumor within the patient, limiting the clinical effectiveness of any purging strategy (10). The degree of leukemic contamination after purging has been compared with that prior to purging. Relapse risk has been found not only to correlate with the post-purge level of contamination but also closely related to the degree of MRD at harvest (11), again suggesting that disease chemosensitivity and residual patient disease rather than re-infused cells contribute to relapse. Nonetheless, there remains the potential risk of infusion of occult residual leukemic cells in the graft. Using gene marking to trace the origin of disease recurrence, a retroviral vector carrying the selectable neomycin resistance gene to mark 5% of the re-infused stem cells demonstrated that in AML at relapse, some leukemic cells did indeed carry the marker gene (12).

Although purging is possible, it can cause difficulties. It is expensive, and its role is controversial. The most common strategy that has been adopted for purging to reduce the leukemic contamination of the graft in AML has been exposure of cells in vitro to the cyclophosphamide derivatives 4-hydroperoxycyclophosphamide (4-HC) or mafosfamide (13). Purging of cells using monoclonal antibodies such as CD33 or CD14/CD15 in AML or CD20 and complement in B-cell lymphoma and leukemia have also been assessed (8,14,15). Positive CD34+ stem-cell selection methods have been employed in lymphoma and breast cancers and with modern devices such as Isolex 300i or Clinimacs, it is possible to purge three to four log-scale of chronic lymphoid leukemia (CLL) cells from fresh leucopheresis products (16,17). This strategy is less relevant in the acute leukemias because these are stem-cell disorders and frequently express CD34 antigen.

There is a lack of clear clinical evidence of a benefit from in vitro purging for acute leukemia and in practice existing techniques would seem to confer a number of disadvantages. In particular, purging may delay engraftment, and this may have contributed to increased death rates in a number of randomized controlled trials (RCTs). A number of studies in AML (predominantly from the United States) have used 4-HC purging. The poorer engraftment associated with this manipulation may have contributed to the higher transplant-related mortality (TRM) seen in the U.S. Intergroup trial and the POG (Pediatric Oncology Group) study (18,19). In spite of rigorous analysis of registry data, no substantial evidence supporting marrow purging has been forthcoming for AML (20–22). Retrospective data analysis of high-risk 2nd CR AML was the only group in which a benefit for 4-HC purging was suggested (21). When a polymerase chain reaction (PCR) negative graft can be obtained in low-grade lymphoma or CLL then a prolonged freedom from progression can be demonstrated following antograft. It may be that this purging efficiency is simply a surrogate marker for those patients with minimal residual disease prior to harvest who would have a better outlook regardless of the antograft procedure (11,23).

An analysis by Appelbaum and Buckner (24) suggested that 135 patients are required in each study arm to detect an effect of purging if it is only effective 50% of the time and if purging were 100% effective then 31 patients would be required in each arm, based on the assumption that:

1. In AML the probability of relapse is 25% for patients in first complete remission who receive syngeneic marrow;
2. The use of untreated autologous marrow increases the probability of relapse by 50% to 63% in this group;
3. Transplant-related mortality occurs in approximately 25% of cases; and
4. Sample sizes are calculated based on an 80% power at the 0.05 significance level.

These large, randomized trials are unlikely to be forthcoming and in the absence of these data, scepticism over this approach persists. Whether emerging leukemia targets will lend themselves to selective removal using monoclonal antibodies remains to be seen.

Another strategy toward purging the stem-cell inoculum is through in vivo purging using, for example, monoclonal antibodies prior to stem cell harvest to attempt collection of HSCs at a time when contamination risk is lowest. This is being evaluated currently in RCTs in low-grade lymphoma.

ASCT: MALIGNANT DISEASES

Autologous Transplantation in AML

It is now over two decades since the feasibility of myeloablative therapy with stem-cell rescue for AML was demonstrated (25,26). This approach is being adopted for an increasing number of patients in the first remission and less commonly as salvage at subsequent remissions. Of adult patients below 60 years of age who develop AML, approximately 60% to 80% can be expected to achieve complete remission following induction chemotherapy with a regimen combining an anthracycline with cytarabine arabinoside (ara-C). In spite of this, the most common cause of treatment failure is relapse. Patients in "morphological complete remission" may still have a burden of up to 10^{10} leukemic cells (27). The aim of repeated courses of intensive chemotherapy is, therefore, to further reduce the tumor load and ultimately eradicate the leukemic clone thus effecting a cure. The principle of administration of ASCT in AML in first complete remission (CR1) is to administer further dose intensification to patients in complete remission to prevent disease relapse. This approach is founded on the principle that administration of chemotherapy leads to a fractional log-cell kill at a given dose regardless of original cell number with the resultant exponential regression of tumor cell number (28,29). This "Skipper–Schabel" model hypothesizes that by administration of a very high dose treatment at the point of MRD there will be a further decrease in cell number by several logs resulting in complete eradication of tumor burden. The induction-intensification strategy proposed for AML aims to give intensified therapy to achieve high log-kill at the point of MRD as post-remission therapy to prevent future relapse with rescue using cryopreserved stem cells collected in remission.

The issue of optimal post-remission therapy in AML remains critical. Important information is now available from RCTs examining ASCT in AML in CR1.

Auto Bone Marrow Transplant vs. Chemotherapy in Adult AML

The first direct comparison of ASCT versus chemotherapy was reported by the EORTC/ GIMEMA (European Organisation for Research and Treatment of Cancer and Gruppo Italiano Malattie Ematologiche Maligne dell'Adulto) (30). This intervention trial entered 941 patients with median age of 33 years. Only 66% of patients attained complete remission (CR) and were eligible for randomization. Patients were allocated alloSCT if they had a consenting human leukocyte antigen (HLA)-matched sibling ($n = 168$). The remaining 254 were randomized to either ASCT using unpurged marrow ($n = 128$) or further chemotherapy ($n = 126$) as a third course. Only 74% of the patients eligible for ASCT actually received this therapy. Analysis by intent-to-treat showed a four-year disease-free survival (DFS) of 55% in the alloSCT arm and 48% in the ASCT arm, which were significantly better than the DFS of 30% in the chemotherapy arm ($P = 0.05$). There was a higher relapse rate (RR) in the chemotherapy arm but many of these patients were salvaged successfully with subsequent ASCT giving rise to a four-year overall survival (OS), which was similar for all three groups. The time taken to treatment was a significant factor in RR prior to the allocated intensification therapy. This factor demonstrates the "time-censoring" effect and highlights the fact that even in randomized intervention trials, biases may be introduced so that analysis of the data on an intent-to-treat basis is critical. Other problems were a suboptimal outcome in the chemotherapy arm, which may have been related to the lack of a high-dose cytarabine arabinoside (HDAC) schedule, and poor engraftment in the ASCT group in whom platelet recovery took up to 20 weeks.

The Medical Research Council (MRC) AML 10 trial was a large collaborative trial designed to examine the value of high-dose therapy after four courses of consolidation treatment compared with no further chemotherapy. It addressed a rather different question to that posed by other RCTs, namely does ASCT confer any further advantage over and above the best available intensity treatment protocol so far designed (31). Of 1622 patients registered, 381 patients below 55 years were randomized after the third course of chemotherapy, representing 38% of those eligible. Of the 190 patients allocated ASCT, only 126 received the intended therapy and five patients in the "stop treatment" arm actually received ASCT in CR1. The reasons for this attrition include patient and physician choice as well as death or relapse before transplant was reached. There was a modest improvement in DFS in the ASCT arm in the standard and high-risk groups when analyzed by intent-to-treat at seven years (DFS 53% vs. 39%, $P = 0.04$). When those patients surviving beyond two years are analyzed, a small OS advantage for ASCT is observed in this group, however, it was not felt that this was sufficient to justify continuing to address the question in further MRC AML trial strategies. The analysis of long-term outcome in the AML 10 cohort will continue to be important as survival advantages may only emerge in the long term.

The U.S. intergroup trial (ECOG/SWOG/CALGB) (18) compared 4-HC-purged bone marrow transplant (BMT) with HDAC. A major flaw in this trial design was the low dosage of treatment prior to the transplant arms of this study; patients proceeded to ASCT after only two intermediate-dose treatments with no HDAC consolidation prior to transplant. There were also features of the transplant protocol that may have contributed to the substantial TRM. Busulfan was used in the conditioning regimen leading to death from hepatic veno-occlusive disease in two patients in the ASCT arm and six patients in the alloSCT arm. Overall mortality from the ASCT procedure was 14%. Four-year DFS was not significantly different in the chemotherapy, ASCT, and alloSCT groups (respectively, 35% chemotherapy, 35% ASCT, and 43% alloBMT, $P > 0.05$). There was actually a weak OS benefit demonstrated in the chemotherapy arm of this trial ($P = 0.1$). It would seem that those patients in the transplant arm suffered the disadvantages of autografting without gaining optimal beneficial effects. Salvage therapy following relapse was better in the chemotherapy group contributing to this difference in overall rate of survival.

Group Ouest Est Leucemies Aigues Myeloblastiques (GOELAM) (32) again showed no difference with transplantation in terms of DFS or OS. This group also used a second course of HDAC (3 g/m^2 for eight doses) in both the chemotherapy and ASCT groups to particularly good effect but not in the allograft group. They were unable to show any benefit for alloSCT in this comparison.

Critical Analysis

A number of general criticisms can be made of these studies. The patients undergoing randomization represent less than half of those achieving remission and therefore eligible for randomization. Of those patients randomized, many failed to receive the assigned therapy because of intercurrent relapse, failure to mobilize, intercurrent infection, neurotoxicity, or patient/physician choice. Many patients relapsing in the chemotherapy arm of these trials are successfully salvaged with ASCT. This attrition seen in the MRC AML 10 trial led to only 66% of randomized patients receiving ASCT in CR1. There were problems, as seen in many studies in the randomization and delivery of assigned treatment particularly in the ASCT arm. Analysis of the reasons for this showed that this was due to both relapse and patient/physician choice. This will lead to an element of selection bias in the patients transplanted, making it difficult to extrapolate the data to all patients and the low proportion of patients actually receiving ASCT calls into question the feasibility of this approach as a standard procedure. The absence of an OS benefit in the majority of studies is probably related to the procedural mortality of 6.5% to 15% in combination with salvage therapy being more successful after chemotherapy. The other factor that has been observed during the course of these trials is a sustained improvement in outcome following chemotherapy over time reflecting better strategies for infectious complications and supportive care.

The varying intensity of the comparison chemotherapy arm also seems to play a critical role. The MRC AML 10 trial would suggest that if the benefit of ASCT is to be seen, then it must be performed after intensive consolidation therapy. The inclusion of a course of HDAC is important whether before ASCT or before further intensification chemotherapy. The MRC trial demonstrated that when ASCT is used as further consolidation compared to stopping treatment there is superior DFS but only a very modest OS benefit. This has been confirmed in a meta-analysis (33).

The question as to whether the function of this fifth course of high-dose therapy could be just as well served by an additional course of consolidation was addressed in the MRC AML 12 study, however, there remains uncertainty as to how many intensive courses are required, and this is being further addressed in randomized studies.

Risk Stratification in AML

The ability to stratify patients by risk group has given a clearer indication as to which patients are most likely to benefit from ASCT. Consistent prognostic indicators in AML include cytogenetics and time to achieve CR1. Other factors include age, disease burden at presentation, presence of the multi-drug resistance gene (MDR-1), French-American-British morphological type, secondary leukemia, and leukemia preceded by myelodysplasia. Favorable karyotypes include t(8;21), inv(16), and t(15;17), whereas in contrast abnormalities of chromosome 5 or 7, 3q, or complex changes forecast a very poor outcome as does >15% blasts in the marrow after course 1 (34–36). It should be possible to identify patients who are likely to benefit from a particular therapeutic approach and thereby refine decisions regarding transplantation so that some patients can be protected from exposure to toxic, expensive treatments. In the MRC AML 12 trial, the five-year survival rate for patients assigned to the good, standard, and poor risk groups was 73%, 44%, and 17%, respectively, and relapse rate was 25%, 52%, and 73%, respectively. Trials prior to the MRC AML 12 have not taken into account cytogenetic data for stratification. This study confirmed that there was no survival advantage for ASCT in the good risk group with data on an additional 330 randomized patients.

Patients with poor prognostic factors seem to benefit little from intensified treatment and still have poor outcome (31). The design of the MRC AML 12 study excluded poor-risk patients defined by cytogenetics or failure to achieve <15% blasts after second course. Patients shown to have a particularly favorable outcome with t(15;17), t(8;21), or inv16 who relapse following chemotherapy have a better chance of achieving second CR. This group should not receive either ASCT or alloSCT in CR1 but ASCT should be held in reserve for those who relapse and in whom an uncontaminated graft determined at a molecular level can be obtained and cryopreserved. Flt-3 gene mutation analysis has retrospectively suggested that the occurrence of a Flt-3 mutation (present in 20% to 27% of AML cases) indicates high relapse risk and adverse prognosis. The impact of MRD at a molecular level at various stages of treatment may be useful to target therapeutic strategies.

In summary, despite the considerable amount of data from RCTs, the optimal treatment of AML remains controversial and the place of ASCT in the management of AML in first remission remains unclear. From the available RCTs, there seems to be a consistent reduction in the RR which is counterbalanced by a procedure-related mortality. The main conclusion that can be drawn from the large RCTs is that dose-intensification, whether delivered as repeated courses of chemotherapy or as single intensive ASCT, improves DFS. Where there is a direct comparison, as in the EORTC/GIMEMA, there is a reduced RR in those receiving ASCT although this has not translated to an improvement in OS due to higher mortality and the significantly better response to subsequent ASCT when relapse occurs after chemotherapy. It may well be that as the outcome from transplantation improves with progress in supportive care and faster marrow recovery following the use of mobilized PBSCs, rather than BM stem cells, the improvement in RRs will follow into better OS. Outcome of good risk AML does not benefit from a transplant approach in CR1.

Transplantation in Second Remission and Relapsed AML

Successful salvage treatment of patients who relapse after chemotherapy is significantly better than that of those relapsing after ASCT (30,31,37). Overall, these patients can attain a 30% leukemia-free survival (LFS) with autografting. There is a better outcome for those with a higher duration of remission. The optimal timing of autologous transplantation is debated. A strategy to delay transplantation to CR2 may be attractive as patients potentially cured by chemotherapy need not receive transplants.

Autologous Transplantation in Acute Lymphoblastic Leukemia

In 75% of cases adult patients with acute lymphoblastic leukemia (ALL) enter CR; however, most will relapse. The strategy to prevent this using consolidation treatment is clearly important, and intensive chemotherapy will lead to five-year DFS rates of 10%

to 42%. Although alloSCT results in better DFS rates of 40% to 63%, it is only available to a minority of patients. What is more controversial is the role of high-dose therapy followed by ASCT in the consolidation phase. ASCT outcomes from registry data (European Bone Marrow Transplant registry, EBMT) in over 200 patients suggest a DFS of 41% at four years for standard risk patients in CR1 with 26% RR at four years. This compares favorably with results from the same group given conventional chemotherapy or allograft. Another study has suggested a 48% DFS in a small cohort at three years (38). No RCT has been of sufficient statistical power to determine whether ASCT can improve patient outcome when compared with consolidation and maintenance chemotherapy. Ongoing trials such as the MRC/ECOG UKALL XII study are aiming to recruit sufficient patients to enable us to answer this important question. It may be that in terms of effectiveness equivalence is demonstrated; however, ASCT may have a role as it offers shorter treatment in comparison with the prolonged consolidation and maintenance used traditionally.

Autologous Transplantation in High-Grade Diffuse Large B-NHL

Patients receiving combination CHOP (cyclophosphamide, hydroxydaunamycin, Oncovin (vincristine), prednisolone) chemotherapy for diffuse large B-cell non-Hodgkin lymphoma (DLBL) can expect long-term remission and probable cure in 30% to 50% of cases. The place of ASCT has been in salvage therapy in patients failing to respond to, or relapsing after, first-line therapy. No study has shown benefit of ASCT in CR1 except for a small population with high-risk features in one subgroup analysis (39), and it is therefore not recommended in this group.

The pivotal RCT that compared chemotherapy vs. ASCT in relapsed DLBL was the PARMA study (40). In this study, 109 patients with relapsed diffuse large B-cell NHL who had shown response to two cycles of DHAP chemotherapy (dexamethasone, cytarabine arabinoside, and cisplatin) were allocated to high-dose therapy and ASCT or continued chemotherapy. A significant difference in failure-free survival (51% vs. 12% at five years) and OS (53% vs. 32% at five years) was observed. This study enrolled chemosensitive, young patients at first relapse. The outcome from this trial has led to the use of ASCT as standard care for salvage in this defined group and is more widely used in relapsed patients able to tolerate autograft.

Those patients refractory to chemotherapy or failing ASCT should be offered allograft where this is feasible as outcome from ASCT is very poor.

Autologous Transplantation in Low-Grade Follicular Lymphoma

Follicular NHL accounts for 15% to 30% of newly diagnosed lymphomas and frequently presents in advanced stage. Although these cases are usually chemosensitive at presentation, they are incurable with standard chemotherapy with high RRs and eventual chemoresistance leading to median survival around 5 to 10 years. Therefore, high-dose therapy and ASCT have been used in an attempt to improve survival in patients with low-grade NHL (LG-NHL).

Initial chemotherapeutic treatment approaches for follicular NHL are safe with low morbidity and therefore the high-dose therapy approach has been used largely at relapse and the majority of the data have been derived from use of ASCT in the salvage setting. Studies demonstrate that although there remains a continued tendency for LG-NHL to relapse after ASCT with no plateau on survival curves, a prolonged DFS can be attained. In 153 patients with recurrent bulky stage III to IV disease, in vitro purging using monoclonal antibodies was used followed by a cyclophosphamide/total body irradiation (TBI)-conditioned ASCT. This was a selected group as patients were only included if they achieved a minimal disease state prior to autograft. Eight-year DFS was 42% and OS was 66% with a TRM under 5%. When compared with historical controls, there was a prolonged DFS (15,23). A similar study reported a two-year DFS of 53% and a four-year DFS of 43% (41). In another trial (the CUP trial—Conventional chemotherapy, Unpurged autograft, Purged autograft), randomization of 140 patients with recurrent LG-NHL between autograft and chemotherapy led to a clear demonstration of statistically significant benefit for ASCT with progression-free survival (PFS) 55% versus 26% and two-year OS 71% versus 46% (42), however, the trial was closed early due to slow accrual, limiting the assessment of purging in this setting. Outcome of ASCT at relapse is influenced by the number of prior chemotherapy regimens and therefore its use earlier in the treatment strategy has been evaluated.

ASCT in CR1 has been examined by Lenz et al. (43) who were able to show a significantly improved PFS (two-year 79% vs. 52%) and five-year survival of 64% versus 33%. This group used interferon (IFN) maintenance therapy following ASCT. They did not show improved OS although follow-up has been short. Preliminary results from the GELF94 study of follicular lymphoma in CR1 reported 401 patients randomized to CHVP (cyclophosphamide, doxorubicin, vindesine, and prednisolone) + IFN versus four courses of CHOP and TBI/etoposide-conditioned ASCT. Event-free survival (EFS) was comparable 45% versus 36% and OS significantly better with ASCT (86% vs. 74%). A recent report compared two approaches in 172 patients with advanced follicular lymphoma as first-line therapy using either CHOP-IFN or ASCT with purged graft. ASCT patients had a higher RR of 81% versus 69% and a longer median EFS (not reached vs. 45 months). This has not yet translated to improved OS however, due largely to the high rate (18.6%) of secondary malignancies in the ASCT group (44). Using the follicular lymphoma prognostic index (FLIPI), they were able to identify a subgroup of patients with a significantly higher EFS rate after ASCT. However, at present autograft cannot be considered standard care as first-line therapy in patients with high tumor burden and should be used only in the context of clinical trials.

Substantial longer-term morbidity occurs following ASCT. Secondary myelodysplastic syndrome (MDS) or secondary AML occur with a 5- to 15-fold increased incidence (45,46) and occurred in 7.4% of patients in the Dana-Farber study (15). The etiology is complex but may be related to the TBI/cyclophosphamide conditioning and the outlook is poor. This has hampered attempts to use ASCT in CR1 to prolong DFS and achieve improved OS. ASCT when used in CR1 improves PFS but confers no survival benefit; it has considerable morbidity and is not therefore recommended in this setting but should be reserved for salvage following disease relapse.

In Vitro and In Vivo Purging in the Setting of Low-Grade NHL

Patients with low-grade NHL frequently present with marrow disease and contamination with occult cells may contribute to a high RR after ASCT. The advantages that may be seen using a purging strategy in low-grade lymphoma have been suggested by the demonstration that those who have *Bcl-2* positivity detected in the collection prior to re-infusion have a higher risk for relapse than those that are *Bcl-2* negative. Therefore, the role of purging may be of importance in this setting. In vivo purging using intensive chemotherapy to achieve PCR-negative grafts has been demonstrated to be effective in improving DFS (47). However, care must be taken with this type of analysis as the ability to attain a molecularly disease-free graft may simply be a marker for chemosensitive disease. Both positive (CD34+) and negative in vitro selection strategies have been examined and the successful generation of a tumor-free graft on molecular analysis correlates with freedom from relapse (11,15,23,48). The CUP RCT was designed to answer this question but was underpowered to demonstrate a difference (42). In an International Bone Marrow Transplant Registry (IBMTR) study using registry data reporting 904 patients in whom 14% received purged ASCT, stem-cell purging was identified as an independent predictor of PFS with five-year recurrence of 43% in purged versus 58% in unpurged and OS (49), as was also demonstrated in another case–control study (50).

In vitro purging strategies can lead to other problems such as increased likelihood of engraftment failure, and current technologies have limited efficacy. An alternative approach using in vivo purging through administration of anti-CD20 monoclonal antibody prior to harvesting has been adopted (51) and is currently being assessed in RCTs.

In summary, ASCT as salvage therapy for follicular lymphoma offers a superior approach in comparison to conventional chemotherapy. The role of in vivo purging is being evaluated. First-line therapy results are promising in high-risk patients but benefits are limited by a high rate of secondary malignancy.

Autologous Transplantation in Hodgkin's Disease

The results of first-line chemotherapy even in advanced-stage Hodgkin's disease (HD) are excellent and can achieve cure in over 50% of patients, so ASCT has a limited role and is not recommended in CR1. No benefit has been shown even in patients with high-risk disease at this stage (39,52), and it should be reserved for the 20% to 50% of patients who relapse.

ASCT has become established as the most effective treatment for relapsed and refractory HD. RCTs have been performed and clearly demonstrated this benefit (53,54). ASCT is considered the standard of care for this group regardless of time to relapse and whether disease is primarily refractory (53,55). Overall five-year survival following ASCT has been 57%. Better prognostic factors are low tumor burden at diagnosis, autograft after a long duration CR, and absence of detectable disease at ASCT (56). Patients with primary refractory HD have had reported 30% to 40% five-year PFS rates.

In spite of the availability of excellent first-line therapy and salvage strategy using ASCT, over 15% of patients still die of progressive HD. For the group of patients with adverse factors such as advanced stage at diagnosis, radiotherapy before ASCT, a short duration first CR, detectable disease at ASCT, detectable disease at ASCT, or extranodal areas involved at ASCT, alternative approaches should be considered, such as reduced-intensity (RIC) alloSCT, which can exert a graft-versus-leukemia (GVL) effect, although this approach has a higher procedural toxicity. The use of fully ablative conditioning is not recommended as this led to high toxicity from GVHD and pneumonitis with transplant-related mortality as high as 50% in some studies (57).

Autologous Transplantation in CLL

CLL is an indolent disease of the elderly with a number of alternative modalities of non-curative therapy available for when a "watch and wait" strategy is no longer appropriate. These include alkylating agents, combination chemotherapy incorporating anthracycline (e.g., CHOP), purine nucleoside analogs such as fludarabine, and monoclonal antibody therapy such as anti-CD52 (Campath). Although some of these agents may offer clinical response and prolonged remission duration, there has been no translation into improved OS.

There are now an increasing number of patients presenting below the age of 55-years. This younger group now comprises 20% of those patients presenting with

CLL. Although median survival is around 10 years for this group, some will have aggressive disease with poor prognosis. Survival with stage B and C disease was 6.5 and 3.5 years in the under 55-year group. Those with the worst prognosis can be predicted using fluorescence in situ hybridization (FISH) analysis of cytogenetic abnormalities, with bad indicators being 17p-, 11q-, and +12q, flow cytometry to detect CD38 positivity and ZAP-70 expression, immunoglobulin heavy-chain variable region mutation status (unmutated V_H gene = high risk) and $p53$ mutation, along with advanced clinical disease stage. These patients require a strategy of aggressive effective therapy targeted toward improving OS with the aim toward cure.

ASCT has been shown to be feasible in CLL with low TRM at 1.5% to 10%, RRs over 80%, and some molecular remissions. In the largest series reported by the Dana-Farber center, 152 patients with advanced CLL underwent TBI/cyclophosphamide and monoclonal antibody-purged autologous BM transplant (ABMT). There was a TRM of 5% and with a median follow-up of 30 months, only 14 patients have relapsed although 63 have detectable residual disease at the molecular level (58). In the multicenter prospective German CLL3 study, a two-year OS rate was 88% in 105 high-risk patients (59). Unfortunately, those risk factors that predict worse OS such as unmutated IgV_H genes also predict a more rapid progression following ASCT (60). There has, however, been some benefit conferred by autograft in this group shown in risk-match analysis but there remains a high rate of clinical relapse with 56% relapsing at three years (61). An update on registry data on 225 patients receiving autografts reported three-year survival at 78% with relapse risk of 45% and TRM of 14% (62).

A recent report of 117 newly diagnosed younger patients receiving fludarabine as first-line treatment followed by stem-cell harvest and autograft showed that 56% of patients were able to proceed to ASCT. The attrition in this study was due largely to progressive CLL, failure to mobilize, and alloSCT favored over ASCT as second-line treatment. There was an increase in patients achieving a CR in the transplanted group and the five-year overall and DFS were 88.6% and 64.7%, respectively (those not able to progress to autograft showed OS 78.6% and DFS 28.8%). This was not a randomized study and limited conclusions can be drawn but these data are promising (63). There is an almost inevitable propensity to relapse following this approach even in those with complete clinical response. No RCT data has yet been able to conclusively show that ASCT does offer a benefit but the MRC CLL 5 phase III trial is aiming to answer this important question.

Autologous Transplantation in Multiple Myeloma

In multiple myeloma (MM), there has been shown to be a clear benefit of dose-intensified therapy. Melphalan dose escalation to 140 mg/m^2 achieved high RRs with CR in 30%. Further escalation to 200 mg/m^2 with ASCT rescue resulted in higher remission rates (57). The first RCT was carried out by the Intergroupe Francais du Myelome (IFM). In an intent-to-treat analysis of 200 patients under 65 years, there was a significant advantage in terms of remission rate (5% vs. 22%), response duration (EFS of 18 months vs. 28 months) and survival, with a median OS of 57 months compared with 44 months in standard arm (64). TRM was only 2.7%. The MRC Myeloma VII trial addressed the same question in 400 patients aged less than 65 years. There was a significant improvement in OS of those treated with intensive therapy, with median survival of 54 months versus 42 months (CR rate 8% vs. 44%, EFS 19 months vs. 31 months). The use of ASCT has therefore become standard of care for patients with newly diagnosed MM up to 65 years (65).

The primary objective of ASCT in MM is to achieve CR and those patients that do this or achieve a very good partial remission (VGPR) after ASCT have improved survival. It was therefore a reasonable strategy to repeated dose-intensive therapy with second ASCT ("tandem" transplants) to try to achieve this in a higher proportion to aim toward further survival benefits.

Tandem Autologous Transplants in MM

The first reports of feasibility of a double ASCT or "tandem transplant" came from the Barlogie group in Arkansas (66). This approach used two transplants conditioned with melphalan 200 mg/m^2. This resulted in a CR of 38% with EFS 43% and overall median survival of 68 months. TRM was 8%. The initial promising results that emanated from this early work led to further assessment in randomized studies.

Tandem transplants have been assessed in an RCT in the IFM94 study. This study randomized 399 patients to single melphalan 140 mg/m^2 and TBI or tandem melphalan 140 mg/m^2 ASCT followed by melphalan 140 mg/m^2 conditioned second transplant. The EFS at seven years following diagnosis was 10% in single and 20% in the double transplant group with OS of 21% and 42%, respectively (67).

Poor prognostic markers in MM include elevated β_2-microglobulin, C-reactive protein or lactate dehydrogenase, and low albumin. Better prognostic indicators are hypo-diploidy and chromosome 13 deletion or t(14;11). In the subgroup analysis of IFM94, all these risk groups benefited. There is a particular benefit shown in those patients who failed to achieve 90% reduction in M-component within three months.

We need to be aware that novel agents such as thalidomide, bortezomib, and revemid may well change the field of ASCT in MM. The exact place of these agents in the pathway of treatment is under evaluation. They may assume prime importance in the first-line treatment or as part of induction prior to ASCT to achieve an MRD state (68). Alternatively, they may assume a more important role in the salvage therapy after autograft failure.

ASCT: NONMALIGNANT DISEASES

Autologous Transplantation in Primary Systemic Amyloidosis

Primary systematic amyloidosis (AL-type) results from a plasma cell dyscrasia and leads to deposition of amyloid deposits in organs including the heart, skin, gut, kidney, and peripheral nervous system. High-dose melphalan and ASCT is the most effective treatment strategy, however this has considerable toxicity in patients whose organs are already compromised by the disease process. Early studies reported a high 30% to 40% TRM, precluding this approach. More recently, analysis of outcomes has led to the development of a risk-adapted approach based on age and organ involvement with stratified melphalan dosing in those groups deemed suitable for ASCT between 140 and 200 mg/m^2 (69). If this approach is adopted, then the TRM is on the order of 13% (70). There remain toxicity issues, however, with cardiac deaths seen and arrhythmias during stem-cell re-infusion due to dimethyl sulfoxide (DMSO) sensitivity and even reports of cardiac deaths during cyclophosphamide for stem-cell mobilization. Benefits are seen with organ response in 34% to 55%, with the potential for improved renal function continuing for up to a year following autograft. A clonal response is seen in 50% to 60% with CR in 30%. Importantly, improved OS and quality of life have been reported (71).

Transplantation for Severe Autoimmune Disorders

The assessment of case reports from patients with co-existent severe autoimmune disease and malignancy receiving ASCT, and promising outcomes from animal studies led to an emergence of this approach. ASCT has been undertaken in the context of phase I/II studies for the treatment of the following autoimmune diseases (72):

- Multiple sclerosis (MS)
- Systemic sclerosis (SSc)
- Rheumatoid arthritis (RA)
- Systemic lupus erythematosus (SLE)
- Juvenile idiopathic arthritis (JIA)
- Autoimmune cytopenias, immune thrombocytopenic purpura, and pure red cell aplasia.

Autograft conditioning is usually with cyclophosphamide ± anti-thymocyte globulin (ATG) although a number of other regimens have been reported and in many cases a CD34+ manipulated graft has been used as part of the immune-modulatory strategy. Overall TRM for these disorders in the EBMT experience was 8.6%, with higher levels seen in SLE, SSc, and JIA.

In SSc, outcomes have shown some success with 70% improvement of skin score and stabilization of lung function. Caution in patients with high pulmonary artery pressure must be taken and where TBI is used, lung shielding may be critical to limit toxicity. Considerable cardiac susceptibility to the high-dose regimens in SSc has been reported, with some mortality even seen during cyclophosphamide stem-cell mobilization. TRM in the order of 8.5% has been reported (73). An EBMT randomized trial has started recruitment (ASTIS).

MS has an unknown etiology but is thought to be autoimmune in nature with auto-reactive T-cells entering the central nervous system (CNS) to cause demyelination. Following animal studies and case reports of patients with malignancies and concomitant MS, formal investigation of the efficacy of ASCT in MS was proposed. From a retrospective study of the EBMT, a PFS of 74% at three years has been reported ($n = 85$) with some resolution of active MRI gadolinium-enhancing or -expanding lesions in all cases examined ($n = 78$) (74). This is encouraging but the data need to be treated with some caution as there is no information on the type of MS treated and expected outcomes vary. In a group of 19 patients with defined non-primary progressive MS with high disease activity on magnetic resonance imaging (MRI), PBSC collection was performed using Cy/G-CSF followed by BCNU, etoposide, cytarabine, melphalan (BEAM)/ATG-conditioned ASCT. All patients showed stabilization or improvement, with MRI-active lesions after ASCT except in a single patient at 4.5 years post-ASCT. Three patients subsequently deteriorated, one below the baseline at median follow-up of 36 months. Importantly, quality of life scores showed a statistically significant improvement following this approach (75). Care must be taken with mobilization as G-CSF when given alone has been demonstrated to cause exacerbations of MS with significant deterioration. This may be ameliorated with the use of corticosteroids. An RCT is planned to further evaluate this approach in MS (EBMT/ASTIMS).

An improvement in rheumatoid arthritis disease-severity parameters in patients who have failed disease-modifying anti-rheumatic drugs has been seen in 67% of undergoing ASCT, although these patients do eventually relapse. At relapse, they have been shown to recover responsiveness to disease-modifying anti-rheumatic drugs. An RCT (ASTIRA) has been started.

Table 2 Indications for Autologous Transplantation

	Disease stage	Comments
AML	CR1—in context of clinical trial	No benefit in good risk group
ALL	CR1—in the absence of matched sibling donor in context of RCT	
CML	Not currently recommended	
High-grade NHL	CR2, relapse	In chemoresponsive disease
Hodgkins disease	CR2, relapse	
Low-grade follicular lymphoma	In context of clinical trials, relapsed disease/salvage	Beneficial as first-line in high FLIPI; high level of MDS reduces effectiveness
CLL	Relapse, CR1 in context of clinical study	
MM	First-line standard therapy	Consider tandem transplant—auto or RIC alloSCT
Amyloid	First-line therapy	
Autoimmune	In context of clinical trial	

Abbreviations: AML, acute myeloid leukemia; ALL, acute lymphoblastic leukemia; CML, chronic myeloid leukemia; NHL, non-Hodgkin's lymphoma; CLL, chronic lymphocytic leukemia; MM, multiple myeloma; FLIPI, follicular lymphoma prognostic index; MDS, myelodysplastic syndrome.

In SLE, responses have been seen in most of the 53 evaluable patients, with relapse in 32% although again responsiveness to conventional drug therapies may be restored even in these patients. A TRM of 12% was seen reflecting disease severity and organ damage in this group (76).

In Europe, these results have resulted in a consensus in the EBMT/EULAR (European League Against Rheumatism), suggesting that HSCT should be offered to a patient suffering severe life-threatening autoimmune disease (AD) if conventional therapies have failed; patients should have active reversible pathology, which will confer good quality of life if reversed. All patients should receive transplants within the context of controlled trials and where possible these should be existing EBMT approved trials with patients registered and data on toxicity collected from mobilization. The uses of autologous transplantation are summarized in Table 2.

ALLOGENEIC STEM-CELL TRANSPLANTATION

General Overview

The benefits of alloSCT are a tumor-free graft and the ability to harness the powerful immunological GVL effect. In the last decade, there have been a number of developments that have made allogeneic transplantation a feasible option for more patients. The establishment of large national and international donor registries and of cord banks has improved the likelihood of identifying a human leukocyte antigen (HLA) match for patients lacking matched family donors. The improvements in tissue typing using molecular probes permit better matching of unrelated donor–recipient pairs decreasing the risk of graft rejection and GVHD. Monoclonal antibodies allow a range of approaches toward GVHD prophylaxis and immune suppression. Poor immune reconstitution remains a problem, but preemptive approaches to cytomegalovirus (CMV) using quantitative PCR has improved outcome.

The other significant change that has impacted on the use of alloSCT is the advent of nonmyeloablative or reduced intensity conditioning (RIC) transplant regimens. The increased safety of this approach with decreased mucosal toxicity, GVHD, and shortened period of cytopenia has extended the group of patients able to safely undergo alloSCT.

Although the studies comparing alloSCT consistently show lower RRs, the indications from several studies addressing long-term quality of life issues demonstrate that recipients of alloSCT experience more short- and long-term complications. These include chronic GVHD, infections, sterility, secondary malignancies, endocrinopathies, and cataracts, but in addition recipients may experience impairment in sexual function and psychosexual difficulties following allogeneic transplantation (77,78). These important issues need to be taken into account when considering therapeutic approaches for individuals.

General Principles in Clinical Practice of ASCT

The conventional approach to allografting has relied on a combination of intensive myeloablative and immunosuppressive regimens as preparative conditioning therapy prior to stem cell rescue. The escalation of chemotherapy to maximally tolerated doses was initially thought to operate via maximal cell kill of residual leukemia in accordance with theories of log-dose–response, with transplanted allogeneic cells acting simply to rescue the ablated hematopoietic system (79). However, murine experiments have shown that, even at maximally intensive 5000 cGy doses of irradiation at which neurotoxicity intervenes, leukemic cells can escape (80). Since the 1970s, evidence has accumulated for a powerful second component of this treatment modality associated with the donor graft itself leading to eradication of tumor cells. This immunological component is now termed the GVL or "graft versus malignancy" effect in the case of other lymphoid and solid tumors.

Conventional high-intensity conditioning regimens may give rise to considerable transplant-related acute and long-term morbidity and mortality. These are due to collateral damage caused by intensive chemoradiotherapy leading to myelosuppression with resultant prolonged neutropenia and risk of infection, mucosal damage in the gastrointestinal tract, neurological damage leading to impaired development in children, cataract formation, sterility, and endocrine adenopathies (81,82). There is also an increased risk of secondary malignancy (83–86). These serious complications, along with the risk of catastrophic GVHD, have largely precluded the use of allografting for patients with prior co-morbid conditions or for older patients (>55 years) as mortality and morbidity exceed benefits. As these older patients represent the majority of those with acute leukemia and have the worst prognosis, novel approaches have been explored. The ability to achieve safer transplantation may also allow engraftment of haploidentical transplants from family members to the 40% of patients who, despite the worldwide registry network, do not have a suitably matched donor. An important goal has been to develop safer allografting procedures and to optimize the achievable benefits of this approach by developing regimens relying less on profoundly toxic-intensive cytotoxic therapy and more on immunological approaches to tumor eradication.

The GVL Effect

The rationale for nonmyeloablative stem-cell transplantation using reduced-intensity preparative conditioning originates from evidence for the therapeutic potential of adoptive transfer of alloreactive donor lymphocytes to eradicate malignant host cells that escape maximally tolerated doses of chemotherapy (87–92). It is now clear that alloSCT offers an important advantage attributable to an immune effect mediated by donor-derived immunocompetent T-lymphocytes.

Initial clinical evidence for this phenomenon was supported indirectly by the observation that transplantation between syngeneic twin pairs was associated with a higher rate of relapse in comparison with sibling allografts and indeed the RR in this situation was similar to that seen following autografting (93). Anecdotal reports of patients achieving complete remission associated with a "flare" of GVHD or following withdrawal of immunosuppressive therapy suggested that there was a link between these phenomena (94,95). Indeed, retrospective analysis of large transplant series has demonstrated that there is an association between GVHD severity and a lower likelihood of relapse of malignancy after allografting (96). An inverse relationship exists between the degree of major histocompatibility (MHC) matching and RRs (97–103). Further evidence for the importance of the immunological power of the allograft was provided when T-cell depletion of the graft, in an attempt to prevent GHVD, led to an increased likelihood of relapse (104,105).

Donor Leucocyte Infusions

The most powerful direct evidence for the GVL effect has come from the successful induction of durable complete cytogenetic and molecular remission following donor leucocyte infusions (DLIs) without any additional therapy in 60% to 80% of patients with CML who relapse after allografting (106,107).

This approach is effective for treatment of relapse of other hematological malignancies although there are major differences between diseases in their susceptibility to the GVL effect. CML is the most sensitive, whereas AML and ALL show lower RRs to the infusion of DLI after alloSCT (108). The reasons for this differential efficacy may be multi-factorial. Relapsed acute leukemia is rapidly progressive, allowing an insufficient temporal window for the development of the alloreactive anti-tumor T-cell response that can take several months to become established. CP CML tends to have a more protracted natural progression and, as such, is less likely to progress at an uncontrollable rate during this phase. Another reason for the susceptibility of CML could be improved tumor antigen presentation by CML because the leukemic cell progeny includes dendritic cells that may be able to effectively stimulate anti-tumor effectors (109–112). Indolent lymphoid malignancies (low-grade lymphoma, CLL, myeloma, and Hodgkin's disease) are also susceptible to GVL effects (113–115) and there is anecdotal evidence that there is activity against solid tumors including breast cancer, renal cell carcinoma, and melanoma (116–118).

The period between DLI infusion and clinical response may reflect the time taken for the proliferating tumor-specific alloreactive cells to reach a critical mass and achieve a favorable ratio between donor immunocompetent effector T-cells and residual target leukemia cells. It is clear from clinical results following DLI that one major predictive factor for the effectiveness of the GVL approach is tumor cell burden, with optimal response when there is an MRD state (106,119–121). It is not fully understood whether the GVL effect is a continuous process of immune surveillance, and as such is required lifelong, or whether there is the ability for this mechanism to eliminate the dormant leukemic progenitor cells. Long-term follow-up is critical to assess the durability of response. It is likely that conversion of mixed donor chimerism to full donor chimerism would be necessary to eliminate the malignant clone in the case of hematological disease (122–126).

Nonmyeloablative Regimens

The success of DLI in inducing remissions in patients who have relapsed after alloSCT suggested the feasibility of achieving long-term disease control by harnessing the GVL potential of the human allogeneic immune system without high-dose induction therapy. A study examining the impact of reducing the conditioning irradiation dose demonstrated no significant difference in OS as any increase in relapse was compensated for by the lower risk of GVHD (127). Using minimally myeloablative regimens with predominantly immunosuppressive therapy, it was possible to achieve engraftment of allogeneic cells, in effect introducing the allogeneic donor cells "by stealth." Indeed there is preliminary data from clinical trials to support this approach, although follow-up is still too short to assess impact on relapse and OS (128–139).

The delayed kinetics of immune reconstitution also limit the GVL effect during the period of extensive immune suppression prior to the establishment of donor immunity (140–142). In spite of the powerful immunological benefit of alloSCT, the most common cause of treatment failure remains disease recurrence (113,120). The ability to further augment the allogeneic immune phenomenon would be an important advance.

Graft Versus Host Disease

GVHD is a major complication following alloSCT, causing considerable morbidity and is fatal in approximately 30% of cases (106). GVHD occurs following recognition by donor T-cells of antigens presented by MHC molecules on the recipient antigen-presenting cells. Clonal expansion of responder T-cells proceeds and an uncontrolled effector response results, involving lymphocytes and cytokines directed toward a broad spectrum of tissues to which lymphocytes migrate (skin, gastrointestinal mucosa, biliary tract, exocrine glands, synovia, lung, and marrow). Classically, clinical GVHD is subdivided into two syndromes: "acute GVHD" occurring in the first two to three months following transplant and "chronic GVHD" occurring later during the post-transplant period. The GVHD phenomenon necessitates the application of intensive post-transplant immunosuppression (IS). This intensive IS can potentially exacerbate post-transplant morbidity through delay in immune reconstitution, leaving the transplant recipient susceptible to potentially fatal infection such as CMV or the reactivation of Epstein Barr Virus (EBV) with associated malignancy (EBV-related lymphoproliferative disorders). It can also suppress the GVL response, increasing the risk of leukemic relapse (84).

The pathophysiological model of GVHD hypothesized is a complex immunological process. Donor T-cells of the type 1 subset are activated following recognition of host allo-determinants and induced to secrete pro-inflammatory cytokines, for example, interleukin (IL)-2 and IFN-γ. This is not only stimulated by allogeneic recognition of donor–host major- and minor-histocompatability differences, but also via induction of substantial tissue damage by conventional fully ablative preparative regimens, which frequently involve high-dose TBI. Donor monocytes, natural killer cells, and macrophages are triggered to release tumor necrosis factor (TNF)-α, mediating morbidity and mortality of GVHD. The "cytokine storm" theory suggests that tissue damage from the intensive conditioning regimen results in the release of endotoxin and lipopolysaccharide from gut flora leading to dysregulated production of inflammatory cytokines such as IL-1, IL-12, TNF-α, and IFN-γ. This acts as an alarm signal and further stimulates and drives the immunologically active donor T-cells thus augmenting the GVHD reaction, triggering the self-perpetuating cascade of the cytokine storm (143–146). Indeed, the severity of acute GVHD is closely related to conditioning intensity (147). The toxic effect of conditioning

Table 3 Advantages and Disadvantages of alloSCT

Advantages	Disadvantages
GVL effect, reduced relapse risk	GVHD
No stem-cell contamination	Slow immune recovery
Ability to manipulate chimerism and immune GVL effect using DLI in event of relapse	High risk of viral infections (CMV, adenovirus)
	Higher mortality and morbidity
	Requires sibling or HLA-matched stem-cell donor

Abbreviations: alloSCT, allogeneic stem-cell transplantation; GVL, graft-versus-leukemia; GVHD, graft-versus-host disease; DLI, donor leukocyte infusion; CMV, cytomegalovirus; HLA, human leukocyte antigen.

leads to gastrointestinal mucosal damage and cytokine release (148–154). GVHD is seen after administration of DLI in a proportion of patients but this is generally less severe and resembles clinically the scenario of chronic GVHD more closely, possibly reflecting the temporal separation of T-cell administration from the toxic conditioning regimen. Preliminary data suggest that by reducing conditioning intensity, this cycle may be broken and the severity of acute GVHD may be decreased (129,130,134,155,156).

The advantages and disadvantages of allogeneic transplantation are summarized in Table 3.

SPECIFIC CLINICAL INDICATIONS FOR ALLOGENEIC TRANSPLANT

Acute Myeloid Leukemia

During the last two decades, outcome from chemotherapy and ABMT has improved, so the important question remains whether alloSCT still has a role in first remission treatment of AML. The main disadvantages of high procedure-related mortality and morbidity from chronic GVHD are counterbalanced by a lower RR, possibly due to a GVL effect.

No high-quality RCTs are available. Most trials adopt a design whereby patients who have an HLA-matched sibling are assigned alloSCT. This "donor-versus-no donor" design may be justified on the basis that only 25% to 30% of individuals in developed countries will have a sibling-matched donor. The remaining patients are allocated to the ABMT/chemotherapy comparison arm acting as a control group. If such a design were not adopted, then a far larger accrual would be necessary to attain sufficient power to detect a statistical difference; however, only 50% to 60% of randomized potentially eligible patients actually receive transplant. In order to address this issue, the question of whether an allogeneic transplant would have greater benefit after only three courses of chemotherapy must be addressed.

The GOELAM, EORTC/GIMEMA, and U.S. Intergroup investigators (18,31,157–159) have all assessed the role of allogeneic transplantation. RR was consistently lower in all these studies. In another study, the value of alloSCT on DFS was confirmed in 107 patients assigned to either HLA-matched allogeneic transplant or chemotherapy (71% DFS vs. 31% DFS; $P = 0.028$) when analyzed on an intent-to-treat basis (160). An early Dutch study randomized 117 patients to either allogeneic BMT or autografting. The results, although not statistically significant, suggest that allogeneic BMT may provide a better DFS than autologous BMT (three-year DFS 35% ASCT vs. 51% alloBMT; $P = 0.12$) although again only 59% of patients received their transplant procedure. Another study prospectively compared alloBMT with ASCT in 94 patients, again showing improved DFS (161). "Time-censoring" effects also apply to allogeneic transplantation and as there tends to be a longer delay to allogeneic transplant this effect is likely to be more prominent.

The MRC AML 10 study showed no benefit for allografting in patients over 35 years, children, or good risk disease. The current MRC trial (AML 15) allows the use of RIC allograft in standard and high-risk patients over 35-years as course 4.

Retrospective analysis of the EBMT registry data 1987–1992 suggested that LFs was significantly better in allogeneic transplantation. Analysis of 516 allogeneic transplants and 598 autologous marrow transplants aged 3 to 40 years showed a significantly higher TRM, lower RR, and better DFS (55% vs. 42%) (162). Other registry-based studies, for example comparison of BGMT (Bordeaux, Grenoble, Marseilles, Toulouse) registry data have shown allografting to be superior to autografting with lower RR, higher TRM but no effect on OS (163).

The data on non-myeloablative transplant in AML CR1 has much shorter follow-up and in general a different cohort of patients; it cannot, therefore, be compared with outcome from alternative strategies.

Haploidentical Transplantation in AML

The results of haploidentical transplantation were reported in 33 patients by the Perugia group. All were at high risk with post-transplant relapse, CR1 poor risk, or in CR2 or later. Positively selected CD34+ cells were used and patients conditioned with TBI/fludarabine. No immunosuppressive therapy was given post-transplant. Leukemia relapse was controlled in that group of patients whose donor was non-killer (NK) alloreactive with only 2/16 relapsing, 72% (13/18) patients who were in CR at transplant survived, whereas 27% (4/15) of those in relapse at transplant survived. NK cell alloreactivity was associated with LFS with 70% LFS achieved vs. 7% of those without this disparity (164–166). This approach should be considered in the context of a trial for young relapsed/refractory patients without a matched donor.

In summary, current evidence suggests that, for the majority of patients under 56 years of age with a sibling donor, allografting should be considered. Only in those with favorable karyotype [t(15;17), t(8;21), inv16] does the balance of risks clearly fall against the use of allogeneic transplant in CR1. In this group, relapse-free survivals of 60% to 83% are seen following conventional chemotherapy and allografting should be considered only at relapse if a molecularly negative harvest cannot be obtained. In patients without a sibling-matched donor who have poor-risk disease or preceding MDS, an unrelated donor search should be initiated. Older patients may also benefit from a transplant approach but this decision is usually made after remission is achieved and post-induction performance status can be assessed.

Allogeneic SCT in ALL

The MRC UKALL XII/ECOG 2993 is an RCT to determine the impact of alloSCT in comparison with other modes of treatment; interim analysis suggests that patients assigned to allograft ($n = 190$) have a significantly reduced risk of relapse compared with those assigned to ASCT or chemotherapy ($n = 253$) and tendency to improved EFS (54% vs. 34% at five years) (167). The French LALA87 found an advantage for alloSCT

particularly in the high-risk group with Ph+, >35 years, presenting WCC >30×
$10^6 \mu L^{-1}$ or time to CR >4 weeks). Ten-year OS in this group was 44% versus 11%
(168,169).

The Philadelphia chromosome occurs in 20% to 30% of adult patients with ALL.
The outcome is poor in these patients. Allogeneic transplantation offers a clear benefit
to this group (170), and this is one of the indications to search for an unrelated donor in
patients who do not have a sibling match as the chance of cure without transplant is
very low. Imatinib, targeting the BCR-ABL tyrosine kinase may have an important
future role in the optimal preparation prior to allograft. It can induce response in refractory
and relapsed patients in 60% of cases but this is rarely sustained but may improve the
outcome of additional transplant therapy (171).

Unfortunately, haploidentical transplants have not mirrored the promising results
seen in AML. This is probably explained in part by the lack of donor NK cell alloreactivity
against ALL cells (172).

Allogeneic SCT in Diffuse High-Grade B-Non-Hodgkin's Lymphoma

Overall results in relapsed or refractory disease have favored ASCT over allografting due
to high toxicity. Some patients do benefit with prolonged DFS from an allogeneic
approach, and there is a GVL effect. The place of mini-allografting remains to be seen
but there seems to be a high rate of relapse in this aggressive disease situation in contrast
to that seen in low-grade disease (120,139).

Allogeneic SCT in Low-Grade Follicular NHL

Allogeneic transplantation may be a curative treatment for follicular NHL; however, the
benefit has been offset overall by high TRM. The IBMTR reported data on 904 follicular
NHL patients who underwent either ASCT or alloSCT between 1990 and 1999. A total of
176 received alloSCT, 131 (14%) received purged ASCT, and 597 (67%) received
unpurged ASCT. The five-year TRM rates were 30%, 14%, and 8% and recurrence
rates were 21%, 43%, and 58% after alloSCT, purged ASCT, and unpurged ASCT,
respectively. Furthermore, five-year OS was 51%, 62%, and 55%, respectively (49).
Decisions regarding whether allograft or ASCT is used are guided by patient and physician
choice as OS is similar. The risk of secondary MDS as seen in ASCT is minimal.

Non-myeloablative transplants show particular promise in this setting with a much
lower TRM (120,139).

Allogeneic SCT in Hodgkin's Lymphoma

This has a limited role in the overall treatment strategy because first-line chemotherapy
will be effective in the majority. A graft-versus-lymphoma effect undoubtedly occurs
against Hodgkin's disease but allograft tends to be reserved for refractory disease or
CR2/3 (57).

Allogeneic SCT in CLL

Early allogeneic transplantation in CLL has resulted in plateaus seen at 40%. The EBMT
registry reported that OS and EFS at 10 years was 41% and 36.6% in a series of 54 patients,
85% of whom were refractory to chemotherapy, undergoing sibling alloSCT. There was

48% TRM (173). TRM has been reduced in patients transplanted early in the course of the disease, and this approach should be targeted to those young patients at high risk early in the disease course.

The evidence for a GVL effect in CLL is strong and reduced-intensity alloSCT for CLL has suggested good efficacy and the potential for molecular remission in a proportion after tapering immune suppression, GVHD, or DLI infusions [78% ($n = 9$) alloSCT vs. 23% ($n = 26$) ASCT] that translated to persistent remission at median of 25 months in the alloSCT group only (61,174). Follow-up is, however, short at present as this approach is in its infancy. In 488 patients with CLL, 228 received RIC, whereas 222 received standard conditioning. Hazard ratio for TRM and OS were 0.5 and 0.56, respectively, significantly in favor of RIC with no increased risk of relapse (175).

Allogeneic SCT in MM

Cures in MM using an autologous transplant are unlikely as there is no plateau seen in the survival curve following this therapy. Allograft offers the advantage of a tumor-free graft and a graft-versus-myeloma effect. Only 10% of MM patients are suitable for sibling allograft due to age and donor availability. The TRM following conventional allograft in MM has been in the region of 20% to 50%, which severely limits the optimal use of allogeneic HST in MM. An analysis of the EBMT registry showed that OS at four years of syngeneic transplant was 77%, ASCT 46%, and allograft 31% (176). However, there have been improvements in outcome with time and an examination of the EBMT registry has suggested median survival of 10 months for patients transplanted between 1983 and 1993 compared with 50 months between 1994 and 1998. However, this approach has the advantage that those who survive the procedure may be cured because the survival curve plateaus at around five years, and DLI are effective in a proportion of patients (around 25%) who relapse.

The advantage of a graft-versus-myeloma effect may be harnessed using an RIC regimen while effecting a reduction in the high TRM rate. Promising results have been seen confirming the safety of this approach with preliminary evidence of efficacy (177,178). The Fred Hutchinson group used 2 Gy conditioning with mycophenolate mofetil and cyclosporine IS following ASCT. TRM at 100 days was 2% but this was due to disease progression. At median follow-up of 552 days post-allograft, 57% of patients have achieved a CR and 26% PR. These results appear particularly impressive as this was a comparatively elderly allogeneic transplant patient group with median age of 52, with range of 29 to 71 years. There remains a considerable risk of GVHD. The benefit of autograft to achieve best residual disease state followed in tandem by mini-allograft and, where appropriate, DLI is being formally evaluated in the MRC Myeloma IX RCT and in poor-risk patients in the IFM 9903-4 study in comparison with tandem autograft. The results of these trials will clearly be of high importance.

Initial results using the mini-allograft approach using unrelated donors ($n = 28$) have been promising with a TRM of 18% reported in one series with OS of 54%, which was significantly better than that seen using fully ablative conditioning (53% and 18%, respectively) (179).

Allogeneic SCT in CML

Imatinib has become the initial treatment of choice for the majority of newly diagnosed patients with CML since the publication of the IRIS RCT comparing imatinib to IFN and cytarabine (180). Current data suggest that 97% of newly diagnosed patients with

CP CML will achieve a complete hematological remission, 87% will have a major cytogenetic reduction, and 76% will have a complete cytogenetic remission following imatinib treatment.

Progression-free survival following imatinib treatment depends on the log-reduction of bcr-abl. Those with a 3-log reduction have a 100% PFS at 30 months compared with 93% for a 2-log reduction and 81% for a 1-log reduction. However, it is still too early to determine if single-agent imatinib will prolong OS compared with allograft. It is unlikely that patients will be cured with imatinib because only 5% have become PCR-negative and allogeneic transplant may remain the most effective curative modality (181). In those patients relapsing after allograft, DLI can accomplish durable remission in 70% to 80%.

The optimal timing of transplantation for patients with a stem-cell donor is an unresolved issue in CML. One potential decision strategy may be to postpone alloSCT until the imatinib response has been assessed and monitored. Patients with suitable donor may reasonably be offered allogeneic transplant if they do not achieve a complete hematological response with three months of imatinib or if they are predominantly Ph-positive at six months, or still have >35% Ph-positive metaphases at 12 months. Patients could also be transplanted at a point when hematological or cytogenetic response fails suggesting resistance.

Alternatively, allogeneic transplantation may be offered as initial therapy in low-risk patients without taking into account imatinib response for those who are under 45 years of age who are in CP and have a sibling donor, or under 35 years of age in those with a molecularly matched unrelated donor.

To refine this risk-adapted strategy, predictive models have been used to determine the success of both imatinib and transplantation approaches. The Sokal score, developed in the busulfan era, also predicts for response to imatinib as well as transplant outcome. Patients at low-risk on the Sokal score had a 94% PFS at 30 months, intermediate-risk had 88% PFS, and high-risk patients had an 80% PFS. There are also predictive factors for success or failure of allogeneic transplants (182). In registry data (EBMT and IBMTR), TRM for patients 45 years of age or under is 15% but can be higher with the presence of one or more adverse risk factors, such as increasing time from diagnosis to transplant, a female donor, increasing age, and CMV positivity and may be up to 40% when more than one risk factor is present. Patients without any of these risk factors have decreased TRM.

Reduced Intensity Allogeneic Transplants in CML

Given the efficacy of the GVL effect in CML following administration of DLI, it was reasonable to use the non-myeloablative strategy in this setting. RIC allograft was used in 24 patients with CML in CR1. A zero TRM at 100 days was seen and at five-years probability of DFS was 85% (183). These promising results may support an examination of this approach in RCT. A second report has suggested that although RIC transplant in CML was safe, out of a cohort of 12 patients, seven patients in CP1 underwent RIC allograft and five achieved CR, however of those transplanted in CP2, only one achieved molecular remission, with the others dying in blast crisis (184).

Allogeneic SCT in MDS

MDS is a heterogeneous group of clonal stem-cell disorders characterized by hypocellular BM, peripheral cytopenias, and dysplastic features. Allogeneic transplantation is the only means of cure for MDS. It is generally reserved for younger patients with more advanced disease, but as the disease is predominantly diagnosed in elderly patients with co-morbidity, a transplant-based approach is generally precluded. A variable number of blasts are present in the marrow in MDS and cytoreductive treatment prior to allograft may be tailored to degree of blast infiltrate. DFS for conventional fully-ablative alloBMT ranges from 29% to 40% due to high TRM and relapse risk.

Reduced-intensity sibling and matched unrelated donor (MUD) transplants have been reported in 62 patients with differing International Prognostic Scoring System (IPSS) scores and median age of 56 years (range 41 to 70 years). Overall TRM was low at 8%. In the low-risk Int-1 group, DFS was 83%, 67% in the Int-2 group, and only 31% in the high-risk group at median follow-up of 524 days (185). These data are not mature, and it remains to be seen whether the low-risk group has a survival benefit when compared with best supportive care or whether the high-risk group have a poorer outcome than would be conferred by fully-ablative regimens.

Allogeneic SCT in Myeloproliferative Disorders

Myelofibrosis with myeloid metaplasia (MMM) is typically a disease of the middle-aged and elderly. Although median survival is 3.5 to 5.5 years, good-prognosis MMM in younger patients for whom allograft may be an option has a median survival of 15 years with a supportive care approach.

Allogeneic transplant aims to eradicate the mutant MMM clone. It is associated with significant morbidity and mortality. Only 14% five-year survival for patients over 44 years was reported in one study (186) and 41% two-year survival in another study have been reported with high levels of chronic GVHD (187,188). Younger patients with good prognosis disease had better outcome overall, and it may be reasonable to consider allograft in patients with high risk (i.e., expected survival <5 years) and under 60 years of age.

RIC has been used in MMM with one-year survival of 77% in a report of 12 patients without transformation (189). A further recent report of 21 patients receiving RIC using fludarabine, busulfan, and ATG achieved zero TRM at day 100 with three-year DFS of 84% (190). Promising results have been reported in a further 21-patient cohort strengthening the evidence for a role of RIC allograft in MMM (191).

Allogeneic SCT in Severe Aplastic Anemia

Acquired aplastic anemia (AA) is life-threatening and is characterized by pancytopenia and aplastic or hypoplastic BM. The exact pathogenesis is unknown but potential mechanisms include intrinsic stem-cell defects with increased apoptosis, shortened telomeres and clonal abnormalities, defective marrow microenvironment, and abnormal immunological control of marrow.

The options for treatment include IS (using ATG, cyclosporine, and corticosteroids) or where a donor is available, allogeneic transplant. The relative benefits of IS versus those of allograft are dependent on age and severity of disease based on the level of neutropenia. Patients under 40 years of age with neutrophils under $0.3 \times 10^9 \, L^{-1}$ show a benefit from an alloSCT approach, whereas older patients (>40 years) benefit from IS from which a 65% to 75% RR may be expected at four to six months. The results of alloSCT in AA have shown steady improvement and recent reports suggest 88% to 94% five-year DFS following allograft (192,193). Chronic GVHD remains a problem and graft rejection is more common in AA as is secondary malignancy. There is a significant morbidity associated with TBI and therefore conditioning regimens have been developed to avoid this. Transplant morbidity may be hampered by a prolonged transfusion history. Patients with AA

Table 4 Indications for Allogeneic Transplantation

	Disease stage	Comments
AML	CR2/3	Consider in CR1 for moderate/ high-risk patients
ALL	High-risk disease	
CML	Failed imatinib	
CLL	Early-stage disease	Studies show persistent remission, but follow-up short
High-grade NHL	CR2/3, relapse	
Hodgkins disease	Relapsed/resistant disease post-ASCT	
Low-grade follicular lymphoma	Patient/physician choice	High transplant-related mortality; studies show no clear OS benefit over ASCT, but may be curative
MM	Reduced-intensity conditioning in clinical trial	
Solid organ tumor	Subject of clinical trials	Encouraging results in renal cell carcinoma and melanoma
MDS/MPD	Myelofibrosis-MDS, risk	Young patients More advanced disease
AA	Failed immunosuppressive	Young patients; WCC2 < 0.3

Abbreviations: AML, acute myeloid leukemia; ALL, acute lymphoblastic leukemia; CML, chronic myeloid leukemia; CLL, chronic lymphocytic leukemia; NHL, non-Hodgkin's disease; MM, multiple myeloma; MDS, myelodysplastic syndrome; AA, aplastic anemia; ASCT, autologous hematopoietic stem-cell transplantation; OS, overall survival.

who fail immunosuppressive therapy and who have a matched sibling donor should be offered SCT.

Unrelated donor transplantation can be considered for patients without sibling donor and failing IS, however the mortality is considerable and the higher risk of rejection and GVHD necessitates the use of TBI. Consequently, morbidity is substantial. The uses and indications for allogeneic transplantation are summarized in Table 4.

CORD BLOOD TRANSPLANTATION

General Overview

One of the major limitations of HSCT is the lack of availability of suitable donors. Less than 40% of patients who could benefit from an allogeneic HSCT have a suitable donor identified (194). Cord blood (CB) provides an attractive alternative source of HSC for transplantation. The ease of collection and potential availability to groups that are under-represented in the BM registries, such as certain racial and ethnic populations, are the advantages of CB compared with the other sources of HSC. Furthermore, CB contains fewer T-cells and/or more naïve T-cells than BM or PBSC, and may permit a greater degree of mismatch with less GVHD.

The first cord blood transplant (CBT) was carried out on a patient with Fanconi's anemia in 1988 (195), and with increasing experience and success rates, it has become

Table 5 Merits of CB Compared with BMT

Advantages	Disadvantages
Rapid availability	Less experience
Lower risk of GVHD	Delayed or failed engraftment
Better representation of minority ethnic groups	No donor recall available for boost or donor lymphocytes
Lack of risk to donor	
Lower CMV transmission	
Less HLA restriction	

Abbreviations: CB, cord blood; BMT, bone marrow transplantation; GVHD, graft-versus-host disease; CMV, cytomegalovirus; HLA, human leukocyte antigen.

an accepted alternative therapy in children who do not have a matched sibling or unrelated donor. There are a number of theoretical advantages of CB compared with adult cells: besides a reduced incidence and severity of acute and chronic GVHD compared with unrelated BMT (196,197), CB cells produce larger hematopoietic colonies in vitro, are able to expand further in long-term culture (198), have longer telomeres (199), and higher content of short-term repopulating cells (200). These characteristics theoretically should go some way in compensating for the limited numbers of cells available. However, in clinical studies, failed engraftment has been reported (201) and when CB is compared with BM, delayed engraftment remains a problem (197) (Table 5).

Clinical Results for CBT

Indications for CBT are those described for allogeneic transplantation.

Related Donor CBT

Rocha et al. (196) compared the outcomes of 2052 HLA-identical sibling donor BMT treated between 1990 and 1997 with 113 HLA-identical sibling donor CBT treated in the same period. Multivariate analysis demonstrated a lower risk of acute and chronic GVHD among recipients of CBT. Compared with BMT, recovery of the neutrophil count and the platelet count was significantly lower in the first month, however no difference was seen in RRs and mortality was similar in the two groups.

Data presented at EBMT March 2004 from the Eurocord registry described 177 patients who received related donor CBT between October 1988 and August 2003. Only 3% were adults, 17% were HLA incompatible, half were carried out for malignant disease, and half for other diseases (SCID, hemoglobinopathies, AA, and metabolic diseases). OS was 48% for the malignant group and 80% to 100% in the non-malignant group.

Thus far, despite the lower incidence of acute and chronic GVHD, the relapse risk with CBT is no higher than that with BMT. Longer follow-up of CBT patients may reveal improved quality of life as a consequence of reduced incidence of chronic GVHD.

Unrelated Donor CBT

Several groups have reported outcomes on unrelated donor CBT (202–206). The importance of nucleated cell dose in predicting engraftment and survival has been demonstrated and HLA disparity has been shown to impact on time to neutrophil recovery and likelihood of graft failure. The New York National Cord Blood Program has reported on 791 patients receiving unrelated donor CBT of whom 80% were children. Of these, 748 had mismatched grafts and the time to neutrophil recovery was 28 days compared with 23 days for those who received matched grafts (207). Overall, 93% of patients showed engraftment, but only 74% of adult patients engrafted by day 42. Three-year survival was only 27% for malignant disease and 48% for genetic diseases. The number of HLA antigen mismatches correlated positively with severity of acute GVHD; however, the incidence of acute GVHD was lower than might be anticipated compared with BMT and considering the degree of donor–recipient HLA mismatch.

Data presented at EBMT March 2004 from the Eurocord registry described 587 patients who had received unrelated donor CBT between October 1988 and August 2003. Thirty-two percent of patients were adults, 86% received HLA incompatible grafts, and four-fifths were transplanted for malignant disease. In this group, survival was inversely related to advanced disease and two-year OS was similar to the New York Registry at 27% for adults (37% for early-stage disease and 22% for advanced-stage disease) and 39% for children. Although the decreased incidence of GVHD has raised concern that the GVL effect might be reduced, there is no evidence, to date, of increased RRs compared with BMT.

CBT allows many patients access to a potentially curative treatment that they would otherwise be denied; however, the findings of relatively high treatment-related mortality and slow engraftment kinetics indicate that it should continue to be performed in specialized centers with a research focus on CB cells. In adults, the procedure is limited by the number of cells obtained from a single CB unit (205) and remains the subject of clinical trials. Consequently, strategies are being developed to extend access to transplantation to many patients who have previously been disqualified on the basis of the available cell dose.

Ex Vivo Expansion

Identification of ex vivo conditions that support the self-renewal and expansion of HSC has the potential to increase enormously the number of patients to whom transplantation is available, and to reduce the associated morbidity and mortality by shortening the time engraftment.

CB cells have been shown to have the greatest capacity for expansion of progenitors and long-term culture-initiating cells (LTC-ICs) in vitro, and have recently been the subject of clinical ex vivo expansion protocols. Four trials have reported the combination of unmanipulated cells with ex vivo expanded CB. Jaroscak et al. (208) reported a phase 1 study of 28, mainly pediatric, patients with malignant and non-malignant disorders who were eligible for BMT, but did not have a sibling or matched unrelated BM donor. Patients received a minimum cell dose of 1×10^7 thawed cells/kg on day 0 and the remainder of the CB graft (typically $1–2 \times 10^8$ total cells) was expanded in perfusion cultures supplemented with 10% fetal bovine serum, erythropoietin (EPO), Flt-3, and PIXY-321. Expanded cells were given as a boost to the unmanipulated graft on day 12. Twenty-one of 27 patients who received expanded cells successfully engrafted; however, augmentation of the CBT did not alter the time to engraftment. The relatively late timing (day 12) of the infusion of expanded cells leads to difficulties in interpreting what influence, if any, the cultured cells had on engraftment and a randomized phase II study is underway to help clarify the issue.

Pecora et al. (209) reported two cases of adult patients with CML who experienced stable engraftment and cytogenetic remission at 19 and 8 months following transplantation of mainly unmanipulated (83% and 89%) CB cells. The remainder of the CB graft was cultured in the same conditions as described above for 12 days, and most of the colony-forming unit-granulocyte macrophage and $CD34^+$ lin^- cells transplanted to the patients were derived from the expanded portion of the graft. It was not possible to assess the relative contributions to hematopoietic recovery of expanded versus unmanipulated cells as gene marking was not used.

One study of six patients who received unmanipulated CB (60% of graft) and $CD34^+$ selected cells (from remaining 40%) cultured for 10 days with thrombopoietin (TPO), stem-cell factor (SCF), and G-CSF suggested a potential hastening of neutrophil engraftment (210). The results of this study were updated the following year and included a second cohort of patients who received CB cells from grafts frozen in two aliquots (211). One aliquot was cultured for 10 days prior to infusing both expanded and unexpanded cells at day 0. Neutrophil engraftment failure was reported in zero of 19 patients (compared with a 15% to 60% failure rate reported for recipients of unexpanded CB transplants). The median time to neutrophil and platelet engraftment rates in adult patients was comparable to that in smaller pediatric patients despite lower CB cell doses.

Together, these studies suggest a potential benefit to ex vivo expansion of CB, as many of these patients received low numbers of cells infused per kg body weight and very few graft failures were observed; however, a major criticism is that the contributions of expanded and unexpanded cells to hematopoietic recovery cannot be assessed.

When one considers the potential utility and future directions of ex vivo expansion of CB cells, it is important to note that the cytokine combinations used in clinical trials of CB expansion reported to date are not those that have achieved maximum expansion of cells and progenitors in vitro and that using other cytokine cocktails might achieve more impressive results.

Transplantation of Two CB Units

Transplantation of two partially HLA-matched CB units is the strategy being adopted by the University of Minnesota for obtaining acceptable numbers of CB cells to transplant adult patients previously denied treatment, on the basis of the available cell dose in a single unit. An initial report described 40 high-risk adult patients, 29 of whom received double unit transplants (212). The early outcome data look encouraging with 91% of patients showing engraftment and no significant difference in GVHD rates in the patients receiving two units compared with those receiving one. The mean time to neutrophil engraftment was 11.5 days and only three patients did not engraft. More recent data from the same group report 23 consecutive patients with malignant hematological disease who received two unit CBT with myeloablative conditioning (213). Twenty-one of 23 patients were able to be evaluated for engraftment, and median time to engraftment was 23 days (range 15 to 41). All 21 patients exhibited donor chimerism with sustained donor engraftment (median follow-up of 10 months) and there was no significant increase in acute GVHD rates compared with single unit CBT. Two of the 21 able to be evaluated patients died of disease relapse, one from a pulmonary hemorrhage and one from *Aspergillus* infection.

The high engraftment rate and low incidence of severe acute GVHD have resulted in a relatively low TRM. Therefore, further investigation of this approach in the context of larger clinical trials is indicated to determine the full impact of double-unit CBT on transplantation outcome in adults and larger adolescents.

Future Developments

HSCs are the best studied of the tissue-specific stem cells. By definition, HSC has long been regarded as restricted to formation of blood cells of both the lymphoid and myeloid lineages. HSCs residing in the BM microenvironment have self-renewal capacity and can repopulate the hematopoietic system of irradiated transplant recipients for the lifetime of the individual. Therefore, HSCs are extremely important targets for gene therapy applications aimed toward the treatment of inherited and acquired blood disorders. However, recent studies have suggested that a subpopulation of HSCs may have the ability to develop into diverse cell types such as hepatocytes, myocytes, and neuronal cells, especially following tissue damage. This raises the possibility that HSC transplants have the potential to provide therapeutic benefit for a wide variety of diseases and contradicts the dogma that adult stem cells are developmentally restricted.

Gene Therapy

Gene therapy has the potential to treat inherited and acquired diseases for which there is little hope of cure by conventional medicine. In 2000, the successful treatment of children with X-linked severe combined immune deficiency (X-SCID) hit the headlines (214). This is a rare immune disease, usually fatal in the first year of life, caused by an abnormal gene on the X chromosome that encodes the common γ C chain. The γ C chain is necessary for the development of T-cells and natural killer cells and affected individuals suffer recurrent life-threatening infections. In each of the patients treated, a retroviral vector was used to introduce a functional copy of the defective gene into BM stem cells before being injected back into the patient. Initially T-cell numbers and repertoire were nearly normal and the treatment was widely accepted as the first true clinical success for gene therapy. Sadly, the excitement came abruptly to an end when two of the 10 children treated in France developed leukemia-like conditions (215). In the two cases of leukemia, genetic analysis of the malignant cells showed that the retroviral vector had inserted into, and activated, an oncogene called LMO2 that is associated with childhood leukemia. Although none of the preclinical studies had shown any evidence of cancer in animals, it had always been known to be theoretically possible that gene insertion could activate oncogenes.

Further concerns were raised by data from a group at Stanford University regarding another method of correcting faulty genes; adeno-associated viruses (AAV) are considered to be safe for gene therapy as they do not cause disease in humans naturally and rarely integrate randomly into the genome. Results of a phase 1 clinical trial using AAV expressing the gene for Factor IX showed encouraging results (216), however, in 2003 a study conducted in mice found the vector used in this clinical trial integrates itself into coding regions of DNA. There was no particular pattern to the integration, but the studies demonstrated that the AAV vector could cause similar problems to those seen with X-SCID patients (217).

In addition to retroviral and AAV vectors, there have also been problems documented with adenoviral vectors. Adenoviruses are the most commonly used viral vectors in clinical trials owing to their ability to transduce dividing and non-dividing cells. Furthermore, they do not introduce their own genome into the host cell, which ensures less chance of harm caused by insertional mutagenesis. However, the death of a young man in the United States was attributed to the toxic effects of an adenoviral vector used to treat ornithine transcarbamylase (OTC) deficiency. His death was caused by massive inflammatory response to the adenoviral vector (218).

After the leukemia cases occurred, scientists and regulatory authorities called for a halt to clinical experiments. In several countries, trials were allowed to resume after a temporary hold, on the basis that the potential benefits to patients outweighed the risks.

Many argue that there is a need to develop new, safer vectors that avoid the problem of insertional mutagenesis, and for more preclinical studies to enable better assessment of the risks. Research must go on, particularly in the area of vector design. Nevertheless, this work may take many years, and even the best animal model may not be able to predict all the possible risk factors when treating patients, as were shown by the X-SCID case.

The varied responses from regulatory authorities add greatly to the uncertainty surrounding gene therapy. By creating a complex web of different rules in different countries, multi-center clinical trials become harder to plan and execute.

Stem-Cell Therapy for Cardiac Repair

Evidence of plasticity of adult human stem cells (the ability to generate cells of different lineage from their organ of origin) has led to investigation into their potential for cellular repair and organ regeneration. Studies in several species have demonstrated that BM-derived stem cells are not simply stromal or HSCs, but they are precursors for many peripheral tissues. Ultimate stem-cell fate and normal growth depend on the environment in which they engraft. In cardiovascular disease, the aim of stem-cell therapy is to transplant cells of non-cardiac origin, such as BM-derived mononuclear cells, to act as a precursor for heart muscle and coronary blood vessels and result in functional improvement of damaged tissue.

After an injury, such as myocardial infarction or as a consequence of cellular damage due to pressure or volume overload of the heart, specific factors including cytokines that stimulate cell replication are produced in the surrounding tissues. The stress of increased mechanical activity in the recipient heart muscle may therefore provide a more favorable environment for stem-cell engraftment than normal tissue. In clinical myocardial infarction, autologous BM cells may regenerate infarcted myocardium and improve perfusion in the infarct zone (219).

However, prior to successful engraftment, stem cells have to be delivered to the target site. To facilitate this, it is desirable to have high concentrations of transplanted cells in the area of interest and low levels of homing to other tissues. Consequently, regional and/or targeted administration is preferred to intravenous delivery of cells (Table 6).

HSCs and mesenchymal stem cells (MSCs) contribute to regenerative processes involved in tissue remodeling, however, other types of stem cells present in the BM may also be involved, for example, hemangioblasts take part in neovascularization and mesodermal progenitor cells differentiate to endothelium. MSCs represent a stem-cell population present in adult tissues that can be isolated, expanded in culture, and characterized in vitro and in vivo. MSCs differentiate readily into chondrocytes, adipocytes, and osteocytes, and they can support HSCs or embryonic stem cells in culture. Evidence suggests MSCs can also express phenotypic characteristics of endothelial, neural, smooth muscle, skeletal myoblasts, and cardiac myocyte cells. When introduced into the infarcted heart, MSCs prevent deleterious remodeling and improve recovery, although further understanding of MSC differentiation in the cardiac scar tissue is still needed. MSCs have been injected directly into the infarction, or they have been administered intravenously and seen to home to the site of injury. Interestingly, examination of the interaction of allogeneic MSCs with cells of the immune system indicates little rejection by T-cells.

Table 6 Routes of Administration of Stem Cells for Cardiac Repair

Route of administration	Potential benefits	Disadvantages
Intracoronary	Effective accumulation and concentration of cells in damaged area	Risk of procedure-related arrhythmia or ischemia
Intramyocardial transendocardial	Injection under visualization allows anatomical identification of target area and even distribution of injections	Risk of procedure-related arrhythmia
Intravenous	Ease of delivery	Low cell concentrations reaching target tissue

Some groups have investigated stem-cell therapy with more homogeneous populations, such as CD34+ and AC133[+] BM cells. However, using highly selected populations of cells may result in less successful outcomes; two recent preclinical studies published in *Nature* demonstrate that murine HSCs do not transdifferentiate into cardiac myocytes in the presence of myocardial ischemia (220,221), which raises the possibility that the improved recovery and cardiac function shown with stem-cell therapy (222,223) may not result from cardiac muscle regeneration, but from impact on left ventricular remodeling and/or angiogenesis. Furthermore, these effects may require a more heterogenous population of stem cells, suggesting that the therapeutic use of mononuclear BM cells may be more promising than single isolated cell fractions alone.

Stem-cell therapy is an exciting new approach to heart disease. Recent clinical trial results have shown the feasibility of adult autologous therapy in acute myocardial infarction. However, there are many outstanding issues and unresolved questions for experimental and clinical research:

- What is the best method of delivery of cells to the damaged heart? Intravenous, intracoronary, transendocardial, or intramyocardial?
- What is the optimum time course of stem-cell therapy after myocardial injury?
- What are the optimal conditions for engraftment of stem cells in the ischemic heart?
- What is the arrhythmic potential of implanted cells?
- Can specific detection of engrafted cells by labeling techniques be achieved?
- Can specific characterization of stem-cell populations be used to predict the therapeutic effect of transplanted cells?

Preclinical and early clinical data look encouraging; however, further studies are required to clarify the potential merits of adult stem-cell transplantation for patients with cardiovascular disease, and, equally importantly, to determine the long- and short-term safety profile of the intervention.

Stem-Cell Therapy in Neurological Repair

The CNS is vulnerable to a variety of illnesses and neurodegenerative diseases. Unlike most other organs, there is little endogenous repair of the CNS and so recovery from injury is usually modest at best. Most therapeutic strategies have focused on maintaining viability of damaged tissue or slowing the progress of degenerative disease. Recent studies have suggested that populations of stem cells previously thought to be lineage restricted can transdifferentiate from one tissue type to another, for example, hematopoietic to neural cells and vice versa (224–226). With advances in the understanding of the biology of stem and progenitor cells, there is increasing hope for stem-cell therapies for structural brain repair and to restore lost neurological function.

Evidence for the Use of HSC in Neurological Repair

Mezey et al. (225) showed that transplanted adult BM stem cells enter the brain of irradiated mice and differentiate into microglia, astrocytes and neurons. They subsequently went on to examine postmortem brain samples from females who had received BMTs from male donors (226). Using a combination of immunocytochemistry and FISH histochemistry to search for Y chromosome-positive cells, they demonstrated that in human transplantation, BM cells can enter the brain and generate neurons. The possibility that this phenomenon could be exploited to prevent the development or progression of neurodegenerative diseases, or to repair damaged tissue, is supported by the functional benefit observed in rodent models of Parkinson's disease (227). Mice received intravenous human CB mononuclear cells without immunosuppression and significant delays were seen in time to onset of symptoms and death compared with control animals.

Interestingly, transplanted cells may not exert their influence simply by replacing lost or damaged cells, but may act indirectly to increase plasticity or resistance to disease. In mouse models of cerebral ischemia where BM cells engrafted in the brain and resulted in functional improvement, the morphology of transplanted cells was atypical despite staining with neural markers (228).

The advantages of autologous HSC therapy are that cells could be harvested electively as an outpatient procedure, purified, expanded, and possibly differentiated before being transplanted back, thus removing the risk of cross-infection and need for IS.

Clearly, further studies are needed to elucidate the mechanisms by which HSCs achieve improved function and repair before large-scale studies can be carried out safely in patients with neurological disease.

Neural Stem Cells

One of the long-held dogmas is that neurogenesis in the adult CNS does not occur. However, recent evidence suggests that this is not the case and that neural precursor cells (NPCs), which are capable of proliferation are present primarily in two areas of the brain: the subependymal layer of the ventricular zone and the dentate gyrus of the hippocampus (229).

Neural stem-cell (NSC) transplantation has the potential to prevent or to restore anatomic or functional deficits associated with injury or disease through cell replacement, release of specific neurotransmitters, and the production of factors that promote neuronal growth and regeneration.

Treatment with NSC can be from immortalized human cell lines such as the teratocarcinoma-derived cell line (230). Transplantation of cultured neuronal cells is safe in animal models and has been shown to improve motor and cognitive deficits in rats with stroke. These observations led to a clinical study in patients with basal ganglia stroke and fixed motor deficits (231). Serial evaluations showed no adverse cell-related serologic or radiological effects, and an improved European Stroke Scale score was observed in 50%

of patients (six of 12). The procedure was considered safe and feasible in patients with motor infarction and warrants further investigation.

The lack of infection risk and the absence of need for immunosuppression might make the use of autologous cells preferable to NSC cell lines. However, in order for autologous NSC to be used therapeutically, endoscopic minimally invasive surgery has to be used to obtain subependymal biopsies. Expansion and differentiation could be carried out in vitro (232) before returning cells to the patient.

Extensive investigation is underway into the difficult ethical area of use of embryonic stem cells and fetal tissue for treatment of neurological diseases, however, as this chapter examines the clinical applications of adult (or at least postnatal) stem cells, this area of research is not explored.

Although in vitro and animal model data look encouraging, there are a number of issues that need to be clarified before stem-cell therapy for neurological disease is widely accepted in clinical medicine:

- Which type of cells are most suitable for replacement therapies (HSC, NSC, clonogenic cell lines, etc.)?
- Should cells be expanded and/or differentiated ex vivo?
- How should cells be administered? Directly into damaged tissue, intravenously?
- Identification of favorable conditions of the recipient brain environment to support transplanted cells.

It is likely that it will be some time before stem-cell therapies for neural disease move out of the research laboratory and into the clinic as a safe and effective therapeutic option. Stem-cell technology does, however, offer real hope for the future treatment of degenerative diseases such as Parkinson's and Huntington's disease, and for stoke sufferers.

Mesenchymal Stem Cells

BM has been known to contain non-hematopoietic cells for many years as was shown by Friedenstein et al. (223), who demonstrated that isolated and cultured cells from guinea-pig BM could form ectopic bone. MSCs are multi-potential, nonhematopoietic progenitor cells of the adult marrow, which are capable of differentiating into various lineages of the mesenchyme (bone, cartilage, fat, muscle, etc.) and are characterized by the absence of hematopoietic markers, such as CD45 and CD34, and by the presence of adhesion molecules such as CD105 and CD106. They are present in low numbers in adult BM, but can be isolated and cultured in therapeutic quantities.

Animal models have shown that MSCs are useful in the repair or regeneration of myocardial tissues (234), damaged bone, cartilage (235), and tendon (236). Furthermore, MSCs provide cytokine and growth factor support for hematopoietic and embryonic stem cells (237,238). As discussed earlier, MSCs have been shown to be of particular interest in the area of cardiac remodeling and regeneration following myocardial injury.

MSC in Treatment of OI

Interestingly MSCs can also be used in the treatment of OI. OI is a genetic disorder of mesenchymal cells in which generalized osteopenia leads to bone deformities, excessive fragility with fracturing, and short stature. The underlying defect is a mutation in one of the two genes encoding type 1 collagen (COL1A1). The most severe type is lethal in utero, and modern medicine provides no cure or effective therapy for individuals who survive through to birth.

Animal models of OI have shown improvement in disease phenotype after mesenchymal cell transplantation and so an initial clinical study was carried out in three children with severe deforming type of disease (type III) (239). Following myeloablative conditioning, each patient received BM from a sibling donor, and all three showed engraftment with hematopoietic donor cells. Three months after engraftment, BM biopsy confirmed 1.5% to 2% donor-derived osteoblasts, improvement in bone architecture, and increase in bone mineral content. Most importantly, the children showed much greater growth after transplantation and reduced fracture rates. The low levels of donor osteoblasts (1.5% to 2%) are thought to be capable of producing such clear clinical improvements because the degree of severity of disease is related to the ratio of normal to mutated polypeptide chains. Therefore, even low levels of MSC engraftment may be sufficient to produce a shift in the balance of normal to mutated protein.

More recently, Chamberlain et al. (240) developed a strategy to inactivate the mutated alleles in BM-derived MSC ex vivo using AAV as a vector. Results using MSC from two patients with OI were encouraging; 31% to 90% of cells demonstrated successful gene-construct insertion and the altered MSC showed improved quality of synthesized bone in vitro. Although insertional mutagenesis still remains a potential problem in the cells that have not been correctly targeted, measures such as the use of different markers and more extensive selection of cells may further reduce this risk.

Immunomodulatory Function of MSC

MSCs have been shown to have immunomodulatory functions in vitro (241) and in allo-transplantation models in vivo (242) by mechanisms that have not yet been fully elucidated. It is, however, known that MSCs are not immunogenic and avoid recognition by alloreactive T-cells and NK cells. Furthermore, they are immunosuppressive and inhibit proliferation of alloreactive T-cells. MSCs occur in small numbers in BMT but not in HSCT derived from peripheral blood. Preliminary studies of co-transplantation of MSCs and HSCs from haploidentical sibling donors suggested that MSC could potentially suppress development of acute and chronic GVHD (243). This observation led to an attempt (244) to treat a boy with severe treatment-resistant GVHD after allogeneic HSCT for ALL with haploidentical MSC from his mother. The patient showed rapid improvement in gut and liver GVHD and at the time of report remained well one year after transplantation. No allo-reactivity was seen when the patient's lymphocytes were co-cultured with donor MSC either before or after treatment.

A more recent study of allogeneic immune response to human MSC has shown that they alter cytokine production by dendritic cells, T-helper cells, and NK cells. In particular, MSCs provoke a reduction in TNF-α and IFN-γ, and an increase in IL-10 and IL-4, and produce elevated prostaglandin E2 (245). These responses induce a more anti-inflammatory or tolerant phenotype by inhibiting or limiting inflammatory responses and promoting the mitigating and anti-inflammatory pathways.

Results suggest that the potential therapeutic applications of MSC could range from autoimmune and allergic diseases to rejection prevention in solid organ transplantation.

CONCLUSION

ASCT and alloSCT are well established in the treatment of patients with leukemia, lymphoma, and other malignancies, and have been critical in the progress toward cure of these diseases. More recently, research has focused on the potential benefits of stem cells for gene therapy, tissue engineering, and the treatment of cardiac, neurological, and other forms of disease. Preclinical and early clinical studies have yielded encouraging results, and yet our knowledge and ability to deliver these forms of therapy in a safe and efficacious manner will require additional advances in the understanding of the basic biology of stem cells.

REFERENCES

1. Juttner CA, To LB, Haylock DN, Branford A, Kimber RJ. Circulating autologous stem cells collected in very early remission from acute non-lymphoblastic leukaemia produce prompt but incomplete hemopoietic reconstitution after high-dose melphalan or supralethal chemoradiotherapy. Br J Haematol 1985; 61(4):739–745.
2. To LB, Roberts MM, Haylock DN, Dyson PG, Branford AL, Thorp D, et al. Comparison of hematological recovery times and supportive care requirements of autologous recovery phase peripheral blood stem cell transplants, autologous bone marrow transplants and allogeneic bone marrow transplants. Bone Marrow Transplant 1992; 9(4):277–284.
3. Henon PR, Liang H, Beck-Wirth G, Eisenmann JC, Lepers M, Wunder E, et al. Comparison of hematopoietic and immune recovery after autologous bone marrow or blood stem cell transplants. Bone Marrow Transplant 1992; 9(4):285–291.
4. Korbling M, Fliedner TM, Holle R, Magrin S, Baumann M, Holdermann E, et al. Autologous blood stem cell (ABSCT) versus purged bone marrow transplantation (pABMT) in standard risk AML: influence of source and cell composition of the autograft on hemopoietic reconstitution and disease-free survival. Bone Marrow Transplant 1991; 7(5):343–349.
5. Barnett D, Janossy G, Lubenko A, Matutes E, Newland A, Reilly JT. Guideline for the flow cytometric enumeration of CD34+ haematopoietic stem cells. Prepared by the CD34+ haematopoietic stem cell working party. General Haematology Task Force of the British Committee for Standards in Haematology. Clin Lab Haematol 1999; 21(5):301–308.
6. Sanz MA, de la Rubia J, Sanz GF, Martin G, Martinez J, Jarque I, et al. Busulfan plus cyclophosphamide followed by autologous blood stem-cell transplantation for patients with acute myeloblastic leukemia in first complete remission: a report from a single institution. J Clin Oncol 1993; 11(9):1661–1667.
7. Mehta J, Powles R, Singhal S, Horton C, Tait D, Milan S, et al. Autologous bone marrow transplantation for acute myeloid leukemia in first remission: identification of modifiable prognostic factors. Bone Marrow Transplant 1995; 16(4):499–506.
8. Voso MT, Hohaus S, Moos M, Pforsich M, Cremer FW, Schlenk RF, et al. Autografting with CD34+ peripheral blood stem cells: retained engraftment capability and reduced tumor cell content. Br J Haematol 1999; 104(2):382–391.
9. Miyamoto T, Nagafuji K, Harada M, Niho Y. Significance of quantitative analysis of AML1/ ETO transcripts in peripheral blood stem cells from t(8;21) acute myelogenous leukemia. Leuk Lymphoma 1997; 25(1–2):69–75.
10. Gale RP, Champlin RE. How does bone-marrow transplantation cure leukemia? Lancet 1984; 2(8393):28–30.
11. Gribben JG, Freedman AS, Neuberg D, Roy DC, Blake KW, Woo SD, et al. Immunologic purging of marrow assessed by PCR before autologous bone marrow transplantation for B-cell lymphoma. N Engl J Med 1991; 325(22):1525–1533.
12. Brenner MK, Rill DR, Holladay MS, Heslop HE, Moen RC, Buschle M, et al. Gene marking to determine whether autologous marrow infusion restores long-term haemopoiesis in cancer patients. Lancet 1993; 342(8880):1134–1137.
13. Yeager AM, Kaizer H, Santos GW, Saral R, Colvin OM, Stuart RK, et al. Autologous bone marrow transplantation in patients with acute nonlymphocytic leukemia, using ex vivo marrow treatment with 4-hydroperoxycyclophosphamide. N Engl J Med 1986; 315(3):141–147.

14. Robertson MJ, Soiffer RJ, Freedman AS, Rabinowe SL, Anderson KC, Ervin TJ, et al. Human bone marrow depleted of CD33-positive cells mediates delayed but durable reconstitution of hematopoiesis: clinical trial of MY9 monoclonal antibody-purged autografts for the treatment of acute myeloid leukemia. Blood 1992; 79(9):2229–2236.

15. Freedman AS, Gribben JG, Nadler LM. High-dose therapy and autologous stem cell transplantation in follicular non-Hodgkin's lymphoma. Leuk Lymphoma 1998; 28(3–4):219–230.

16. Scime R, Indovina A, Santoro A, Musso M, Olivieri A, Tringali S, et al. PBSC mobilization, collection and positive selection in patients with chronic lymphocytic leukemia. Bone Marrow Transplant 1998; 22(12):1159–1165.

17. Paulus U, Schmitz N, Viehmann K, von Neuhoff N, Dreger P. Combined positive/negative selection for highly effective purging of PBPC grafts: towards clinical application in patients with B-CLL. Bone Marrow Transplant 1997; 20(5):415–420.

18. Cassileth PA, Harrington DP, Appelbaum FR, Lazarus HM, Rowe JM, Paietta E, et al. Chemotherapy compared with autologous or allogeneic bone marrow transplantation in the management of acute myeloid leukemia in first remission. N Engl J Med 1998; 339(23):1649–1656.

19. Ravindranath Y, Yeager AM, Chang MN, Steuber CP, Krischer J, Graham-Pole J, et al. Autologous bone marrow transplantation versus intensive consolidation chemotherapy for acute myeloid leukemia in childhood. Pediatric Oncology Group. N Engl J Med 1996; 334(22):1428–1434.

20. Reiffers J, Stoppa AM, Attal M, Michallet M, Marit G, Blaise D, et al. Allogeneic vs autologous stem cell transplantation vs chemotherapy in patients with acute myeloid leukemia in first remission: the BGMT 87 study. Leukemia 1996; 10(12):1874–1882.

21. Gorin NC, Labopin M, Meloni G, Korbling M, Carella A, Herve P, et al. Autologous bone marrow transplantation for acute myeloblastic leukemia in Europe: further evidence of the role of marrow purging by mafosfamide. European Co-operative Group for Bone Marrow Transplantation (EBMT). Leukemia 1991; 5(10):896–904.

22. Gorin NC, Aegerter P, Auvert B. Autologous bone marrow transplantation for acute leukemia in remission: an analysis of 1322 cases. Haematol Blood Transfus 1990; 33:660–666.

23. Freedman AS, Neuberg D, Mauch P, Soiffer RJ, Anderson KC, Fisher DC, et al. Long-term follow-up of autologous bone marrow transplantation in patients with relapsed follicular lymphoma. Blood 1999; 94(10):3325–3333.

24. Appelbaum FR, Buckner CD. Overview of the clinical relevance of autologous bone marrow transplantation. Clin Haematol 1986; 15(1):1–18.

25. Dicke KA, McCredie KB, Stevens EE, Spitzer G, Bottino JC. Autologous bone marrow transplantation in a case of acute adult leukemia. Transplant Proc 1977; 9(1):193–195.

26. Dicke KA, Zander A, Spitzer G, Verma DS, Peters L, Vellekoop L, et al. Autologous bone-marrow transplantation in relapsed adult acute leukemia. Lancet 1979; 1(8115):514–517.

27. Campana D, Pui CH. Detection of minimal residual disease in acute leukemia: methodologic advances and clinical significance. Blood 1995; 85(6):1416–1434.

28. Skipper HE, Schabel FM Jr, Wilcox WS. Experimental evaluation of potential anti-cancer agents. XIII. On the criteria and kinetics associated with "curability" of experimental leukaemia. Cancer Chemother Rep 1964; 35:1–111.

29. Skipper HE. The effects of chemotherapy on the kinetics of leukemic cell behavior. Cancer Res 1965; 25(9):1544–1550.

30. Zittoun RA, Mandelli F, Willemze R, de Witte T, Labar B, Resegotti L, et al. Autologous or allogeneic bone marrow transplantation compared with intensive chemotherapy in acute myelogenous leukemia. European Organization for Research and Treatment of Cancer (EORTC) and the Gruppo Italiano Malattie Ematologiche Maligne dell'Adulto (GIMEMA) Leukemia Cooperative Groups. N Engl J Med 1995; 332(4):217–223.

31. Burnett AK, Goldstone AH, Stevens RM, Hann IM, Rees JK, Gray RG, et al. Randomised comparison of addition of autologous bone-marrow transplantation to intensive chemotherapy for acute myeloid leukemia in first remission: results of MRC AML 10 trial. UK

Medical Research Council Adult and Children's Leukemia Working Parties. Lancet 1998; 351(9104):700–708.

32. Harousseau JL, Cahn JY, Pignon B, Witz F, Milpied N, Delain M, et al. Comparison of autologous bone marrow transplantation and intensive chemotherapy as postremission therapy in adult acute myeloid leukemia. The Groupe Ouest Est Leucemies Aigues Myeloblastiques (GOELAM). Blood 1997; 90(8):2978–2986.

33. Suciu S. The value of BMT in AML patients in first remission: a statistician's viewpoint. Ann Hematol 1991; 62(2–3):41–44.

34. Grimwade D, Walker H, Oliver F, Wheatley K, Harrison C, Harrison G, et al. The importance of diagnostic cytogenetics on outcome in AML: analysis of 1612 patients entered into the MRC AML 10 trial. The Medical Research Council Adult and Children's Leukemia Working Parties. Blood 1998; 92(7):2322–2333.

35. Keating MJ, Smith TL, Kantarjian H, Cork A, Walters R, Trujillo JM, et al. Cytogenetic pattern in acute myelogenous leukemia: a major reproducible determinant of outcome. Leukemia 1988; 2(7):403–412.

36. Burnett AK. Karyotypically defined risk groups in acute myeloid leukemia. Leuk Res 1994; 18(12):889–890.

37. Chopra R, Goldstone AH, McMillan AK, Powles R, Smith AG, Prentice HG, et al. Successful treatment of acute myeloid leukemia beyond first remission with autologous bone marrow transplantation using busulfan/cyclophosphamide and unpurged marrow: the British autograft group experience. J Clin Oncol 1991; 9(10):1840–1847.

38. Carey PJ, Proctor SJ, Taylor P, Hamilton PJ. Autologous bone marrow transplantation for high-grade lymphoid malignancy using melphalan/irradiation conditioning without marrow purging or cryopreservation. The Northern Regional Bone Marrow Transplant Group. Blood 1991; 77(7):1593–1598.

39. Gianni AM, Bregni M, Siena S, Brambilla C, Di Nicola M, Lombardi F, et al. High-dose chemotherapy and autologous bone marrow transplantation compared with MACOP-B in aggressive B-cell lymphoma. N Engl J Med 1997; 336(18):1290–1297.

40. Philip T, Armitage JO, Spitzer G, Chauvin F, Jagannath S, Cahn JY, et al. High-dose therapy and autologous bone marrow transplantation after failure of conventional chemotherapy in adults with intermediate-grade or high-grade non-Hodgkin's lymphoma. N Engl J Med 1987; 316(24):1493–1498.

41. Rohatiner AZ, Freedman A, Nadler L, Lim J, Lister TA. Myeloablative therapy with autologous bone marrow transplantation as consolidation therapy for follicular lymphoma. Ann Oncol 1994; 5(suppl 2):143–146.

42. Schouten HC, Qian W, Kvaloy S, Porcellini A, Hagberg H, Johnson HE, et al. High-dose therapy improves progression-free survival and survival in relapsed follicular non-Hodgkin's lymphoma: results from the randomized European CUP trial. J Clin Oncol 2003; 21(21):3918–3927.

43. Lenz G, Dreyling M, Schiegnitz E, Forstpointner R, Wandt H, Freund M, et al. Myeloablative radiochemotherapy followed by autologous stem cell transplantation in first remission prolongs progression-free survival in follicular lymphoma: results of a prospective, randomized trial of the German Low-Grade Lymphoma Study Group. Blood 2004; 104(9):2667–2674.

44. Deconinck E, Foussard C, Milpied N, Bertrand P, Michenet P, Cornillet-LeFebvre P, et al. High-dose therapy followed by autologous purged stem-cell transplantation and doxorubicin-based chemotherapy in patients with advanced follicular lymphoma: a randomized multicenter study by GOELAMS. Blood 2005; 105(10):3817–3823.

45. Metayer C, Curtis RE, Vose J, Sobocinski KA, Horowitz MM, Bhatia S, et al. Myelodysplastic syndrome and acute myeloid leukemia after autotransplantation for lymphoma: a multicenter case–control study. Blood 2003; 101(5):2015–2023.

46. Armitage JO, Carbone PP, Connors JM, Levine A, Bennett JM, Kroll S. Treatment-related myelodysplasia and acute leukemia in non-Hodgkin's lymphoma patients. J Clin Oncol 2003; 21(5):897–906.

47. Ladetto M, Corradini P, Vallet S, Benedetti F, Vitolo U, Martelli M, et al. High rate of clinical and molecular remissions in follicular lymphoma patients receiving high-dose sequential chemotherapy and autografting at diagnosis: a multicenter, prospective study by the Gruppo Italiano Trapianto Midollo Osseo (GITMO). Blood 2002; 100(5):1559–1565.

48. Fouillard L, Laporte JP, Labopin M, Lesage S, Isnard F, Douay L, et al. Autologous stem-cell transplantation for non-Hodgkin's lymphomas: the role of graft purging and radiotherapy posttransplantation—results of a retrospective analysis on 120 patients autografted in a single institution. J Clin Oncol 1998; 16(8):2803–2816.

49. van Besien K, Loberiza FR Jr, Bajorunaite R, Armitage JO, Bashey A, Burns LJ, et al. Comparison of autologous and allogeneic hematopoietic stem cell transplantation for follicular lymphoma. Blood 2003; 102(10):3521–3529.

50. Bierman PJ, Sweetenham JW, Loberiza FR Jr, Taghipour G, Lazarus HM, Rizzo JD, et al. Syngeneic hematopoietic stem-cell transplantation for non-Hodgkin's lymphoma: a comparison with allogeneic and autologous transplantation—The Lymphoma Working Committee of the International Bone Marrow Transplant Registry and the European Group for Blood and Marrow Transplantation. J Clin Oncol 2003; 21(20):3744–3753.

51. Flinn IW, O'Donnell PV, Goodrich A, Vogelsang G, Abrams R, Noga S, et al. Immunotherapy with rituximab during peripheral blood stem cell transplantation for non-Hodgkin's lymphoma. Biol Blood Marrow Transplant 2000; 6(6):628–632.

52. Hasenclever D, Diehl V. A prognostic score for advanced Hodgkin's disease. International Prognostic Factors Project on Advanced Hodgkin's Disease. N Engl J Med 1998; 339(21):1506–1514.

53. Schmitz N, Pfistner B, Sextro M, Sieber M, Carella AM, Haenel M, et al. Aggressive conventional chemotherapy compared with high-dose chemotherapy with autologous haemopoietic stem-cell transplantation for relapsed chemosensitive Hodgkin's disease: a randomised trial. Lancet 2002; 359(9323):2065–2071.

54. Linch DC, Winfield D, Goldstone AH, Moir D, Hancock B, McMillan A, et al. Dose intensification with autologous bone-marrow transplantation in relapsed and resistant Hodgkin's disease: results of a BNLI randomised trial. Lancet 1993; 341(8852):1051–1054.

55. Constans M, Sureda A, Terol MJ, Arranz R, Caballero MD, Iriondo A, et al. Autologous stem cell transplantation for primary refractory Hodgkin's disease: results and clinical variables affecting outcome. Ann Oncol 2003; 14(5):745–751.

56. Sureda A, Constans M, Iriondo A, Arranz R, Caballero MD, Vidal MJ, et al. Prognostic factors affecting long-term outcome after stem cell transplantation in Hodgkin's lymphoma autografted after a first relapse. Ann Oncol 2005; 16(4):625–633.

57. Milpied N, Fielding AK, Pearce RM, Ernst P, Goldstone AH. Allogeneic bone marrow transplant is not better than autologous transplant for patients with relapsed Hodgkin's disease. European Group for Blood and Bone Marrow Transplantation. J Clin Oncol 1996; 14(4):1291–1296.

58. Provan D, Bartlett-Pandite L, Zwicky C, Neuberg D, Maddocks A, Corradini P, et al. Eradication of polymerase chain reaction-detectable chronic lymphocytic leukemia cells is associated with improved outcome after bone marrow transplantation. Blood 1996; 88(6):2228–2235.

59. Dreger P, Montserrat E. Autologous and allogeneic stem cell transplantation for chronic lymphocytic leukemia. Leukemia 2002; 16(6):985–992.

60. Dreger P, Stilgenbauer S, Benner A, Ritgen M, Krober A, Kneba M, et al. The prognostic impact of autologous stem cell transplantation in patients with chronic lymphocytic leukemia: a risk-matched analysis based on the VH gene mutational status. Blood 2004; 103(7):2850–2858.

61. Ritgen M, Lange A, Stilgenbauer S, Dohner H, Bretscher C, Bosse H, et al. Unmutated immunoglobulin variable heavy-chain gene status remains an adverse prognostic factor after autologous stem cell transplantation for chronic lymphocytic leukemia. Blood 2003; 101(5):2049–2053.

62. Michallet M, Thiebaut A, Dreger P, Remes K, Milpied N, Santini G, et al. Peripheral blood stem cell (PBSC) mobilization and transplantation after fludarabine therapy in chronic lymphocytic leukaemia (CLL): a report of the European Blood and Marrow Transplantation (EBMT) CLL subcommittee on behalf of the EBMT Chronic Leukemias Working Party (CLWP). Br J Haematol 2000; 108(3):595–601.

63. Milligan DW, Fernandes S, Dasgupta R, Davies FE, Matutes E, Fegan CD, et al. Results of the MRC pilot study show autografting for younger patients with chronic lymphocytic leukaemia is safe and achieves a high percentage of molecular responses. Blood 2005; 105(1):397–404.

64. Attal M, Harousseau JL, Stoppa AM, Sotto JJ, Fuzibet JG, Rossi JF, et al. A prospective, randomized trial of autologous bone marrow transplantation and chemotherapy in multiple myeloma. Intergroupe Francais du Myelome. N Engl J Med 1996; 335(2):91–97.

65. Child JA, Morgan GJ, Davies FE, Owen RG, Bell SE, Hawkins K, et al. High-dose chemotherapy with hematopoietic stem-cell rescue for multiple myeloma. N Engl J Med 2003; 348(19):1875–1883.

66. Barlogie B, Jagannath S, Desikan KR, Mattox S, Vesole D, Siegel D, et al. Total therapy with tandem transplants for newly diagnosed multiple myeloma. Blood 1999; 93(1):55–65.

67. Attal M, Harousseau JL, Facon T, Guilhot F, Doyen C, Fuzibet JG, et al. Single versus double autologous stem-cell transplantation for multiple myeloma. N Engl J Med 2003; 349(26):2495–2502.

68. Cavo M, Zamagni E, Tosi P, Cellini C, Cangini D, Tacchetti P, et al. First-line therapy with thalidomide and dexamethasone in preparation for autologous stem cell transplantation for multiple myeloma. Haematologica 2004; 89(7):826–831.

69. Comenzo RL, Gertz MA. Autologous stem cell transplantation for primary systemic amyloidosis. Blood 2002; 99(12):4276–4282.

70. Gertz MA, Lacy MQ, Dispenzieri A, Gastineau DA, Chen MG, Ansell SM, et al. Stem cell transplantation for the management of primary systemic amyloidosis. Am J Med 2002; 113(7):549–555.

71. Skinner M, Sanchorawala V, Seldin DC, Dember LM, Falk RH, Berk JL, et al. High-dose melphalan and autologous stem-cell transplantation in patients with AL amyloidosis: an 8-year study. Ann Intern Med 2004; 140(2):85–93.

72. Burt RK, Traynor AE, Pope R, Schroeder J, Cohen B, Karlin KH, et al. Treatment of autoimmune disease by intense immunosuppressive conditioning and autologous hematopoietic stem cell transplantation. Blood 1998; 92(10):3505–3514.

73. Binks M, Passweg JR, Furst D, McSweeney P, Sullivan K, Besenthal C, et al. Phase I/II trial of autologous stem cell transplantation in systemic sclerosis: procedure-related mortality and impact on skin disease. Ann Rheum Dis 2001; 60(6):577–584.

74. Fassas A, Passweg JR, Anagnostopoulos A, Kazis A, Kozak T, Havrdova E, et al. Hematopoietic stem cell transplantation for multiplesclerosis: a retrospective multicenter study. J Neurol 2002; 249(8):1088–1097.

75. Saccardi R, Mancardi GL, Solari A, Bosi A, Bruzzi P, Di Bartolomeo P, et al. Autologous HSCT for severe progressive multiple sclerosis in a multicenter trial: impact on disease activity and quality of life. Blood 2004.

76. Jayne D, Passweg J, Marmont A, Farge D, Zhao X, Arnold R, et al. Autologous stem cell transplantation for systemic lupus erythematosus. Lupus 2004; 13(3):168–176.

77. Watson M, Wheatley K, Harrison GA, Zittoun R, Gray RG, Goldstone AH, et al. Severe adverse impact on sexual functioning and fertility of bone marrow transplantation, either allogeneic or autologous, compared with consolidation chemotherapy alone: analysis of the MRC AML 10 trial. Cancer 1999; 86(7):1231–1239.

78. Zittoun R, Suciu S, Watson M, Solbu G, Muus P, Mandelli F, et al. Quality of life in patients with acute myelogenous leukaemia in prolonged first complete remission after bone marrow transplantation (allogeneic or autologous) or chemotherapy: a cross-sectional study of the EORTC-GIMEMA AML 8A trial. Bone Marrow Transplant 1997; 20(4):307–315.

79. Thomas ED. The role of bone marrow transplantation for the eradication of malignant disease. Cancer 1969; 10:1963–1969.

80. Burchenal J, Oettgen H, Holmberg E, Hemphill S, Reppert J. Effect of total body irradiation on the transplantability of mouse leukaemias. Cancer Res 1960; 20:425.

81. Cohen A, Rovelli A, Bakker B, Uderzo C, van Lint MT, Esperou H, et al. Final height of patients who underwent bone marrow transplantation for hematological disorders during childhood: a study by the Working Party for Late Effects-EBMT. Blood 1999; 93(12):4109–4115.

82. Socie G, Salooja N, Cohen A, Rovelli A, Carreras E, Locasciulli A, et al. Non-malignant late effects after allogeneic stem cell transplantation. Blood 2003; 2:2.

83. Curtis RE, Rowlings PA, Deeg HJ, Shriner DA, Socie G, Travis LB, et al. Solid cancers after bone marrow transplantation. N Engl J Med 1997; 336(13):897–904.

84. Curtis RE, Travis LB, Rowlings PA, Socie G, Kingma DW, Banks PM, et al. Risk of lymphoproliferative disorders after bone marrow transplantation: a multi-institutional study. Blood 1999; 94(7):2208–2216.

85. Kolb HJ, Guenther W, Duell T, Socie G, Schaeffer E, Holler E, et al. Cancer after bone marrow transplantation. IBMTR and EBMT/EULEP Study Group on Late Effects. Bone Marrow Transplant 1992; 10(suppl 1):135–138.

86. Kolb HJ, Poetscher C. Late effects after allogeneic bone marrow transplantation. Curr Opin Hematol 1997; 4(6):401–407.

87. Drobyski WR, Keever CA, Roth MS, Koethe S, Hanson G, McFadden P, et al. Salvage immunotherapy using donor leukocyte infusions as treatment for relapsed chronic myelogenous leukaemia after allogeneic bone marrow transplantation: efficacy and toxicity of a defined T-cell dose. Blood 1993; 82(8):2310–238.

88. Giralt S, Hester J, Huh Y, Hirsch-Ginsberg C, Rondon G, Seong D, et al. CD8-depleted donor lymphocyte infusion as treatment for relapsed chronic myelogenous leukaemia after allogeneic bone marrow transplantation. Blood 1995; 86(11):4337–4343.

89. Johnson BD, Drobyski WR, Truitt RL. Delayed infusion of normal donor cells after MHC-matched bone marrow transplantation provides an antileukaemia reaction without graft-versus-host disease. Bone Marrow Transplant 1993; 11(4):329–336.

90. Kolb HJ, Mittermuller J, Clemm C, Holler E, Ledderose G, Brehm G, et al. Donor leukocyte transfusions for treatment of recurrent chronic myelogenous leukaemia in marrow transplant patients. Blood 1990; 76(12):2462–2465.

91. Kolb HJ, Mittermuller J, Gunther W, Bartram C, Thalmaier K, Schumm M, et al. Adoptive immunotherapy in human and canine chimeras. Bone Marrow Transplant 1993; 12(suppl 3):S61–S64.

92. Guglielmi C, Arcese W, Dazzi F, Brand R, Bunjes D, Verdonck LF, et al. Donor lymphocyte infusion for relapsed chronic myelogenous leukaemia: prognostic relevance of the initial cell dose. Blood 2002; 100(2):397–405.

93. Gale RP, Horowitz MM, Ash RC, Champlin RE, Goldman JM, Rimm AA, et al. Identical-twin bone marrow transplants for leukaemia. Ann Intern Med 1994; 120(8):646–652.

94. Collins RH Jr, Rogers ZR, Bennett M, Kumar V, Nikein A, Fay JW. Hematologic relapse of chronic myelogenous leukaemia following allogeneic bone marrow transplantation: apparent graft-versus-leukaemia effect following abrupt discontinuation of immunosuppression. Bone Marrow Transplant 1992; 10(4):391–395.

95. Higano CS, Brixey M, Bryant EM, Durnam DM, Doney K, Sullivan KM, et al. Durable complete remission of acute nonlymphocytic leukaemia associated with discontinuation of immunosuppression following relapse after allogeneic bone marrow transplantation: a case report of a probable graft-versus-leukaemia effect. Transplantation 1990; 50(1):175–177.

96. Porter DL, Roth MS, McGarigle C, Ferrara JL, Antin JH. Induction of graft-versus-host disease as immunotherapy for relapsed chronic myeloid leukaemia. N Engl J Med 1994; 330(2):100–106.

97. Apperley JF, Mauro FR, Goldman JM, Gregory W, Arthur CK, Hows J, et al. Bone marrow transplantation for chronic myeloid leukaemia in first chronic phase: importance of a graft-versus-leukaemia effect. Br J Haematol 1988; 69(2):239–245.

98. Weiden PL, Flournoy N, Thomas ED, Prentice R, Fefer A, Buckner CD, et al. Antileukemic effect of graft-versus-host disease in human recipients of allogeneic-marrow grafts. N Engl J Med 1979; 300(19):1068–1073.

99. Weiden PL, Flournoy N, Sanders JE, Sullivan KM, Thomas ED. Antileukemic effect of graft-versus-host disease contributes to improved survival after allogeneic marrow transplantation. Transplant Proc 1981; 13(1 Pt 1):248–251.

100. Weiden PL, Sullivan KM, Flournoy N, Storb R, Thomas ED. Antileukemic effect of chronic graft-versus-host disease: contribution to improved survival after allogeneic marrow transplantation. N Engl J Med 1981; 304(25):1529–1533.

101. Weisdorf DJ, McGlave PB, Ramsay NK, Miller WJ, Nesbit ME Jr, Woods WG, et al. Allogeneic bone marrow transplantation for acute leukaemia: comparative outcomes for adults and children. Br J Haematol 1988; 69(3):351–358.

102. Weisdorf DJ, Nesbit ME, Ramsay NK, Woods WG, Goldman AI, Kim TH, et al. Allogeneic bone marrow transplantation for acute lymphoblastic leukaemia in remission: prolonged survival associated with acute graft-versus-host disease. J Clin Oncol 1987; 5(9):1348–1355.

103. Horowitz MM, Gale RP, Sondel PM, Goldman JM, Kersey J, Kolb HJ, et al. Graft-versus-leukaemia reactions after bone marrow transplantation. Blood 1990; 75(3):555–562.

104. Drobyski WR, Ash RC, Casper JT, McAuliffe T, Horowitz MM, Lawton C, et al. Effect of T-cell depletion as graft-versus-host disease prophylaxis on engraftment, relapse, and disease-free survival in unrelated marrow transplantation for chronic myelogenous leukaemia. Blood 1994; 83(7):1980–1987.

105. Goldman JM, Gale RP, Horowitz MM, Biggs JC, Champlin RE, Gluckman E, et al. Bone marrow transplantation for chronic myelogenous leukaemia in chronic phase: increased risk for relapse associated with T-cell depletion. Ann Intern Med 1988; 108(6):806–814.

106. Kolb HJ, Schattenberg A, Goldman JM, Hertenstein B, Jacobsen N, Arcese W, et al. Graft-versus-leukaemia effect of donor lymphocyte transfusions in marrow grafted patients. European Group for Blood and Marrow Transplantation Working Party Chronic Leukaemia. Blood 1995; 86(5):2041–2050.

107. Porter DL, Collins RH Jr, Shpilberg O, Drobyski WR, Connors JM, Sproles A, et al. Long-term follow-up of patients who achieved complete remission after donor leukocyte infusions. Biol Blood Marrow Transplant 1999; 5(4):253–261.

108. Korngold R, Leighton C, Manser T. Graft-versus-myeloid leukaemia responses following syngeneic and allogeneic bone marrow transplantation. Transplantation 1994; 58(3):278–287.

109. Clark RE, Dodi IA, Hill SC, Lill JR, Aubert G, Macintyre AR, et al. Direct evidence that leukemic cells present HLA-associated immunogenic peptides derived from the BCR-ABL b3a2 fusion protein. Blood 2001; 98(10):2887–2893.

110. Bosch GJ, Joosten AM, Kessler JH, Melief CJ, Leeksma OC. Recognition of BCR-ABL positive leukemic blasts by human CD4+ T cells elicited by primary in vitro immunization with a BCR-ABL breakpoint peptide. Blood 1996; 88(9):3522–3527.

111. ten Bosch GJ, Kessler JH, Joosten AM, Bres-Vloemans AA, Geluk A, Godthelp BC, et al. A BCR-ABL oncoprotein p210b2a2 fusion region sequence is recognized by HLA-DR2a restricted cytotoxic T lymphocytes and presented by HLA-DR-matched cells transfected with an Ii(b2a2) construct. Blood 1999; 94(3):1038–1045.

112. Molldrem JJ, Lee PP, Wang C, Felio K, Kantarjian HM, Champlin RE, et al. Evidence that specific T lymphocytes may participate in the elimination of chronic myelogenous leukaemia. Nat Med 2000; 6(9):1018–1023.

113. Branson K, Chopra R, Kottaridis PD, McQuaker G, Parker A, Schey S, et al. Role of nonmyeloablative allogeneic stem-cell transplantation after failure of autologous transplantation in patients with lymphoproliferative malignancies. J Clin Oncol 2002; 20(19):4022–4031.

114. Marks DI, Lush R, Cavenagh J, Milligan DW, Schey S, Parker A, et al. The toxicity and effi-
     cacy of donor lymphocyte infusions given after reduced-intensity conditioning allogeneic
     stem cell transplantation. Blood 2002; 100(9):3108–3114.
115. Perez-Simon JA, Kottaridis PD, Martino R, Craddock C, Caballero D, Chopra R, et al.
     Nonmyeloablative transplantation with or without alemtuzumab: comparison between 2
     prospective studies in patients with lymphoproliferative disorders. Blood 2002;
     100(9):3121–3127.
116. Childs RW, Clave E, Tisdale J, Plante M, Hensel N, Barrett J. Successful treatment of
     metastatic renal cell carcinoma with a nonmyeloablative allogeneic peripheral-blood
     progenitor-cell transplant: evidence for a graft-versus-tumor effect. J Clin Oncol 1999;
     17(7):2044–2049.
117. Morecki S, Moshel Y, Gelfend Y, Pugatsch T, Slavin S. Induction of graft versus tumor effect
     in a murine model of mammary adenocarcinoma. Int J Cancer 1997; 71(1):59–63.
118. Childs R, Drachenberg D. Allogeneic stem cell transplantation for renal cell carcinoma. Curr
     Opin Urol 2001; 11(5):495–502.
119. Collins RH Jr, Shpilberg O, Drobyski WR, Porter DL, Giralt S, Champlin R, et al. Donor leu-
     kocyte infusions in 140 patients with relapsed malignancy after allogeneic bone marrow
     transplantation. J Clin Oncol 1997; 15(2):433–444.
120. Robinson SP, Goldstone AH, Mackinnon S, Carella A, Russell N, de Elvira CR, et al. Che-
     moresistant or aggressive lymphoma predicts for a poor outcome following reduced-intensity
     allogeneic progenitor cell transplantation: an analysis from the Lymphoma Working Party of
     the European Group for Blood and Bone Marrow Transplantation. Blood 2002;
     100(13):4310–4316.
121. Raiola AM, Van Lint MT, Valbonesi M, Lamparelli T, Gualandi F, Occhini D, et al. Factors
     predicting response and graft-versus-host disease after donor lymphocyte infusions: a study
     on 593 infusions. Bone Marrow Transplant 2003; 31(8):687–693.
122. Mackinnon S, Barnett L, Heller G, O'Reilly RJ. Minimal residual disease is more common in
     patients who have mixed T-cell chimerism after bone marrow transplantation for chronic
     myelogenous leukaemia. Blood 1994; 83(11):3409–3416.
123. Antin JH, Childs R, Filipovich AH, Giralt S, Mackinnon S, Spitzer T, et al. Establishment of
     complete and mixed donor chimerism after allogeneic lymphohematopoietic transplantation:
     recommendations from a workshop at the 2001 Tandem Meetings of the International Bone
     Marrow Transplant Registry and the American Society of Blood and Marrow Transplantation.
     Biol Blood Marrow Transplant 2001; 7(9):473–485.
124. Baurmann H, Nagel S, Binder T, Neubauer A, Siegert W, Huhn D. Kinetics of the graft-
     versus-leukaemia response after donor leukocyte infusions for relapsed chronic myeloid leu-
     kaemia after allogeneic bone marrow transplantation. Blood 1998; 92(10):3582–3590.
125. Mackinnon S, Papadopoulos EB, Carabasi MH, Reich L, Collins NH, Boulad F, et al. Adop-
     tive immunotherapy evaluating escalating doses of donor leukocytes for relapse of chronic
     myeloid leukaemia after bone marrow transplantation: separation of graft-versus-leukaemia
     responses from graft-versus-host disease. Blood 1995; 86(4):1261–1268.
126. Orsini E, Alyea EP, Chillemi A, Schlossman R, McLaughlin S, Canning C, et al. Conversion
     to full donor chimerism following donor lymphocyte infusion is associated with disease
     response in patients with multiple myeloma. Biol Blood Marrow Transplant 2000;
     6(4):375–386.
127. Clift RA, Buckner CD, Appelbaum FR, Bearman SI, Petersen FB, Fisher LD, et al. Allo-
     geneic marrow transplantation in patients with acute myeloid leukaemia in first remission:
     a randomised trial of two irradiation regimens. Blood 1990; 76(9):1867–1871.
128. Childs R, Epperson D, Bahceci E, Clave E, Barrett J. Molecular remission of chronic myeloid
     leukaemia following a non-myeloablative allogeneic peripheral blood stem cell transplant: in
     vivo and in vitro evidence for a graft-versus-leukaemia effect. Br J Haematol 1999;
     107(2):396–400.
129. Giralt S, Estey E, Albitar M, van Besien K, Rondon G, Anderlini P, et al. Engraftment of
     allogeneic hematopoietic progenitor cells with purine analog-containing chemotherapy:

harnessing graft-versus-leukaemia without myeloablative therapy. Blood 1997; 89(12):4531–4536.

130. Khouri IF, Keating M, Korbling M, Przepiorka D, Anderlini P, O'Brien S, et al. Transplant-lite: induction of graft-versus-malignancy using fludarabine-based nonablative chemotherapy and allogeneic blood progenitor-cell transplantation as treatment for lymphoid malignancies. J Clin Oncol 1998; 16(8):2817–2824.

131. Slavin S, Nagler A, Naparstek E, Kapelushnik Y, Aker M, Cividalli G, et al. Nonmyeloablative stem cell transplantation and cell therapy as an alternative to conventional bone marrow transplantation with lethal cytoreduction for the treatment of malignant and nonmalignant hematologic diseases. Blood 1998; 91(3):756–763.

132. Kroger N, Perez-Simon JA, Myint H, Klingemann H, Shimoni A, Nagler A, et al. Relapse to prior autograft and chronic graft-versus-host disease are the strongest prognostic factors for outcome of melphalan/fludarabine-based dose-reduced allogeneic stem cell transplantation in patients with multiple myeloma. Biol Blood Marrow Transplant 2004; 10(10):698–708.

133. Kottaridis PD, Gale RE, Frew ME, Harrison G, Langabeer SE, Belton AA, et al. The presence of a FLT3 internal tandem duplication in patients with acute myeloid leukaemia (AML) adds important prognostic information to cytogenetic risk group and response to the first cycle of chemotherapy: analysis of 854 patients from the United Kingdom Medical Research Council AML 10 and 12 trials. Blood 2001; 98(6):1752–1759.

134. Chakraverty R, Peggs K, Chopra R, Milligan DW, Kottaridis PD, Verfuerth S, et al. Limiting transplantation-related mortality following unrelated donor stem cell transplantation by using a nonmyeloablative conditioning regimen. Blood 2002; 99(3):1071–1078.

135. Kottaridis PD, Milligan DW, Chopra R, Chakraverty RK, Chakrabarti S, Robinson S, et al. In vivo CAMPATH-1H prevents graft-versus-host disease following nonmyeloablative stem cell transplantation. Blood 2000; 96(7):2419–2425.

136. Maris M, Boeckh M, Storer B, Dawson M, White K, Keng M, et al. Immunologic recovery after hematopoietic cell transplantation with nonmyeloablative conditioning. Exp Hematol 2003; 31(10):941–952.

137. Wong R, Giralt SA, Martin T, Couriel DR, Anagnostopoulos A, Hosing C, et al. Reduced-intensity conditioning for unrelated donor hematopoietic stem cell transplantation as treatmen for myeloid malignancies in patients older than 55 years. Blood 2003; 102(8):3052–3059.

138. Corradini P, Tarella C, Olivieri A, Gianni AM, Voena C, Zallio F, et al. Reduced-intensity conditioning followed by allografting of hematopoietic cells can produce clinical and molecular remissions in patients with poor-risk hematologic malignancies. Blood 2002; 99(1):75–82.

139. Faulkner RD, Craddock C, Byrne JL, Mahendra P, Haynes AP, Prentice HG, et al. BEAM-alemtuzumab reduced-intensity allogeneic stem cell transplantation for lymphoproliferative diseases: GVHD, toxicity, and survival in 65 patients. Blood 2004; 103(2):428–434.

140. D'Sa S, Peggs K, Pizzey A, Verfuerth S, Thuraisundaram D, Watts M, et al. T- and B-cell immune reconstitution and clinical outcome in patients with multiple myeloma receiving T-cell-depleted, reduced-intensity allogeneic stem cell transplantation with an alemtuzumab-containing conditioning regimen followed by escalated donor lymphocyte infusions. Br J Haematol 2003; 123(2):309–322.

141. Morris EC, Rebello P, Thomson KJ, Peggs KS, Kyriakou C, Goldstone AH, et al. Pharmacokinetics of alemtuzumab (CAMPATH-1H) used for in vivo and in vitro T-cell depletion in allogeneic transplants: relevance for early adoptive immunotherapy and infectious complications. Blood 2003.

142. Chakrabarti S, Mackinnon S, Chopra R, Kottaridis PD, Peggs K, O'Gorman P, et al. High incidence of cytomegalovirus infection after nonmyeloablative stem cell transplantation: potential role of Campath-1H in delaying immune reconstitution. Blood 2002; 99(12):4357–4363.

143. Antin JH, Ferrara JL. Cytokine dysregulation and acute graft-versus-host disease. Blood 1992; 80(12):2964–2968.

144. Ferrara JL. The cytokine modulation of acute graft-versus-host disease. Bone Marrow Transplant 1998; 21(suppl 3):S13–S15.

145. Ferrara JL, Levy R, Chao NJ. Pathophysiologic mechanisms of acute graft-versus-host disease. Biol Blood Marrow Transplant 1999; 5(6):347–356.

146. Krijanovski OI, Hill GR, Cooke KR, Teshima T, Crawford JM, Brinson YS, et al. Keratino-cyte growth factor separates graft-versus-leukaemia effects from graft-versus-host disease. Blood 1999; 94(2):825–831.

147. Deeg HJ, Spitzer TR, Cottler-Fox M, Cahill R, Pickle LW. Conditioning-related toxicity and acute graft-versus-host disease in patients given methotrexate/cyclosporine prophylaxis. Bone Marrow Transplant 1991; 7(3):193–198.

148. Hill GR, Cooke KR, Brinson YS, Bungard D, Ferrara JL. Pretransplant chemotherapy reduces inflammatory cytokine production and acute graft-versus-host disease after allogeneic bone marrow transplantation. Transplantation 1999; 67(11):1478–1480.

149. Hill GR, Ferrara JL. The primacy of the gastrointestinal tract as a target organ of acute graft-versus-host disease: rationale for the use of cytokine shields in allogeneic bone marrow trans-plantation. Blood 2000; 95(9):2754–2759.

150. Hill GR, Krenger W, Ferrara JL. The role of cytokines in acute graft-versus-host disease. Cytokines Cell Mol Ther 1997; 3(4):257–266.

151. Holler E, Ertl B, Hintermeier-Knabe R, Roncarolo MG, Eissner G, Mayer F, et al. Inflamma-tory reactions induced by pretransplant conditioning—an alternative target for modulation of acute GvHD and complications following allogeneic bone marrow transplantation? Leuk Lymphoma 1997; 25(3–4):217–224.

152. Holler E, Kolb HJ, Eissner G, Wilmanns W. Cytokines in GvH and GvL. Bone Marrow Transplant 1998; 22(suppl 4):S3–S6.

153. Panoskaltsis-Mortari A, Lacey DL, Vallera DA, Blazar BR. Keratinocyte growth factor admi-nistered before conditioning ameliorates graft-versus-host disease after allogeneic bone marrow transplantation in mice. Blood 1998; 92(10):3960–3967.

154. Schwaighofer H, Kernan NA, O'Reilly RJ, Brankova J, Nachbaur D, Herold M, et al. Serum levels of cytokines and secondary messages after T-cell-depleted and non-T-cell-depleted bone marrow transplantation: influence of conditioning and hematopoietic reconstitution. Transplantation 1996; 62(7):947–953.

155. Slavin S, Nagler A, Naparstek E, Kapelushnik Y, Aker M, Cividalli G, et al. Nonmyeloabla-tive stem cell transplantation and cell therapy as an alternative to conventional bone marrow transplantation with lethal cytoreduction for the treatment of malignant and nonmalignant hematologic diseases. Blood 1998; 91(3):756–763.

156. Kottaridis PD, Milligan DW, Chopra R, Chakraverty RK, Chakrabarti S, Robinson S, et al. In vivo CAMPATH-1H prevents GvHD following nonmyeloablative stem-cell transplantation. Cytotherapy 2001; 3(3):197–201.

157. Keating S, de Witte T, Suciu S, Willemze R, Hayat M, Labar B, et al. The influence of HLA-matched sibling donor availability on treatment outcome for patients with AML: an analysis of the AML 8A study of the EORTC Leukemia Cooperative Group and GIMEMA. European Organization for Research and Treatment of Cancer. Gruppo Italiano Malattie Ematologiche Maligne dell'Adulto. Br J Haematol 1998; 102(5):1344–1353.

158. Stevens RF, Hann IM, Wheatley K, Gray RG. Marked improvements in outcome with che-motherapy alone in pediatric acute myeloid leukaemia: results of the United Kingdom Medical Research Council's 10th AML trial. MRC Childhood Leukemia Working Party. Br J Haematol 1998; 101(1):130–140.

159. Burnett AK, Wheatley K, Goldstone AH, Stevens RF, Hann IM, Rees JH, et al. The value of allogeneic bone marrow transplant in patients with acute myeloid leukaemia at differing risk of relapse: results of the UK MRC AML 10 trial. Br J Haematol 2002; 118(2):385–400.

160. Ferrant A, Doyen C, Delannoy A, Cornu G, Martiat P, Latinne D, et al. Allogeneic or auto-logous bone marrow transplantation for acute non-lymphocytic leukaemia in first remission. Bone Marrow Transplant 1991; 7(4):303–309.

161. Mitus AJ, Miller KB, Schenkein DP, Ryan HF, Parsons SK, Wheeler C, et al. Improved sur-vival for patients with acute myelogenous leukaemia [see comments]. J Clin Oncol 1995; 13(3):560–569.

162. Gorin NC, Labopin M, Fouillard L, Meloni G, Frassoni F, Iriondo A, et al. Retrospective evaluation of autologous bone marrow transplantation vs allogeneic bone marrow transplantation from an HLA identical related donor in acute myelocytic leukaemia: a study of the European Cooperative Group for Blood and Marrow Transplantation (EBMT). Bone Marrow Transplant 1996; 18(1):111–117.

163. Reiffers J, Gaspard MH, Maraninchi D, Michallet M, Marit G, Stoppa AM, et al. Comparison of allogeneic or autologous bone marrow transplantation and chemotherapy in patients with acute myeloid leukaemia in first remission: a prospective controlled trial. Br J Haematol 1989; 72(1):57–63.

164. Ruggeri L, Capanni M, Mancusi A, Perruccio K, Burchielli E, Martelli MF, et al. Natural killer cell alloreactivity in haploidentical hematopoietic stem cell transplantation. Int J Hematol 2005; 81(1):13–17.

165. Ruggeri L, Capanni M, Urbani E, Perruccio K, Shlomchik WD, Tosti A, et al. Effectiveness of donor natural killer cell alloreactivity in mismatched hematopoietic transplants. Science 2002; 295(5562):2097–2100.

166. Ruggeri L, Capanni M, Casucci M, Volpi I, Tosti A, Perruccio K, et al. Role of natural killer cell alloreactivity in HLA-mismatched hematopoietic stem cell transplantation. Blood 1999; 94(1):333–339.

167. Avivi I, Rowe JM. Acute lymphocytic leukaemia: role of hematopoietic stem cell transplantation in current management. Curr Opin Hematol 2003; 10(6):463–468.

168. Sebban C, Lepage E, Vernant JP, Gluckman E, Attal M, Reiffers J, et al. Allogeneic bone marrow transplantation in adult acute lymphoblastic leukaemia in first complete remission: a comparative study. French Group of Therapy of Adult Acute Lymphoblastic Leukaemia. J Clin Oncol 1994; 12(12):2580–2587.

169. Thiebaut A, Vernant JP, Degos L, Huguet FR, Reiffers J, Sebban C, et al. Adult acute lymphocytic leukaemia study testing chemotherapy and autologous and allogeneic transplantation. a follow-up report of the French protocol LALA 87. Hematol Oncol Clin North Am 2000; 14(6):1353–1366, x.

170. Dombret H, Gabert J, Boiron JM, Rigal-Huguet F, Blaise D, Thomas X, et al. Outcome of treatment in adults with Philadelphia chromosome-positive acute lymphoblastic leukaemia—results of the prospective multicenter LALA-94 trial. Blood 2002; 100(7):2357–2366.

171. Shimoni A, Kroger N, Zander AR, Rowe JM, Hardan I, Avigdor A, et al. Imatinib mesylate (STI571) in preparation for allogeneic hematopoietic stem cell transplantation and donor lymphocyte infusions in patients with Philadelphia-positive acute leukaemias. Leukaemia 2003; 17(2):290–297.

172. Aversa F, Terenzi A, Felicini R, Carotti A, Falcinelli F, Tabilio A, et al. Haploidentical stem cell transplantation for acute leukaemia. Int J Hematol 2002; 76(suppl 1):165–168.

173. Michallet M, Archimbaud E, Bandini G, Rowlings PA, Deeg HJ, Gahrton G, et al. HLA-identical sibling bone marrow transplantation in younger patients with chronic lymphocytic leukaemia. European Group for Blood and Marrow Transplantation and the International Bone Marrow Transplant Registry. Ann Intern Med 1996; 124(3):311–315.

174. Ritgen M, Stilgenbauer S, von Neuhoff N, Humpe A, Bruggemann M, Pott C, et al. Graft-versus-leukaemia activity may overcome therapeutic resistance of chronic lymphocytic leukaemia with unmutated immunoglobulin variable heavy-chain gene status: implications of minimal residual disease measurement with quantitative PCR. Blood 2004; 104(8):2600–2602.

175. Dreger P, Brand R, Hansz J, Milligan D, Corradini P, Finke J, et al. Treatment-related mortality and graft-versus-leukaemia activity after allogeneic stem cell transplantation for chronic lymphocytic leukaemia using intensity-reduced conditioning. Leukaemia 2003; 17(5):841–848.

176. Gahrton G, Svensson H, Cavo M, Apperly J, Bacigalupo A, Bjorkstrand B, et al. Progress in allogenic bone marrow and peripheral blood stem cell transplantation for multiple myeloma: a comparison between transplants performed 1983–1993 and 1994–1998 at

European Group for Blood and Marrow Transplantation centres. Br J Haematol 2001; 113(1):209–216.

177. Kroger N, Schwerdtfeger R, Kiehl M, Sayer HG, Renges H, Zabelina T, et al. Autologous stem cell transplantation followed by a dose-reduced allograft induces high complete remission rate in multiple myeloma. Blood 2002; 100(3):755–760.

178. Maloney DG, Molina AJ, Sahebi F, Stockerl-Goldstein KE, Sandmaier BM, Bensinger W, et al. Allografting with nonmyeloablative conditioning following cytoreductive autografts for the treatment of patients with multiple myeloma. Blood 2003; 102(9):3447–3454.

179. Shaw BE, Peggs K, Bird JM, Cavenagh J, Hunter A, Alejandro Madrigal J, et al. The outcome of unrelated donor stem cell transplantation for patients with multiple myeloma. Br J Haematol 2003; 123(5):886–895.

180. O'Brien SG, Guilhot F, Larson RA, Gathmann I, Baccarani M, Cervantes F, et al. Imatinib compared with interferon and low-dose cytarabine for newly diagnosed chronic-phase chronic myeloid leukaemia. N Engl J Med 2003; 348(11):994–1004.

181. Lange T, Bumm T, Otto S, Al-Ali HK, Kovacs I, Krug D, et al. Quantitative reverse transcription polymerase chain reaction should not replace conventional cytogenetics for monitoring patients with chronic myeloid leukaemia during early phase of imatinib therapy. Haematologica 2004; 89(1):49–57.

182. Gratwohl A, Hermans J, Goldman JM, Arcese W, Carreras E, Devergie A, et al. Risk assessment for patients with chronic myeloid leukaemia before allogeneic blood or marrow transplantation. Chronic Leukaemia Working Party of the European Group for Blood and Marrow Transplantation. Lancet 1998; 352(9134):1087–1092.

183. Or R, Shapira MY, Resnick I, Amar A, Ackerstein A, Samuel S, et al. Nonmyeloablative allogeneic stem cell transplantation for the treatment of chronic myeloid leukaemia in first chronic phase. Blood 2003; 101(2):441–445.

184. Sloand E, Childs RW, Solomon S, Greene A, Young NS, Barrett AJ. The graft-versus-leukaemia effect of nonmyeloablative stem cell allografts may not be sufficient to cure chronic myelogenous leukaemia. Bone Marrow Transplant 2003; 32(9):897–901.

185. Ho AY, Pagliuca A, Kenyon M, Parker JE, Mijovic A, Devereux S, et al. Reduced-intensity allogeneic hematopoietic stem cell transplantation for myelodysplastic syndrome and acute myeloid leukaemia with multilineage dysplasia using fludarabine, busulphan, and alemtuzumab (FBC) conditioning. Blood 2004; 104(6):1616–1623.

186. Guardiola P, Anderson JE, Bandini G, Cervantes F, Runde V, Arcese W, et al. Allogeneic stem cell transplantation for agnogenic myeloid metaplasia: a European Group for Blood and Marrow Transplantation, Societe Francaise de Greffe de Moelle, Gruppo Italiano per il Trapianto del Midollo Osseo, and Fred Hutchinson Cancer Research Center Collaborative Study. Blood 1999; 93(9):2831–2838.

187. Daly A, Song K, Nevill T, Nantel S, Toze C, Hogge D, et al. Stem cell transplantation for myelofibrosis: a report from two Canadian centers. Bone Marrow Transplant 2003; 32(1):35–40.

188. Deeg HJ, Gooley TA, Flowers ME, Sale GE, Slattery JT, Anasetti C, et al. Allogeneic hematopoietic stem cell transplantation for myelofibrosis. Blood 2003; 102(12):3912–3918.

189. Devine SM, Hoffman R, Verma A, Shah R, Bradlow BA, Stock W, et al. Allogeneic blood cell transplantation following reduced-intensity conditioning is effective therapy for older patients with myelofibrosis with myeloid metaplasia. Blood 2002; 99(6):2255–2258.

190. Kroger N, Zabelina T, Schieder H, Panse J, Ayuk F, Stute N, et al. Pilot study of reduced-intensity conditioning followed by allogeneic stem cell transplantation from related and unrelated donors in patients with myelofibrosis. Br J Haematol 2005; 128(5):690–697.

191. Rondelli D, Barosi G, Bacigalupo A, Prchal JT, Popat U, Alessandrino EP, et al. Allogeneic hematopoietic stem cell transplantation with reduced intensity conditioning in intermediate or high-risk patients with myelofibrosis with myeloid metaplasia. Blood 2005.

192. Storb R, Blume KG, O'Donnell MR, Chauncey T, Forman SJ, Deeg HJ, et al. Cyclophosphamide and antithymocyte globulin to condition patients with aplastic anemia for allogeneic

marrow transplantations: the experience in four centers. Biol Blood Marrow Transplant 2001; 7(1):39–44.

193. Locatelli F, Bruno B, Zecca M, Van-Lint MT, McCann S, Arcese W, et al. Cyclosporin A and short-term methotrexate versus cyclosporin A as graft versus host disease prophylaxis in patients with severe aplastic anemia given allogeneic bone marrow transplantation from an HLA-identical sibling: results of a GITMO/Ebone marrowT randomised trial. Blood 2000; 96(5):1690–1697.

194. McNiece I, Briddell R. Ex vivo expansion of hematopoietic progenitor cells and mature cells. Exp Hematol 2001; 29(1):3–11.

195. Gluckman E, Broxmeyer HA, Auerbach AD, Friedman HS, Douglas GW, Devergie A, et al. Hematopoietic reconstitution in a patient with Fanconi's anemia by means of umbilical-cord blood from an HLA-identical sibling. N Engl J Med 1989; 321(17):1174–1178.

196. Rocha V, Wagner JE Jr, Sobocinski KA, Klein JP, Zhang MJ, Horowitz MM, et al. Graft-versus-host disease in children who have received a cord-blood or bone marrow transplant from an HLA-identical sibling. Eurocord and International Bone Marrow Transplant Registry Working Committee on Alternative Donor and Stem Cell Sources. N Engl J Med 2000; 342(25):1846–1854.

197. Gluckman E. Current status of umbilical cord blood hematopoietic stem cell transplantation. Exp Hematol 2000; 28:1197–1205.

198. Hows JM, Bradley BA, Marsh JC, Luft T, Coutinho L, Testa NG, et al. Growth of human umbilical-cord blood in long-term hemopoietic cultures. Lancet 1992; 340(8811):73–76.

199. Noort WA, Falkenberg JHF. Haematopoietic content of cord blood. In: Cohen SBA, Madrigal A, eds. Cord Blood Characteristics: Role in Stem Cell Transplantation. London: Martin Dunitz, 2000; 13–37.

200. Wang JC, Doedens M, Dick JE. Primitive human hematopoietic cells are enriched in cord blood compared with adult bone marrow or mobilized peripheral blood as measured by the quantitative in vivo SCID-repopulating cell assay. Blood 1997; 89(11):3919–3924.

201. Cairo MS, Wagner JE. Placental and/or umbilical cord blood: an alternative source of hematopoietic stem cells for transplantation. Blood 1997; 90(12):4665–4678.

202. Gluckman E, Rocha V, Boyer-Chammard A, Locatelli F, Arcese W, Pasquini R, et al. Outcome of cord-blood transplantation from related and unrelated donors. Eurocord Transplant Group and the European Blood and Marrow Transplantation Group. N Engl J Med 1997; 337(6):373–381.

203. Wagner JE, Rosenthal J, Sweetman R, Shu XO, Davies SM, Ramsay NK, et al. Successful transplantation of HLA-matched and HLA-mismatched umbilical cord blood from unrelated donors: analysis of engraftment and acute graft-versus-host disease. Blood 1996; 88(3):795–802.

204. Kurtzberg J, Laughlin M, Graham ML, Smith C, Olson JF, Halperin EC, et al. Placental blood as a source of hematopoietic stem cells for transplantation into unrelated recipients. N Engl J Med 1996; 335(3):157–166.

205. Wagner JE, Barker JN, DeFor TE, Baker KS, Blazar BR, Eide C, et al. Transplantation of unrelated donor umbilical cord blood in 102 patients with malignant and nonmalignant diseases: influence of CD34 cell dose and HLA disparity on treatment-related mortality and survival. Blood 2002; 100(5):1611–1618.

206. Laughlin MJ, Eapen M, Rubinstein P, Wagner JE, Zhang M-J, Champlin RE, et al. Outcomes after transplantation of cord blood or bone marrow from unrelated donors in adults with leukaemia. N Engl J Med 2004; 351(22):2265–2275.

207. Rubinstein P, Stevens CE. Placental blood for bone marrow replacement: the New York Blood Center's program and clinical results. Baillieres Best Pract Res Clin Haematol 2000; 13(4):565–584.

208. Jaroscak J, Goltry K, Smith A, Waters-Pick B, Martin PL, Driscoll TA, et al. Augmentation of umbilical cord blood (UCB) transplantation with ex vivo-expanded UCB cells: results of a phase 1 trial using the AastromReplicell System. Blood 2003; 101(12):5061–5067.

209. Pecora AL, Stiff P, Jennis A, Goldberg S, Rosenbluth R, Price P, et al. Prompt and durable engraftment in two older adult patients with high-risk chronic myelogenous leukaemia (CML) using ex vivo expanded and unmanipulated unrelated umbilical cord blood. Bone Marrow Transplant 2000; 25(7):797–799.

210. Shpall EJ, Champlin R, Glaspy JA. Effect of CD34+ peripheral blood progenitor cell dose on hematopoietic recovery. Biol Blood Marrow Transplant 1998; 4(2):84–92.

211. Shpall EJ. The utilization of cytokines in stem cell mobilization strategies. Bone Marrow Transplant 1999; 23(suppl 2):S13–S19.

212. Barker JN, Weisdorf DJ, Defor TE, O'Brien MR, Wagner JE. Non-myeloablative umbilical cordblood transplantation in 40 high-risk adults: extending access to transplant with high rate of engraftment, low risk of severe acute GVHD and demonstration of graft-versus-malignancy effect. Blood 2003; 2003:858a.

213. Barker JN, Weisdorf DJ, DeFor TE, Blazar BR, McGlave PB, Miller JS, et al. Transplantation of 2 partially HLA-matched umbilical cord blood units to enhance engraftment in adults with hematologic malignancy. Blood 2005; 105(3):1343–1347.

214. Cavazzana-Calvo M, Hacein-Bey S, de Saint Basile G, Gross F, Yvon E, Nusbaum P, et al. Gene therapy of human severe combined immunodeficiency (SCID)-X1 disease. Science 2000; 288(5466):669–672.

215. Hacein-Bey-Abina S, Von Kalle C, Schmidt M, McCormack MP, Wulffraat N, Leboulch P, et al. LMO2-associated clonal T cell proliferation in two patients after gene therapy for SCID-X1. Science 2003; 302(5644):415–419.

216. Kay MA, Manno CS, Ragni MV, Larson PJ, Couto LB, McClelland A, et al. Evidence for gene transfer and expression of factor IX in haemophilia B patients treated with an AAV vector. Nat Genet 2000; 24(3):257–261.

217. Check E. Harmful potential of viral vectors fuels doubts over gene therapy. Nature 2003; 423(6940):573–574.

218. Lehrman S. Virus treatment questioned after gene therapy death. Nature 1999; 401:517–518.

219. Strauer BE, Brehm M, Zeus T, Kostering M, Hernandez A, Sorg RV, et al. Repair of infarcted myocardium by autologous intracoronary mononuclear bone marrow cell transplantation in humans. Circulation 2002; 106(15):1913–1918.

220. Balsam LB, Wagers AJ, Christensen JL, Kofidis T, Weissman IL, Robbins RC. Haematopoietic stem cells adopt mature haematopoietic fates in ischaemic myocardium. Nature 2004; 428(6983):668–673.

221. Murry CE, Soonpaa MH, Reinecke H, Nakajima H, Nakajima HO, Rubart M, et al. Haematopoietic stem cells do not transdifferentiate into cardiac myocytes in myocardial infarcts. Nature 2004; 428(6983):664–668.

222. Orlic D, Kajstura J, Chimenti S, Limana F, Jakoniuk I, Quaini F, et al. Mobilized bone marrow cells repair the infarcted heart, improving function and survival. PNAS 2001; 98(18):10344–10349.

223. Orlic D, Kajstura J, Chimenti S, Jakoniuk I, Anderson SM, Li B, et al. Bone marrow cells regenerate infarcted myocardium. Nature 2001; 410(6829):701–705.

224. Bjornson CR, Rietze RL, Reynolds BA, Magli MC, Vescovi AL. Turning brain into blood: a hematopoietic fate adopted by adult neural stem cells in vivo. Science 1999; 283(5401):534–537.

225. Mezey E, Chandross KJ, Harta G, Maki RA, McKercher SR. Turning blood into brain: cells bearing neuronal antigens generated in vivo from bone marrow. Science 2000; 290(5497):1779–1782.

226. Mezey E, Key S, Vogelsang G, Szalayova I, Lange GD, Crain B. Transplanted bone marrow generates new neurons in human brains. PNAS 2003; 100(3):1364–1369.

227. Buske C, Feuring-Buske M, Abramovich C, Spiekermann K, Eaves CJ, Coulombel L, et al. Deregulated expression of HOXB4 enhances the primitive growth activity of human hematopoietic cells. Blood 2002; 100(3):862–868.

228. Zhong JF, Zhan Y, Anderson WF, Zhao Y. Murine hematopoietic stem cell distribution and proliferation in ablated and nonablated bone marrow transplantation. Blood 2002; 100(10):3521–3526.

229. Kim HM, Qu T, Kriho V, Lacor P, Smalheiser N, Pappas GD, et al. Reelin function in neural stem cell biology. Proc Natl Acad Sci USA 2002; 99(6):4020–4025.

230. Kondziolka D, Wechsler L, Goldstein S, Meltzer C, Thulborn KR, Gebel J, et al. Transplantation of cultured human neuronal cells for patients with stroke. Neurology 2000; 55(4):565–569.

231. Kondziolka D, Wechsler L, Goldstein S, Meltzer C, Thulborn KR, Gebel J, et al. Transplantation of cultured human neuronal cells for patients with stroke. Neurology 2000; 55(4):565–569.

232. Smith PM, Blakemore WF. Porcine neural progenitors require commitment to the oligodendrocyte lineage prior to transplantation in order to achieve significant remyelination of demyelinated lesions in the adult CNS. Eur J Neurosci 2000; 12(7):2414–2424.

233. Friedenstein AJ, Chailakhjan RK, Lalykina KS. The development of fibroblast colonies in monolayer cultures of guinea-pig bone marrow and spleen cells. Cell Tissue Kinet 1970; 3:393–403.

234. Pittenger MF, Martin BJ. Mesenchymal stem cells and their potential as cardiac therapeutics. Circ Res 2004; 95(1):9–20.

235. Noel D, Djouad F, Jorgense C. Regenerative medicine through mesenchymal stem cells for bone and cartilage repair. Curr Opin Investig Drugs 2002; 3(7):1000–1004.

236. Awad HA, Boivin GP, Dressler MR, Smith FN, Young RG, Butler DL. Repair of patellar tendon injuries using a cell-collagen composite. J Orthop Res 2003; 21(3):420–431.

237. Majumdar MK, Thiede MA, Haynesworth SE, Bruder SP, Gerson SL. Human marrow-derived mesenchymal stem cells (MSCs) express hematopoietic cytokines and support long-term hematopoiesis when differentiated toward stromal and osteogenic lineages. J Hematother Stem Cell Res 2000; 9(6):841–848.

238. Haynesworth SE, Baber MA, Caplan AI. Cytokine expression by human marrow-derived mesenchymal progenitor cells in vitro: effects of dexamethasone and IL-1 alpha. J Cell Physiol 1996; 166(3):585–592.

239. Horwitz EM, Prockop DJ, Fitzpatrick LA, Koo WW, Gordon PL, Neel M, et al. Transplantability and therapeutic effects of bone-marrow-derived mesenchymal cells in children with osteogenesis imperfecta. Nat Med 1999; 5(3):309–313.

240. Chamberlain JR, Schwarze U, Wang PR, Hirata RK, Hankenson KD, Pace JM, et al. Gene targeting in stem cells from individuals with osteogenesis imperfecta. Science 2004; 303(5661):1198–1201.

241. Beyth S, Borovsky Z, Mevorach D, Liebergall M, Gazit Z, Aslan H, et al. Human mesenchymal stem cells alter antigen-presenting cell maturation and induce T-cell unresponsiveness. Blood 2005; 105(5):2214–2219.

242. Bartholomew A, Sturgeon C, Siatskas M, Ferrer K, McIntosh K, Patil S, et al. Mesenchymal stem cells suppress lymphocyte proliferation in vitro and prolong skin graft survival in vivo. Exp Hematol 2002; 30(1):42–48.

243. Frassoni F, Labopin M, Bacigalupo A. Expanded mesenchymal stem cells (MSC), co-infused with HLA identical hematopoietic stem cell transplants, reduce acute and chronic GVHD: a matched pair analysis. Bone Marrow Transplant. 2002; 29(suppl 2):S2(abstr).

244. Le Blanc K, Rasmusson I, Sundberg B, Gotherstrom C, Hassan M, Uzunel M, et al. Treatment of severe acute graft-versus-host disease with third-party haploidentical mesenchymal stem cells. Lancet 2004; 363(9419):1439–1441.

245. Aggarwal S, Pittenger MF. Human mesenchymal stem cells modulate allogeneic immune cell responses. Blood 2005; 105(4):1815–1822.

Index

Milton Keynes UK
Ingram Content Group UK Ltd.
UKHW052022071024
449327UK00027B/2382

9 780367 390921